Environmental Degradation of the Black Sea: Challenges and Remedies

NATO Science Series

A Series presenting the results of activities sponsored by the NATO Science Committee. The Series is published by IOS Press and Kluwer Academic Publishers, in conjunction with the NATO Scientific Affairs Division.

A. Life Sciences	IOS Press
B. Physics	Kluwer Academic Publishers
C. Mathematical and Physical Sciences	Kluwer Academic Publishers
D. Behavioural and Social Sciences	Kluwer Academic Publishers
E. Applied Sciences	Kluwer Academic Publishers
F. Computer and Systems Sciences	IOS Press

1. Disarmament Technologies	Kluwer Academic Publishers
2. Environmental Security	Kluwer Academic Publishers
3. High Technology	Kluwer Academic Publishers
4. Science and Technology Policy	IOS Press
5. Computer Networking	IOS Press

NATO-PCO-DATA BASE

The NATO Science Series continues the series of books published formerly in the NATO ASI Series. An electronic index to the NATO ASI Series provides full bibliographical references (with keywords and/or abstracts) to more than 50000 contributions from internatonal scientists published in all sections of the NATO ASI Series.
Access to the NATO-PCO-DATA BASE is possible via CD-ROM "NATO-PCO-DATA BASE" with user-friendly retrieval software in English, French and German (WTV GmbH and DATAWARE Technologies Inc. 1989).

The CD-ROM of the NATO ASI Series can be ordered from: PCO, Overijse, Belgium

2. Environmental Security – Volume 56

Environmental Degradation of the Black Sea: Challenges and Remedies

edited by

Sükrü T. Beşiktepe

Institute of Marine Sciences, METU,
Erdemli, İçel, Turkey

Ümit Ünlüata

Institute of Marine Sciences, METU,
Erdemli, İçel, Turkey

and

Alexandru S. Bologa

Romanian Marine Research Institute,
Constanta, Romania

Kluwer Academic Publishers

Dordrecht / Boston / London

Published in cooperation with NATO Scientific Affairs Division

Proceedings of the NATO Advanced Research Workshop on
Environmental Degradation of the Back Sea: Challenges and Remedies
Constanta-Mamaia, Romania
6–10 October 1997

A C.I.P. Catalogue record for this book is available from the Library of Congress.

ISBN 0-7923-5675-6 (HB)
ISBN 0-7923-5676-4 (PB)

Published by Kluwer Academic Publishers,
P.O. Box 17, 3300 AA Dordrecht, The Netherlands.

Sold and distributed in North, Central and South America
by Kluwer Academic Publishers,
101 Philip Drive, Norwell, MA 02061, U.S.A.

In all other countries, sold and distributed
by Kluwer Academic Publishers,
P.O. Box 322, 3300 AH Dordrecht, The Netherlands.

Printed on acid-free paper

Printed in the Netherlands.

TABLE OF CONTENTS

PREFACE

The Black Sea presently faces severe ecological disequilibrium due primarily to eutrophication and other types of contaminants, from atmospheric, river and landbased sources. Major contaminants include nutrients, pesticides, hydrocarbons and heavy metals. Among the most critical contemporary concerns are eutrophication and associated deterioration of water quality, plankton blooms, hypoxia and anoxia, loss of biodiversity and decline of living resources.

A better understanding of conditions leading to eutrophication and of the associated changes during the last four decades, is being carried out at national , regional and international levels. High quality scientific research has been conducted in all Black Sea riparian countries (Bulgaria, Georgia, Romania, Russian Federation, Turkey, Ukraine). In addition, several successful regional research programmes (e.g., CoMSBlack, NATO-TU Black Sea, NATO-TU Waves, EC-EROS 2000 Phase III, IOC-Black Sea Regional Center with Pilot Projects 1/2) and one major environmental management program (GEF-BSEP) have been successfully launched.

New international efforts like the Black Sea Commission, the Black Sea Program Coordination Unit, the Black Sea Economic Cooperation (all situated in Istanbul), together with the Convention for the Protection of the Black Sea against Pollution (Bucharest, 1992) and the Odessa Interministerial Decleration (1993) attest to the economic and political importance of these problems and the attention presently paid to this endangered sea.

The purpose of the ARW was to synthesize the present knowledge based on the past and the ongoing research, technology and management programs, the most recent field and laboratory results, and the progress achieved in regional communication, networking and the creation of data base management systems for the Black Sea. The transboundary environmental issues were highlighted by the key speakers who reviewed the most relevant environmental questions, sharing data and approaches, and discussed ways to promote interdisciplinary and international efforts for identifying remedies for the near future. The ARW identified future research needs and a framework for a continuous ocean observing system and forecast capabilities for the Black Sea in parallel with developments in modern ocean science.

Each presentation have focused on different aspect of the Black Sea oceanography and therefore wide range of topics from oil spills to geology, from observational ecological processes to modeling were covered. Papers presented in the workshop can be divided into two groups as scientific papers addressing environmental degradation of the Black Sea and reports on the achievements of the existing international programs

Thanks are expressed to all participants of the workshop and the contributors to this volume. All ao the participants contributed to the working group discussions and prepared reports at the end of this book. We thank the NATO Science Committee for giving us opportunity to carry out this Workshop. We should like to express our sincere appreciation to Dr. L. Veiga da Cunha, Director for the Priority Area on Environment of the Scientific and Environmental Affairs Division of the NATO, for his support

The editors

ORIGIN OF THE BLACK SEA

NACİ GÖRÜR
ITU Maden Fakultesi,
Jeleoloji Bolumu,
80626 Ayazaga, Istanbul, Turkey

Abstract

Origin of the Black Sea has long been a matter of discussion. Many tectonic models have been proposed to explain how and when this mini ocean was created. One model regarded it as a remnant ocean. Another one attributed its formation to the basifcation of a continental crust. A third model suggested that it resulted from a continuous uplift and erosion of a landmass. A fourth model indicated that it resulted from strike-slip fault activities. The most favourite model is the back-arc opening. In these models various age estimates, ranging from pre-Cambrian to Quaternary have been proposed, although mid-Cretaceous has gained a wide support.

1. Introduction

The Black Sea is an inland marine basin located north of Turkey. It is connected with the Mediterranean by the Thracian Bosporus, the Sea of Marmara, and the Dardanelles. It is about 900- km- long and 300- km- wide with a 460 000 km^2 surface area and 534000 km^3 volume. It contains low-salinity water resulted from restricted exchange with oceans, large freshwater input, and low evaporation rate [32]. The surface waters are oxygeneted, whereas deep waters (generally deeper than 150 m) are anoxic and rich in H_2S [10, 29].

Geomorphological features, such as continental shelf, continental slope, continental rise, and abyssal plain are well-developed in the Black Sea Basin. The continental shelf on the Turkish side is narrow; 30 to 50 km wide along the Thrace and the Kocaeli Peninsula and about 10 km along the Eastern Pontides. The continental slope here are steep with a gradient of 1:40 and cut by many turbidite canyons. Continental rise generally has a smooth surface with a gradient between 1:40-1:1000. The abyssal plain usually has a gradient of less than 1:1000 [32, 20, 12].

Geophysical studies indicate that the Black sea consist of two deep basins: the West and the East Black Sea Basins. These basins are floored by oceanic crust and are separated by the mid-Black Sea ridge, a region of thinned continental crust. The Moho is about 20 km deep in the West Black Sea Basin and 25 km deep in the East Black Sea

S. Beşiktepe et al. (eds.),
Environmental Degradation of the Black Sea: Challenges and Remedies, 1–8.

Basin [27, 23, 24, 8, 34, 6, 13, 4]. Each basin shows different stratigraphic and structural features [21, 40, 13, 28]. The West Black Sea Basin trends E-W and has more than 14 km sedimentary fill, whereas the East Black Sea Basin trends NW-SE and contains less than 12 km infill [12, 44]. The infill of both basins ranges in age from Cretaceous to Holocene and are cut particularly in the East Black Sea Basin by numerous faults [13].

Formation of the Black Sea has long been debated. Many contradictory models of the origin with various age estimates have been proposed. Most of these models are reviewed below to show their similarities and dissimilarities. The purpose of doing this is to see what consensus of opinion exists about timing and kinematics of opening of the Black Sea.

2. Timing of Opening Proposed

A wide range of age has been attributed to the opening of the Black Sea. Two extreme age estimations bracket these proposals. One is pre-Cambrian [22] and the other is early Quaternary [26]. The rest ranges in between [11] regarded the Black Sea as being Palaeozoic. Sorokhtin[35] and Vardapetyan [41] thought that it may be early Mesozioc. Adamia *et al.* [2], Hsü *et al.* [19] and Letouzey *et al.* [21] suggested that it opened during the latest Cretaceous to early Cenozoic time. Goncharov and Neprochnov [15] and Muratov [24] proposed a late Cenozoic age. Zonenshain and Le Pichon [44] suggested that the Black sea formed during three episodes in the middle Jurassic, late Jurassic, and late Cretaceous. Later authors, such as Görür [16], Görür *et al.* [17], Okay *et al.* [28], and Robinson *et al.* [31] specified this age particularly for the West Black Sea Basin as Aptian (the mid-Cretaceous).

3. Models of the Origin Proposed

Many authors, including Dewey *et al.* [11], Sorokhtin [35], and Vardapetyan [41], suggested that the Black Sea is a remnant of an ancient ocean basin. Zonanshain and Le Pichon [44] proposed that it is a remnant of an originally much greater marginal sea developed during three separate episodes in the middle Jurassic, late Jurassic, and the late Cretaceous. Following Beloussov's [5] view, Subbotin *et al.* [38], Muratov [25], Yanshin *et al.* [43], and Shlezinger [33] argued against the idea about the oceanic nature of the Black Sea. They claimed that the oceanic crust beneath the Black Sea was not an originally oceanic one, but had formed later by basification of a continental crust. Brinkmann [7] accepted the former presence of a landmass, so called Pontian massif of Frech [14], in the Black Sea area until late in the Mesozoic. According to him this continental area was eroded and subsequently submerged to form the Black Sea in the mid-Cretaceous. Existence of such landmass was also indicated by earlier workers, such as Wilser [42] and Stille [36, 37]. Petrascheck [30] rejected this suggestion. According to him the conspicuous similarities of the Jurassic and the Cretaceous in the Crimea to those in the central Pontides contradict such hypothesis. In accordance with Petrascheck's [30] views, Adamia *et al.* [2], Hsü *et al.* [19], and Letouzey *et al.* [21]

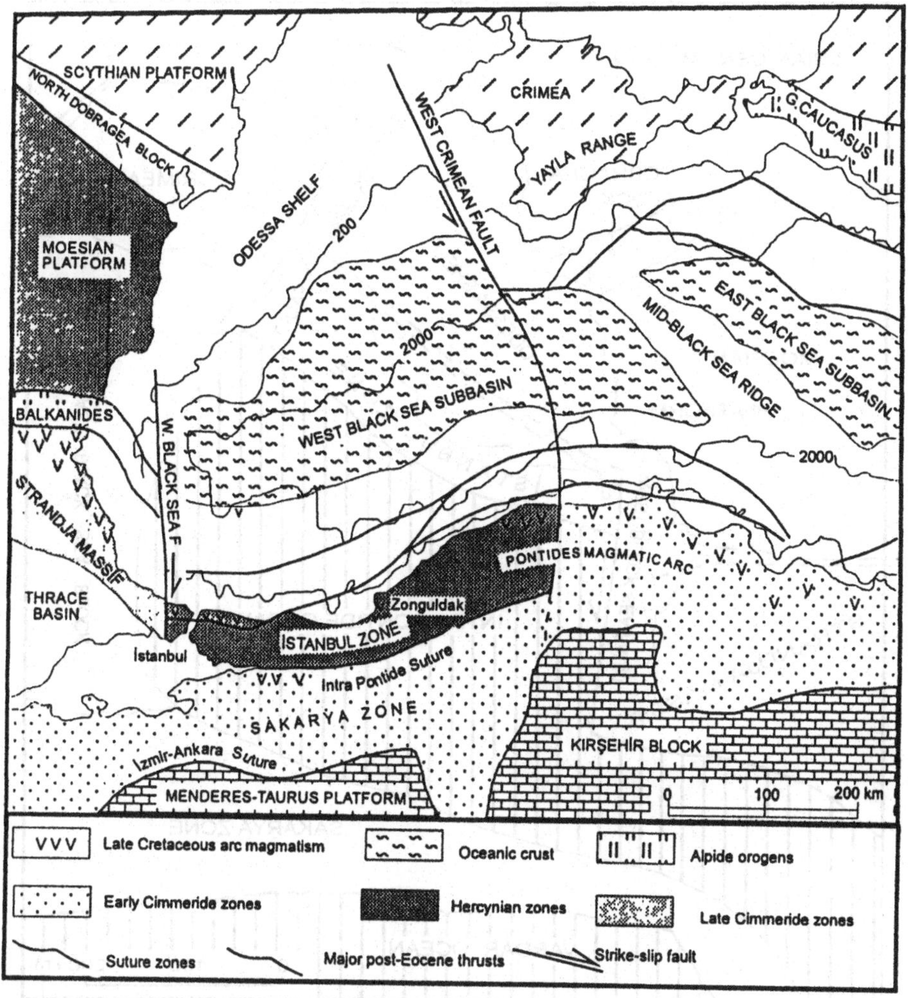

Figure 1 Tectonic setting of the Black Sea basin (after Okay *et al* [30])

suggested that the Black Sea opened as a result of a back-arc rifting behind the Pontide magmatic arc. Apolskiy [3], however, thought that this marine basin resulted from sinistral strike slip fault movement between the Kopet Dag and the Carpathians, rather than the rifting. Later studies, i.e. Görür [16] and Görür *et al.* [17], are in general agreement with the back-arc rifting model. Data reviewed in these studies showed that

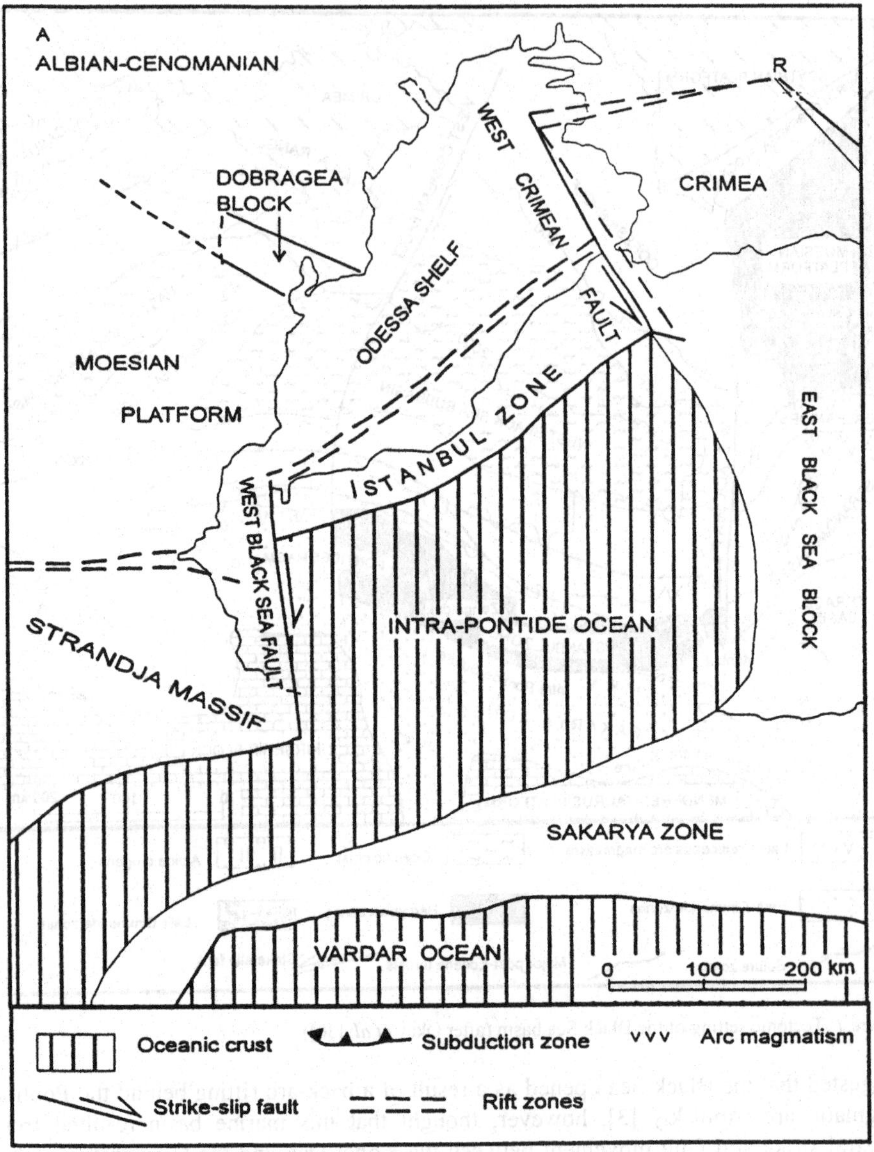

Figure 2a. Albian-Cenomanian palinspastic palaeogeographic map of the West Black Sea region (after Okay *et al* [30])

the Black Sea began opening as a back-arc basin by the rifting of a young continental margin magmatic arc (Pontic arc) during Aptian-Cenomanian time. Kinematic model for this opening was given by Okay *et al.* [28]. According to this model the West and the East Black Sea basins have separate origins. The West Black Sea basin opened during the Aptian-Albian by back-arc rifting of a Hercynian continental sliver (the Istanbul zone of Okay, 1989) from what is now the Odessa shelf (Fig. 1). In contrast, the East Black Sea opened as a result of the counterclockwise rotation of an east Black Sea block, comprising Crimea and the eastern Black Sea, around a pole located northeast of Crimea. Robinson *et al.* [31] share a similar view, but they believe that the East Black Sea probably formed by separation of the mid-Black Sea ridge from the Shatsky Ridge during the Palaeocene to Eocene (Fig. 1).

4. Conclusions

Recent research indicates that the back-arc basin model and its kinematic details, which are presented by Adamia *et al.* [2], Hsü *et al.* [19], Letouzey *et al.* [21], and Okay *et al.* [28], have won general support for the opening of the Black Sea. Görür's [16] age estimation also seems to have been widely accepted. Formation of this sea will be summarized below on the base of these studies, because this topic is also critical in unraveling the geological history of the Mesozoic-Cenozoic palaeotectonics of the central Tethyan area.

The Black Sea formed during the Aptian to Cenomanian time as a result of a complex interplay between rifting, transform faulting and block rotation. The West Black Sea Basin opened by rifting the Istanbul Zone from the southern continental margin of Laurasia. This zone, which formed from a sedimentary sequence of Ordovician to Cretaceous age [1, 18], was located until the mid-Cretaceous along the Odessa shelf between the Moesian Platform and the Crimea (Fig. 2a). It moved south during the late Cretaceous to Palaeocene along two transform faults (the dextral west Black Sea and the sinistral west Crimean faults, Dachev *et al.* [9] and Finetti *et al.* [13], respectively), opening in its wake the West Black Sea Basin (Fig. 2b). This movement was ended at the end of early Eocene by the collision of the İstanbul Zone with a Cimmeride continental unit (the Sakarya Zone, Okay, 1989), obliterating the arm of the Neo-Tethys Ocean in between (the Intra-Pontide Ocean, Sengör and Yilmaz [39]) (Figs. 2a and 2b). The East Black Sea Basin resulted from anticlockwise rotation of an east Black Sea block around a rotation pole located north of the Crimea (Fig. 2b). The east Black Sea block, which comprised the Crimea and the future East Black Sea Basin, was bounded by the Karkinisky Basin in the north, by the West Crimean fault in the west, by the Black Sea margin in the south, and by the Slate Diabase Zone Ocean in the east (Fig. 2b). The rotation of the east Black Sea block was coeval with the opening of the West Black Sea, but lasted until the Miocene resulting in both closure of the Slate-Diabase Ocean and continuous compression along the Greater Caucasus [28].

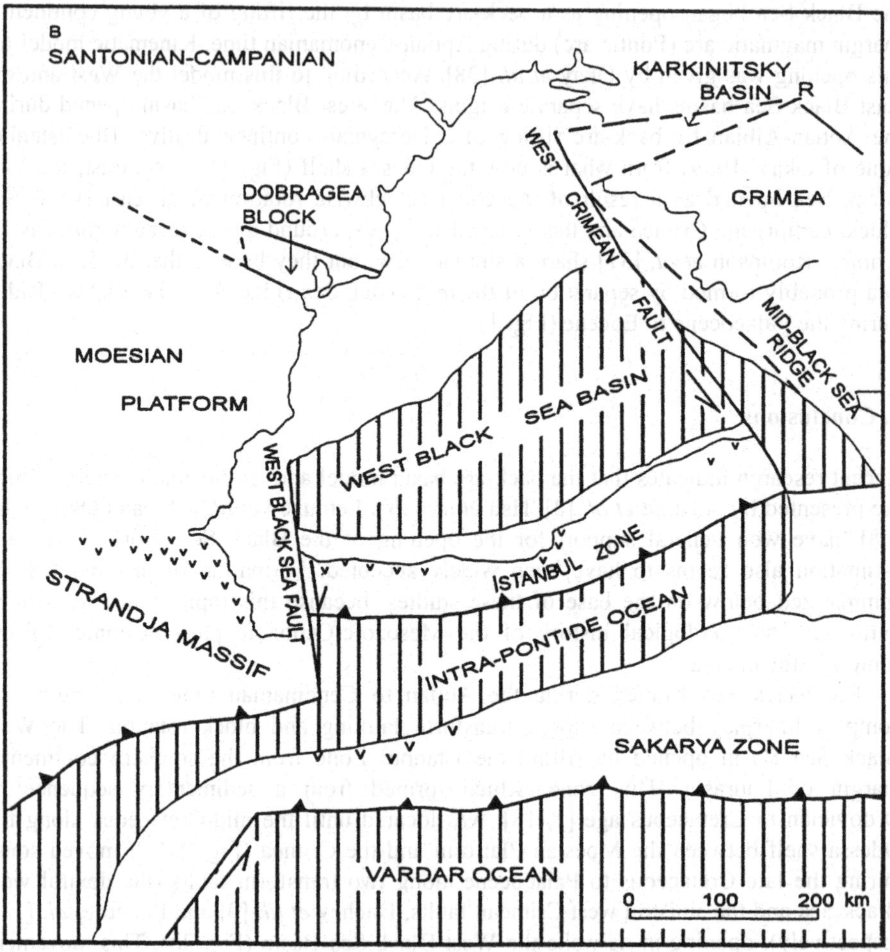

Figure 2b Santonian-Campanian palinspastic palaeogeographic map of the West Black Sea region (after Okay *et al* [30])

Timing of rifting of the West Black Sea Basin can be clearly demonstrated by its well-developed southern continental margin sequence. The stratigraphic record of the East Black Sea Basin is poor in the eastern Pontides, because this region is underlain by the Upper Cretaceous to Eocene Pontide magmatic arc (Fig. 1). The continetal margin sediments of this basin must be below the sea. The syn-rift sediments of the West Black Sea Basin is characterized by a deepening-upward clastic sequence accumulated during the Aptian to Cenomanian time interval. The post-rift sequence is represented by the post-Cenomanian volcaniclastic turbidites, shales, and sandy limestones. The syn breakup sequence is formed from red pelagic micrites and marls of the Upper Cenomanian age [16, 17].

References

1 Abdüsselamoğlu, Ş. (1977) The Palaeozoic and Mesozoic in the Gebze region Explanatory text and excursion guidebook 4th Colloquium on Geology of the Aegean Region, Excursion 4 Western Anatolia and Thrace, ITÜ Maden Fak , Istanbul, 16 pp

2 Adamia, Sh A , Gamkrelidze, I P , Zakariadze, G S and Lordkipanidze, M.B (1974) Adjaro Trialetsky progib i problema formirovaniya glubokovodnoi vpadiny Chernogo morya *Geotektonika* **1**, 78-94

3 Apolskiy, O P (1974) O proiskhozhdenu vpadin Chernogo Morya i Yuzhnogo Kaspiya *Geotektonika* **5**, 94-97

4 Artyushkov, E V (1992) Role of crustal streching on subsidence of the continental crust *Tectonophysics* **215**, 187-207

5 Beloussov, V V. (1967) Some problems of the evolution of the Earth's crust and upper mantle of the oceans *Geotectonics* **1**, 3-14

6 Beloussov, V V , Volvovsky, B S , Arkhipov, I V , Buryanova, B.V , Evsyukov, Y D , Goncharov, V P , Gordienko, V V , Ismagilov, D F , Kislov, G K , Kogan, L I , Kondyurin, A V , Kozlov, V N , Lebedev, L I , Lokholatnikov, V M , Malovitsky, Y.P , Moskalenko, V N , Neprochnov, Y P , Otisty, B K , Rusakov, O M , Shimkus, K M , Schlezinger, A E , Sochelnikov, V V , Sollogub, V B , Solovyev, V D , Starostenko, V.I , Starovoitov, A F , Terekhov, A A , Volvovsky, I S , Zhigunov, A S and Zolotarev, V G. (1988) Structure and evolution of the Earth's crust and upper mantle of the Black Sea *Boll Geofis Teor Appl* **30**, 109-196

7 Brinkmann, R (1974) Geologic relations between Black Sea and Anatolia In. E.T Degens and D A Ross (eds), *The Black Sea-Geology, Chemistry and Biology*, AAPG Memoir 20, Tulsa, Okla, pp. 63-76

8 Bulandje, Y D , editor (1976) Kompleksnoye Issledovaniye Chernomorksy Vpadini *Akad Nauk*, SSSR, Moscow, 98 pp

9 Dachev, C , Stanev, V and Bokov, P (1988) Structure of the Bulgarian Black Sea *Bollettino di Geofisica Teorica ed Applicata* **30**, 79-107

10 Deuser, W G (1974) Evolution of anoxic conditions in Black Sea during Holocene In E T Degens and D A Ross (eds), *The Black Sea-Geology, Chemistry and Biology*, AAPG Memoir 20, Tulsa, Okla, pp 133-136

11 Dewey, J F , Pitman, W C III , Ryan, W B F and Bonnin, J (1973) Plate tectonics and the evolution of the Alpine System *Geol Soc Am Bull* **84**, 3137-3180

12 Erinç, S (1984) Geomorphological and structural characteristics of the Black Sea basin and its morphometry *V I S.I İstanbul* **1**, 15-22

13 Finetti, I , Bricchi, G , Del Ben, A , Pipan, M and Xuan, Z (1988) Geophysical study of the Black Sea area *Boll Geofis Teor Appl* **30**, 197-324

14 Frech, F. (1899) *Lethaea Geognostica Vol 2,* Schweizerbart, Stuttgart, 788 pp

15 Goncharov V P and Neprochnov, Yu P (1960) Geomorphology of the bottom and tectonic problems in the Black Sea, in *International Dictionary of Geophysics*, Pergamon, New York, N Y , pp 1-6

16 Görür, N (1988) Timing of opening of the Black Sea basin, *Tectonophysics* **147**, 247-262

17 Gorür N , Tuysuz, O., Aykol, A , Sakinç, M , Yiğitbaş, E and Akkök, R (1993) Cretaceous red pelagic carbonates of northern Turkey Their place in the opening history ofthe Black Sea *Eclogae Geologica Helvetica* **86**, 819-838

18. Görür, N , Monod, O , Okay, A I , Sengör A M C , Tüysuz, O , Yigitbaç, E , Sakmç, M and Akkök, R (1997) Palaeogeographic and tectonic position of the Carboniferous rocks of the western Pontides (Turkey) in the frame of the Variscan belt *Bull Soc Geol France* **168(2)**, 197-205

19 Hsü, K J , Nacev, I K and Vucev, V T (1977) Geologic evolution of Bulgaria in the light of plate tectonics *Tectonophysics* **40**, 245-256

20 Ketin, I (1983) *Turkiye jeolojisine genel bir bakış* Istanbul Teknik Universitesi Vakfi, Istanbul, 595 pp

21 Letouzey, J , Bijou-Duval, B , Dorkel, A , Gonnard, R , Kristchev, K , Montadert, L and Sungurlu, O (1977) The Black Sea a marginal basin, geophysical and geological data, in B Bijou-Duval and L

Montadert (eds.), *Internatıonal Symposıum on the Structural Hıstory of the Medıterranean Basıns* Editions Technip, Parıs, pp. 363-376
22 Milanovskiy, Ye. (1967) Problema proıskhozdenıya Chernomorskoy vpadıny ı yeye mesto v strukture al'piyskogo poyasa Moscow Unıv, *Vestn Ser Geol* **22,** 27-43
23 Mındelı, P Sh, Neprochnov, Yu P and Pataraya, Ye I (1965) Opredelenıye oblastı otsutstvıya granıtnogo sloya v Chernomorskoy v padine po dannym GSZ ı seysmologıı *Akad Nauk SSSR Izv Ser Geol* **2**, 2-15 Engl transl ,1966, Granıte-free area ın Black Sea trough from seısmıc data *Internat Geology Rev* **8(1)**, 36-43.
24. Muratov, M V and Neprochnov, Yu P. (1967) Stroyenıye dna Chernomorskoy kotlovıny I yeye proıskho zhdenıye (Structure of Black Sea basın and ıts origın) *Moskov Obshch. Ispytatelye Prırody Byull Otdel. Geol* **42(5)**, 40-59.
25 Muratov, M V. (1972) Istorıya formirovanıya glubokovodnoı kotlovıny Chernogo Morya v sravnenuıs vpadınami Sredızemnogo *Geotektonıka* **5,** 22-41
26 Nalıvkin, D V (1960) *The geology of the USSR* Pergamon, New York, N Y ,170 pp
27. Neprochnov, Yu.P (1959) Glubınoye stroyenıye zemnoy kory pod Chernym morem v yugozapady ot Kryma po seysmıcheskım dannym (Deep structure of Earth's crust underneath Black Sea, southwest of Crımea, from seismıc data) *Akad Nauk SSSR Doklady* **126(5)**, 1119-1122
28 Okay, A I., Şengör, A M C and Gorür, N (1994) Kınematıc hıstory of the openıng of the Black Sea and ıts effect on the surrounding regıons *Geology* **22**, 267-270
29 Östlund, G (1974) Expedıtion " Odysseus 65" Radıocarbon age of Black Sea deep water ın E T Degens and D A. Ross (eds), *The Black Sea- Geology, Chemıstry and Bıology* AAPG Memoır 20, Tulsa, Okla, pp 127-132
30 Petrascheck, W.E. (1960) Über ostmedıterrane Gebırgzusammenhange *Abh. Dtsch Akad Wıss Berlın,* **K1. III, 1** (Kraus-Festschrıft). 9-18
31 Robınson, A.G , Rudat, J H , Banks, C J and Wıles, R L F (1996) Petroleum geology of the Black Sea, *Marıne and Petroleum Geology* **13(2)**, 195-223
32 Ross, D A., Uchupı, E , Prada, K E and MacIlvaıne, J C (1974) Bathymetry and mıcrotopography of Black Sea, ın: E T. Degens and D A Ross (eds), *The Black Sea- Geology, Chemıstry and Bıology* AAPG Memoır 20, Tulsa, Okla, pp 1-10
33 Schlezinger, A E. (1981) Struktura osadochnogo chekhla chernomorskogo basseına, ın A V Peıve and Yu N. Pushcharovsky (eds), *Problemy Tektonıkı Zemnoı Kory* Nauka, Moscow, pp 237-262
34 Sidorenko, A V (1978) *Karta razlomov terrıtorıı SSSR ı sopredelnıch stran,1/2 500 000*, Akad Nauka, SSSR, Moscow
35 Sorokhtın, O G (edıtor) (1979) *Geodınamıka*, ın Okeanologıya Geofizıka Okeanskogo Dna Nauka, 2, Moscow, 375 pp
36 Stılle, H. (1928) Uber europäisch-zentralasıatısche Gebırgzusammenhänge Nachr Ges. Wıss Gottıngen, *Math -Phys* **Kl**, 173-201.
37 Stılle, H ,1953 Der geotektonısche Werdegang der Karpaten *Beıh Geol Jahrb* **8**, 239
38 Subbotin, C I , Sollogub, V B., Posen, D , Dragasevıc, T , Mıtuch, E and Posgay, K (1968) Junctıon of deep structures of the Carpatho-Balkan regıon wıth those of the Black and Adrıatıc Seas *Can J Earth Scı* **5**, 1027-1035
39 Şengör, A M C and Yılmaz, Y (1981) Tethyan evolutıon of Turkey A plate tectonıc approach *Tectonophysıcs* **75**, 181-241.
40 Tugolesov, D.A , Gorshkov, A.S., Meysner, L.B , Solov yev, V V and Khakhalev, Y M (1985) The tectonıcs of the Black Sea trough *Geotectonıcs* **19**, 435-445
41 Vardapetyan, A N (1981) Pozdnekaınozoıskaya tektonıka plıt chernomorsko Kaspuskogo regıon Ph D Thesıs Inst of the Oceanology, Moscow, 24 pp
42 Wılser, J L (1928) Dıe geotectonısche Stellung des Kaukasus und dessen Bezıehungen zu Europe (Geologıe der Schwarzmeer- Umrandung und Kaukasıens II. Stuck) Z *Dtsch Geol Ges* **80**, 153-194
43 Yanshın, A L , Bassenyants, Sh A , Pılıpenko, A I and Shlezınger, A E (1980) Novye dannye o vremenı obrazovanıya glubokovodnoı Chernomarskoı vpadıny *Dokl Akad Nauk SSSR* **252**, 223-227
44 Zonenshaın, L P and Le Pıchon, X (1986) Deep basıns of the Black Sea and Caspıan Sea as remnants of Mesozoıc back-arc basıns. *Tectonophysıcs* **123**, 181-212

GEOCHEMISTRY OF THE LATE PLEISTOCENE-HOLOCENE SEDIMENTS OF THE BLACK SEA: AN OVERVIEW

M. NAMIK ÇAĞATAY
Institute of Marine Sciences and Management,
Istanbul University
Müşküle Sokak, Vefa 34470 Istanbul, Turkey

Abstract

The Black Sea sediments deposited during the last 30,000 yr consist of three units which can be traced over the deep Black Sea basin. The geochemical composition of these sediments reflects the paleogeographic, oceanographic and biological evolution of the basin during this period. The youngest unit, Unit 1, is a microlaminated coccolithophore mud that has been depositing since ~3,000 yr BP, after the invasion of the Black Sea by coccolithophore *Emiliania Huxleyi*. This unit is enriched in Ca and Sr because of its high biogenic carbonate content and contains organic matter of mixed marine/terrestrial origin.

Unit 2 is a sapropel unit that was formed after the influx of Mediterranean waters at ~7,000 yr BP. Compared to the other units, this unit is enriched in organic carbon (C_{org}), Mo, U, B, Cu, Ni, Co, V and Ba. The elements enriched in Unit 2 are either mainly associated with the organic matter (U, V, B), sulfides (Cu, Co, Mo), or with biologically secreted barite (Ba). Organic matter in this unit appears to be heterogeneous in composition, but probably mainly of marine origin. Low Mn contents of Unit 2, which are comparable with that of Unit 1, imply deposition of this unit under unoxic bottom waters. High Ba contents in Unit 2 indicate high rate of organic productivity during its formation.

Unit 3 is a laminated clayey mud deposited during a freshwater lacustrine phase of the basin. The organic matter in this unit is of terrestrial origin. Compared to the other two units, this unit contains higher contents of Mn, Zr and Ti. The high Mn contents imply deposition through oxic waters and high Zr and Ti point to a detrital aluminosilicate source.

1. Introduction

The Black Sea is the largest anoxic basin of the world with a maximum depth of 2250 m (Figure 1). It is connected to the Mediterranean and the world ocean system via the Sea of Marmara and the Straits of Bosphorus and Dardanelles. The present sill depths of the Boshorus and Dardanelles straits are 35 and 65 m, respectively. The Black Sea has a

S. Beşiktepe et al. (eds.),
Environmental Degradation of the Black Sea: Challenges and Remedies, 9–22.

pycnocline at a depth of about 150-200 m, separating aerated brackish waters (~18 ‰) from anaerobic, H_2S-rich more saline waters (~22.5 ‰). In recent years oceanographic studies have established the presence of a suboxic layer close to the pycnocline, which is important for biogeochemical and redox reactions (e.g., [4, 22, 57, 58]).

The Black Sea basin consists of four physiographic parts: continental shelf, continental slope, basin apron and abyssal plain [68, 69]. The shelf is wide in the northwest whereas it is narrow (<20 km) along the Anatolian coastline and

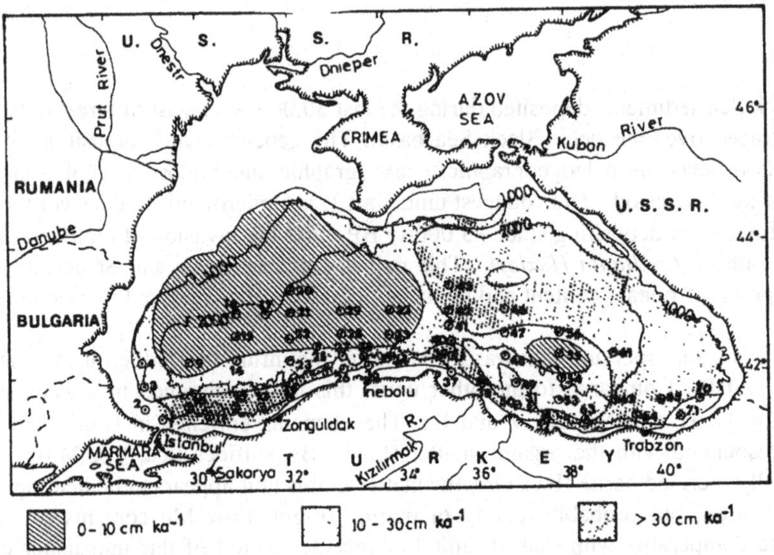

Figure 1 Location of the cores used by Çağatay et al (1987, 1990) on a simplified bathymetric sedimentation map of the Black Sea (batymetry and sedimentation rates during the last 3 kyr from [67]

Caucasus (Figure 1). The southern shelf and slope has been dissected by some canyons trending at roughly right angles to the coast. These canyons are important in transportation of riverine sediment loads to the abyssal plains that cover large areas in the central part.

Because of its special oceanographic characteristics and geological setting, the Black Sea sediments contain sensitive records of the sea level, tectonic, past climatic changes. The shallow sill depth at Bosphorus caused the connection of the Black Sea with the world ocean system to be intermittently cut off especially during the glacial lowstands. In the last glacial epoch and deglaciation, the Black Sea was a fresh water lake, and a lacustrine clay unit (Unit 3) was deposited (Figure 2) [24, 66]. The last connection with the Mediterranean was established through the Bosphorus at 7,150 yr BP [70]. Following this connection, high organic productivity and restricted circulation conditions caused deposition of a sapropel unit (Unit 2). With the invasion of the basin by the coccolithophore *Emiliania Huxleyi* at about 3000 yr BP [66] and the establishment of the present oceanographic conditions, a microlaminated coccolith mud (Unit 1) started depositing in the basin.

Black sea has been cited as a present-day model for the formation of petroleum source rocks (e.g., [28, 39, 83, 84]), syngenetic Kupferschiefer-type Cu-Pb-Zn deposits [88] and black-shale-type uranium deposits [8]. Therefore, the late Quaternary Black Sea sediments have been the subject of numerous mineralogical and organic/inorganic geochemical studies (e.g., [6, 7, 16, 17, 18, 19, 26, 28, 31, 33, 37, 39, 42, 45, 56, 60, 65, 71, 73, 74, 78, 86]).

The purpose of this paper is to overview these studies concerning the salient organic and inorganic geochemical characteristics of the late Quaternary Black Sea sediments, and together with some new data, relate these characteristics to the paleogeographic, oceanographic and biological evolution of the Black Sea.

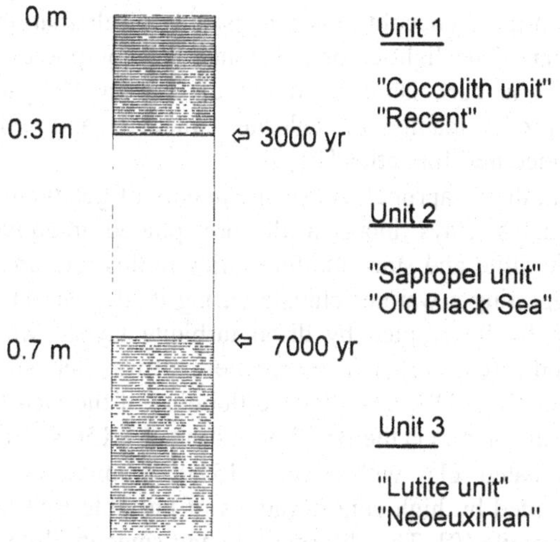

Figure 2. Late Pleistocene-Holocene stratigraphy of the deep Black Sea basin

2. Stratigraphy of the late Quaternary sediments

The general stratigraphic and sedimentological features of the late Quaternary sediments in the Black Sea were studied extensively by the Soviet workers [1, 3, 79, 80] and by the R/V Atlantis II expedition [66, 67]. These studies clearly established the presence of three distinct stratigraphic units in the top 1 m in the abyssal plain (Figure 2). These units have been deposited in the last 30,000 yr and named from top to bottom as Unit 1 (coccolith), Unit 2 (sapropel) and Unit 3 (lutite). They roughly correspond to the "Recent", "old Black Sea" and "Neoeuxinian" units of the Russian workers (e.g., [79, 80]). The shelf areas are mostly composed of pelecypod, gastropod, foraminifer, coccolithophore and ostracod-bearing sand and mud [16]. These shelf sediments locally contain layers consisting of shell accumulations and black organic-rich lamina.

The Unit 1/Unit 2 and Unit 2/Unit 3 boundaries on the R/V Atlantis II cores were dated to be 3450 and 7090 BP, respectively, using ^{14}C dating [66]. Later, Jones (1990) [47] and Jones *et al.* (1994) [48] obtained 2720 and 7540 yr BP, respectively, for the same boundaries, using the accelerator mass spectrometer ^{14}C technique on the R/V Knorr core samples. Some workers used varve chronology to date the unit boundaries, assuming that one light and one dark laminae represent one year [26, 30, 41]. On the basis of varve chronology the ages obtained for the base of Unit 1 and Unit 2 in the deep basin range between 1000-1600 and 4000-5100 yr, respectively. Calvert *et al.* (1987) [19] placed the boundaries at 1600 and 6600 yr BP, respectively, after allowing for a correction to the radiocarbon ages by the varve chronology established by Degens et al. (1978) [25]. However, dating by varve counting in the Black Sea sediments is now known to be incorrect on the bases of ^{210}Pb mass accumulation rate calculations [14] and sediment trap studies [15, 43]. Therefore, the radiocarbon ages of ~3000 and ~7000 yr BP for the base of Unit 1 and Unit 2 are accepted in the present paper.

Unit 1 is ~30-cm-thick coccolith marl, consisting of alternations of light- and dark-colored microlaminae. The light-colored laminae are composed mainly of calcareous coccolith remains. The dark laminae consist of clays and organic matter. The clay minerals consist predominantly of chlorite, smectite and illite [56], with high chlorite/illite and smectite/illite ratios [18].

Unit 2 is ~40 cm thick sapropel, consisting mainly of gelatinous organic matter with some coccolith remains, clays, inorganically precipitated aragonite, iron monosulfides and pyrite. Chlorite/illite and smectite/illite ratios in this unit are lower than the other two units [18], suggesting a warmer climate during its deposition than the present one. Unit 2 is occasionally interrupted by distal turbidite layers, which sometimes occur between Units 1 and 2 (e.g., [18]). The sapropel unit was deposited during a period of high plankton productivity [18, 61] after the flooding of the lacustrine Black Sea basin by the Mediterranean waters via the Bosphorus strait at 7,150 yr BP [70].

Unit 3 is a laminated, clay with a low (~15%) carbonate content. It includes dark laminae that are formed by high concentrations of unstable iron mono-sulfides such as mackinawite and greigite [9]. The chlorite/illite and smectite/illite ratios in this unit are high as in Unit 1 [18]. Unit 3 was deposited under freshwater lacustrine conditions during ~30,000-7,000 yr BP [29, 66] when the water level of the Black Sea was about 100 m lower than the present sea level [70].

3. Geochemistry

3.1 ORGANIC GEOCHEMISTRY

3.1.1 Organic carbon content and distribution

The organic carbon (C_{org}) content of the sapropel unit 14.25% and averages ~10% (Table 1). The average C_{org} contents of Units 1 and 3 are ~3% and ~1%, respectively. The surface sediments from all the physiographic regions of southern part of the Black Sea Basin contain an average of 2.3% C_{org}. The areal distribution of C_{org} contents of composite samples comprising all the three units shows an increase from <1% on the shelf to >5% on the abyssal plain [17]. This distribution closely approximates that of the sedimentation rates given by Ross *et al.* (1970) [67].

TABLE 1. Calcium carbonate and organic carbon contents of the Black Sea sediments [16]

Stratigraphic unit	n	CaCO3 (%)		Organic carbon (%)	
		Average	Range	Average	Range
Unit 1	34	39.4	22.0-65.4	3.1	2.0-5.4
Unit 2	51	12.0	7.0-15.1	10.1	6.2-14.3
Unit 3	24	15.1	10.1-18.0	1.0	0.4-1.5
Surface sediments[+]	65	30.0	15.5-65.4	2.3	1.1-5.4

n= number of samples analyzed

[+] include shelf, slope and Unit 1 sediments of the abyssal plain.

The high C_{org} values in Unit 2 are a result of the high organic productivity and organic mass accumulation rates in the Black Sea during the sapropel formation (e.g., [18, 21]). The relatively high C_{org} content of Unit 1 is partly the result of high organic matter preservation under unoxic water-column conditions. High influx of terrigenous aluminosilicate material (sedimentation rates=40-90 cm/kyr; [66], together with low preservation rates in an oxic water column, has resulted in the low C_{org} content of Unit 3.

3.1.2 Organic matter characterization

The composition of the organic matter in the three sediment units, and especially in the sapropel unit (Unit 2), has been extensively studied by gas chromatography-mass spectrometry (GC/MS) [62, 74, 75, 87], infra-red spectroscopy [44], pyrolysis-gas chromatography-mass spectrometry (Py/GC/MS) [31, 33, 35, 85], ^{13}C isotope analysis – [19, Çağatay, Carrigan and Egesel, unpubl. Data], ^{13}C magnetic resonance spectrometry (Ediger et al., in press), CHN elemental analysis [Egesel and Çağatay, unpubl. Data], Rock-Eval pyrolysis [Çağatay, Carrigan and Egesel, unpubl. Data] and reflectance microscopy [33]. These various studies have often produced contradictory results concerning the origin of the organic matter in these sediments. The discrepancy between the GC/MS and other methods is probably due to the extraction of a small amount of organic matter that may not be representative of the bulk material [19] and/or to spatial and temporal variability of the organic matter source. The results of the organic geochemical studies for each stratigraphic unit are summarized below.

According to the GC/MS studies by Peake *et al.* (1974) [60], the organic matter in Unit 1 is mainly of planktonic origin, consisting of fatty acids, sterols, chlorins and porphyrins. A similar study by Pelet and Debyser (1977) [62], however, shows that the organic matter in this unit is of mixed origin, with the marine component being more dominant than the terrestrial one. The organic carbon in Unit 1 has a $\delta^{13}C$ range from - 24.0 to - 25.0 ‰ [19]. Considering that the Black Sea plankton has a $\delta^{13}C$ value of - 23 ‰ [29] and the temperate zone C_3-type land plants -27 ‰ Deines [27], Calvert and Fontugne [19] concluded that the organic carbon of Unit 1 contains a significant fraction of terrestrial carbon.

Results of organic geochemical studies on the sapropel unit are more contradictory. According to the early studies the organic matter is mainly of terrestrial origin and consists of spores, pollen and other plant parts containing sterols and fatty acids [74, 75] and bitumen, humic and fulvic acids [86]. Similarly, Hunt [45] concluded that the hydrocarbon composition of the organic matter is enriched in aromatic and asphaltic compounds of terrestrial origin, but poor in paraffin compounds of marine organic affinity. However, the results of Pelet and Debyser [62] and Lee *et al.* [54] showed that the marine component of the organic matter is more important than the terrestrial one. Furthermore, infra-red spectroscopic studies by Huc *et al.* [44] indicate that the sapropel is composed predominantly of aliphatic hydrocarbons.

Elemental C/N of Unit 2 in Core 46 in the eastern Black Sea (Figure 1) ranges from 12.0 to 16.4 and averages 14.5 (Table 2). Considering the C/N of marine phytoplankton and zooplankton lies between 5 and 8 and that of land plants between 20 and 200 [34], the C/N values of the Black Sea sapropel suggest that the organic carbon is probably mainly of marine origin, with a significant contribution of terrestrial material.

According to Calvert and Fontugne [19], the $\delta^{13}C$ of Unit 2 range from -23.1 and -26.5 ‰, with the heaviest values corresponding to the highest organic carbon contents in the cores. However, recent unpublished $\delta^{13}C$ data obtained by Çağatay, Carrigan and Egesel (Table 2) suggest a significant contribution of terrestrial organic matter. Although stable carbon isotope analyses are widely applied to infer the origin of organic matter, $\delta^{13}C$ values of different organic matter have overlapping wide ranges [27], thus making the interpretation difficult.

TABLE 2 Elemental C/N, stable carbon isotope and Rock-Eval analyses of Black Sea sapropel (Unit 2) from Site 46 in eastern Black Sea basin (Çağatay, Carrigan and Egesel, unpubl. data)

Paramete r	C/N n = 17	$^{13}\delta$ (‰ PDB) n = 20	HI n = 7
Average	14.49	-25.78	1071
Range	12.01-16.43	-24.43 to -27.35	400-1800

n = number of samples analyzed

Recent studies, involving Py/GC/MS analysis [31, 33, 35, 85], Rock-Eval pyrolysis (Table 2; Çağatay, Carrigan and Egesel, unpublished data) also indicate the predominance of marine organic matter in Unit 2. These various analyses, together with ^{13}C magnetic resonance spectrometry [33], have further demonstrated that the sapropel is a Type 1 kerogen, having similar organic structures to oil shales. Rock-Eval analysis of Unit 2 has produced very high hydrogen index values (Table 2) that also strongly suggest Type 1 kerogen of marine algal origin [36, 83]. Py/GC/MS analysis shows that the principal organic components in the sapropel are alkanes, alkenes, alkyl benzenes, alkyl naphthalenes and a few phenols [33]. In view of these recent results, it can be concluded that the organic matter in Unit 2 is mostly of marine planktonic origin, with local and variable contribution of a terrestrial component causing some variation in the composition.

The results of the various methods of analyses, such as GC/MS and stable C isotope, show without any ambiguity that the organic matter in Unit 3 consists predominantly of terrestrial remains [19, 74]. This is to be expected from the paleogeographic setting of the Black Sea during the deposition of Unit 3. At this time, with the lake water level being 100 m lower than the present sea level and most of the shelf areas being subareally exposed [70], the rivers emptied their loads of terrestrial organic matter directly into the deep basin.

3.2 INORGANIC GEOCHEMISTRY

3.2.1 Calcium Carbonate and Organic Carbon Distributions

The $CaCO_3$ contents of the sediments are highest in Unit 1, ranging up to 65% and averaging 39% (Table 1). The $CaCO_3$ concentrations of Units 2 and 3 are much lower, averaging 12% and 15%, respectively. Most of the $CaCO_3$ content of Unit 1 is made up of coccolithophore tests that are present as light laminae in this unit. However, on the shelf areas there is an important contribution of $CaCO_3$ from mollusk, foraminifer and ostracod shells. This makes the surface sediments in the whole of the Black Sea basin relatively rich in $CaCO_3$.

3.2.2 Minor and trace element composition

A comparison of the minor and trace element concentrations in the three units shows enrichments of Mo, U, B, Cu, Ni, Co, V, and Ba, in Unit 2 relative to the other two units (Figure 3; Table 3). The Mo and U contents of Unit 2 are about four times and Cu, Ni, Co, B, and V contents are twice those of Unit 3. Unit 1 contains strontium and some of the elements enriched in Unit 2, such as U, Co and Mo, in higher concentrations than Unit 3. Unit 3 is enriched in Ti, Zr and Mn relative to the other two units.

The elements enriched in Unit 2 are either associated with the organic matter or precipitated as authigenic minerals. Separation and selective extraction of the organic and sulfide fractions of Unit 2 sediments by Volkov and Fomina [86] and Philipchuk and Volkov [63] suggest that Mo, Cu, Ni, Co and V are mostly concentrated in both the fractions, with V being present mainly in the organic, and Mo and Co in the sulfide fractions. They also conclude that Co, Cu and Mo are coprecipitated with iron sulfides in their sulfide phases. In anoxic basins sulfide precipitation starts in the anoxic water layer [12, 46, 52, 77], whereas in oxic basins it takes place in sediment pores of the sulfate reduction zone (e.g., [72]). In Unit 2, V enrichment probably occurs by surface adsorption of vanadyl ions (VO^{2+}) on organic matter c‹ by precipitation of V_2O_3 or $V(OH)_3$ [20], although V (and Ni) could also be present as metal porphyrins [55].
Compared to Unit 3, uranium is one of the most enriched elements in Units 1 and 2 (Figure 3). The association of this element with sedimentary organic matter is well known (e.g.,[11, 49, 51]. Although according to the thermodynamic stability relations [53], U(IV) species should be the most favoured ones in the anoxic water column of the Black Sea, U occurs almost entirely as U(VI) [2]. Experimental studies by Szalay [81], Baturin *et al.* [7], Kochenov *et al.* [50] and Nakashima *et al.* [59], and thermodynamic calculations by Andreyev and Chumachenko [5] and Langmuir [53] show that under the Eh-pH conditions of Units 1 and 2, U(VI) species are reduced to U(VI) and uraninite could precipitate. In the Black Sea cores pore water profiles [2], together with high levels of solid phase U, indicate the removal of U from the pore waters. A selective extraction study by Çağatay *et al.* [17] on the Unit 1 and 2 sediments suggest that U is mainly associated with the organic matter in ion-exchangeable and organo-uranium forms. This study, however, does not rule out the presence of uraninite in the Black Sea

sediments and demonstrates that high solid phase U always occurs in sediments with >2% organic carbon - a possible requirement for the reduction and fixation of uranium.

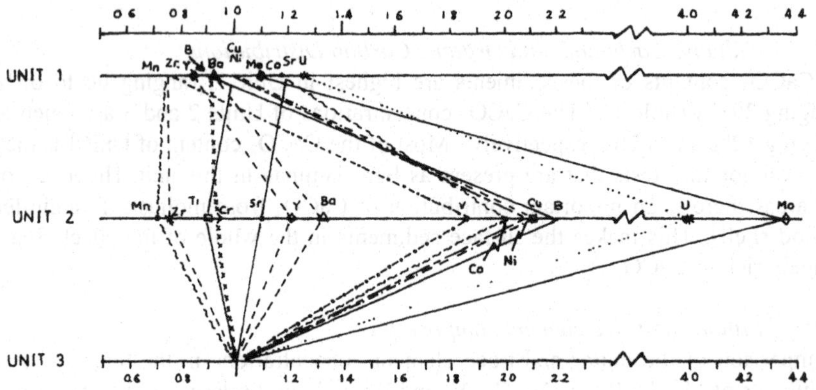

Figure 3 Relative enrichment of elements in different sediment units of the Black Sea. The numbers along the horizontal axes represent the ratios of average element concentrations in Unit 1 or Unit 2 to those in Unit 3 [16]

TABLE 3 Minor and trace element composition of late Quaternary sediments in the Black Sea [16] (Number of samples analyzed in Units 1, 2 and 3 is 335, 47, and 95, respectively)

Unit	Sr	Ba	Cu	Mo	U	V	B	Mn	Co	Cr	Ni
Unit 1											
Average	246	207	30	12	3.9	47	32	501	14	80	39
Minimum	205	50	25	5	1.7	10	10	350	5	5	10
Maximum	2100	1900	75	60	18.7	250	150	950	70	305	180
Unit 2											
Average	224	297	62	48	11.5	124	75	469	24	81	81
Minimum	200	50	35	10	2.5	40	30	290	5	40	20
Maximum	710	1510	150	150	21.7	300	205	790	70	410	215
Unit 3											
Average	207	223	30	11	3.6	63	34	680	12	87	39
Minimum	210	50	20	8	1.7	10	10	415	5	40	10
Maximum	700	1100	80	25	13.6	175	85	2950	40	200	120

The Ba enrichment in Unit 2 indicates increased primary organic productivity during the sapropel deposition [18]. Barium is used as a proxy for paleoproductivity, with high Ba contents being often found in organic-rich sediments, including sapropels [13, 18, 32, 40]. In such sediments, Ba occurs as biologically secreted 1-3 μm barite crystals [10, 82]. Although no barite crystals have yet been reported from the Black Sea sapropel, it is reasonable to assume that Ba occurs as micro barite crystals in this organic-rich sediment unit [18, 42].

Boron shows more than two-fold enrichment in Unit 2 compared to Units 1 and 3 (Table 3; Figure 3). At low concentrations, B is a nutrient for most plants and enriched in the marine shales and plants [55].

Manganese contents of Units 1 and 2 are comparable, and lower than those of Unit 3 (Table 3; Figure 3). This is contradictory with the results of Calvert [18] who found higher contents of this element in Unit 2 than in Unit 1 in a core from the central part of the Black Sea basin. Manganese exists as Mn(IV) oxyhydoxide colloids in the oxic and as dissolved Mn(II) in the anoxic water layers of the Black Sea basin [76]. In the anoxic sediment column, it precipitates as Mn(II) carbonate, provided sufficient alkalinity is reached [18]. In most surface oxic sediments of deep basins and shelves, Mn is highly enriched as Mn-oxyhdroxide crusts and nodules [23, 72]. This enrichment occurs by the upward diffusion of dissolved Mn(II) from the sulfate reduction zone and its precipitation as oxyhydroxides in the surface oxic layer [38]. The recycling of Mn between the oxic and unoxic zones of a sediment column and its enrichment in the oxic layer can obviously occur only under oxic water column conditions (e.g., [18]). Using this criteria, Calvert [18], Pederson and Calvert [61] and Calvert and Pederson [20] interpreted the relatively high levels of Mn in Unit 2 in the core from the central Black Sea to infer that the basin waters were oxygenated during the formation of the sapropel unit. However, on a basin-wide scale [16; Table 3] found no significant difference between the Mn contents of Units 1 and 2, suggesting that Unit 2 was formed under similar euxinic conditions as Unit 1. The relatively high values of Mn in Unit 3 imply that this unit was deposited through oxic waters [16, 18].

Compared to their contents in Unit 3, the lower concentrations of authigenic elements, such as U, Mo, Ni, Cu and Co, in Unit 1 is partly due to the carbonate dilution effect. High Sr concentrations in the Unit 1 are associated with the coccolithic carbonate debris of low Mg-calcite composition. Elevated contents of Ti and Zr in Unit 3 point to a predominantly detrital aluminosilicate source.

4. Conclusions

Three sediment units have been deposited in the deep Black Sea during the last 30,000 yr. These, from top to bottom, are the microlaminated coccolith mud (Unit 1), the sapropel (Unit 2) and the laminated lutite (Unit 3). Unit 3 was deposited during a freshwater lacustrine phase of the Black Sea. Unit 2 deposition startad at ~7,000 yr BP, just after the influx of Mediterranean waters via the Bosphorus Strait. This influx caused a two-layer water body and high surface organic productivity, which in turn, initiated the sapropel deposition. Unit 1 has begun depositing at ~3,000 y BP, after the invasion of the Black Sea by coccolithophore *Emiliania Huxleyi* and establishment of the present oceanographic conditions.

Organic geochemical studies suggest that the organic matter in Unit 1 is mixed marine/ terrestrial and in Unit 3 terrestrial origin. The results for unit 2 are conflicting, but most recent studies show a predominantly marine origin for the organic carbon in this unit. However, the contradictory results suggest a temporally and spatially variable source of organic matter for this unit.

Unit 1 is enriched in Ca (and $CaCO_3$) and Sr; Unit 2 in C_{org}, Mo, U, B, Cu, Ni, Co, V and Ba; and Unit 3 in Mn, Zr and Ti.. The high Ca and Sr contents in Unit 1 are the result of the coccolithic carbonate debris, which has caused the dilution of the other element concentrations in this unit. The elements enriched in Unit 2 are either mainly associated with the organic and biogenic matter (U, V, B, Ba) and/or authigenic sulfides (Cu, Co, Mo). High contents of Ba in Unit 3 indicate high organic productivity during its formation. The Ti and Zr enrichment in Unit 3 point to a detrital source. Elevated Mn content of Unit 3 indicates deposition through an oxic water column, whereas relatively low Mn contents of Units 1 and 2 imply anoxic bottom waters during their deposition.

References

1. Andrusov, N I (1980) Preliminary account of participation in the Black Sea deep-water expedition of 1890, *Isvest Vsesov Geogr. Obshch* **26**, 380-409 (in Russian)
2. Anderson, R F., Fleischeri M Q and LeHuray, A P (1989) Concentration, oxidation state, and particulate flux of uranium in the Black Sea, *Geochim. Cosmochim. Acta* **53**, 2215-2224
3. Arkangel'sky, A D. (1927) On sediments of the Black Sea and their significance in sedimentology, *Bvul. Mosk. Obsch. Ispyt Prir* **5**, 199-289 (in Russian).
4. Baştürk, Ö., Saydam, C. Salihoğlu, I, Eremeeva, L V, Konovalov, S K, Stoyanov, A., Dimitrov, A., Cociasu, A, Dorogan, L. and Altabet, M (1994) Vertical variation in the principle chemical properties of the Black Sea in the autumn of 1991, *Marine Chemistry* **45**, 149-165
5. Andreyev, P F and Chumachenko, A.P (1964) Reduction of uranium by natural organic substances, *Geochemistry International* **1**, 3-7
6. Baturin, G.N. and Kochenov, A V (1968) Relation between some rare metals and organic material in marine sediments, *Oceanology* **7(6)**, 792-809
7. Baturin, G N., Kochenov, A.V and Kovaleva, S.A (1965) Some features of uranium distribution in Black Sea water. Dokl Akad Nauk U S S R **166**, 172-174
8. Bell, R T (1978) Uranium in black shales - a review, in M M Kimberley (Ed), *Short Course in Uranium Deposits*, Mineral. Assoc Can., Toronto, Ontario, pp 307-329
9. Berner, R A (1974) Iron sulfides in Pleistocene deep Black Sea sediment and their paleo-oceanographic significance, in E T Degens and D A Ross (Eds), *The Black Sea - Geology, Chemistry and Biology*, Am. Assoc Pet. Geol, Mem 20, pp 524-531
10. Bishop, J K B (1988) The barite-opal-organic carbon association in oceanic particulate matter, *Nature* **332**, 341-343
11. Breger, I A (1974) The role of organic matter in accumulation of uranium, the organic geochemistry of the coal-uranium association, in *Formation of Uranium Ore Deposits*, I.A.E A, Vienna, IAEA-SM-183/29, pp 99-123
12. Brewer, P G and Spencer, D W (1974) Distribution of some trace elements in Black Sea and their flux between dissolved and particulate phases, in E.T Degens and D.A. Ross (Eds), *The Black - Geology, Chemistry and Biology*, Am Assoc. Pet Geol, Mem 20, pp 137143
13. Brumsack, H J (1986) The inorganic geochemistry of Cretaceous black shales (DSDP Leg 41) in comparison to modern upwelling sediments from the Gulf of California, in C P. Summerhayes and N J Shackleton (eds), *Geol Soc Spec Publ* No 21, pp 447-462
14. Buesseler, K O and Benitez, C R (1993) Determination of mass accumulation rates and sediment radionuclide inventories in the deep Black Sea, *Deep-Sea Research* **41**, 1605-1615
15. Buesseler, K O, Livingston, H D, Honjo, S, Hay, B J, Manganini, S J, Degens, E ,Ittekot, E, Izdar, E, and Konuk, T (1987) Chernobyl radionuclides in a Black Sea sediment trap, *Nature* **329**, 825-828
16. Çağatay, M N, Saltoğlu, T. and Gedik, A (1987) Geochemistry of the recent lack Sea sediments (in Turkish), *Geological Engineering*, Ankara **30-31**, 47-64
17. Çağatay, M N, Saltoğlu, T and Gedik, A (1990) Geochemistry of uranium in the late Pleistocene-Holocene sediments from the southern part of the Black Sea basin, *Chemical Geology* **82**, 129-144

18 Calvert, S.E (1990) Geochemistry and origin of the Black Sea sapropel, in V. Ittekkot, S Kempe, W Michaelis and A Spitzy (eds.), *Facets of Marine Biogeochemistry*, Springer, Berlin, pp 326-352

19 Calvert, S.E and Fontugne, M R (1987) Stable carbon isotopic evidence for the marine origin of the organic matter in the Holocene Black Sea sapropel, *Chemical Geology (Isotope Geoscience Section)* **66**, 315-322

20 Calvert, S.E. and Pederson,T F (1993) Geochemistry of recent oxic and anoxic marine sediments Implications for the geological record, Marine Geology **113**, 67-88

21 Calvert, S.E , Vogel, C S and Southon, J R (1987) Carbon accumulation rates and the origin of the Holocene sapropel in the Black Sea, *Geology* **15**, 818-921

22. Codispoti, L.A Friederich, G.E , Murray, J.W , and Sakamato, C M (1991) Vertical variability in the Black Sea. Implications of continuous vertical profiles that penetrated the oxic/anoxic interface, *Deep-Sea Res* **38(2A)**, 691-710

23 Cronan, D S (1980) *Underwater Minerals*, Academic Press. London, 362 pp

24 Degens, E T and Ross, D A. (Eds) (1972) Chronology of the Black Sea over the last 25,000 years, *Chemical Geology* **10**, 1-16

25 Degens, E T ,Stoffers, P , Golubic, S , and Dickman, M D (1978) Varve chronolgy Estimated rates of sedimentation in the Blak Sea deep-basin, in D A Ross, Y P Neprochanov et al (eds), *Initial Reports of the Deep Sea Drilling Project, Leg 42B*, Deep Sea Drilling Project, Washington DC, pp 499-508

26 Degens, E T Michaelis, W , Garrasi, C , Mopper, K , Kempe ,S and Ittekot, V.A. (1980) Warven-chronolgie und fruhdiagenetische Umsetzungen organischer substanzen Holozaner sedimente des Scwarzen Meeres, *Neues Jahrb Geol Paleontol Monatsch* **1980/2**, 65-86

27 Deines, P (1980) The isotopic composition of reduced organic carbon, in P Fritz and J Ch Fontes (eds), Handbook of Environmental Isotope Geochemistry, Vol 1, The Terrestrial Environment, Elsevier, Amsterdam, pp 329-406

28 Demaison, G J and Moore, G T (1980) Anoxic environments and oil source bed genesis, *Organic Geochemistry* **2**, 9-31

29 Deuser, W G (1974) Evolution of anoxic conditions in Black Sea during Holocene, in E T Degens and D A Ross (Eds), *The Black Sea - Geology, Chemistry and Biology*, Am Assoc Pet Geol , Mem 20, pp 133-136

30 Duman, M (1994) Late Quaternary chronology of the southern Black Sea basin, *Geo-Marine Letters* **14**, 272-278

31 Didyk, B M., Simoneit, B R T , Brassell, S C and Eglinton, G (1978) Organic geochemical indicators of palaeoenvironmental conditions of sedimentation, *Nature* **272**, 216-222

32 Dymond, J , Suess, E and Lyle, M (1992) Barium in deep-sea sediment A geochemical proxy for paleoproductivity, *Paleoceanography* **7**, 163-181

33 Ediger, V , Gaines, A Karayığıt, A I , Galletti, G , Fabbri, D , Snape, C and Sirkecioğlu, O (in press), Chemical structure of Black Sea sapropels, *Marine Geology*

34 Emerson, S and Hedges, J I (1988) Processes controlling the organic carbon content of open ocean sediments, *Paleoceanography* **3**, 139-162

35 Ergin, M , Gaines, A , Galetti, G C , Chiavari, G , Fabbri, D. and Yucesoy-Yilmaz, F (in press) Early diagenesis of organic matter in recent Black Sea sediments characterization and source assessment, *Applied Geochemistry*

36 Espitalié, J , Laporte, J L , Madec, M , Marquis, F , Leplat, P and Paulet, J (1977) Métode rapide de charactérisation des roches méres, de leur potential pétrolier et de leur degré évoultion, *Revue Inst France du Pétr* **32**, 23-43

37 Florovskaya, V N and Gurskiy, Yu No (1966) Organic material in deep-water sediments of Black Sea, *Geochemistry* **3 (1)**,78-83

38 Froelich, P N , Klinkhammer, G P , Bender, Bender, M L , Luedtke, N A , Heath, G R , Cullen, D , Dauphin, P , Hammond, D , Hartman, B and Maynard, V (1979) Early oxidation of organic matter in pelagic sediments of the eastern equatorial Atlantic suboxic diagenesis, *Geochim Cosmochim Acta,* 43 1075-1090 Glagoleva, M A (1961) Regularities in distribution of chemical elements in modern sediments of the Black Sea *Am Geol Inst* **135**, 1-6

39 Glen, C R and Arthur, M A (1985) Sedimentary and geochemical indicators of productivity and oxygen contents in modern and ancient basins The Holocene Black Sea as the "type anoxic basin, *Chemical Geology* **48**, 325-354

40 Goldberg, E D (1958) Determination of opal in marine sediments, *J Marine Research* **17**, 178-182

41. Hay, B J Honjo, S , Kempe, S., Ittekkot, V A , Degens, E T , Konuk, T. and Izdar, E (1990) Interannual variability in particle flux in southwestern Black Sea, *Deep Sea Research* **38 (Suppl 2)**, S1211-S1235
42 Hirst, D.M (1974) Geochemistry of sediments from eleven Black Sea cores, in E.T. Degens and D.A Ross (Eds), *The Black Sea - Geology, Chemistry and Biology*, Am Assoc. Pet. Geol , Mem. 20, pp 430-455
43 Honjo, S , Hay, B J , Manganini, S J , et al., 1987, Seasonal cyclicity of lithogenic particle fluxes at a southern Black Sea sediment trap station, in E T Degens, E Izdar and S. Honjo (eds), *Particle Flux in the Oceans*, Mitteilungen aus des Geologisch-Palaeontologischen Institute der, Univesitat Hamburg, 62, pp 19-39.
44. Huc, A Y , Durand, B and Bonin, J G (1978) Humic compounds and kerogens in cores from Black Sea sediments, in D.A. Ross, Y.P. Neprochanov et al (eds), *Initial Reports of the Deep Sea Drilling Project, Leg 42B*, Deep Sea Drilling Project, Washington DC, pp 749-753
45 Hunt, J M. (1974) Hydrocarbon geochemistry of Black Sea, in E.T Degens and D.A Ross (Eds), *The Black Sea - Geology, Chemistry and Biology*, Am. Assoc Pet Geol., Mem 20, pp 499-504
46. Jacobs, L , Emerson, S and Skei, J. (1985) Partitioning and transport of metals across the O2/H2S in a permanently unoxic basin. Framvaren Fjord, Norway, *Geochim Cosmochim Acta* **49**, 1433-1444
47. Jones, G A. (1990) AMS radiocarbon dating of sediments and waters from the Black Sea, *EOS* **71**, 152
48 Jones, G.A. and Gagnon, A.R (1994) Radiocarbon chronology of Black Sea sediments, *Deep Sea Research* **41(3)**, 531-557
49 Klinkhammer, G P and Palmer, M R (1991) Uranium in the oceans Where it goes and why, *Geochim Cosmochim Acta* **55**, 1799-1806
50 Kochenov, A.V , Korolev, K G , Dubinchuk, V T. and Medvedev, Yu.L (1977) Experimental data on the conditions of precipitation of uranium from aqueous solutions *Geochemistry International* **14 (4)**, 82-87
51 Kornfeld, J.A.(1964) Geochemistry of uranyl oxides in Devonian marine black shales of North America, in U Colombo and G D Hobson (Eds), *Advances in Organic Geochemistry*, Pergamon, Oxford, pp 261-262
52 Landing, W M and Lewis, B L (1991) Thermodynamic modelling of trace metal speciation in the Black Sea, in J.W Murray and E Izdar (eds), *Black Sea Oceanography*, NATO advanced Study Institute, **Kluwer Acad. Publ., Dordrecht, pp 125-160.**
53 Langmuir, D (1978) Uranium solution-mineral equilibria at low temperatures with applications to sedimentary ore deposits, in M M Kimberley (Ed), *Short Course in Uranium Deposits, Their Mineralogy and Origin*, Mineral Assoc Can , Toronto, Ont , pp 17-55
54 Lee, C., Gagosian, R B and Ferrington, J W (1980) Geochemistry of sterols in sediments from Black Sea and South West Africa shelf and slope, *Organic Geochemistry* **2**, 103-113
55 Mason, B and Moore C B (1982) *Principles of Geochemistry*, John Wiley & Sons, New York, 350 p
56 Muller, G and Stoffers, P (1974) Mineralogy and petrology of Black Sea sediments In E T Degens and D A Ross (Eds), *The Black Sea - Geology, Chemistry and Biology*, Am Assoc Pet Geol , Mem 20, pp 200-248
57 Murray, J W , Jannasch, H W , Honjo, S , Anderson, S , Reeburgh, W S , Top, Z , Friederich, G E , Codispoti, L A and Izdar, E (1989) Unexpected changes in the oxic/unoxic interface in the Black Sea, *Nature* **338**, 411-413
58 Murray, J W , Friederich, G.E , Codispoti, L A (1993) The suboxic zone in the Black Sea, in C P Huang, C R O'Melia and J J Morgan (eds), *Aquatic Chemistry*, Advances in Chemical Series No 244, Am Chem Soc.,Washington DC,
59 Nakashima, S , Disnar, J R , Peruchot, A and Trichet, J (1984) Experimental study of mechanism of figation and reduction of uranium by sedimentary organic matter under diagenetic and hydrothermal conditions, *Geochim Cosmochim Acta* **48**, 2321-2329
60 Peake, E , Casagrange, D J and Hodgson, G W (1974) Fatty acids chlorins hydrocarbons, sterola, and carotenoids from a Black Sea core, in E T Degens and D A Ross (Eds), *The Black Sea - Geology, Chemistry and Biology*, Am Assoc Pet Geol , Mem 20, pp 505-523
61 Pedersen, T F , and Calvert, S E , (1990) Anoxia vs Productivity What controls the formation of organic -carbon - rich sediments and sedimentary rocks *AAPG Bull* **74**, 454-466
62 Pelet, R and Debyser, Y (1977) Organic geochemistry of Black Sea cores *Geochim Cosmochim Acta* **41**, 1575-1586

63 Philipchuk, M F. and Volkov, I I (1974) Behaviour of molybdenum in processes of sediment formation and diagenesis in Black Sea. In E T Degens and D.A. R,oss (Eds), *The Black Sea - Geology, Chemistry and Biology*, Am. Assoc Pet Geol , Mem 20, pp 542-553
64 Rona, E and Joensu, O (1974) Uranium geochemistry in Black Sea, in E T Degens and D.A Ross (Eds), *The Black Sea - Geology, Chemistry and Biology*, Am Assoc Pet Geol , Mem 20, pp 570-572
65 Rosanov. A G , Volkov, I.I and Yagodinskaya, T A (1974) Forms of iron in surface layer of Black Sea sediments. In. E T Degens and D A Ross (Eds), *The Black Sea - Geology, Chemistry and Biology*, Am Assoc Pet Geol , Mem 20 pp. 542-553
66 Ross, D A and Degens, E T (1974) Recent sediments of the Black Sea, in E.T Degens and D A Ross (Eds), *The Black Sea - Geology Chemistry and Biology*, Am Assoc. Pet. Geol , Mem. 20, pp.183-199.
67 Ross, D A , Degens, E T and MacIlvaine, J.C (1970) Black Sea recent sedimentary history, *Science* **170** 163-165
68 Ross, D A , Uchpi, E , Prada, K E and MacIlvaine, J C (1974) Bathimetry and microtopography of Black Sea, in E T Degens and D A Ross (eds), *The Black Sea - Geology, Chemistry and Biology*, Am Assoc Pet Geol , Mem 20, pp 1-10
69 Ross, A D , Stoffers, P and Trimonis, E S (1978) Black Sea sedimentary framework, in *Initial Reports of the Deep Sea Drilling Project*, Vol 42, Part 2 U S Gov Print Off, Washington, D C , pp 359-363
70 Ryan, W B F , Pitman III, W C , Major, C O , Shimkus, K , Moskalenko, V , Jones, G A , Dimitrov, P , Gorur, N , Sakınç, M and Yüce. H (in press) An abrupt drowning of the Black Sea shelf, *Marine Geology*
71 Sevast'yanov, V F and Volkov, I I (1966) Chemical composition of iron-manganese concretions of the Black Sea, *Am Geol Inst* **166**, 174-176
72 Shaw, T J , Gieskes, J M and Jahnke, R A (1990) Early diagenesis in differing depositional environments The response of transition metals in pore waters, *Geochim Cosmochim Acta* **54**, 1233-1246
73 Shimkus, K M and Trimonis, E S (1974) Modern sedimentation in Black Sea, in E T Degens and D A Ross (eds), *The Black Sea - Geology, Chemistry and Biology*, Am Assoc Pet Geol , Mem 20, pp 249-278
74 Simoneit, B R (1974) Organic Analysis of Black Sea cores, in E T Degens and D A Ross (eds), *The Black Sea - Geology, Chemistry and Biology*, Am Assoc Pet Geol , Mem 20, pp 477-488
75 Simoneit, B R (1978) Organic chemistry of terrigenous muds and various shales from the Black Sea, in D A Ross, Y P Neprochanov et al.(eds), *Initial Reports of the Deep Sea Drilling Project. Leg 42B*, Deep Sea Drilling Project, Washington DC, pp. 749-753
76 Spencer, D W and Brewer, P G (1971) Vertical advection, diffusion and redox potentials as controls on the distribution of manganese and other trace metals dissolved in waters of the Black Sea, J Geophys Res **76**, 5877-5892
77 Spencer, D W , Brewer, P G and Sachs, P L (1972) Aspects of the distribution and composition of suspended matter in the Black Sea, *Geochim Cosmochim Acta* **36**, 71-86
78 Starikova, N D (1961) Organic substances in the liquid phase of sea and ocean deposits, *Akad Nauk S S S R Inst Okeanol* Tr **50**, 130-169
79 Strakhov, N M (1967) *Principles of Lithogenesis*, Vol 1, Oliver&Boyd, p 245
80 Strakhov, N M (1969) *Principles of Lithogenesis*, Vol 2, Oliver&Boyd, p 609
81 Szalay, A (1964) Cation egchange properties of humic acids and their importance in the geochemical enrichment of $(UO_2)^{2+}$ and other cations, *Geochim Cosmochim Acta* **28(10)**, 1605-1614
82 Thomson, J Higgs, N C , Wilson, T R S , Croudace, I W , de Lange, G J and van Santvoort, P J M (1995) Redistribution of geochemical behaviour of redox-sensitive elements around S1, the most recent eastern Mediterranean sapropel, *Geochim Cosmochim Acta* **17**, 3487-3501
83 Tissot, B P and Welte, D H (1978) *Petroleum Formation and Occurrence*, Springer, Berlin, 538 pp
84 Tourtelot, H A (1979) Black shale - its deposition and diagenesis, *Clays Clay Miner* **27**, 313-321
85 van de Meent, D , Brown, S C , Philp, R P and Simoneit, B R T (1980) Pyrolysis-high resolution gas chromatography and pyrolysis gas chromatography-mass spectrometry of kerogens and kerogen precursors, *Geochim Cosmochim Acta* **44**, 999-1013
86 Volkov, I I and Fomina, L.S (1974) Influence of organic material and processes of sulfide formation on distribution of some trace elements in deep-water sediments of Black Sea, in E T Degens and D A Ross (eds), *The Black Sea - Geology, Chemistry and Biology*, Am Assoc Pet Geol , Memoir 20, pp 456-476

87. Wakeham, S F., Beier, J.A and Clifford, C H (1991) Organic matter sources in the Black Sea as inferred from hydrocarbon distributions, in J W Murray and E Izdar (eds.), *Black Sea Oceanography*, NATO advanced Study Institute, Kluwer Acad. Publ., Dordrecht, pp 319-341.
88 Wedepohl, K.H , Delevaux, M H and Doe, B R (1978) The potential source of lead in the Permian Kupferschiefer bed of Europe and some selected Paleozoic mineral deposits in the Federal Republic of Germany, *Contrib. Mineral. Petrol* **65**, 273-281

IMPORTANCE OF SEDIMENTARY PROCESSES IN ENVIRONMENTAL CHANGES: LOWER RIVER DANUBE – DANUBE DELTA – WESTERN BLACK SEA SYSTEM

N. PANIN, D.C. JIPA, M. T. GOMOIU, D. SECRIERU
National Institute for Marine Geology and Geoecology – Geoecomar
23-25 Dimitrie Onciul St., 70318 Bucharest, Romania

1. Introduction

Draining a territory almost twice the Black Sea area, the Danube River is the most important sediment supplier of this marine basin. Presently the Danube River determines the sedimentation on the northwestern Black Sea shelf area. The Danube influence extends down to the deep sea floor. In turn the Black Sea strictly controls the sedimentation of the Danube material through the sea level variations. Danube Delta is

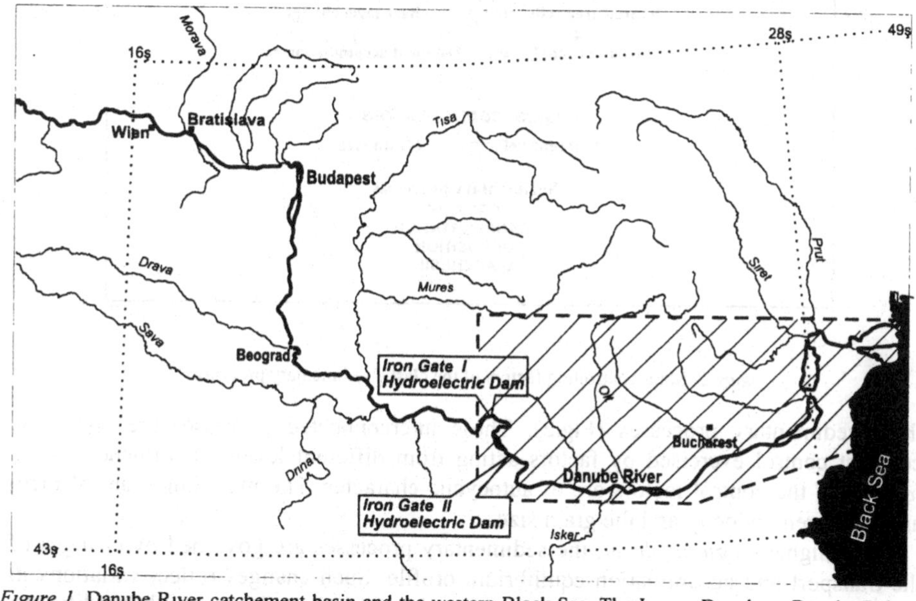

Figure 1 Danube River catchement basin and the western Black Sea The Lower Danube – Danube Delta - western Black Sea area and the Iron Gates dams locations are shown

acting as a regulator of the sediment transfer between the Danube River and the Black Sea. The three factors -river, delta and marine basin- make up a well-established geo-system. Our present interest will concentrate on the sedimentary cycle including the

S. Beşiktepe et al. (eds.),
***Environmental Degradation of the Black Sea: Challenges and Remedies*, 23–41.**

Romanian Carpathians source area, the Lower Danube transport agent and the western Black Sea accumulation basin (Fig. 1).

1.1. SEA LEVEL VARIATION: FOCAL POINT OF SEDIMENTATION – ENVIRONMENT ANALYSIS

Sedimentary terrigenous processes, acting in the Danube – Danube Delta – Black Sea System are represented by the classical *erosion – transportation – deposition – diagenesis* quartet. Biological and anthropogen processes are closely associated with

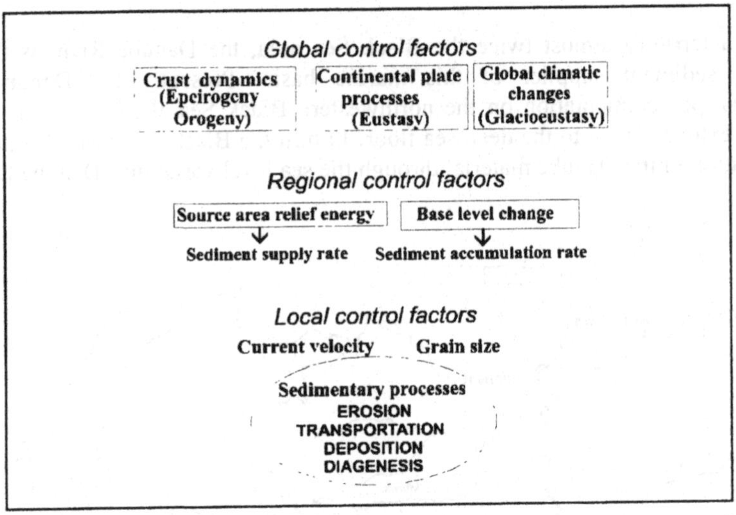

Figure 2 Main controlling factors of the clastic sedimentary processes

these sedimentary processes (Fig.2). These interconnected processes are under the constant control exercised by factors acting from different levels of influence. At a *local level* the control factors are of autocyclic character. The most important of them are the current velocity and the grain size.

At a higher, *regional level*, the sedimentary processes are governed by changes of the transport and accumulation equilibrium profile. Such changes reflect variations in the relief energy of the source area and/or base level position which modify the sediment supply and sediment accumulation rates.

The highest level influence is exerted on the sedimentary processes by the *global control.* The dynamic topography of the earth's crust influences primarily the sediment source areas. Eustatic -and glacioeustatic- sea level variations emerge as another mighty category of factors dominating the sedimentary processes.

More than the influence exercised by the source areas, the sea level variation is known as the driving element of the Black Sea terrigenous sedimentation. Consequently our sedimentary environments analysis will be built around the Black Sea water level variations. Sedimentary processes and their environmental effects will be presented for the high and low sea level scenarios. Considering the available data, the Recent environment can be regarded in the center of the high sea level scenario. The low sea level scenario will be based mostly on the data from the Pleistocene sediments.

2. Sea level rise controlled sedimentary processes and environments of the Lower Danube – Danube Delta – Black Sea System

2.1. DANUBE RIVER

On its way to the Black Sea, Danube River crosses and/or drains territories of twelve European countries. The Danube represents a vital element for about 76 million people. From the second half of the 19th century Danube began to endure aggressive anthropogenic changes, aiming at the more extensive use of its potential. Since that time, while the Black Sea level was high, the anthropogen activities have caused significant modifications of the Danube behavior as a sediment transporting agent and a clastic material supplier.

Along the 2860km long Danube reach and within its catchment area various regulation projects have been realized. The hydroelectric power stations at Aschach, Ottensheim and Linz have been built up on the upper and middle Danube reach. Rather recently the Gabcikovo power station has been constructed on the Danube River, at the Hungarian – Slovakian border.

In Romania the anthropogen changes of the Lower Danube started in 1857 when Sulina branch, the Danube Delta middle distributary, was rectified for navigation purposes. Promptly, the river started to deepen its channel, significantly altering the distribution of water and sediment discharge between the Delta distributaries.

TABLE 1. Recent evolution of the Lower Danube River sediment discharge
Data from National Institute of Metereology and Hydrology, Bucharest and Bondar et al (1991)[3]

BEFORE 1970 Average annual sediment discharge: ~ **65 Mt/y** (of which ~10% sandy fraction)
AFTER 1971 (building of Iron Gates I dam) Compared to the average value before 1971, sediment discharge dropped by: ➢30-40% at Zimnicea hydrographic station ➢30-50% at Vadul Oii hydrographic station ➢15-20% at Isaccea hydrographic station
AFTER 1985 (building of Iron Gates II dam) Compared to the average value before 1971 **Sediment discharge dropped by 50-70% at all hydrographic stations**
PRESENT-TIME ➢ **~30 Mt/y sediments discharged in Iron Gates I reservoir** ➢ small amounts of sediments discharged in Iron Gates II reservoir ➢ **~30 Mt/y present-time sediment discharge**; of which 10-12% sandy fraction

Since 1960 hydrotechnical works have been carried out along the Romanian tributaries of the Danube River, leading to numerous dam building. An important moment for the Lower Danube sedimentary history is the year 1971, when building up the Iron Gates I power station dammed the river. Soon afterward a similar dam (Iron Gates II) was built up not far downstream (Fig. 1).

The Iron Gates dams have produced important anthropogen changes, causing the Danube sediment discharge to diminish by 50 to 70% (Table 1). Due to the many dams constructed on the Romanian Danube distributaries, their sediment contribution also decreased at about the same rate.

A grain size study of the riverbed sediments was carried out right after the completion of the Iron Gate I dam [10]. The investigation revealed the downstream increase of the silt fraction through progressive sorting, with the simultaneous reduction of the sand grain size fraction. No improvement in the quartz sand grains roundness was observed throughout the 1100km long transport distance of the Lower Danube [13].

Both Romanian and Bulgarian hydrologic data indicate that Danube is getting part of its detrital load by eroding its own bed [2]. Moreover Danube riverbed is being dredged for building material, which further intensifies the natural erosion of the Lower Danube bed.

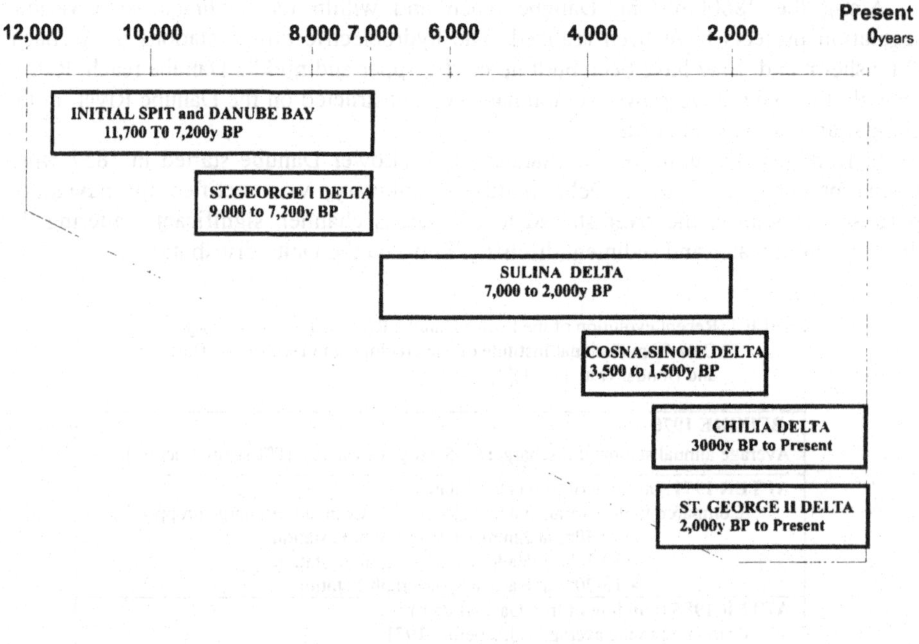

Figure 3 Chronology (C14 ages) of the Danube Delta development phases

2.2. DANUBE DELTA

Danube Delta is mainly a Holocene sediment accumulation developed during the last 12000 years [14, 15, 16, 18]. The deltaic development was accomplished through several main phases as indicated by the relationships between the sets of littoral sandbars making up the Danube Delta body (Figs. 3, 4).

Most of the Danube deltaic units show different petrographic composition on the two deltaic wings. The northern wing deltaic – littoral bars are made up of more mature, quartzose sand of northern origin. The southern delta wings consist of more lithic sand with heavy minerals, supplied by the Danube River discharge.

2.2.1. Blocked Danube Delta phase

C14 dating [18] indicated that the oldest beach ridge deposits on the delta territory (known as the *Letea-Caraorman Initial Spit*) is 12.000 to 11.000 years old. This spit was almost closing a large bay *(the Danube Bay*) where the Danube River was discharging its sedimentary load. The spit formed under the action of the north to south littoral drift, fed by rivers flowing out from the area now represented by

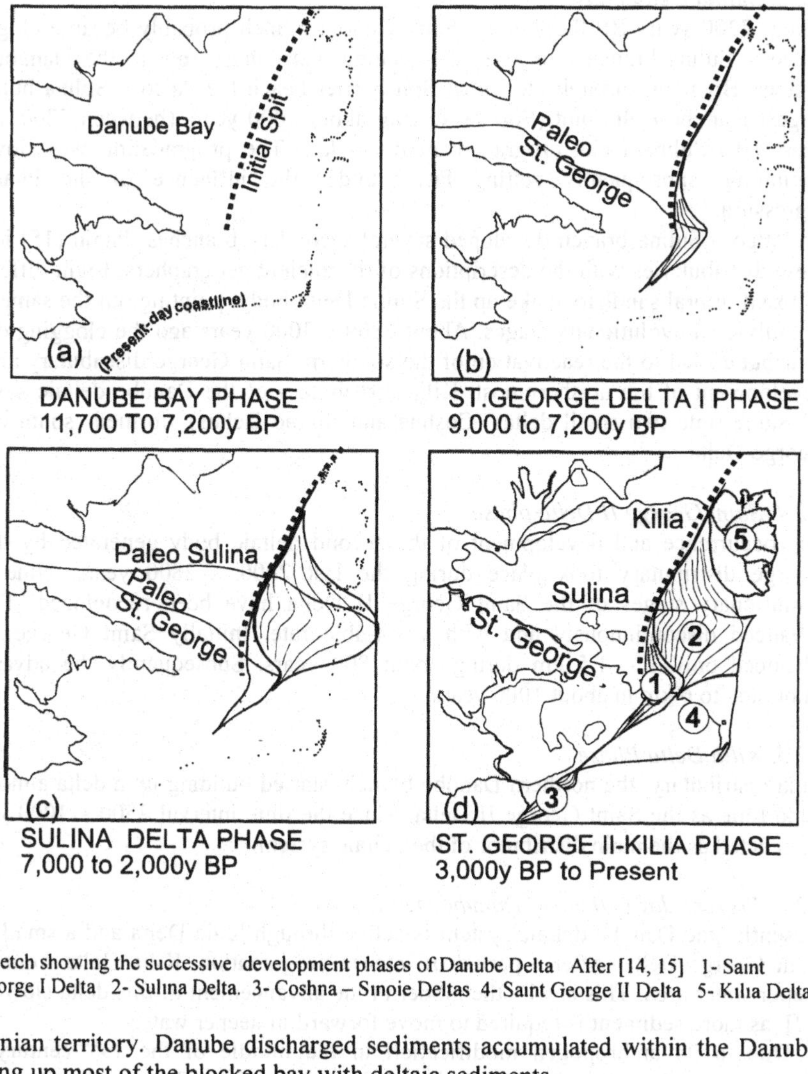

Figure 4 Sketch showing the successive development phases of Danube Delta After [14, 15] 1- Saint George I Delta 2- Sulina Delta. 3- Coshna – Sinoie Deltas 4- Saint George II Delta 5- Kilia Delta

the Ukrainian territory. Danube discharged sediments accumulated within the Danube Gulf, filling up most of the blocked bay with deltaic sediments.

2.2.2. Saint George I Delta phase

The deltaic accumulation, known as the Saint George I Delta, developed between 9000 and 7200 years BP in the area crossed by the present-day Saint George Distributary (Fig. 4). The paleo-Saint George branch flowed through a passage at the southern end of the Initial Spit, or broke through the spit. Presently only the northern wing of the Saint George I Delta is visible, showing the seaward arching of the littoral bars.

The structure of the littoral sand bars assemblage points out that the deltaic progradation acted in three distinctive stages. On the whole Saint George I Delta advanced about 10 km during 2000 years.

2.2.3. Sulina Delta phase

About 7200 years BP the Paleo – Saint George branch probably became clogged and Paleo – Sulina branch took over the leading distributary role in the Danube deltaic system. Breaking through the Initial Spit barrier beach the Paleo – Sulina built up the largest Danube Delta unit (Fig. 4). During about 5000 years (between 7200 and 2000 years BP) Sulina Delta prograded about 30 km. The progradation was slow at the beginning, subsequently getting faster under the influence of the Phanagorian regression.

Paleo – Sulina branch developed several secondary branches. Panin [15] identified these distributaries with the descriptions of the ancient geographers. Eight different sets of fossil littoral sandbars make up the Sulina Delta body, pointing out the same number of individual evolutionary stages. About 2800 – 3000 years ago the clogging of Sulina distributary led to the reactivation of the southern, Saint George distributary and to the development of the northernmost, Kilia distributary of the Danube deltaic system. At the same time two small deltas (Coshna and Sinoie Deltas) formed south of Saint George Delta.

2.2.4. Saint George II Delta phase

The occurrence and development of the second deltaic body generated by the Saint George distributary took place during the last 2000 – 2800 years. Nine distinct evolutionary stages of the Saint George II Delta have been recognized. The delta advanced discontinuously and with a variable rate. Initially Saint George II Delta advanced only 1 – 1.5 km during about 800 years. Subsequently the advancement amounted to 6 km in about 1000 years.

2.2.5. Kilia Delta Phase

Kilia distributary, the northern Danube branch, started building up a delta almost at the same time as the Saint George II Delta. Since the time interval 2500 – 1000 years BP Kilia acted as the main distributor of the deltaic system.

2.2.6. Present-day activity of Danube Delta

Presently the Danube deltaic system is active through Kilia Delta and a small modern Saint George Delta. After a period of intense progradation, Kilia Delta front reached deeper water area. This is why the modern time advancement of this delta slowed down [22], as more sediment is required to move forward in deeper water.

Due to its anthropogen modification in the middle of the 19th century Sulina distributary took away part of the sediment load from Saint George distributary. This was practically the end of the Saint George II Delta. The deltaic activity continued at small scale through a secondary distributary of the Saint George branch. Through successive branch division this secondary distributor began to accumulate deltaic deposits (Fig. 5). After an exceptionally high flood (1897) a several hundred meters long sand bar occurred in front of the small delta. This bar (the Sahalin Island) grew out (up to 17 km long) as a barrier island protecting the deltaic accumulation. During its evolution Sahalin Island displayed backstepping migration through overwashing, with the tendency to get attached to the dryland. This is probably a model for the development of the fossil littoral bars making up most of the Danube deltaic unit.

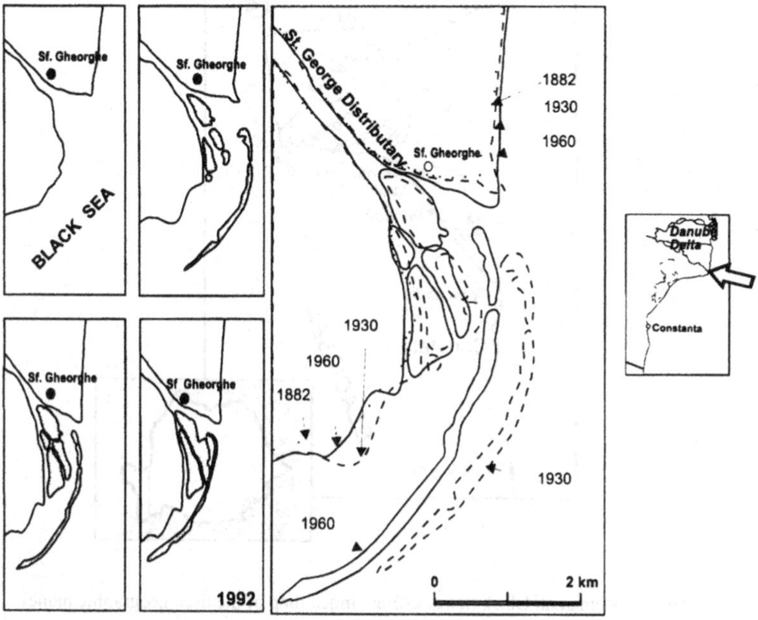

Figure 5 Development of the modern Saint George Delta and Sahalin barrier island.
Data from Popp [22] for 1882, 1930 and 1960 years and Gâstescu, Driga[6] for 1992 year

2.3. ROMANIAN BLACK SEA LITTORAL ZONE.

From genetic and geomorphologic viewpoints the Romanian Black Sea shore is divided into two main units:

- northern shoreline unit (between Kilia Delta and Cape Midia) is of accumulative type, with low relief, built up in front of Danube Delta;
- southern shoreline unit (from Cape Midia to Vama Veche village) is of erosional origin (especially its southern part) with a higher relief and loess cliffs.

2.3.1. Erosional and accretional littoral environments

The shoreline in front of the Danube Delta is subjected to both accretional and erosional migrations. Constant shoreline advances occur in connection with Kilia Delta progradation. Local accretional shores are located south of deltaic tributary mouths (Sulina and Saint George). Engineering works (for example Sulina and Midia jetties) also produce littoral accretion.

Shoreline retreat is dominant in the area in front of the Danube Delta. The process is purely erosional in the case of the dryland attached beaches. The barrier beaches of the Danube Delta area migrate through a combined erosional – constructional, overwashing mechanism.

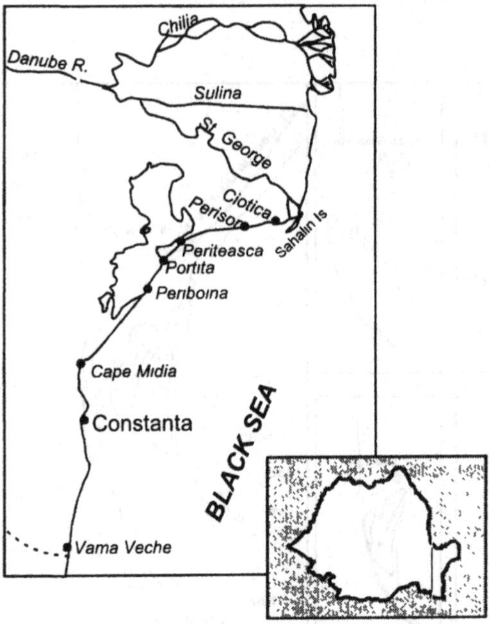

Figure 6 Romanian Black Sea coastline Index map of the used geographic names

Important littoral erosion processes occur on the shore located between Sulina and Saint George mouths. Since 1882 [22] the shoreline several kilometers south of Sulina jetties has undergone an average erosional migration of about 8 m/y (Fig. 5). Another high erosional shoreline is Ciotica – Perisor littoral area (south of Sahalin Island). During the last 50 years this shoreline retreated more than 500 m, converting the former Zãtonu Mare Lake into a bay.

Sahalin barrier island represents a spectacularly retreating shoreline. On the 1910 Danube Delta map [1] the northern end of the island is located two km offshore from the modern Saint George Delta front. Migrating faster than the rest of the island, the northern extremity of the Sahalin Island is presently attached to the delta front.

The 30 km long littoral bar Periteasca-Portita–Periboina closes the Razelm lagoonal system located in the southern part of of the Danube Delta area (Fig. 6). This sandy barrier displays retrograding migration through overwashing. Since the end of the 19th century the Periteasca – Periboina barrier beach has moved about 1 km westward [22].

Protected by a loess cliff up to 12 m high the southern coast of Romania is more stable as compared with the northern deltaic coast. Erosional processes are also active in the southern Romanian littoral area. Under the influence of the marine abrasion and gravitational processes the cliff line retreats with the average rate of 50 cm/year (Selariu, 1971 and Serbănescu, 1969, in Kosyan and Panin,1996)[12]. The erosion is more active on the accumulative beaches interspersed in between the cliffs. In view of their tourist value, hydrotechnical barriers have been constructed for the protection of these shores.

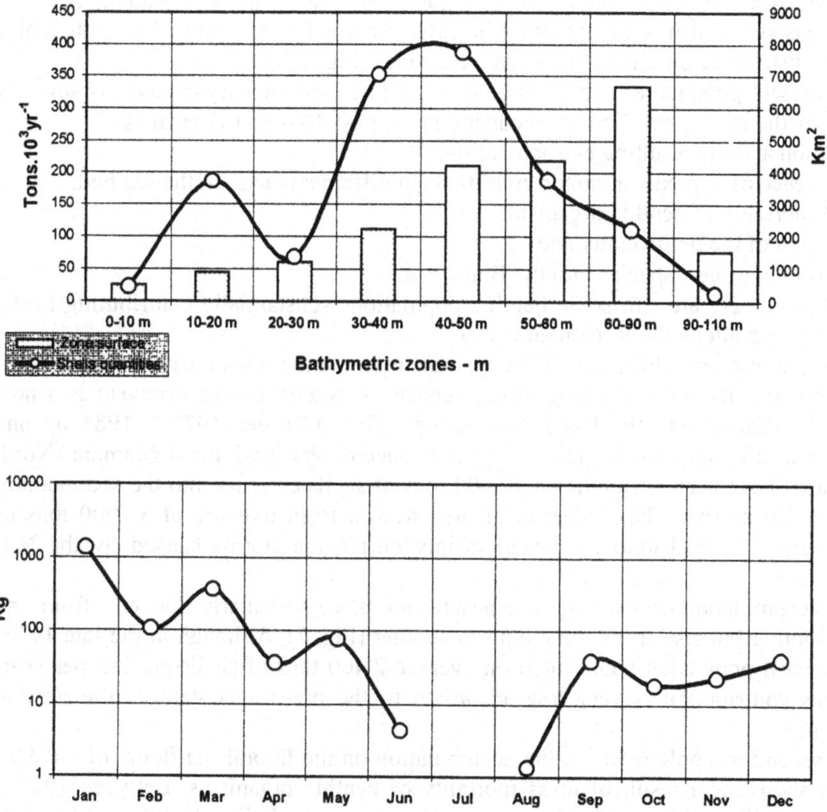

Figure 7. Potential production of Molluscs shells for sedimentogenesis on the Romanian Black Sea shelf (upper diagram). *Mya arenaria* shells deposited on one meter length of beach per month at Mamaia Beach (lower diagram)

The southernmost Romanian shores, south of Constanta, receive no Danube originated sandy sediments. As there is no other source to supply siliciclastic sand material, the shore deposits of this area are of biogenic origin, derived from the mechanical working of the littoral shells.

2.3.2. Biologic sediment source of the southernmost Romanian shores

Many marine planktonic and benthic organisms play an important role in the sedimentary deposits. Benthic invertebrates are known to contribute to the sedimentary processes, enriching the deposits, reworking and mechanically modifying the sediments through burrowing, tube building and deposit feeding [7, 25]. Among them the Molluscs are most representative in the Black Sea.

It is well known that one of the specific features of the Black Sea is the presence of biofiltering forms on its bottoms. These forms dominated by mussels and clams, play at the same time an important role in sedimentary processes.

The potential of Molluscs population in forming the bottom deposits on the Romanian Black Sea continental shelf (≈ 22,000 km^2) was in the 1980s about 1.3 – 1.6 million tons yr^{-1}, which is about half of the value recorded in the 1960s. This potential is unevenly distributed on different bathymetric zones (Fig. 7).

The present potential of benthic organisms to form sedimentary deposits is severely affected by the ecological changes occurring in the past decades [8], such as:

pollution and the resulting eutrophication,
occurrence of hypoxic or anoxic conditions in different places on the sea bed,
mass mortality of benthic organisms,
reduction of the biodiversity and
penetration of new species into the Black Sea.

However, there are still some benthic populations substantially contributing to the littoral deposits allong the Romanian coasts.

Two examples are illustrative. First example refers to the role of *Mya arenaria* (soft-shell clam) population in forming beach deposits at Mamaia. *Mya arenaria* is a new species penetrating into the Black Sea in the 1970s. Between 1977 – 1981 on one kilometer length and several meters or tens of meters beach width, at Mamaia (North Constantza) have been accumulated 12,000 tons of shells entering into the sedimentary processes. These shells have been originated from a fresh biomass of 35,000 tons of *Mya arenaria*, carried onto the beach mainly during the storms caused by the N-E winds.

The accumulation of shells on the beach took place irregularly not only from one year to another but also from one month to another (Fig. 7). Although in the late 1970s *Mya arenaria* population contributed on average 2,400 tons of shells per km per year, today this contribution is very low according to the precarious state of the benthic ecosystem.

The second example refers to the accumulation on the littoral sea floor of the *Mya arenaria* shells, as a result of mass mortality of benthic organisms. Between 1972 – 1981 on the littoral zone stretching from Constantza to Sulina and from the very shallow to 30 m deep zone (about 1,765 km^2), *Mya arenaria* population was lost, as a consequence of mass mortality and 3.5 million tons of fresh biomass was produced. This means about 1.2 tons of deposited shells, besides the organic substance entering into the sedimentary processes.

The loss of *Mya arenaria* biomass varies within the different bathymetric or geographical zones lying at the Romanian Black Sea littoral. Bottoms most affected by mortality are those lying in the coastal shallow zone between 0 – 10 m and 10 – 20 m depth at Sulina, Portitza, Periboina and Vadu. Mass mortality occurs more frequently than the mortality which affects only partially the *Mya arenaria* population, but both of them represent a major source for shell deposits. Because of the *Mya arenaria* deposits both the grain size and chemistry of sediments have changed, thus changing the biotope and further influencing the structure of the bottom associations [7].

2.3.3. Main anthropogen influence on the littoral sedimentary processes

Significant anthropogen influence is exercised on the Romanian littoral sedimentary environment. The operations carried out for navigation purposes at the mouth of the Sulina distributary represent the best known case. The intense activity of Kilia Delta always represented a threat for the navigation at the Sulina branch mouth. This is the reason that jetties, continuously advancing into the sea, have been built up at the Sulina mouth. Sulina mouth bar is also continuously dredged in order to keep clear the navigation channel; the dredged sand being dumped away offshore. The combined effect of these two anthropogen activities deeply disturbed the littoral sedimentation processes in the area (Table 2).

TABLE 2. Sulina jetties technogenous impact on littoral sedimentation

SULINA JETTIES PROJECT 1856, Danube European Commission (Charles Hartley project)	
DEVELOPMENT	
1861	1412 m (jetty length, seaward extension)
1925	3180 m
1939	4150 m
1956	5773 m
1995	~ 8000 m
IMPACT ON LITTORAL SEDIMENTATION PROCESS. ➢ cuts the natural north to south sedimentation flux, ➢ influences the local littoral sedimentation balance removing an important amount of sediment (dredged sand) from the littoral sedimentary system, ➢ modifies the littoral longshore current structure (creates a large eddy-like littoral circulation cell, south of the jetty)	

Harbor constructions represent another technogenic factor disturbing the sedimentation process on the southern Romanian coast. The harbor development led to significant seaward extension of the protected basins (up to 23 m water depth at Constanta harbor and 13 m depth at Midia harbor). Consequently the already sediment under-saturated longshore current was forced to shift away from the beach.

2.4. ROMANIAN BLACK SEA SHELF

2.4.1. Sediment-fed and sediment-starved shelf environments

On the Romanian shelf (part of the northwestern Black Sea shelf area) the dispersal pattern of the Danube sediment supply points out to the existence of two main areas with different depositional processes: the Danube sediment fed internal shelf and the sediment starved external shelf (Fig. 8).

The internal, western zone of the Romanian shelf stands out as the shallow marine area which receives clayey and silty sediments supplied by the Danube River. Moving as suspended load, the sediment flux goes beyond the area in front of the Danube Delta but does not reach the eastern external shelf zone. Under the influence of the dominant currents the clayey sediment flux moves southward toward the Bulgarian shelf, keeping closer to the shoreline.

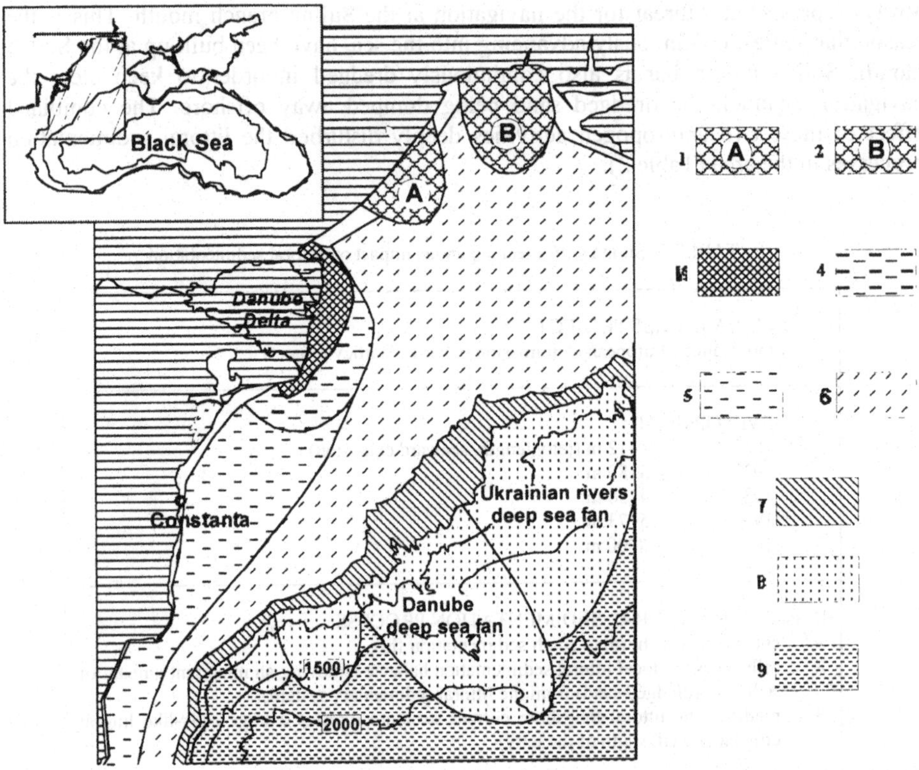

Figure 8 Main sedimentary environments in the northwestern Black Sea area
1-2. Areas under the influence of the Ukrainian rivers (Dnieper and Dniester) sediment discharge 2.Danube delta front area 3 Danube prodelta area 5-6 Western Black Sea shelf areas (5-area under the influence of the Danube originated sediment flux, 6-sediment starved area). 7.Shelfbreak and uppermost continental slope zone 8.Deep sea fan area. 9 Deep sea floor area.

Situated outside the area covered by the Danube fed sediment flux the external, eastern part of the Romanian continental shelf represents an area practically deprived of clastic material. Within this sediment-starved shelf area the condensed sediment accumulation is of biogenic origin, consisting of organic pellets on relict sediments or shell concentrations (Fig. 9).

The Danubian sediments seldom reach the shelf area north and northwest of the Danube mouths. Dniester and Dniepr, the main rivers north of Danube Delta, are themselves no significant suppliers of sediment for the northwestern Black Sea shelf. These Ukrainian rivers discharge their sedimentary load into lagoons, separated by beach barriers from the Black Sea. Consequently the sediment-starved status characterizes almost the whole Black Sea continental shelf west of the Crimean Peninsula.

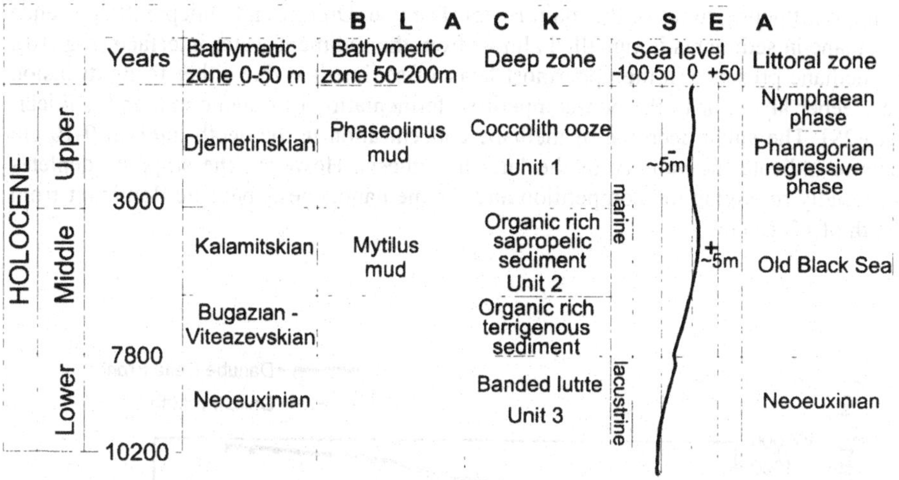
B L A C K S E A
Years
Bathymetric zone 0-50 m
Bathymetric zone 50-200m
Deep zone
Sea level
-100 50 0 +50
Littoral zone
HOLOCENE
Upper
Middle
Lower
3000
7800
10200
Djemetinskian
Kalamitskian
Bugazian - Viteazevskian
Neoeuxinian
Phaseolinus mud
Mytilus mud
Coccolith ooze
Unit 1
Organic rich sapropelic sediment
Unit 2
Organic rich terrigenous sediment
Banded lutite
Unit 3
marine
lacustrine
-
~5m
+
~5m
Nymphaean phase
Phanagorian regressive phase
Old Black Sea
Neoeuxinian

Fig 9 Black Sea Holocene sea level variations and the main organic rich sediment types
From Panin [15] (after Scerbakov *et al*, 1979, Fedorov, 1972, Degens and Ross, 1972;references in Panin, 1983)

2.4.2. *Anthropogen-induced diagenetic changes on the western Black Sea shelf*

The most important changes of the sedimentary regime with diagenetic consequences, affecting the entire northwestern sector of the Black Sea, are a reduction of the coarse grain-size fractions weight in the riverine sedimentary material, superimposed on an overall decrease of the river transported terrigenous material and an increase of the organic matter quantity reaching the bottom .

As a consequence of the grain size decrease of the sediment on the western Black Sea shelf, the porosity of the sediments increases but their permeability show a significant decrease, hindering the diffusion of oxygen from the overlaying water.

Combined with the augmentation of the organic matter quantity reaching the bottom because of the increased biological productivity of an eutrophicated Black Sea, the above mentioned modifications has lead to the rapid installation of a highly reducing environment in sediment. The oxic layer of the sediments has became thinner, the zero mV Eh limit occurring usually at a depth less than 5 cm in the sediments, occasionally reaching at this depth to values as low as -200 mV (Panin et al, 1992 a, b)[19, 20]. The high reducing capacity of the sediments induces a very shallow sulphate reduction zone, the sulphate concentration in the interstitial water decreasing from 12 - 19 mmol/l in the upper 15 cm of sediment to 1.5 - 5.7 mmol/l at a depth of 30 - 40 cm (Pimenov and Roussanov, 1997)[21]. The shallow H_2S production has caused spreading of the H_2S contaminated superficial sediments, extending now over almost the entire zone of the river influence. In favourable conditions of restricted water circulation and/or extremely high inputs of organic matter due to episodes of algal blooming or of disposal of highly organic waste waters, the oxic-anoxic boundary can rise above the sediment-water interface. The oxic layer acting as a H_2S blocking screen, H_2S diffuses in the bottom waters. This is the case of the depressionary zones from the Odessa Bay where the H_2S concentrations in the bottom water, at water depths between 17 and 31 m, reach up to 1.50 ml/l (Panin et al., 1992 a, b)[19, 20] .

In the influence zones of the main rivers (Danube, Dniestr and Dniepr) the presence of methane in sediments is signalled almost from the sediment-water interface (Fig. 10). The methane presence in the uppermost layer of sediment is due either to its diffusion from deeper layers or to the non-competitive fermentation of organic matter (Whiticar, 1996)[29]. The slight decrease of methane concentration in the depth interval 0-10 cm seems to indicate the validity of the later hypothesis. However, the sulphate depletes very rapidly removing the competition and the methanogenesis become dominant from a depth of 15-20 cm.

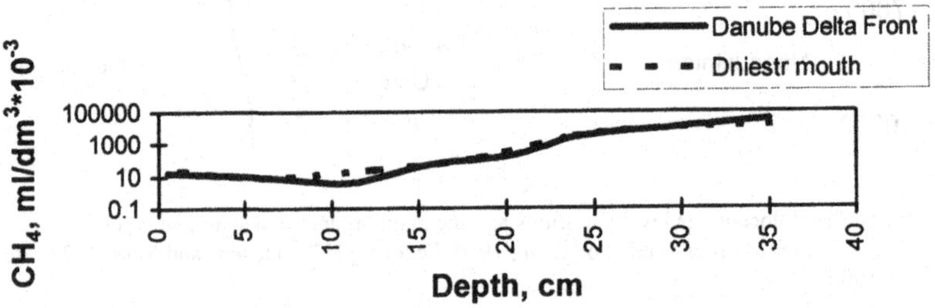

Figure 10 Vertical distribution of CH_4 in sediments from Danube Delta Front and Dniestr mouth (data from Pimenov, Roussanov, 1997)[21]

The bacterial mineralization of organic matter leads to the recycling of the main nutrients (P, N). As a result of increased nutrient concentrations in interstitial waters out-bound nutrient fluxes from sediments were recorded almost everywhere on the Black Sea northwestern shelf, with higher values in Danube Delta Front area and at Dniestr mouth (Table 3).

Another consequence of the high organic matter concentration in sediments and consequently their high reducing capacity is the reduction of the higher oxidation forms of iron and manganese and their dissolution in interstitial waters. Ensuing fluxes of iron and manganese out of sediments were recorded on the entire Black Sea northwestern continental shelf (Table 3, data from Friedl et al., 1996)[5]. The highest iron flux accompanied by a quite high manganese flux was recorded at the limit of the outer shelf, on the shelf edge. High manganese and iron fluxes were also recorded at the Dniestr mouth. Favourable conditions (oxidizing conditions at the sediment-water interface, presence of adequate substrata – mollusc shells, especially *Modiolus* for the outer shelf and a very low sedimentation rate made even lower by the decrease of terrigenous material input) lead to iron and manganese reprecipitation at the interface and formation of iron-manganese crusts and nodules.

TABLE 3 Benthic fluxes on the Black Sea North-Western shelf (in mmolxm^{-2}xd^{-1})(Friedl et al , 1996)[5]

Station	NH_4^+	PO_4^{3-}	Fe^{2+}	Mn^{2+}
Danube Delta Front 45° 12 210 N; 29° 50.760 E	2.75	0.060	0 50	0 43
Danube Prodelta 44° 34 980 N; 29° 11 370 E	0 13	0.025	0 45	0.15
Dniestr mouth 46° 03 400 N, 30° 29 120 E	3 68	0 460	1 85	2 10
Shelf edge 43° 41 830 N, 30° 03 490 E	0 09	0 050	16 35	1 15
Varna Bay 43° 15 950 N; 28° 07 760 E	0 15	0 070	0 40	0 16

Both oxidized iron and manganese are known as very good scavengers for a whole series of trace elements, including some of the most toxic. Their reduction determines the dissolution of these elements in interstitial waters and in the absence of reprecipitation reactions such as sulphides formation they diffuse and contaminate the bottom waters. Iron reduction can contribute especially substantial quantities of phosphate to bottom waters (Williams et al., 1976)[28].

2.5. DEEP SEA ZONE OF THE WESTERN BLACK SEA

The modern, higher sea level epoch is marked in the deep-sea zone of the western Black Sea by the cessation of the Danube Deep-Sea Fan sedimentary accumulation. Without the Danube supply this area became sediment starved. The existing accumulation is represented by coccolith ooze overlying a sapropelic or organic-rich sediment unit (Ross et al., 1970)[23] (Fig. 9); pointing out the domination of the organic component over the detrital one.

2.6. MAIN SEDIMENTARY FEATURES OF THE LOWER DANUBE – DANUBE DELTA – BLACK SEA SYSTEM IN THE HIGH SEA LEVEL ENVIRONMENT

The drastic decrease of the clastic material supply stands out as the most important sedimentary feature of the high sea level Danube – Danube Delta – Black Sea System. The sedimentary load of the Danube River, the fluvial component of the system, is significantly diminished. This appears as a consequence of the sea level rise, but the anthropogenic activity carry a good share of responsibility for this state. Presently Danube Delta concentrates most of its constructive sedimentary activity along a single distributary. A large shelf interfere between the deltaic depocenter and the deep-sea area.

With the littoral currents under-saturated in sediment the beach erosion is a widespread process. Some anthropogen interventions intensify this situation. The southernmost beaches of the Romanian littoral receive no Danube born sand sediment, relying on biogenic sources. Barrier beaches of the Danube Delta area show backstepping migration through overwashing.

Sediment starved conditions prevail over a large area, from the external shelf to the deep-sea zone. Reflecting the clastic material deficiency the sedimentary accumulations are organic-rich and occur in condensed sections. Danube Deep-Sea Fan is presently inactive.

2.7. SEDIMENTARY ENVIRONMENTS OF THE DANUBE – DANUBE DELTA – BLACK SEA SYSTEM IN THE SEA LEVEL RISE PERSPECTIVE

Sea level rise is a process in progress. The expected sea level rise will aggravate the scarcity of clastic material affecting the Danube – Danube Delta – Black Sea System.

In the Danube Delta area the sea level rise will induce a deeper upstream penetration of the salt wedge (Panin, 1992)[17]. Consequently the transfer process of the sedimentary load to the marine basin will be disturbed.

A possible 20 – 30 cm sea level rise by the years 2020 – 2030 would lead to a transgressive shoreline migration, the sea covering a good part of the present deltaic area. Under the conjugated action of the base level rising and the diminishing sediment supply the shoreline retreat will be extensive. The erosional marine processes could redesign the shoreline, converting into bays some present–day lagoons or intradeltaic lakes.

The sediment starved status will get more severe on the shelf as well as in deep water, down to the abyssal floor. Without the buffer protection of the sediment flux these areas will be more exposed to contamination produced by anthropogen activities.

In comparison with some past Black Sea level rises the modern rise is slow, limited and previsible. According to Ryan et al. (1997)[26] 7,500 years ago the sea level rose 140 m in about two years.

3. Low sea level sedimentary processes and environments in the western Black Sea

Sea level fluctuations determined by glaciations and deglaciations controlled the geo – environmental state of the Danube River – Danube Delta – Black Sea System. The Black Sea sedimentary regime alternated between marine and lacustrine, according to the temporary linkage with or cut-off from the Mediterranean Sea. The low sea level scenario is based on data regarding the Pleistocene deposits and takes into account quite extensive sedimentary and environmental changes.

3.1. LOWER DANUBE IN THE LOW SEA LEVEL SCENARIO

The Quaternary and pre-Quaternary history of the Danube River – Black Sea relationship is still obscure. The only fact about this subject is that Danube Delta evolution started 12000 years ago (Panin, 1983) [15].

The need to call upon the Danube supply came up in order to explain the occurrence of the Black Sea terrigenous sedimentation unit during the Middle Quaternary to Present (Hsu, Trimonis and Shimkus, Stoffers and Muller, in Ross, 1978) [24]. The beginning of the terrigenous sedimentation was interpreted by Hsu (1978) [9] as reflecting the penetration of the large Danube detrital influx into the Black Sea basin. Other authors belive that Danube River did not entered Black Sea until 12,000 years ago; previously the paleo-river system discharging its sediments within the Dacian Basin.

The Danube supply problem is complicated due to the fact that between the Black Sea and the Carpathian source area (acting through the Danube River) there was another Paratethys basin, the Dacian Basin. Sedimentological investigation, now in progress, of the Late Neogene – Quaternary evolution of the Dacian Basin (Jipa, 1995)[11] revealed that during the Romanian stage (4.2 to 1.8 Ma) the sedimentation in the Dacian Basin was entierly fluvial. Consequently a paleo lower Danube River could have existed long before the Quaternary. It is not known yet where this paleo-river discharged its sediment load.

3.2. COASTAL AND SHELF ENVIRONMENTS IN THE LOW BLACK SEA WATER LEVEL

During the dropping of the sea level the shoreline and the deltaic depocenters migrate basinward. In contrast with the highstand environment, during the lowstand the shelf is getting narrower and disappears; becoming completely subaerially exposed if the shoreline reaches the shelfbrake.

Neoeuxinian paleoshorelines have been recognized near the present western Black Sea shelfbreak zone. These features are evidenced by a number of paleo – beach ridges, locally associated with paleocliffs (Kosyan and Panin, 1996)[12]. On the present-day Romanian continental shelf a complex system of apparently meandering buried channels have been observed at 50–75m water depth interval (Kosyan and Panin, 1996) [12].

Recently Winguth et al. (1997)[32] identified within the western Black Sea shelf sediments the equivalents of the Pliocene sequences recognized in the Danube and Dnieper deep sea fan area. Based on on seismic profiles investigations the authors pointed out the sea level changes effects on the shelf area. The shelf sediment sequences are progradational, retrogradational and aggradational. The shelfbreak line migrated landward and seaward, the shifting amplitude varying between 0.3 and 10.6 km.

3.3. DANUBE DEEP SEA FAN ENVIRONMENT

The abyssal fan environment is the most important depositional center in the low sea level scenario. A complex system of abyssal fans occurs in the northwestern Black Sea area. The deep sea fan area was intensely studied during the last two decades. Six research teams have been active in this area between 1979 and 1988 (see references in Wong at al., 1994)[30]. Since 1993 a German – Romanian team, collaborating with Ukrainian and Russian scientists (Wong et al., 1994;Wong et al., 1997;Winguth at al., 1997))[30] [31] [32], produced important new data on the structure and evolution of the western Black Sea abyssal fan complex.

Several distinct, interfingering abyssal fans occur in the deep, western area of the Black Sea (Fig. 8). According to seismic and bathymetric data the largest, northern fan was built up under the influence of the Ukrainian Dniepr and Bug Rivers. Danube Deep Sea Fan is next to the south of the Ukrainian Rivers Fan. Another two smaller fans, south of the Danube Fan, are also assigned to the Danube River activity.

The Danube Deep Sea Fan extends up to the shelfbreak zone, approximately at the 200 m isobath. The selfbreak zone is deeply cut by a submarine valley which created the Viteaz Canyon. This morphologic feature represents one of the way the sediment supply reached the submarine fans area. Within the present-day physiography the Viteaz Canyon is visible on 30 km. It continues with the most recent central channel of the Danube Fan. Seismic records suggest the connection of the Viteaz Canyon with a Danube paleo valley on the continental shelf.

The upper part of the Danube Fan extends down to about 1500m water depth. The central channel of the fan displays a 15km wide levee system. Near the shelf break zone the central channel (Viteaz Canyon) is 350m deep.

On the middle Danube Fan the width of the levee system increases to about 60 km. The main channel bifurcates and the secondary channels are strongly meandering. The central channel is shallower (50 – 100m) but its width extends to 4.5km. Four other, older channels have been recognized on side scan, subbottom profiler and seismic records.

The passage to the lower fan area begins at about 2000m. The levees of the lower fan are not any more important morphologic features. Eight seismic sequences of the Danube Deep Sea Fan have been recognized (Wong et al., 1994) [30]. All the lowstand system tract features have been observed in the six upper sequences: channel, overbank and mass transport sediments. These sequences, representing the deep sea fan development accumulated during the past 480 ka, with an average sedimentation rate between 2.4 and 7.2 m/ka (Wong et al,1997) [31].

4. Conclusions

During the Neogene – Quaternary evolution of the Black Sea the balance within the Danube – Danube Delta – Black Sea sedimentary system was governed by the influence of the Carpathian source area versus the sea level variations.

The source area influence was strictly regulated. This was realized through the existence of an intermediary accumulation area (the Dacian Basin) between the Carpathians and the Black Sea.

The sea level variations controlled with authority the sedimentary processes and environment within the Danubian – Euxinic System. Main differences between the rising versus dropping sea level sedimentary conditions within the Danube – Danube Delta - Black Sea System consisted in (1) the volume of the sediment influx, (2) the location of the depositional centers within the accumulation area as well as in (3) the character (erosional versus sediment starved) of the sediment deficient areas.

References

1 Antipa, G (1915) Wissenschaftliche und wirtschaftliche Probleme des Donaudeltas, *Ann Inst Geol Roumanie* VI/1

2.Behr, O , Bondar, C , Modev, S , Prohaska, S (1977) Morphological changes and abatement of the negative effects in a selected part of the Danube River, Environmental Progr Danube River Bassin, Danube Applied Research Conference, Sinaia (Romania), Abstract

3.Bondar, C., State, I., Cernea, D , Harabagiu, E. (1991) Water flow and sediment transport of the Danube at its outlet into the Black Sea, *Meteorology and Hydrology* **21**, 21 – 25

4 Drever, J. I (1988) *The Geochemistry of Natural Waters,* Prentice Hall, Englewood Cliffs, New Jersey

5 Friedl, G , Dinkel, C., Wehrli, B (1996) Benthic fluxes across the sediment – water interface in the northwestern Black Sea, Black Sea Pollution Assesment Workshop, Istanbul.

6 Gâstescu, P , Driga, B (1995) *Delta Dunării*, Harta turistică, Editura Sport - Turism, Bucuresti
7 Gomoiu, M.T. (1983) The role of *Mya arenaria* L. populations in coastal sedimentary processes, *Cercetări Marine* **16**, 7-24
8 Gomoiu, M T (1992) Marine eutrophiccation syndrome in the nortwestern part of the Black Sea, *Science of the Total Environment, Suppl.*, 683-692.
9 Hsu, K. J. (1978) Stratigraphy of the lacustrine sedimentation in the Black Sea. In D. A. Ross, Yn P. Neprochnov et al (eds), *Initial Reports of the Deep Sea Drilling Project*, 42 (2)
10.Jipa, D. (1974) Grain size of the Recent Lower Danube sediments (in Romanian) *Dări de seamă, Inst. Geol*, H5, 61 – 74.
11 Jipa, D (1995) Upper Neogene evolution of sedimentary environments in the littoral zone of Dacic Basin, *Romanian Journal of Stratigraphy*. Abstract, **76**, 131-132
12 Kosyan, R D and Panin, N (1996) Black Sea coastal zone, Technical report on coastal erosion Economic effects and their evaluation, GEF Black Sea Environmental Programme. Bucharest
13.Mihesan, M , Jipa, D (1978) Roundness of the arenitic quartz particles - Recent alluvial deposits of Lower Danube (in Romanian) *Dări de Seamă, Inst Geol*, LXVI, 291-305
14.Panin, N (1972) Histoire Quaternaire du Delta du Danube Essai d'interprétation des facies des depôts deltaiqes. *Cercetări marine* **4**, 5 –15
15.Panin, N (1983) Black Sea coast line changes in the last 10,000 years A new attempt at identifying the Danube mouth as described by the ancients, *Dacia*, N S , XXVII, 175 -184,
16.Panin, N (1989) Danube Delta Genesis, evolution and sedimentology, *Rev Roumaine Géol. Géophys Géogr , Ser Géographie* **33**, 25 –36.
17 Panin, N (1992) Impact of expected climate change and sea level rise on the Danube Delta area, UNEP (OCA) WG 19, Istanbul, Inf 8
18 Panin, N , Panin, S , Herz, N , Noakes, J E., (1983) Radiocarbon dating of Danube Delta deposits, *Quaternary Research* **19**, 249 - 255
19 Panin, N , Secrieru, D , Manoleli, D , Reznik, V , Fesiunov, O and Nazarenko, M F (1992a) The ecological conditions in the N - W part of the Black Sea, *Rapp Comm Int Mer Medit* **33**, 146
20 Panin, N , Secrieru, D , Manoleli, D., Reznik, V., Fesiunov, O and Nazarenko, M.F (1992 b) The Ecological Conditions in the NW Part of the Black Sea A brief description of the Romanian - Ukrainian geo-ecological research in 1990, *Mediul Inconjurator*, III, 27-35
21 Pimenov, N V , Roussanov, I I (1997) Anaerobic microbial processes in the Black Sea sediments (preliminary results), EROS 21 Black Sea Cruise Workshop, Kastaniembaum (Switzerland)
22 Popp, N (1965) Present-day changes of the littoral zone morphology in the Danube Delta (in Romanian), *Studii de Hidraulică*, IX, 412 – 439
23 Ross, D A , Degens, E T. and MacIlvaine, J (1970) Black Sea. recent sedimentary history *Science*, **170/3954**, 103 – 165
24 Ross, D A. (1978) Summary of results of Black Sea drilling, *Initial Reports of the Deep Sea Drilling Project* **42**, 1149- 1178
25 Row, G T (1974) The effects of the benthic fauna on the physical properties of deep-sea sediments In A L Inderbitzen (ed), *Deep Sea Sediments*, Plenum Publishing Corporation, 381-400
26 Ryan, W B F , Pitman, W C III, Major, C O , Shimkus, K , Moskalenko, V , Jones, G A , Dimitrov, P Gorur, N , Sakinc, M , Emelyanov, E M., Trimonis, E S (1997) An abrupt drowning of the Black Sea shelf at 7 5 kyr BP, *Marine Geology* **138**, 119 – 126
27 Shimkus, K M and TRIMONIS, E S (1977) Modern sedimentation in Black Sea, in E T Degens and D A Ross(eds), *The Black Sea - Geology, Chemistry, and Biology*, AAPG Memoir 20, 249 - 279
28 Williams, J D H , Jauet, J M and Thomas, R L (1976) Forms of phosphorus in the surficial sediments of Lake Erie - *J Fish. Res Can* **33**, 413 - 429
29 Whiticcar, M J (1996) Isotope tracking of microbial methane formation, in Adams, D D , S P Seitzinger, and P M Crill, (eds) *Cycling of reduced gases in the hydrosphere*, Mitt. Int Ver Limnol **25**, 39 - 55
30 Wong, H K , Panin, N , Dinu, C , Georgescu, P and C Rahn, C (1994) Morphology and post - Chaudian (Late Pleistocene) evolution of the submarine Danube fan complex *Terra Nova* **6**, 502 - 511
31 Wong, H K , Winguth, C , Wollschilager, M , Panin , Dinu C , Georgescu, P , Ungureanu, G , Krugliakov, U V , Podshuveit, V (1997) The Danube and Dniepr Fans Morphostructure and evolution *Geo-Eco-Marina* **2**, 77 - 101
32 Winguth, C , Wong, H.K , Dinu, C., Panin, Georgescu, P and Uugureanu, G , (1997) Upper Quaternary sea level changes in the northwestern Black Sea Preliminary results *Geo-Eco–Marina* **2**, 103 – 113

CHARACTERISTIC CHEMICAL FEATURES AND BIOGEOCHEMICAL CYCLES IN THE BLACK SEA

Ö. BAŞTÜRK[1], E. YAKUSHEV[2], S. TUĞRUL[1] and İ. SALIHOĞLU[1]
[1]*Middle East Technical University-Institute of Marine Sciences, Erdemli-İçel, Turkey*
[2]*P.P. Shirshov - Institute of Oceanology, Moscow, Russia*

Abstract

In this report, temporal and spatial variabilities in the biochemical and biological sectors of the Black Sea are reviewed, based on the long-term chemical and biological data collected by the Black Sea riparian countries and USA. Past and present biochemical data, starting from R/V J. Elliott Pillsbury August-1965 cruise till R/V Bilim July-1997 cruise, were compared in terms of density dependent profiles for the dynamically different regions of the Black Sea. Observed changes in the ecosystem of the Black Sea during recent decades are reviewed in conjunction with the alterations and modifications in the wide range of environmental controls. Recent physical and biochemical measurements such as nutrients, dissolved oxygen, hydrogen sulfide have also been evaluated for describing the possible routes and mechanisms of transport of coastally trapped pollutants into the offshore, deep-basin waters through meandering rim current system, and the redox mechanisms controlling the sub-oxic zone located between oxic and anoxic layers.

1. Introduction

The Black Sea, the most isolated and the largest inland sea permanently occupied by anoxic water below 100-200 meters, possesses various oceanographic features which make it distinctively different from other basins. What make the Black Sea different from the other basins are its topography in which the interacting coastal and deep ecosystems are contained in the same enclosure, and the presence of cyclonically meandering rim current along the peripheries of the basin. The Black Sea is further unique in having very narrow shelf zone along more than half of its margin, except northwestern shelf area. Detrital particulate organic and inorganic matter of both allochthonous and autochthonous origin is partly assimilated within the coastal waters and partly transported to the offshore waters through transient eddies and jet filaments derived by the abrupt changes in the marginal topography. Because of the absence of bioturbation, delicate laminae of sediments are generally well preserved in the abyssal

S. Beşiktepe et al. (eds.),
***Environmental Degradation of the Black Sea: Challenges and Remedies*, 43–59.**

bottom sediments as a result of the establishment of anoxic conditions in the water column throughout the last 5,000 years [1]. Detectably thick transition zone, so-called suboxic zone where DO <10 μM) and H_2S <1μM [2], located between the upper oxygenated and the deep anoxic layers, is another unique feature that makes it distinct from other anoxic basins, such as Framvaren Fjord [3] and Tyro and Bannock Basins of the Mediterranean [4] where there is a sharp boundary between oxic and anoxic waters. Complicated biogeochemistry and related redox processes within the suboxic zone (SOZ) has been of great interest to chemical oceanographers and modellers.

The upper, oxic layer, comprising about 10-15 % of the total volume of the Black Sea [5], supports the necessary energy to derive the biological cycles ranging from microbiological up to higher trophical levels. It receives wide spectrum of industrial, agricultural and domestic wastes of about 160 million population occupying a catchment area of $2x10^6$ km^2, nearly five times larger than its total surface area of $423x10^3$ km^2 (Figure 1).

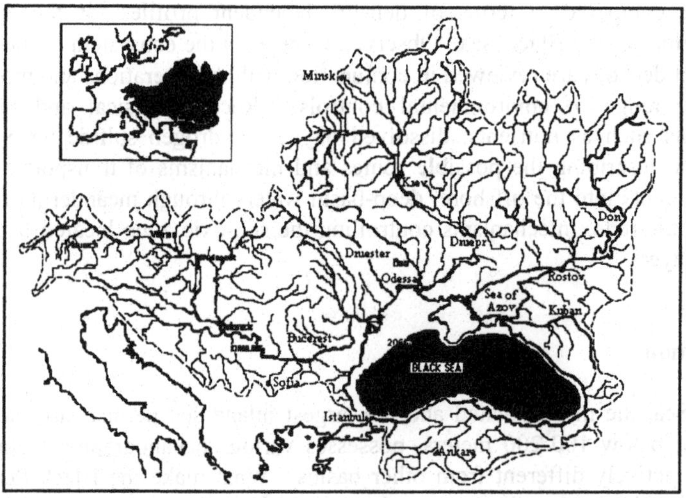

Figure 1 Total catchment area of the rivers discharging into the Black Sea basin

Majority of the riverine discharges are located on the northern and north-western shelf area which comprises 35% of the total surface area of the basin [6]. The Danube River alone, with its about 200 km^3/y water discharge, represents 3/4 of the north western river run off and 2/3 of the total riverine input (370 km^3/y) into the basin [7, 8], and carries wastes amounting more than the total discharge to the North Sea [9]. Dnyester (9.3 km^3/y) and Dnyeper (54 km^3/y) are the second group of fresh water inputs to the basin which were strongly regulated after 1955 [10, 11]. The total fresh water contribution of Turkish rivers, none of which are regulated, is being only 35 km^3/y [12]. However, annual sedimentary loads of Turkish rivers, being $32.1x10^6$ tons, exceed all other rivers discharging into the basin, except Danube which brings $83x10^6$ tons of sediments. On the other hand, annual particulate organic matter load of Turkish rivers ($0.275x10^3$ tons/yr) is 1/3 of that of Danube ($0.913x10^3$ tons/y) and discharges of all other rivers on the northern part of the Black Sea($0.977x10^3$ tons/y) [13].

2. Materials and methods

The data obtained during the R/V J. Elliott Pillsbury August-1965, R/V Atlantis Aprıl-March-1969 and R/V Knorr August-1988 cruises were taken from technical reports [14, 15, 16]. Other data were taken during the various cruises of R/V Bilim since September-1991 till July, 1997. Water samples were collected by a Sea Bird model CTD probe, equipped with a Rosette type sampling system and insitu light transmissometer, at selected density surfaces. Nutrients were analysed on board R/V Bilim by using a two channel Technicon Autoanalyser, whereas dissolved oxygen was measured by Winkler method modified for low level oxygen concentrations. Hydrogen sulfide concentrations for September-1991 period were measured by iodometric technique whereas those for other cruises by colorimetric methylene blue technique. Positions of stations and transects used in the discussions are displayed in Fig. 2.

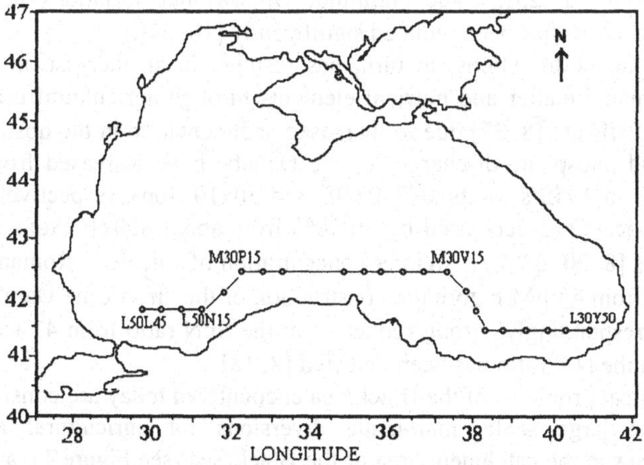

Figure 2 Locations of stations along the southern transect for September,1991

3. Results and Discussions

It is well known that coastal ecosystems of semi-enclosed seas act as natural filters for the open ocean waters by assimilating the major fraction of organic and inorganic inputs into the basin via absorption/desorption, fast sedimentation, and/or complex redox processes in the water column before their detrimental effects are fully transmitted to offshore waters. The remaining fraction of the materials of terrestrial origin are transported to the offshore areas of the basin where they are further converted into particulate organic and inorganic matter. 10-15% of the particulate organic matter in the open ocean is exported to the deep benthic layer where it is permanently buried in sediments for further diagenesis. However, when the coastal ecosystems of such basins are forced to survive under diverse and intense man-made stress, firstly their ecosystem shifts from natural, well balanced state to a less stabilised state which is more sensitive to further perturbations. Present ecosystem of the Black Sea is suffering such modifications and considered to have little resistance to destabilisation.

3.1.EUTROPHICATION PROBLEMS AND THEIR ECOLOGICAL IMPLICATIONS

Scientific evidence clearly indicated that profound changes have occurred in the Black Sea ecosystem during last three decades [2, 6, 9, 17, 18, 19, 20, 21]. The most dramatic changes have been detected on the north-western shelf area in the form of a decrease in taxonomic diversity of hydrobionts, as very intense blooms of some species to the detriment of others, decline in water transparency, and hypoxia in the benthic layer which influenced the benthic ecosystem [22]. Deterioration of the Black Sea ecosystem has been the consequence of the increased organic pollution through river discharges, even though the inflow of fresh water by the major rivers has been reduced by 25-50% since late 1970's [6, 8, 23]. The amplitude of seasonal variations in the water discharges of these rivers has been reduced significantly [10, 24].

These external modifications, in turn, has resulted in an increase in the annual loads of labile organic matter and nutrient elements through agricultural usages, but a decrease in that of silicate [8, 27] due to increased sedimentation in the upstream dams. Annual nitrate and phosphate discharges of the Danube have increased from $143x10^3$ and $12.5x10^3$ tons in 1950's to about $740x10^3$ and $30x10^3$ tons, respectively, whereas that of reactive silicate has decreased by 30-50% from about $800x10^3$ tons to $330x10^3$ tons in late 1980's [8, 20, 23, 27]. Winter concentration of silicate in Romanian coastal waters decreased from 55 μM before the construction of the "Iron Gate Dam" to 20 μM in 1992 [8]. Correspondingly, strong reduction in the Si:N ratio from 42 to about 2.8, and an increase in the N:P ratio has been detected [8, 18].

Main ecological problems of the Black Sea encountered today are considered to be induced by these large scale, man-made diversions for agricultural and power production purposes in the catchment area of the Black Sea (see Figure 1), and possibly by the decadal variations in the climatic field [24]. Relative contributions of these factor are not clearly defined yet. Most probably, man-made modifications are imposed on the climatic variations as suggested by Niermann and Greve [25] and Tiit [26]. Artificial interferences together with the introduction of opportunistic invaders to the Black Sea ecosystem [9] forced the system to attain a new ecological state in which regular late winter-early spring and autumn blooms have been replaced by late spring-summer blooms with extended periods during last 2-3 decades. In the mean time, the mean value of secchi disk decreased from 20-21 m in 1920s down to 6-8 m in 1990-92 after which its value started to increase gradually to about 14 m in 1995 [28].

3.2. CHANGES IN THE BIOCHEMICAL PROPERTIES OF THE UPPER LAYER

It has long been known that the depth of the oxic-anoxic interface in the Black Sea varies both in space and time, being shallower (70-80m) in the interior of cyclonic gyres in the central basin, and deeper (150-200m) at the peripheries of the main cyclonic circulation and in the anti-cyclonic eddies located between the cyclonically meandering rim current and the coastal zone [29, 30, 31]. September-1991 data collected by R/V Bilim (IMS-METU) were used to show the affects of water circulation on the vertical distributions of biochemical properties of the southern Black along an east-west transect (Fig. 2). Water circulation characteristics of the basin, in terms of biochemical properties, for this period were discussed in some detail by Baştürk et al. [32, 33]. Some of these stations, along the transect, fall into the rim current frontal zone (RCFZ), some in peripheries of the cyclonic gyres and some in the anti-cyclonic eddies.

Depth dependent variation of the potential density surfaces along the E-W transect (Figure 3) implies that the upper (σ_θ = 15.8) and lower (σ_θ = 16.15) boundaries of the SOZ approach to each other at the frontal zone between cyclonic gyres and the RCFZ. On both sides of these regions SOZ thickness increases again. 5 μM H_2S isoline approaches to that of 10 μM DO surface (Fig. 4) where the suboxic layer becomes very thin (≈10 m) as compared to other regions (30-40 m). In parallel to the deepening of isopycnal surfaces towards the east-end of the transect, DO and H2S contours deepen. Primary PO_4 minimum within the SOZ, as was shown by Baştürk et al. [32, 34] for the cyclonic gyres, is clearly evident at these locations (Fig. 5a). However, at the peripheries of these stations, deep PO_4 minimum layer disappears due to the increased vertical and diapycnal mixing between oxic and anoxic surfaces. PO4 concentrations attain higher values (1.0-1.2 μM) compared to those measured at the central stations (0.2-0.3 μM).

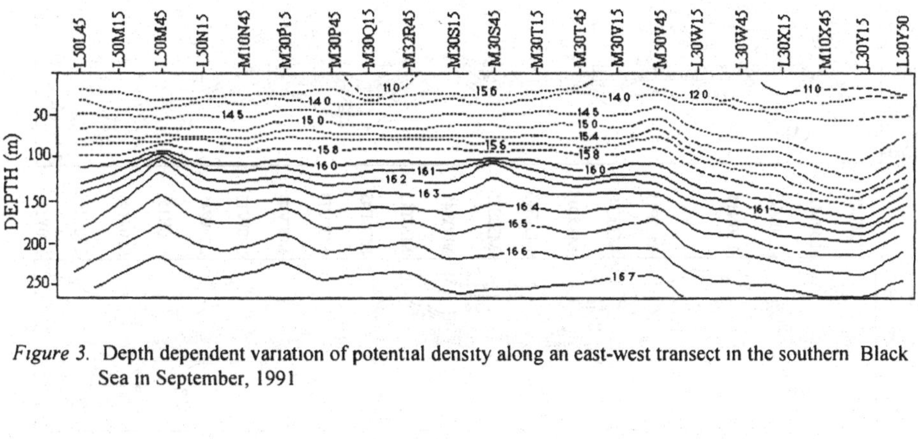

Figure 3. Depth dependent variation of potential density along an east-west transect in the southern Black Sea in September, 1991

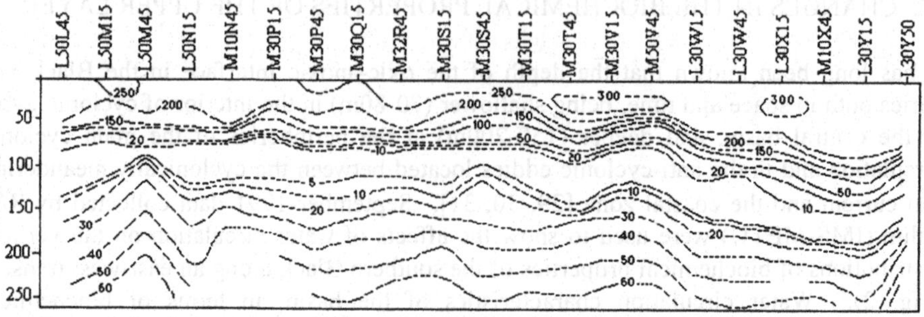

Figure 4. Depth dependent variations of dissolved oxygen and hydrogen sulfide, in μM, along the east-west transect in southern Black Sea in September, 1991.

Similarly, the peak concentrations of TNO_x (NO_3+NO_2) at the peripheral stations are higher (≈ 7-8 μM) than those of stations located in the cyclonic gyre (≈ 5 μM) where isopycnal surfaces are sandwiched in a narrow zone (Fig. 5b). Details of vertical profiles of biochemical properties were given for cyclonic, anticyclonic and rim current regions by Baştürk et al. [32]. All these features imply that the sandwiching of the isopycnal surfaces, which is observed frequently in the regions where abrupt changes in the bottom topography couples with the RCFZ, increases the concentration gradients of biochemical elements within the SOZ, which ,in turn, increases the material diffusion between oxic and anoxic layers. These redox sensitive elements are then transported laterally and cyclonically towards the peripheries of gyres. This process is important for the introduction of reduced elements into the upper layer through SOZ.

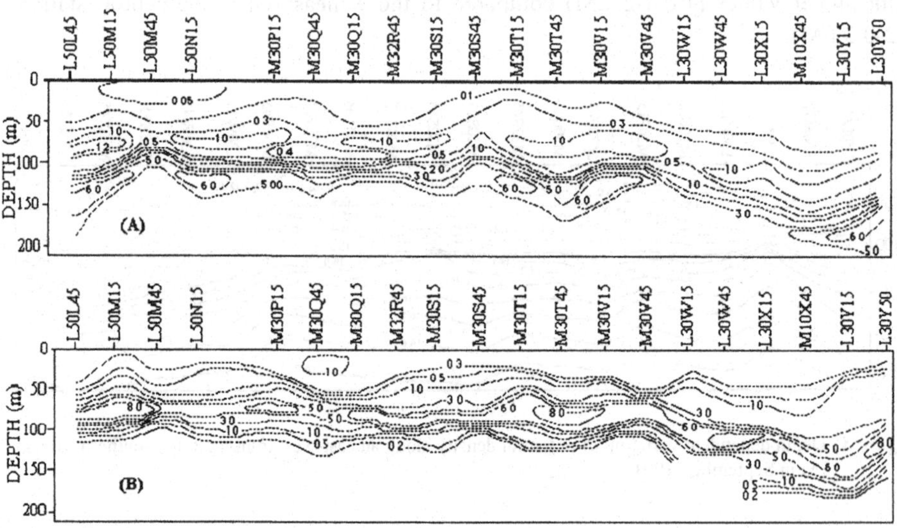

Figure 5 Depth dependent variations of phosphate (A) and total oxidized nitrogen (B) along the southern E-W transect in Black Sea (all are in units of μM)

Difficulties in determining the trace levels of dissolved oxygen and hydrogen sulfide in the transition zones between oxic, suboxic and anoxic zones together with the spatial and temporal variability in the biochemical structure of the upper layer, have led to conflicting conclusions when the comparisons were made in terms of depth profiles even from the similar locations. Researches also derived conflicting conclusions about the long-term changes in the depth of chemical boundaries of the oxic/anoxic interface [2, 5, 8, 19, 25, 36, 37, 38, 39, 40, 41, 42]. The chemistry of the entire basin as a function of water density - rather than depth - has been examined by numerous investigators recently [19, 42, 43] to define the chemical boundaries of the transition zone. All these investigators have shown that the distinct chemical features in the oxic/anoxic transition layer are consistently formed at certain density surfaces, independent of geographical location.

Ecological changes in the photic layer of the Black Sea observed in the form a shift from productive diatom based state to consumptive, organic rich state have led an upward expansion of the upper boundary of the suboxic zone, but not that of sulfidic layer [2, 19, 40, 43] in the Black Sea. The first redox sensitive element that had responded to these shifts is the dissolved oxygen within the oxycline. When the density dependent variation of dissolved oxygen for different years within the western cyclonic gyre of the Black Sea are examined, it will be recognised that the DO gradient zone in 1960's was much broader than in 1990's in terms of density surfaces (Fig. 6). Ventilation of the oxycline down to 15.3-15.4 density surfaces during the winter mixing and horizontal advection processes is also evident. It is also evident seen that 25 μM DO concentration measured at about 15.8-15.9 σ_θ surface in 60's has shifted upward by 0.3-0.4 density units in 1990's. All these changes clearly indicated that in parallel to changes in the coastal zones, the offshore regions of the Black Sea has also undergone detectable biochemical and ecological deteriorations.

Inter-annual as well as intra-annual variations within the basin can be triggered by different driving forces which affect the general circulation and hence the coastal-offshore interactions, horizontal advection and the vertical mixing processes between oxic/suboxic/anoxic layers [33]. These interactions influence the depth dependent profiles of biochemical properties down to the lower boundary of SOZ [33]. However, episodic events, like the one observed in July-1992, may induce pronounced changes in the vertical distributions of chemical properties due to changes in the climatology and circulation characteristics which affected the vertical mixing intensity and hence induced intense planktonic blooms in mid-summer (Fig. 7).

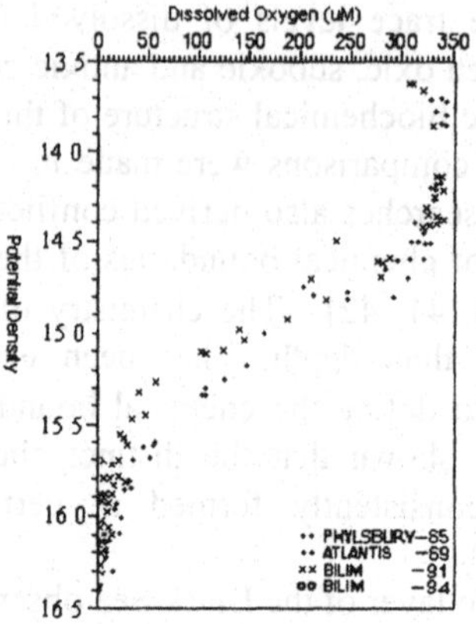

Figure 6 Potential density dependent variations of dissolved oxygen in the western interior of the Black Sea for different periods.

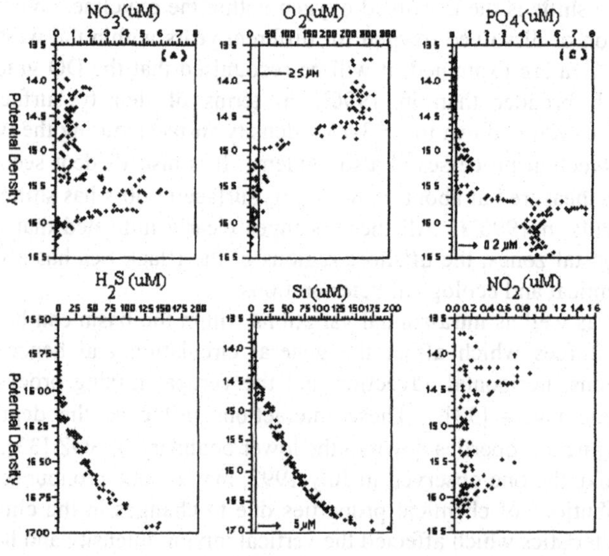

Figure 7 Potential density dependent plots of biochemical properties during the abnormal July,1992 period

During July-1992 period, DO, PO_4 and TNO_x concentrations in the oxycline were greatly reduced down to 15.4 surface where normal, well established TNO_x peak was positioned. This erosion of upper nutricline can not be due to simple dilution effect of vertical mixing, because there should be an increase in the DO concentration in parallel to the reduction in TNO_x concentration due to the entrainment of oxic waters. If these types events are repeating regularly or randomly at about decadal frequency, then the low TNO_x concentrations measured in 1960s can be attributed to such episodic events which may influence nearly the whole basin in 1960's; even though the event observed in July-1992 was confined to offshore waters of the southern Black Sea. The hydrochemical details of these events and the factors and mechanism(s) responsible for this type events are still poorly understood and debates more process oriented, detailed surveys.

Upper and lower gradients (μM property/m) of the total oxidised nitrogen (TNO_x) were calculated from the depth integrals of chemical properties given by Baştürk et al. [34] for September-1991. When the measured and the calculated gradients for the water layer between 14.2 and 15.4 σ_θ surfaces, which denote the nitracline onset and nitrate maximum, are compared, practically no difference is observed between two gradients. However, 0.03-0.04 unit differences between the values for anticyclonic region and rim current frontal zone implies that more nitrogen species accumulated in these zone, probably due to less intense denitrification. On the other hand, measured lower gradient of NO_x was found to be lower than the gradient calculated from *in situ* O:N:P ratios measured within the sub-oxic layer of the Black Sea for September, 1991. The difference, in contrast to upper gradient, becomes larger in lower gradient of cyclonic gyres (-0.354 μM/m). This large difference between the measured and calculated gradients of NO_x, suggests that nearly half of the NO_3 is utilized by the reduced species diffusing into the narrow sub-oxic zone of the cyclonic gyres.

3.3. SUB-OXIC LAYER

The layer of co-existence (C-layer), in which DO and H_2S were suggested to coexist [35, 44] was shown to be analytical artefacts in sampling and analysis procedures [2, 19, 40]. Co-existence of DO and H_2S is proved to be unrealistic when one considers the rapid dynamics of the reaction between them [45, 46]. At least 10 times faster H_2S oxidation rates in deep Black Sea compared to surface waters with added H_2S, and the similarities between the rates of oxidation of filtered and unfiltered samples under oxic (150-200 μM DO) [47] and suboxic oxygen levels (DO<50 μM) [48] imply that the oxidation of H_2S is mostly abiotic in nature and O_2 is not a direct oxidant for H_2S. Instead, it may be oxidized by the coupling of Mn and Fe cycles involving dissolved and particulate forms of them.

Insitu H_2S oxidation rates within the suboxic zones of the dynamically different regions of the Black Sea were shown to be different [48]; faster removal rates ($t_{1/2}$=2.5-3.4 hr) within the suboxic layers of the western cyclonic gyre compared to those in anticyclonic eddies ($t_{1/2}$=2.3-15.4 hr), and even faster rates within the RCFZ ($t_{1/2}$=1.2-1.3 hr) due to the catalytic actions of particulate forms of iron and manganese oxides resuspended from the shelf-slope zone by strong rim current. Tebo [49] measured nearly two orders of magnitude higher Mn(II) residence time in the SOZ of the central parts of the basin (30-90 days) compared to that at the near-shore waters (0.6-1.0 day). At offshore stations no overlapping of NO_3 and NH_4 gradients was evident whereas they overlapped at the near-shore stations [49]. 5 μM DO concentration was shown to penetrate down to 16.0 density surface at the near-shore stations whereas the same level was observed at 15.4 density surface of the interior of the basin.

In order to understand the redox processes and mechanisms of hydrogen sulfide removal within the sub-oxic zone of coastal frontal zones, a process oriented joint survey was conducted during July, 1997 cruise of R/V Bilim along a transverse transect where stations were located 5 nm apart in the offshore direction in the region close to Sakarya Canyon which is located at the south western Black Sea. This area is characterised by an abrupt change in the topography from about 200 m to more than 1000 m along the coast and covers an anticyclonic eddy and strong rim current at its offshore boundary. A detailed sampling strategy was followed along a transect vertical to the coastline. Here we will present anomalous distributions of some chemical parameters at a station (St. L29M46) which was located within the Sakarya Canyon (Fig. 8).

Deep light transmission minimum layer, called as Fine Particle Layer (FPL) [50], weakly observed within the lower boundary of the SOZ of offshore waters intensifies within that of coastal shelf zone but at deeper density surfaces (down to 16.6 density surface). This layer commences at 15.9 density surface and expands down to the 16.4 surface in coastal shelf zone. At the Sta. L29M46 (Fig. 8), the onsets of NH_4 and Mn(II) were observed at the same density surface (15.85) and have similar profiles. However, dissolved oxygen concentration less than 10 μM, on contrary to its frequently observed basin-wide position (DO< 10 μM at σ_θ = 15.6-15.8), was measured down to 16.4 density surface below which sulfidic layer has known to be started (H_2S> 1 μM at σ_θ = 16.15-16.20). About 20 μM DO concentration was measured at 16.2 density surface which was accepted as the onset of sulfidic layer. The well established SOZ layer is very narrow at this station compared to previous boundaries [43]. Even under the intense erosion conditions within the upper zone of the sulfidic layer and thus oxygenation down to the 16.4 density surface, oxic and anoxic layers still do not overlap with each other. In other words, the C-layer does not exist even under intense mixing conditions.

At the station at about 5 nm north of the coastal station (not shown here), the FPL starts to be sliced intermittently by horizontal intrusions of sulfidic water masses from the surrounding in the offshore direction. There is no single and broad FPL as was observed at coastal station, but rather a series of minimum. Additionally, lower boundary of the FPL deepens from 16.4 surface down to 16.7 density surface at this station. Vertical distributions of DO and H_2S at that station followed sequentially each other, never being overlapped. When there is a DO peak, H_2S concentration drops to below detection limits (<1 μM) or wise versa. Additionally, 50 μM DO concentration was detected at around 16.4 surface, followed again by a sulfidic zone. At the interfaces of these oxic/anoxic zones no overlapping of the layers are detectable.

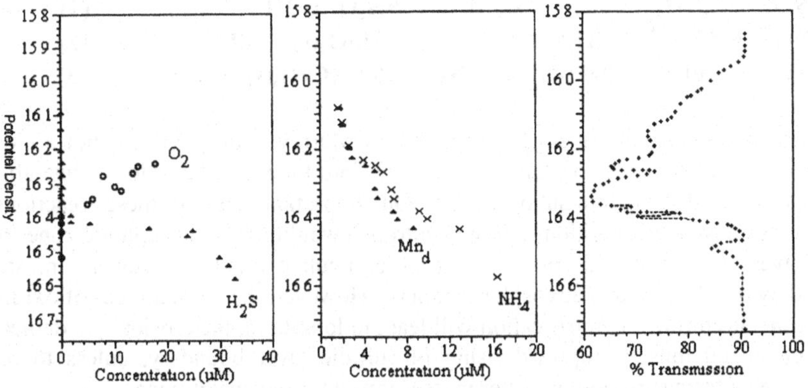

Figure 8 Anomalous distributions of biochemical parameters within deep fine particle layer observed at Sta L29M46 located within the Sakarya Canyon region (R/V Bilim, IMS-METU July,1997 cruise)

Comparison of the vertical O_2 and H_2S gradients in the suboxic zone of the Black Sea, and the insufficiency of the O_2-electron gradient to compensate the electron concentration required for the oxidation of H_2S diffusing upward, have led Murray et al. [43] to suggest three alternative processes.

1. Sulfide is oxidized by the settling particulate MnO_2(s) [47, 51], and possibly by iron oxyhydroxides [FeOOH(s)] formed by bacteria in shallower oxic waters [47]. Manganese cycle may be the key process for maintaining the broad suboxic zone; oxidised Mn may oxidised reduced sulfur while reduced Mn(II) may consume dissolved oxygen, as was suggested by Tebo [49]. However, vertical electron equivalent gradients of Mn(II) and Fe(II) were shown to be much less than that of the sulfide gradient [43].

2). Another explanation is the anaerobic oxidation of sulfide associated with the phototrophic reduction of CO_2 to organic carbon as suggested by Jannash [52] and Jorgensen et al. [53], as also supported by the findings of Repeta et al. [54] who measured considerable quantities of bacteriochlorophyll pigments below the suboxic zone .

3). Horizontal as well as vertical ventilation processes may supply sufficient DO. However, it was shown by Lewis and Landing [55] that horizontal mixing of oxygen could provide the necessary oxygen to account for Mn oxidation, but probably not that of sulfide and other reduced species.

Since NO_3, NH_4, Mn(II) and Fe(II) all decrease to low concentrations at about the same density levels (σ_θ =15.95, 15.95, 15.85 and 16.00, respectively), equivalence of upward and downward electron gradients have led Murray et al. [43] to hypothesise that NO_3 is the main account for the oxidation of reduced species diffusing upward from the anoxic layer. The following reactions were suggested by Murray et al. [43] to occur within the suboxic-anoxic boundary:

$$3NO_3^- + 5NH_4^+ \Leftrightarrow 4N_2 + 9H_2O + 2H^+ \quad (1)$$
$$2NO_3^- + 5Mn^{2+} + 4H_2O \Leftrightarrow N_2 + 5MnO_2(s) + 8H^+ \quad (2)$$
$$2NO_3^- + 10Fe^{2+} + 24H_2O \Leftrightarrow N_2 + 10\,Fe(OH)_3(s) + 18H^+ \quad (3)$$

When one examines the suggested model of Murray *et al.* [43], it is not a closed redox cycle since it needs constant input of NO_3 into the redox layer to oxidise all the reduced species diffusing into suboxic zone. One important output of theses reactions is that, under steady-state conditions , no new nitrogen would reach the euphotic zone from deeper layer, and hence the sinking flux of particulate organic nitrogen, must be sustained by the riverine and atmospheric inputs. However, the constant loss of oxidised nitrogen species through denitrification will lead, in long-term, the erosion and depletion of nitracline and thus the upward rising of sulfidic layer boundary, unless there is constant inputs from surrounding water masses through isopycnal mixing.

Luther *et al* [56] have suggested two thermodynamically favourable reactions between manganese and nitrogen species: the first one is the reduction of NO_3^- to N_2 through the oxidation of Mn(II) to $MnO_2(s)$ (rxn 4), the second one is the oxidation of Mn(II) in the presence of O_2 to $MnO_2(s)$ (rxn 5) which oxidises NH_4^+ and organic nitrogen to N_2 (rxn 6). Mass-balance calculations show that the oxidation of NH_4 and organic-nitrogen by MnO_2 may be the dominant process producing N_2 in Mn-rich continental margin sediments. Above pH 6.8, the reaction between Fe(III) species and NH_4^+ to form N_2 was shown to be thermodynamically unfavourable [56]. On the other hand, Rozanov [57] has stated that oxygen is the main and direct oxidant of all reduced species diffusing from the anoxic layer in the suboxic zone.

$$5\,Mn^{2+} + 2\,NO_3^- + 4\,H_2O \Rightarrow 5\,MnO_2 + N_2 + 8\,H^+ \quad (4)$$
$$2\,Mn^{2+} + 4\,OH^- + O_2 \Rightarrow 2\,MnO_2 + 2\,H_2O \quad (5)$$
$$2\,NH_3 + 2\,MnO_2^- + 6\,H^+ \Leftrightarrow 3\,Mn^{2+} + N_2 + 6\,H_2O \quad (6)$$

NO_2 maximum ($\sigma_t = 15.85$) was shown almost always to coincide with the zone of denitrification and with the zone of Particulate Mn maximum ($\sigma_t = 15.85$) [43]. Coincidence of Mn-Oxide maximum surface with that of NO_2 , and positioning of particulate manganese layer between the layer of NO_3 minimum, suboxic levels of DO, and particulate Fe oxides suggest that particulate manganese oxide couples the redox processes in the upper layer of the suboxic zone with those in the lower layer by carrying the oxidation potential of nitrate and oxygen to the lower section of SOZ where ammonia is oxidised to N_2 and/or to NO_2 plus reduces NO_3 to NO_2 and/or N_2 while itself is oxidised. On the other hand, it was shown that Fe(II) in the form of pyrite reduce MnO_2 rapidly (rxn 7) to form iron(hydr)oxides, which in turn oxidises H_2S to S^o and $S_2O_3^=$ (rxn 8) [58, 59]. Particulate manganese is also suggested to be involved in the oxidation of hydrogen sulfide by Nealson *et al.* [60] (rxn 13).

$$MnO_2 \quad + 2\ Fe^{2+} + 2H_2O \Rightarrow Mn^{2+} + 2\ FeOOH + 2\ H^{+.} \qquad (7)$$
$$2FeOOH + H_2S \quad \Leftrightarrow 2\ Fe^{2+} + S^o + 4\ OH^- \qquad (8)$$

Field studies and surveys suggested that NH_4 is oxidised basically by the particulate $Mn0_2$, probably, to N_2 or NO_2. Observed decrease in the Mn(II) concentration within the oxygenated water layers (see Fig. 8) is proposed to be due to the oxidation of Mn(II) back to $Mn0_2(s)$ by oxygen, which in turn oxidises NH_4+ to N_2 or NO_2, and indirectly the H_2S by coupling with iron oxidation.

By considering the reaction types suggested by different researchers, the following reaction schemes are suggested as possible redox reactions within the oxic/sub-oxic/anoxic interfaces. In the given order, particulate manganese plays a key role and acts as an electron carrier, or a catalyst, for the electrons given by O_2 and NO_3^- to the NH_4^+ and H_2S diffusing from the anoxic layer.

$$5Mn(II) + 2\ NO_3^- + 4\ H_2O \Rightarrow 5MnO_2\ (s) + N_2 + 8\ H^+ \qquad (9a)$$
$$4Mn(II) + 2\ O_2 + 4\ H_2O \Rightarrow 4MnO_2\ (s) + 8\ H^+ \qquad (9b)$$
$$3Mn0_2(s) + NH_4^+ + 4H^+ \Leftrightarrow 3\ Mn^{2+} + N0_2^- + 4H_20 \qquad (10a)$$
$$3Mn0_2\ (s) + 2NH_4^+ + 4H^+ \Rightarrow 3Mn^{2+} + N_2 + 6H_20 \qquad (10b)$$
$$MnO_2 + 2Fe^{2+} + 2H_2O \Rightarrow Mn^{2+} + 2FeOOH + 2H^{+.} \qquad (11)$$
$$2FeOOH + H_2S \Leftrightarrow 2Fe^{2+} + S^o + 4OH^- \qquad (12)$$
$$MnO_2(s) + HS^- + 3H^+ \Rightarrow Mn^{2+} + S^o + 2H_2O \qquad (13)$$

4. Conclusions and Suggestions for Future Studies

The chronology of the catastrophic events related to the ecological deteriorations in the basin during the last 2-3 decades in the Black Sea can be summarised as;

a - Substantial increase in annual nutrient loads (N,P) even though the river discharges has been reduced on contrary to that observed for silicate load.

b - Changes in the relative proportions of nutrient elements within the photic layer have induced a shift from regular late winter-early spring and autumn blooms to late-spring-early summer blooms with extended periods and remarkable changes in the taxonomic composition of bloom-producing planktonic species.

c - Collapse of the Black Sea pelagic and benthic ecosystems has been followed by a decrease in relative abundance of diatoms, and an increase in the low-food value planktons, especially dinoflagellates during 1980-1990 period [20, 62, 63]. Formation of large hypoxic areas was recorded in 1973 on NW shelf which later became frequent due to the increased sedimentation of organic matter produced in the photic layer or introduced from land-based sources.

d - These changes, together with the invasion by opportunistic organisms such as *Mnemiopsis leidyi* have put additional stress on the ecosystem of the Black Sea which induced drastic changes in the taxonomic composition of the zooplankton species which were serving as food for the fish larvae. Thus a detectable decrease has experienced in fish stocks and catches in conjunction with the reductions in the diets of fish larvae.

e - modifications of the organic matter cycle within the upper layer of the water column have induced changes in the vertical distributions of the biochemically active elements which were detected as an upward expansion of oxic/anoxic interface with respect to density surfaces. Nearly two-fold increase in the nitrate content of the oxycline, detectable decrease in that of silicate, and removal of ammonia which was at detectable levels in 1960s were the overall outcome of all these modifications.

f- In coastal zones, removal of ammonia, hydrogen sulfide and dissolved iron(II) and manganese(II) are probably following biochemical cycles that differ from those in the interior of the basin as suggested by Murray *et al.* [43]. Therefore, manganese cycle and its impact on the distributions of reduced species diffusing into suboxic zone should be studied in future.

Acknowledgements

This work was supported by the Turkish Scientific and Technical Council (TÜBİTAK) and the NATO-Science for Stability Program. The authors wish to thank all oceanographers interested in the Black Sea and supplied the valuable information through publications related with the Black Sea.

References

1. Degens, E T and Stoffers, P (1980) Environmental events recorded in Quaternary sediments of the Black Sea. *Jr. Geol Soc London* **137**, 131-138.
2. Murray, J. M., Jannasch, H. W, Honjo, S., Anderson, R. F., Reeburgh, W S., Top, Z., Friderich, G E, Codispoti, L A. and Izdar, E. (1989) Unexpected changes in the oxic/anoxic interface in the Black Sea. *Nature* **338**, 411-413
3. Emerson, S (1980) Redox species in a reducing fjord·The oxidation rate of Manganese-II In H J Freeland, D M. Farmer, C D Levins and Victoria, B J (eds.), *Fjord Oceanography*, pp·680-692, NATO Conference Series Vol.4.
4. Bregant, D, Catalano, G., Civitarese, G. and Luchetta,A (1990) Some chemical characteristics of the brines in Bannock and Tyro Basins salinity, sulfur compunds, Ca^{+2} ,F^- ,pH, A_t, PO_4^{3-} SiO_2, NH_3 *Marine Chemistry* **31**, 35-62
5. Sorokin, Yu I. (1983) The Black Sea. In *Estuaries and Enclosed Sea Ecosystem of the World.* Vol 26, B. H Ketchum (ed), Elsevier, Amsterdam, pp:253-292.
6. Bologa, A.S. (1986) Planktonic primary productivity of the Black Sea: A review *Thalassia Jugoslavica* **21**, 1-22.
7. Cociasu, A, Dorogan, L, Humborg, C, and Popa, L. (1996) Long-term Ecological Changes in the Romanian Coastal Waters of the Black Sea. *Marine Pollution Bulletin* **32(1)**, 32-38.
8. Humborg, C., Ittekkok, V, Cociasu, A and v Bodungen, B (1997) Effect of Danube River dam on Black Sea biogeochemistry and ecosystem structure *Nature* **386**,*385-388*
9. Mee, L D. (1992) The Black Sea in Crisis The need for concerted international action *Ambio* **21(3)**, *278-286*
10. Simonov, A I and Al'tman, E N (1991) Hydrometeorology and hydrochemistry of the USSR seas V 4, Black Sea. Hydrometeoizdat, St Peterbug (in Russian)
11. Tolmazin, D (1985) Changing coastal oceanography of the Black Sea, I Northwest Shelf. *Prog in Oceanogr.* **15**, *217-276*

12 Balkas, T , Dechev, G , Mıhnea, R., Sebanescu, O and Unluata, U (1990) State of Marine Envıronment ın the Black Sea Regıon UNEP Regıonal Seas Reports and Studıes No 124

13 Izdar, E , Konuk, T , Ittekkok, V Kempe, S and Degens, E T (1987) Partıcle flux ın the Black Sea· Nature of the Organıc Matter In *Partıcle Flux ın the Ocean* SCOPE/UNEP Sonderband, Heft 62 June 23-28, 1986 Izmır

14 NOAA-NODC World Ocean Data Center-A, USA, (1997)

15 Brewer, P G. (1971) Hydrographıc and chemıcal data from the Black Sea, Woods Hole Oceanographıc Inst Technıcal Report, Reference No.71-65

16 Frıederıch, G E , Codispotı, L. A and Sakamato, C M (1990) Bottle and pump cast data from the Black Sea expeditıon, Monterey Aquarıum Res Inst Tech. Report No·90-3, 224 pp

17 Shuskına, Eh A and Musaeva, Eh I (1990) Structure of planktonic community from the Black Sea epıpelagıcal and its change as the result of ıntroduction of a ctenophore specıes Oceanology **30(2)**, 306-310

18. Smayda, T. J (1990) Novel and nuısance phytoplankton blooms ın the sea Evıdence for a global epıdemıc In E Granelli and D. M. Anderson (eds) *Proc of 4 th Int Conf, Toxıc Marına Phytoplankton.* Lund, Sweden, 26-30 June, 1989, pp 29-40

19 Tuğrul, S , Baştürk, O., Saydam, C and Yılmaz, A (1992) Changes ın the hydrochemıstry of the Black Sea ınferred from water densıty profiles, *Nature* **359**, *137-139*

20 Bodeanu, N. (1989) Les developements massıfs du phytoplancton des eaux du lıttoral Roumanıan de la Mer Noıre au cours de l'annee *Cercet Mar* **22**, *127-146*

21 Gomoıu, M. T (1992) Marıne eutrophıcatıon syndrome ın the northwestern part of the Black Sea In· *R A Wollenweıder, R Marchettı and R V Vıvıanı (eds) Marıne Coastal Eutrophıcatıon* Elsevıer Publ Amsterdam, pp.683-692

22 Zaıtsev, Yu P and Alexandrov, B G (1997) Recent man-made changes ın the Black Sea ecosystem In *E Özsoy and A Mıkaelyan (eds), Sensıtıvıty to change Black Sea, Baltıc Sea and North Sea,* 25-31 Kluwer Academıc Publıshers, Netherlands

23 Dorogan, L , Popa, L , Cocıasu, A and Voınescu, I (1985) L'Apport de sels nutrıtıfs des eaux de Danube dans la Mer Noıre *Rapp Comm. Int Mer Medıt* **29**, *57-59*

24. Polansky A , Voskresenskaya and Belokopytov, V (1997) Variabilıty of northwestern Black Sea hydrography and river dıscharges as part of global Ocean-Atmosphere fluctuatıons. In· *E.Özsoy and A Mıkaelyan (eds), Sensıtıvıty to change Black Sea, Baltıc Sea and North Sea,* 11-24 Kluwer Academic Publishers, Netherlands.

25 Nıermann, U and Greve, W (1997) Dıstrıbutıon and fluctuatıon of dominant zooplankton species ın the southern Black Sea ın comparıson to the North Sea and Baltıc Sea In. *E Özsoy and A Mıkaelyan (eds), Sensıtıvıty to change Black Sea, Baltıc Sea and North Sea,* 65-77 Kluwer Academıc Publıshers, Netherlands

26 Tıtt, R (1997) The effect of hydrologıcal condıtions on the state of herrıng stocks of the Baltıc Sea. In *E Özsoy and A Mıkaelyan (eds), Sensıtıvıty to change Black Sea, Baltıc Sea and North Sea,* 139-147 Kluwer Academic Publishers, Netherlands

27 Cocıasu, A., Dıaconu, V , Popa, L , Buga, L , Nae, I , Dorogan, L and Malciu, V. (1997) The nutrient stock of the Romanıan shelf of the Black Sea durıng the last three decades. In *E Özsoy and A. Mıkaelyan (eds.), Sensıtıvıty to change Black Sea, Baltıc Sea and North Sea,* 49-63 Kluwer Academic Publishers, Netherlands

28 Vladimirov, V. L., Mankovsky, V I , Solov'ev, M V and Mıshonov, A V (1997) Seasonal and long-term variabılıty of the Black Sea optıcal parameters In *E Özsoy and A. Mıkaelyan (eds), Sensıtıvıty to change Black Sea, Baltıc Sea and North Sea,* 33-48. Kluwer Academic Publishers, Netherlands

29 Grasshoff, K (1975) Hydrochemıstry of land-locked basıns and Fjords, In J P Rıley and G Skırrow (eds), *Chemıcal Oceanography,* Academıc Press, New York, NY, vol 2, pp.455-579

30 Oğuz, T , La Vıolette, P. E and Ünluata, U (1992) The upper layer circulatıon of the Black Sea Its varıabılıty as inferred from hydrographıc and Satellite observatıons *Jr Geophys Res.* **97(C8)**, *569-584*

31 Oğuz, T , Latun, V S., Latıf, M A., Vladımırov, V V , Sur, H. I , Markov, A A , Özsoy, E , Kotovshchıkov, B B , Eremeev, V V. and Unlüata, Ü (1993) Circulatıon in the surface and ıntermediate layers of the Black Sea , *Deep-Sea Res* **40(8)**, *1597-1612*

32. Baştürk,, Ö , Tuğrul, S , Konovalov, S and Salıhoğlu, I (1997a) Varıatıons ın the vertıcal chemıstry of water masses wıthın hydrodynamıcally different regıons of the Black Sea NATO-ARW Meetıng,

Sensitivity of North Sea, Baltic Sea and Black Sea to anthropogenic and climatic changes, 14-18, Nov., 1995, Varna, Bulgaria

33. Baştürk, Ö., Tuğrul, S , Konovalov, S and Salihoğlu, I (1997b) Effects of circulation on the spatial distributions of principle chemical properties and unexpected short- and long-term changes in the Black Sea. NATO-ARW Meeting, 1-10 June, 1997, Sebastopol, Ukraine.
34. Baştürk,, Ö., Tuğrul, S., and Salihoğlu, I. (1996) Vertical chemistry of the three dynamically different regions of the Black Sea. *Tr.Jr of Marine Sciences* **2**, 35-50.
35. Fashchuk, D. Y and Ayzatullin, T. A. (1986) A possible transformation of the anaerobic zone of the Black Sea. *Oceanology* **26(2)**, 171-173
36 Leonov, A V. and Ayzatullin, T. A (1987) Mathematical modelling of the oxidation of hydrogen sulfide in connection with calculations of the dynamics of the hydrogen sulfide-oxygen co-existence layer and process conditions for obtaining sulfur from the Black Sea water. *Oceanology* **27(2)**, 174-178.
37 Bryantsev, V. A., Faschuk, D Ya , Ayzatullin, T A , Bagotskiy, S V and Leonov, A. V. (1988) Variation in the upper boundary of the hydrogen sulfide zone in the Black Sea. Analysis of field observations and modelling results *Oceanology* **28**, *180-185*.
38 Buesseller, K. O , Livingston, H. D , Ivanov, L. and Romanov, A. (1994) Stability of oxic/anoxic interface in the Black Sea. *Deep-Sea Res.* **41(2)**, *283-296*.
39. Codispoti, L A., Friederich, G. E , Murray, J W and Sakamoto, C M. (1991) Chemical variability in the Black Sea: Implications of continuous vertical profiles that penetrated the oxic/anoxic interface. *Deep-Sea Res.* **38(2)**, S691-S710.
40 Bastürk ,Ö., Saydam, C , Salihoğlu, I., Eremeev, L. V , Konovalov, S., Stoyanov, A , Dimitrov, A , Cociasu, A , Dorogan, L. and. Altabet, M. (1994) Vertical variations in the principle chemical properties of the Black Sea in the autumn of 1991. *Marine Chemistry* **45**, 149-165.
41. Oğuz, T , Latif, M A , Sur, H. I., Özsoy, E.and Ünlüata, Ü (1991) On the dynamics of the southern Black Sea. In: E.Izdar and J W Murray (eds.) *Black Sea Oceanography. 43-63*, Kluwer Academic Publishers. The Netherlands
42. Saydam, C., Tuğrul, S., Baştürk, Ö. and Oğuz, T (1993) Identification of the oxic/anoxic interface by isopycnal surfaces in the Black Sea. *Deep-Sea Res.* **40(7)**, 1405-1412.
43. Murray, J M., Codispoti, L. A. and Freiderich, G. E. (1995) Oxidation-reduction Environments The suboxic zone in the Black Sea In. *Aquatic Chemistry*, C. P Huang, C R. O'Melia and J. J. Morgan (eds), ACS Advances in Chemistry Series No 244, pp 157-176.
44 Vinogradov, M Ye and Nalbandov, YU R (1990) Effect of changes in water density on the profiles of physicochemical and biological characteristics in the pelagic ecosystem of the Black Sea. *Oceanology* **30**, 567-573.
45 Millero, F. J., Hubinger, S , Fernandez, M and Garnett, S. (1987) Oxidation of H_2S in seawater as a function of temperature, pH and ionic strength *Jr. Environ Sci and Tech* **21**, 439-443.
46. Gökmen, S (1996) A comparative study for the determination of hydrogen sulfide in the suboxic zone of the Black Sea. Ms. Thesis, Inst. of Marine Science, Erdemli-Icel, Turkey, 156 pp
47. Millero, F. J. (1991) The oxidation of H_2S with O_2 in the Black Sea. In. *Black Sea Oceanography*, E İzdar and J W. Murray (eds), NATO ASI Series C: Volume 351 pp 205-227.
48. Gökmen, S. and Baştürk, Ö (1997) Some remarks on the H_2S removal rates within the suboxic zone of dynamically different regions of the Black Sea. NATO TU-Black Sea Project· Symposium on Scientific Results, Extended Abstracts, p 64 15-19 June, 1997, Crimea-Ukraine
49 Tebo, B. M. (1991) Manganese(II) oxidation in the suboxic zone of the Black Sea. *Deep-Sea Res* **38(2A)**, s883-s905.
50. Kempe, S., Dierks, A. R , Liebezeit, G. and Prange, A (1991) Geochemical and structural aspects of the pycnocline in the Black Sea (R V. Knorr 134-LEG-1,1988). In: *Black Sea Oceanography*, E. İzdar and J. W Murray (eds), Kluwer Academic Publ Netherlands, pp 89-110.
51 Luther III, G. W., Church, T M and Powell, D (1991) Sulfur speciation and sulfide oxidation in the water column of the Black Sea. *Deep-Sea Res.* **38(2A)**, 1121-1137
52. Jannasch, H. W. (1991) Microbial processes in the Black Sea water column and top sediment An overview. In. *Black Sea Oceanography*, E. Izdar and J. W. Murray (Eds), p:271-281, Kluwer Academic Publ , Netherlands.
53. Jorgensen, H. W , Fossing, H., Wirsen, C O.and Jannasch, H W (1991) Sulfide oxidation in the anoxic Black Sea chemocline, *Deep-Sea Res* **38(2A)**, S1083-S1104.

54 Repeta, D J, Simpson, D J, Jorgensen, B B. and Jannash, H. W. (1989) Evidence for anoxygenic photosynthesis from the distribution of bacteriochlorophylls in the Black Sea. *Nature, 342·69-72*

55 Lewis, B L. and Landing, W. M (1991) The biogeochemistry of manganese and iron in the Black Sea *Deep-Sea Res.* **38(2A)**, S773-S803

56. Luther III, G. W, Sundby, B., Lewis, B L, Brendel, P J and Silverberg, N (1997) Interactions of manganese with the nitrogen cycle: Alternative pathways to dinitrogen *Geochim. Cosmochim Acta* **61(19)**, 4043-4052.

57 Rozanov, A G., Neretin, L. N. and Volkov, I I. (1997) Redox Nepheloid Layer (RNL) of the Black Sea. Its location, consumption and origin. NATO TU-Black Sea Project. Symposium on Scientific Results, Extended Abstracts, p·61. 15-19 June, 1997, Crimea-Ukraine

58 Burdige, D. J. and Nealson, K. H (1986) Chemical and microbiological studies of sulfide-mediated manganese reduction *Geomicrobiol, J.* **4**, 361-387

59. Lovely, D R and Phillips, E. J. P. (1988) Manganese inhibition of microbial iron reduction in anaerobic sediments. *Geomicrobiol J.* **6**, 145-155

60 Nealson, K H., Myers, C R. and Wimpee, B. B (1991) Isolation and identification of manganese-reducing bacteria and estimates of microbial Mn (IV)-reducing potential in the Black Sea. *Deep-Sea Res* **38(2A)**, S907-S920.

61 Porumb, F. (1992) On the development of *Noctiluca scintillans* under eutrophication of Romanian Black Sea waters In. R A Wollenweider, R Marchetti and R V Viviani (eds) *Marine Coastal Eutrophication.* Elsevier Publ Amsterdam, pp 907-920

62. Bodeanu, N. (1992) Algal blooms and developments of the main phytoplanktonic species at the Romanian Black Sea littoral in conditions of intensification of the eutrophication processes In. R A Wollenweider, R. Marchetti and R V. Viviani (eds) *Marine Coastal Eutrophication.* Elsevier Publ Amsterdam, pp·891-906.

EUTROPHICATION: A PLAUSIBLE CAUSE FOR CHANGES IN HYDROCHEMICAL STRUCTURE OF THE BLACK SEA ANOXIC LAYER

S.K. KONOVALOV[1], L.I. IVANOV[1], J.W. MURRAY[2], L.V. EREMEEVA[1]
[1]*Marine Hydrophysical Institute, NASU, 2 Kapitanskaya St., Sevastopol, 335000, Ukraine*
[2] *School of Oceanography, Box 357940, University of Washington, Seattle WA 98195-7940, USA*

Abstract

Field observations have been used to analyze changes in chemical properties (distribution of nutrients and sulfide) of the Black Sea anoxic layer from 1960 to 1995. The results reveal notable changes (increase in the inventories of nutrient and sulfide) in the chemical structure of the anoxic zone, which has been considered for a long time as a conservative layer of the Black Sea. It is inferred that intensive eutrophication is the main reason of the change. The results of the analysis of long-term variations in chemical structure demonstrate not only intensive degradation of the ecosystem of the Black Sea in 70's and 80's, but also relative improvement of the situation in 90's.

1. Introduction.

The chemistry of the Black Sea has been a subject of scientific investigations for about a century. The results have been published in a great number of publications, since the first one to appear in 1890 [1]. Recently published bibliography [2] contains 4256 references for the period from 1974 to 1994. There are several publications [3,4,5] where various problems pertinent to Black Sea chemical oceanography have been discussed in detail.

Though much has been achieved in the past in understanding the structure of the oxic/anoxic transition zone (suboxic zone) in the Black Sea, its spatial and temporal variability as well as the specific biogeochemical processes related to interactions of the oxic and anoxic environments, there are still certain gaps in our knowledge about Black Sea biogeochemistry. Since the sulfide containing waters are an adverse environment for most biological species (except some forms of bacteria), recent research activities have focused on studying biochemical processes within the transition zone [6,7], seeking proof for possible changes in the position of the onset of sulfide [8,9,10] or challenging

S. Beşiktepe et al. (eds.),
Environmental Degradation of the Black Sea: Challenges and Remedies, 61–74.

such claims [6,11]. Less attention has been paid to possible alterations within the anoxic layers of the Black Sea.

One of the issues that was beyond the scope of the earlier studies was climatic variability and trends for biochemical variables within the anoxic layer. This layer of the sea was considered to be a conservative water mass with a residence time on the scale of ten years to centuries. To partially fill the gap, we focus in this paper on the assessment of long term variability of nutrients and sulfide within the anoxic layer of the Black Sea. Another specific objective is to determine the relationship between changes in sulfide concentration and variations in nitrate, ammonia, silica and phosphate. Characterization of such relationships will be also valuable in terms of understanding the processes within the suboxic as well as oxic layers.

The premise of this work is that the basin average for the position of the sulfide onset in the Black Sea, within a certain accuracy, may be considered as stationary on a long time scale. Though this has been proven in several recent publications [6,11,12], at this point, it is worth inserting a comment on this assumption.

Temporal long term variations in the position and structure of the pycnocline [13], which is defined in density coordinates, might virtually determine the limits for the accuracy of determination of the chemocline position, which is also defined in density coordinates. Indeed, the uncertainty in identifying sulfide onset in sigma-t coordinates (0.1 sigma-t) [11] is of the same order of magnitude as the observed variations in sigma-t values at a fixed depth level [13]. When comparing this information with published data [14,15,16] we may come to conclusion that possible changes in the position of the sulfide onset versus density scale are not larger than 0.1 - 0.15 sigma-t. A more accurate analysis is restricted by quality of the data. However, the results of the last 20 to 30 years observations reveal appreciable changes in biochemical structure of the anoxic zone and this will be discussed in the paper. Moreover, it will be shown that variations in all of the above mentioned variables, in general, might result from the process of eutrophication.

2. Data and methodology.

The data from 38 cruises of the Marine Hydrophysical Institute / National Academy of Sciences of Ukraine for the period from 1960 to 1995 as well as the data of the American cruises in the Black Sea, R/V ATLANTIS (1969) and R/V KNORR (1988), have been included in the analysis. Since, in general, the raw data appear to be noisy because of their coarse resolution and due to appreciable seasonal as well as large- and mesoscale space-time variability, the mean values were used for comparison and calculations of the concentration ratios.

Most of the historical data were collected by means of Nansen bottles, therefore, we assume that, within the pycnocline, temperature/salinity was measured with much better accuracy than depth. Hence, to avoid obvious errors related to uncertainty in determining the depth and to eliminate uncertainties related to variations in the position of the chemocline means were calculated for selected isopycnal interfaces in the pycnocline layer [11,12,17]. In the deep water, where vertical sigma-t gradients are small, and precision in measuring temperature/salinity is crucial, averaged sulfide concentrations were calculated for fixed depth levels (1000 and 2000 m, in particular).

Unfortunately, due to coarse vertical resolution and/or poor spatial coverage, comparison of estimates of the mean position of isopycnal surfaces are practically useless for almost all former cruises. However, mean positions of the interfaces can be derived with high confidence from several recent high resolution basin wide surveys (HydroBlack 91, CoMSBlack 92). Therefore, vertical gradients for sulfide concentration and total inventories of sulfide and nutrients were calculated using averaged positions of the specific sigma-t interfaces derived from these recent hydrographic surveys. These estimates are based on the assumption of a relative stationarity of temperature/salinity structure in the pycnocline which is not true in general [13]. However, such an approach is considered to be justified in this case since, virtually, all variations in chemical parameters considerably exceed variations in the pycnocline structure.

To make an estimate of variations in terrestrial inputs of nutrients in the Black Sea, we used published data on the annual amount of nutrients discharged with the waters of Danube in different years [18,19].

3. Results.

3.1. TIME VARIATIONS IN NUTRIENTS.

3.1.1 Inorganic Nitrogen.

The overall range of changes that occurred over two decades in both nitrate and ammonia concentration within the oxic, suboxic and anoxic layers are shown in Figure1,a,b. The average maximum nitrate concentration increased, as has been documented by Codispoti, et al. [17] and Tugrul, et al. [14], by two to three times by the end of 1980's, but then decreased notably after 1993 (Figure 2,a.). The position of nitrate maximum shoaled in terms of sigma-t from, approximately, σ_t=15.7 in 1969 to σ_t=15.5 for recent cruises. Variations in nitrate concentrations within the oxic and suboxic layers are highly correlated with variations in values of apparent oxygen utilization (AOU) [15], supporting the hypothesis that these changes occurred due to eutrophication.

Over the same period, ammonia concentrations decreased dramatically within the oxic and suboxic layers. However, it is obvious from Figure 1,b that the sharp increase in ammonia commenced at a lower sigma-t (16.3) interface in 1969 (R/V ATLANTIS II cruise) than in 1988 (R/V KNORR cruise) when it was at σ_t=16.1.

Within the anoxic layer, ammonia concentrations increased considerably from 1969 to 1988. This increase is especially noticeable for the layer of σ_t=16.2 to 16.8, where the difference can be estimated as 5 to 10 μM/l. Below σ_t=17.1 depth level average values of ammonia concentration for 1969 and 1988 are very close to each other. Unfortunately, the general lack of data for ammonia and, in particular, the lack of data below σ_t=16.8 from the R/V KNORR (1988) cruise, as well as scattering of values for

R/V ATLANTIS (1969) cruise, is the main reason of uncertainties in average profiles and limits the analysis of long-term changes in the distribution of this nutrient element.

Though there is an apparent need for more data in order to reveal magnitude of temporal variations of ammonia concentrations in the Black Sea and to reconstruct the nitrogen cycle and budget, the existing data is sufficient to allow us to conclude that ammonia concentrations have increased in the layer below the onset of sulfide since the 1960's. Besides tracing time variations of nutrients concentrations on specific sigma-t surfaces, we have also calculated total amount or inventory of inorganic nitrogen (ammonia, nitrate, and nitrite) for 1969 (data of R/V ATLANTIS II) and for 1988 (R/V KNORR data). Estimated difference in the total inventory of inorganic nitrogen above σ_t=17.1 (approximately, 650 m), basically in the form of ammonia, is $8.0 \cdot 10^{11}$ M for this 19 year

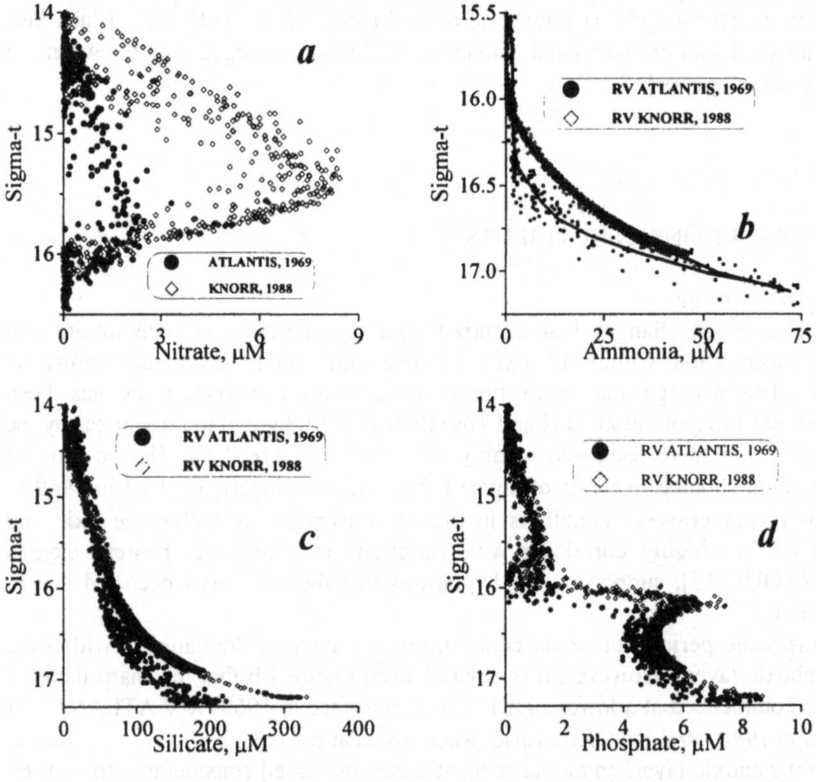

Figure 1 Nitrate (*a*), ammonia (*b*), silica (*c*), phosphate (*d*) versus sigma-t scale in 1969 (R/V ATLANTIS II cruise) and 1988 (R/V KNORR cruise)

period. For the same time interval, the increase in the amount of nitrogen within the oxic and suboxic layers (basically in the form of nitrate) was only $6.4 \cdot 10^{10}$ M, and accounts for less than 8% of the total increase in the inventory of inorganic nitrogen (down to

σ_t=17.1) in the Black Sea. However, the out flux of nitrogen from the oxic layer increased, approximately, three fold in accordance with increase in vertical gradient. The sum of nitrogen increment and estimated additional losses is $10 \cdot 10^{11}$ M for the same period.

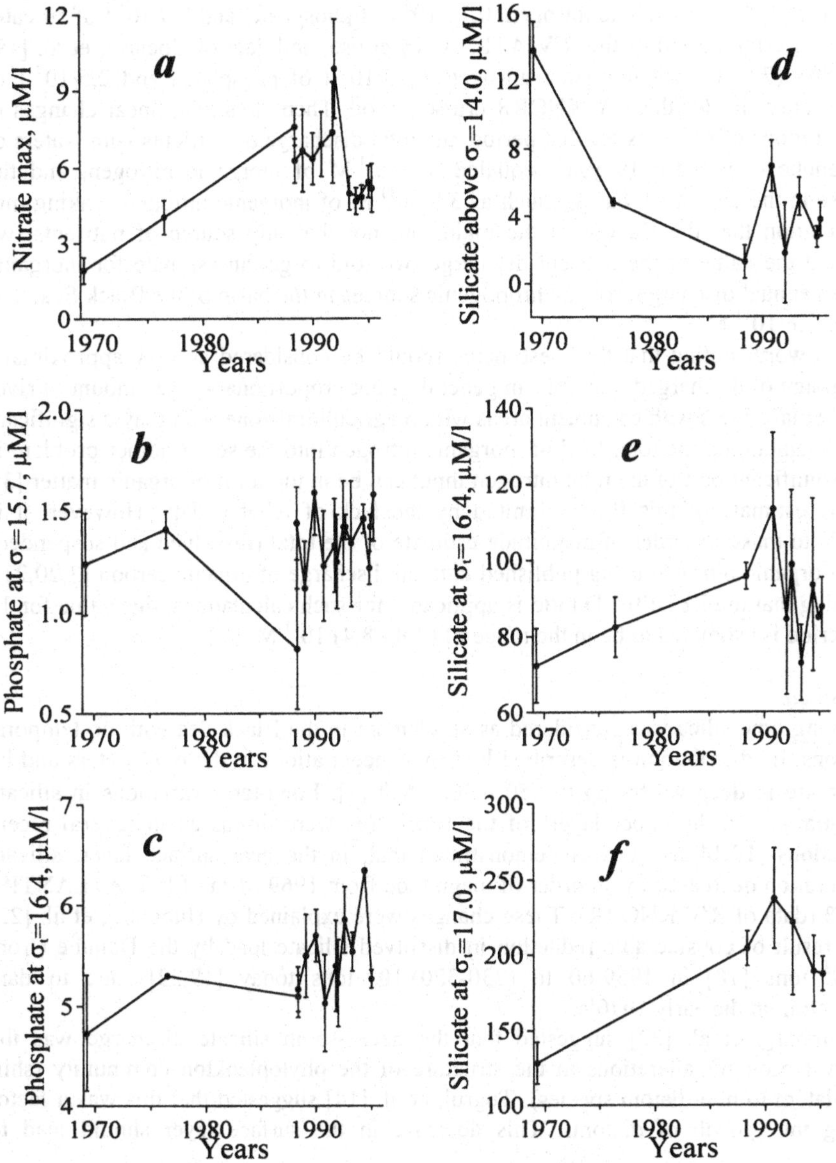

Figure 2. Temporal variations of concentrations for Nitrate within the layer of maximum(a), Phosphate at σ_t=15 7(b) and 16 4(c) interfaces, Silicate above σ_t=14.0(d), and at 16.4(e), 17 0(f) interfaces

This increase in the inventory of inorganic nitrogen above σ_t=17.1 roughly coincides with the increase in the amount of inorganic nitrogen discharged into the Black Sea from land based sources, which we also estimate to be about $8 \cdot 10^{11}$ M for the same period. To calculate this figure, we used published data for the annual discharge of nutrients with the waters of Danube [18,19]. We assumed that the data published by Almazov [18] for 1960's ($1.4 \cdot 10^5$ t of inorganic nitrogen, $0.12 \cdot 10^5$ t of phosphate, and $7.9 \cdot 10^5$ t of silicate) are valid for the period of the R/V ATLANTIS cruise, and data of Cociasu, et al. [19] for 1980's ($7.0 \cdot 10^5$ t of inorganic nitrogen, $0.25 \cdot 10^5$ t of phosphate, and $2.5 \cdot 10^5$ t of silicate) are valid for the R/V KNORR cruise period. Then, assuming linear changes in concentrations of nutrients for that period, the total discharge of nutrients with waters of the Danube over these 19 years equaled $5.7 \cdot 10^{11}$ M of inorganic nitrogen, and the increase in the discharge was as much as $3.8 \cdot 10^{11}$ M of inorganic nitrogen. Taking into consideration that the Danube is the main, but not the only source of nutrients, we increased the value of the nutrient discharge two-fold to get an estimate for inorganic nitrogen annual discharge from anthropogenic sources in the basin of the Black Sea. The value is $7.6 \cdot 10^{11}$ M.

It is worth noting that these estimates should be considered as very approximate. The amount of discharged nutrients, in general, is not proportional to the amount of river runoff. Small rivers with catchment areas within agricultural zones can play a significant role in maintaining the total load of inorganic nitrogen into the sea. Another problem is that a significant part of the total nitrogen input can be in the form of organic matter [3], and any estimate of this flux is limited by the lack of reliable data. However, it is possible to make an order of magnitude estimate of the total (dissolved and suspended) load of organic nitrogen using published data on discharge of organic carbon [3,20,21]. Assuming that the REDFIELD ratio is applicable for such calculations, this value for 19 year period is estimated to be in the range of $(4.8 - 8.4) \cdot 10^{11}$ M.

3.1.2 Silica.

For a long time silica was considered as an element in the Black Sea with no temporal variations. Its distribution is described by low concentrations in the oxic waters and by an increase in deep waters up to 250 - 300 μM/l [3]. Long-term variations in silicate concentrations in the upper layers of the Black Sea were discussed in several recent publications [19,14,22]. It was demonstrated that, in the near surface layer, silicate concentration decreased by an order of magnitude from 1969 (data of R/V ATLANTIS) to 1988 (data of R/V KNORR). These changes were explained by Humborg, et al. [22] as the result of considerable reduction in dissolved silicate load by the Danube (from $800 \cdot 10^3$ tons [18] in 1959-60 to $(230\text{-}320) \cdot 10^3$ tons today [19,22]) due to dam construction in the early 1970's.

Humborg, et al. [22] suggested that the decrease in silicate discharge was the primary reason for alterations in the structure of the phytoplankton community (shift from diatom to non-diatom species). Tugrul, et al. [14] suggested that this was a factor limiting the growth of diatoms. This decrease in the surface layer should lead to decrease in the silica flux from the oxic to anoxic waters and result in a reduction of silicate concentrations in the sulfide bearing waters.

However, data of Figures 1,c and 2,d-f demonstrate that while there was a decrease in silicate concentrations in the oxic waters (from 13.8±1.7 μM/l in 1969 to 1.3±1.0 μM/l in 1988) there was a two fold increase of the silicate inventory in the anoxic waters from the beginning of 1970's to the beginning of 1990's. After the early 1990's, silicate concentrations increased in the surface waters and decreased within the anoxic zone.

For silicate, the total increase in the inventory for the layer between σ_t=14.5 and 17.1 was equal to $9.1 \cdot 10^{12}$ M, in spite of the fact that the discharge of silicate from land based sources decreased during this period due to dam constructions and regulations of the riverine discharge.

3.1.3 Inorganic Phosphorus.

Distribution of phosphate (plotted vs sigma-t, Figure 1,d) has been previously discussed in detail [11,14,23,24,25]. A broad phosphate maximum is centered at approximately σ_t=15.50, sharp minimum in the suboxic zone is centered at σ_t=15.85, and the lower maximum is located at about σ_t=16.20.

Taking into consideration distinct seasonal and spatial variability in phosphate concentration in the minimum layers at σ_t=15.85 and the maximum at σ_t=16.20 [12], not these levels but σ_t=15.70 and σ_t=16.40 have been chosen to analyze long-term variations in phosphate concentrations (Figure 2,b,c). It is seen that the phosphate concentrations have not changed much. The observed increase in concentrations is less than 20% and it is close to the uncertainties in the average values for individual cruises.

As was shown for nitrogen and silica, the total increase in the inventory of phosphate in the layer between σ_t=14.5 and 17.1 equals $1.3 \cdot 10^{11}$ M for the period between the ATLANTIS II to KNORR surveys. This value is 18 times higher than the increase in the discharge of inorganic phosphorus from known land based sources for the same period (the additional source for phosphate loaded with river runoff can be estimated as $0.07 \cdot 10^{11}$ M). However, if we exclude the data from 1969 the increase in phosphate concentrations is not so prominent, and time variations in total amount of inorganic phosphate are within confidence limits for each estimate. On the other hand, increase in the inventory of individual nutrients from 1969 to 1988 in the upper part of the anoxic zone fits the REDFIELD ratio (the difference between theoretical and calculated value is only about 10%) for N:P and demonstrates good correlation for N:Si [15]. This suggests that the observed increase in the stock of individual nutrients could be the result of a more intensive downward flux of particulate organic matter and consumption of this organic matter within the anoxic layer and, consequently, the observed increase in phosphate concentrations could not be higher.

3.2 RECENT TRENDS FOR SULFIDE.

The relative stability of the oxic/anoxic interface in the Black Sea [6,16,25,26] is in some contradiction with alterations in nutrients distribution below the pycnocline. When analyzing mean sulfide profiles from individual surveys we observe that mean sulfide concentrations at 1000 and 2000 m have increased appreciably since the second half of the 1970's (Figure 3). Before that time values for individual cruises showed a random scatter rather than a systematic trend. While the absence of a regular trend can be assumed as an evidence of steady state situation in Black sea environment before the middle of the 1970's, any quantitative estimates of augmentation in the inventory of sulfide in the 1980's, that are based on data of Figure 3, are hardly possible. The reason comes from our knowledge about different samplers that were used before and after 1980. Metallic samplers were used before 1980 and plastic bottles - after that time. Systematic difference was registered between the results when metallic and plastic samplers were applied to analyze water from the same depth [27,28]. On the average, concentration of sulfide was 17% [28] higher when plastic bottles were in use.

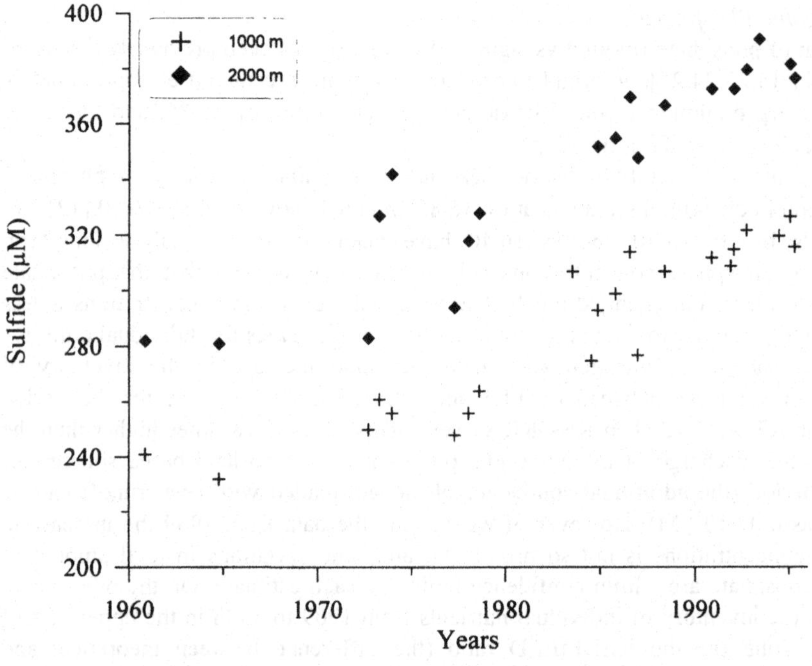

Figure 3. Mean hydrogen sulfide concentrations at 1000 and 2000 m for individual cruises.

An attempt to apply 1.17 coefficient to the "old" data has revealed that 3/4 of the difference between the data for 1960-70's and for 1980-90's is systematic error. Thus, 65μM/l of increase in the average concentration of sulfide from the 1970's to the 1990's in the layer below 1000m, that comes from data (Figure 3), include 50μM/l of systematic error. The rest, that does not exceed 15μM/l, is close to range of scattering in data for individual cruise data set. For these reasons all the data on the increase in sulfide concentration from 1960-70's to 1980's must be considered qualitative rather than quantitative. On the other hand, the only plastic samplers and the same analytical procedure were reported to be in use after 1980. Nevertheless, an increase in sulfide concentration for the period from 1985 to 1995 is seen from data of Figure 3 as well. Unfortunately, this period is short to make quantitative estimates of trend in sulfide concentrations, but the presence of changes in chemical structure of the anoxic zone triggered by intensive eutrophication is evident.

To make an assessment of possible changes in sulfide distribution in deep part of the anoxic zone we have used the value of annual upward sulfide flux at 700m depth that is close to $5 \cdot 10^{10}$ M per year. The last value is based on the average profile of sulfide distribution in 1985 to 1995 and coefficient of turbulent diffusion and vertical advection derived from the model [29]. Assuming a balance in sinks and sources of sulfide for 1960-70's and a 4-fold increase in the rate of sulfide production in the 1980's [21] we get 8 μM/l as the upper limit of possible changes in sulfide concentrations in 20 years in the layer below 700 m. Such increase in sulfide concentration should correspond to about 3 μM/l increase in ammonia concentration. While these values of the possible increase in sulfide and ammonia concentrations are not in any contradiction with the results of observations, they are very low to be verified from the existing data sets.

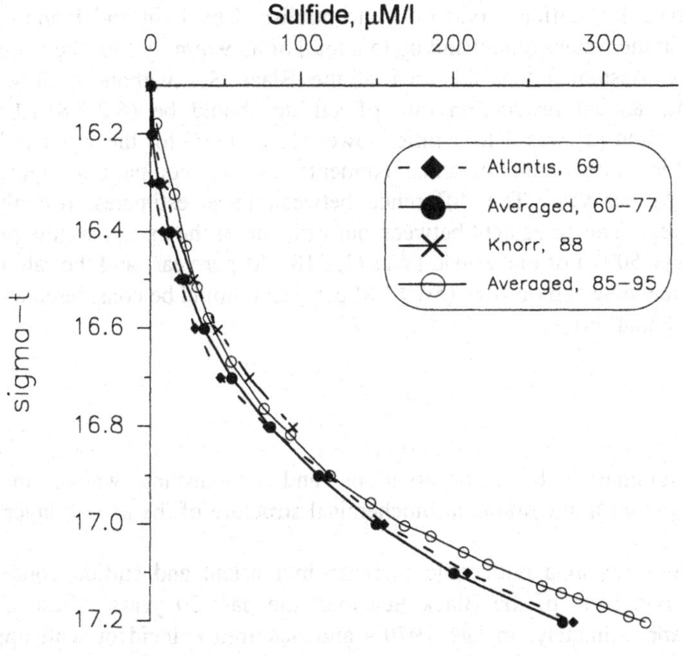

Figure 4. Hydrogen sulfide mean profiles in sigma-t coordinates for 1960-77, 1985-1995, ATLANTIS II (1969) and KNORR (1988) surveys.

To make an assessment of possible changes in sulfide distribution in the upper part of anoxic layer, where quality of existing data is much better, we use the average sulfide profiles for the period from 1960 to 1977 and for the period from 1985 to 1995 (Figure

4). In that manner, mean profile for the ATLANTIS II (1969) cruise coincides well with the average profile for the earlier period (1960-1977), and KNORR (1988) data are close to the mean values for the later period (1985-1995). The results, shown in Figure 4, reveal that the increase in sulfide concentrations was fairly constant for the upper layer of the anoxic zone, above σ_t=16.9, where it is close to 5-6 μM/l.

The increase in total sulfide inventory in the layer above σ_t=17.1, the layer where we observe increase in ammonia concentrations (Figure 1,b), was $2.3 \cdot 10^{12}$ M, or about $12 \cdot 10^{10}$ M per year. Very close value of the increase in the sulfide inventory, that is as much as $2.5 \cdot 10^{12}$ M, we have calculated from the data on ammonia. This annual increase in the sulfide inventory may be considered as an estimate of an excess (relative to a steady state production level in the 1960's) sulfide production.

It is interesting to compare this value of the excess sulfide production rate with the same estimates but based on the flux of particulate organic carbon (POC) and the rate of sulfate reduction in anoxic waters. For such a comparison we used the data published by Lein and Ivanov [21] and Sorokin [5]. Sorokin used results of measurements carried out in late 1960's - beginning of 1970's [5] while Lein and Ivanov used data collected in 1980-88 [21].

The estimate for sulfide production rate published by Lein and Ivanov [21] was 65.9 $g \cdot m^{-2} \cdot y^{-1}$ in the water column and up to a total of 83.9 $g \cdot m^{-2} \cdot y^{-1}$ for the water column and sediments. Assuming that the area of the Black Sea without shallow areas is ~$3 \cdot 10^{11}$ m^2, the annual production rate of sulfide should be $(6.2\text{-}7.8) \cdot 10^{11}$ M. The estimate of Sorokin [5] was 3 to 5 times lower: 13 $g \cdot m^{-2} \cdot y^{-1}$ for the water column and 25 $g \cdot m^{-2} \cdot y^{-1}$ for water column and sediments or, in compatible figures, it is $(1.2\text{-}2.4) \cdot 10^{11}$ M per year. The difference between these estimates, roughly, equals $5 \cdot 10^{11}$ M per year. The agreement between our estimate of the excess sulfide production rate in the upper 500 m of the anoxic layer ($1.2 \cdot 10^{11}$ M per year) and the rate of sulfide reduction in the whole anoxic zone ($5 \cdot 10^{11}$ M per year) should be considered as good for such coarse calculations.

4. Discussion.

Below, we summarize basic observations and conclusions which, in general, characterize long term alterations in biochemical structure of the anoxic layer over the past 30 years.

Observations reveal a noticeable increase in nutrient and sulfide concentrations within the anoxic layer of the Black Sea over the past 20 years. These alterations commenced, approximately, in late 1970's and occurred coincident with appreciable changes in the hydrochemical structure of the upper layer [19,14,22] as well as with an obvious decline in water transparency [30]. Within the oxic layer, the total inventory of nitrate increased two to three fold [14], while concentrations of ammonia and silica decreased considerably [14,22]. It should be also emphasized that variations in the total inventory of silica did not correlate with variations in the amount of silica input with riverine discharge, as was suggested by Humborg, et al. [22]. Actually, the increase in the inventory of nutrients and sulfide in the anoxic zone and the high correlation between the increase in inventory of individual nutrients supports the idea that excess in discharge of nitrate over other nutrients [22] is the main reason of: (a) increased production and downward flux of particulate organic matter, (b) essential decrease in the concentrations of silicate, phosphate and ammonia in the oxic layer of the sea, (c) intensive uptake of silicate and phosphate from sediments.

The last phenomena helps to explain the discrepancy between the amount of silica and phosphorus discharged from land based sources and the increase in the inventory of these nutrients in the anoxic zone. Thus, excess discharge of nitrate over the other nutrients [22] led to intensive consumption of phosphate and silicate in the upper layer of the sea due to removal of these nutrients in the form of sinking particulate organic matter. To balance the excess nitrate an increasingly intensive uptake of silicate and phosphate from sediments took place. That means that discharge of enormous amount of inorganic nitrogen was enough to activate sources of phosphate and silicate other than riverine discharge and to give rise to additional production and increase in the amount of sulfide in the deep water.

Quantitatively, increase in total inventory of sulfide is supported by direct measurements revealing at least three fold increase in the sulfide production rate. On the other hand, decrease in ammonia, phosphate and silicate concentrations in the oxic layer along with dramatic increase in nitrate concentration supports the idea that the Black Sea ecosystem has transformed from the system with nitrogen limitation of primary productivity to another type with phosphorus and/or silicate limitation.

Most of the changes are in accord with a general scheme of plausible results from eutrophication.

1. Increase in discharge of inorganic nitrogen [22] (that, probably, was the limiting factor for phytoplankton growth) from land-based sources resulted in increase in the level of primary production and, consequently, in increase of downward flux of particulate organic matter. At the same time, relative increase in discharge of nitrate versus phosphate and silicate [22] resulted in decrease of ammonia, silicate and phosphate concentrations in the euphotic zone and the whole oxic layer [14,22] thus activating other sources of phosphorus and silica different from coastal based sources.

2. Increase in the flux of particulate organic matter caused depletion of oxygen in the layer beneath the euphotic zone and / or beneath the depth of direct ventilation of the pycnocline due to penetrative convection [15] and increase in nitrate concentrations [14] as the result of POM decomposition in the process of sulfate reduction.

3. An increased amount of organic matter sinks into the anoxic zone. It is utilized in that layer resulting in an increase in both nutrients and sulfide concentration (Figures 1-4).

The results of this work can be summarized in the following basic conclusions:

1. There are evident changes in the chemical structure and these changes may be attributed to intensive eutrophication of the sea. Increase in inventory of nutrients and sulfide in the anoxic zone, and dramatic variations in nutrients and oxygen distribution in the oxic zone are among these results.
2. Residence time of the anoxic waters assessed from changes in chemical properties is short, relative to values calculated from physical properties. This

can be explained by the short time needed for particulate organic matter to sink below the pycnocline. This demonstrates an important role of particulate organic matter, an agent responsible for fast conveyance of ecosystem response from the upper layer into the deep layers, in evolution of the Black Sea ecosystem.

3. The effect of eutrophication is superimposed on 'natural' variability resulting from changes in external forcing. This means that the Black Sea is "small" enough to respond clearly at a time scale of a decade to anthropogenic impact.

The analysis of long-term changes in the chemical structure of the Black Sea in the most recent years revealed positive tendencies: increase in nutrient concentrations decreased in the last three to four years (concentrations of nitrate decreased considerably, and concentrations of silicate increased in the oxic and decreased in the anoxic zone of the sea). These changes are in agreement with data on temporal variations in optical properties [30] and, all together, this supports the idea of improving ecological situation in the Black Sea. Unfortunately, these good news do not come from scientific forecast, but from the results of monitoring. This fact demonstrates the importance of both monitoring for tracing of current changes in the Black Sea environment and process oriented studies in order to understand the key processes and their role in evolution of the Black Sea ecosystem in order to qualitatively improve forecasting capabilities.

Acknowledgments

This work has been made possible through the support of National Research Council of the United States, NATO and TUBITAK.

References

1. Andrusov, N I (1890) Preliminary report on the Black Sea cruise. *Izvestiya imperatorskogo Russkogo geographicheskogo obschestva (News of the Russian Royal geographic society)*, **26**, No 5, 398-409. (in Russian)
2. Mamaev, V O., D. Aubrey and V N Eremeev, editors (1996) *Black Sea Bibliography 1974 - 1994*
3. Skopintsev, B.A. (1975) *Formirovanie sovremennogo khimicheskogo sostava vod Chernogo morya (Formation of the present chemical structure of the Black Sea)*, Leningrad, Gidrometeoizdat, 336pp (in Russian)
4 Izdar, E. and J.W. Murray, editors (1991) *Black Sea Oceanography*. NATO ASI Series, Series C Mathematical and Physical Sciences, **351**, 487pp
5 Sorokin, Yu I (1982) *Chernoe more· priroda i resursi (The Black Sea. the Nature and the resources)* Moscow, Nauka, 217pp (in Russian)
6 Bezborodov, A.A. and V N. Eremeev (1993) *Chernoe more. Zona vzaimodeistviya aerobnikh i anaerobnikh vod (Black Sea The oxic/anoxic interface)*. Sevastopol, MHI ASU, 299pp (in Russian)
7. Murray, J.W., L.A. Codispoti and G.E. Friederich (1995) Oxidation-Reduction Environments. The suboxic zone in the Black Sea In: *Aquatic Chemistry. Interfacial and Interspecies Processes*, Chin Pao Huang, Charles R. O'Melia and James J. Morgan, editors, ACS Advances in Chemistry Series, No.244, 157-176

8 Boguslavskiy, S G , V A. Zhorov and A.A Novoselov (1985) O problemakh serovodorodnoy zoni Chernogo morya (On the problems of the Black Sea anoxic zone). *Morskoy gidrophizicheskiy zhurnal (Marine Hydrophysical journal)*, Sevastopol, MHI ASU, **1**, 54-58. (in Russian)

9. Faschuk, D.Ya. and T.A. Aizatulin (1986) O vozmozgnosti transformatsii anaerobnoy zoni Chernogo morya (On the possibility of transformation of the Black Sea anaerobic zone). *Okeanologiya (Oceanology)*, **26**, 171-178. (in Russian)

10 Faschuk, D.Ya., T.A. Aizatulin and L K Sebakh (1987) Osobennosti sovremennogo sostoyaniya sloya sosuschestvovaniya kisloroda I serovodoroda v Chernom more (On the peculiarities of the present state of the C-layer in the Black Sea). *Sovremennoe sostoysnie ekosistemi Chernogo morya (The present state of the Black Sea ecosystem)*. Moscow, Nauka, 29-41.

11 Buesseler, K O., H.D Livingston, L.I. Ivanov and A.S Romanov (1994) Stability of the oxic-anoxic interface in the Black Sea *Deep-Sea Research I*, **41**, No 2, 283-296

12 Konovalov, S.K., S Tugrul, O Basturk, I Salihoglu (1997) Spatial isopycnal analysis of the main pycnocline chemistry of the Black Sea: Seasonal and interannual variations. In: *Sensitivity to change Black Sea, Baltic Sea and North Sea*, E. Ozsoy and A. Mikaelyan, editors, Kluwer Academic Publishes, Dordrecht, NATO ASI, Series 2: Environment, **27**, 197-210

13. Ivanov, L.I , S Besiktepe, E. Ozsoy (1997) Physical Oceanography Variability in the Black Sea Pycnocline, in A. Mikaelyan and E. Ozsoy (eds.), *Sensitivity to change: Black Sea, Baltic Sea and North Sea*, Kluwer Academic Publishes, Dordrecht, NATO ASI Series, 265-274

14 Tugrul, S, O Basturk, C Saydam and A Yilmaz (1992) Changes in the hydrochemistry of the Black Sea inferred from water density profiles *Nature*, **359**, 137-139

15 S K Konovalov, J W Murray, L I Ivanov and A.S. Samodurov Temporal variations in the basic hydrochemical properties of the Black sea. (Submitted to Progress in Oceanography).

16 Saydam, C., S. Tugrul, O. Basturk and T. Oguz (1993) Identification of the oxic/anoxic interface surfaces in the Black Sea. *Deep-Sea Research I*, **40**, No.7, 1405-1412.

17 Codispoti, L.A , G E Friederich, J W Murray and C M Sakamoto (1991) Chemical variability in the Black Sea. implications of continuos vertical profiles that penetrated the oxic/anoxic interface *Deep-Sea Research*, **38**, No 2a, 691-710

18 Almazov, N M. (1961) Discharge of dissolved nutrients by rivers of the USSR to the Black Sea. *Naukovi Zapiski Odesskoy Biologicheskoy Stantsii (Scientific notes of the Odessa biological observatory)*, Kiev, **3**, 99-107 (in Russian)

19 Cociasu, A., L Dorogan, C. Humborg, and L. Popa (1996) Long-term ecological changes in the Romanian coastal waters of the Black Sea. Marine Pollution Bulletin, **32**, 32-38.

20 Deuser, W.G (1971) Organic-carbon budget of the Black Sea *Deep-Sea Research*, **18**, 995-1004

21 Lein, A Yu and M V Ivanov (1991) On the sulfur and carbon balances in the Black Sea In: *Black Sea oceanography*, Izdar, E and J W Murray, editors, Kluwer Academic Publishers, 307-318.

22 Humborg, C , V Ittekkot, A Cociasu and B v.Bodungen (1997) Effect of Danube River dam on Black Sea biogeochemistry *Nature*, **386**, 385-388

23 Shaffer, G (1986) Phosphate pumps and shuttles in the Black Sea *Nature*, No 321, 515-517

24 Basturk, O , S. Tugrul, S. Konovalov, and I. Salihoglu (1997) Variations in vertical structure of water chemistry within the three hydrodynamically different regions of the Black Sea. In: *Sensitivity to change Black Sea, Baltic Sea and North Sea*, E. Ozsoy and A Mikaelyan, editors, Kluwer Academic Publishes, Dordrecht, NATO ASI, Series 2 Environment, **27**, 183-196

25 Vinogradov, M.E and Yu.R Nalbandov (1990) Vliyanie izmeneniya plotnosti vodi na raspredelenie phizicheskikh, khimicheskikh i biologicheskikh kharakteristik ekosistemi Chernogo morya (The influence of the variations in density of water at the distribution of physical, chemical, and biological properties of the Black Sea ecosystem) *Okeanologiya (Oceanology)*, **30**, No.5, 769-777 (in Russian)

26. Eremeev, V N., A M. Suvorov, A Kh Khaliulin, E.A Godin (1996) O sootvetstvii polozgeniya verkhney granitsi serovodorodnoy zoni opredelennoy izopiknicheskoy poverkhnosti v Chernom more po mnogoletnim dannim (On the question of correlation between the depth of hydrogen sulfide onset and an individual density value in the Black Sea on the basis of multiannual data set) - *Okeanologiya (Oceanology)*, **36**, No 2, 235-240 (in Russian)

27. Ivanenkov, V.N. (1975) Zaschita metallicheskikh batometrov ot korrozii I kachestvo rezul'tatov khimicheskikh opredeleniy v serovodorodnoi zone Chernogo morya (Special painting of metal samplers and the quality of chemical analysis of water from the anoxic zone of the Black Sea). Morskoy Gidrofizicheskiy Zhurnal (Marine Hydrophysical Journal), MHI UAS, Sevastopol, No.1, 182-190.

28 Novoselov, A.A. and A.S Romanov (1985) Nekotorie metodıcheskie problemi pri issledovanii serovodorodnoi zonı Chernogo morya (Analytical problems ın chemical analysis of water from the anoxic zone of the Black Sea). Sovershenstvovanie upravleniya razvitiem rekreatsionnikh sistem (Development in management of the coastal zone). VINITI, No.7791-B855, 359-372.

29 Samodurov, A.S., L.I. Ivanov (1998) Processes of ventılation of the Black Sea related to water exchange through the Bosporus. In: L.Ivanov, T.Oguz (eds.) *Nato TU-Black Sea project: Ecosystem Modellıng as a Management Tool for the Black Sea, Symposıum on Scientıfic Results*. Kluwer Academic Publıshes, Dordrecht, NATO ASI Series. Vol 2, p.p 221-236.

30 Vladimirov, V.L., V.I Mankovskiy, M.V. Solov'ev, and A V. Mishonov (1997) Seasonal and long-term variability of the Black Sea optıcal parameters. Sensitivity to Change. Black Sea, Baltic Sea and North Sea. (Edited by E. Ozsoy and A. Mikaelyan), NATO ASI, Kluwer Academic Publishers, Series 2: Environment, **27**, 33-48.

THE SUBOXIC ZONE OF THE BLACK SEA

JAMES W. MURRAY, BING-SUN LEE
School of Oceanography Box 357940
University of Washington
Seattle WA 98195-7940, USA

JOHN BULLISTER
Pacific Marine Environmental Lab/NOAA
7600 Sand Point Way NE
Seattle WA.98115-0070, USA

GEORGE W. LUTHER, III
College of Marine Studies
University of Delaware
Lewes, DE 19958, USA

Abstract

One of the most fascinating and unexpected discoveries during the US-Turkish Expedition to the Black Sea in 1988 was the detection of a suboxic zone at the oxic-anoxic interface. The suboxic zone was defined as the region where the concentrations of oxygen and sulfide were both extremely low and had no perceptible vertical or horizontal gradients. The suboxic zone is a site of intensive redox cycling of species of sulfur, nitrogen, manganese and iron. Further analysis showed that the features of the vertical profiles occurred on characteristic density surfaces, which means they can be reproducibly sampled. Similar reactions appear to occur in hemipelagic marine sediments but are more difficult to study. In the oxygen minima of the world's oceans the extent of the redox reactions is less complete, Thus, the Black Sea is a natural laboratory to study these reactions because the complete progression of redox reactions is well resolved on characteristic density surfaces within a suitable depth range for resolution by pump-type sampling. In this paper we review our knowledge of the suboxic zone and present five hypotheses for future study.

1. Introduction

The Black Sea has long been a major site for studying extreme anoxic oceanographic conditions [10, 62]. It is the world's largest stable anoxic basin and is thought to provide

S. Beşiktepe et al. (eds.),
***Environmental Degradation of the Black Sea: Challenges and Remedies*, 75–91.**

a quasi-steady state system. The water column is characterized by the absence of oxygen and elevated concentrations of hydrogen sulfide and methane from about 100m to the bottom at depths greater than 2000m. The water column is anoxic (e.g. sulfide containing) because oxygen is consumed by oxidation of sinking organic matter and a shallow, sharp salinity-determined density gradient prevents exchange of oxygen between the surface and deep water. The boundaries between oxic and anoxic environments, such as this one, are fascinating study sites because they contain a rich population of oxidation-reduction reactions which result in a change of speciation of many elements.

Though they are minor in terms of ocean area and volume, it is important to study anoxic marine conditions because they represent an end-member of the spectrum of ocean redox environments. In addition, anoxic conditions are thought to have existed in the ocean's past (e.g. Berger and Roth [3]; Schlanger and Jenkyns [57]) and understanding present day anoxic environments should help us understand their origin and influence on the geochemistry of particle fluxes and sediment accumulation in the geological record.

There have been periodic visits by US research vessels to the Black Sea. These include the R.V. Pillsbury in 1965 [50, 59], R/V Atlantis 11 in 1969 [6, 13, 63], R/V Chain in 1975 [19] and R/V Knorr in 1988 [24, 41]. The 1988 US - Turkish R/V Knorr Expedition provided an especially comprehensive set of chemical and biological studies using modern oceanographic techniques. One of the most fascinating discoveries in 1988 was the detection of a suboxic zone at the oxic-anoxic interface. The suboxic zone was defined as the region where concentrations of oxygen and sulfide were both extremely low and had no perceptible gradients. This observation was facilitated by use of an in situ pumping system that provided high resolution sampling with minimal contamination. Subsequent observations by Ukrainian (Konovalov and Romanov, personal communication, 1991) and Turkish [69] chemists confirmed these distributions.

This suboxic zone is a site of dynamic oxidation and reduction reactions readily accessible on known density surfaces in a predictable and reproducible way. Other anoxic interfaces and suboxic environments have been observed, but they are either less predictable because of temporal variability (not at steady state, e.g. Saanich Inlet; [15, 65] or Chesapeake Bay; [32, 39]) or do not appear to have a suboxic zone (e.g. Cariaco Trench; [22, 73]). Suboxic conditions exist in the oxygen minima of the world's oceans (especially the northern and southern eastern tropical Pacific and the Arabian Sea) but they do not extend through the entire sequence from aerobic to anoxic conditions (e.g. [11, 44, 55]). In the Arabian Sea and eastern tropical Pacific, for example, only about a third of the expected NO_3 is missing [5]. Nevertheless, they are only a slight nudge away from becoming anoxic themselves. One difficulty in studying oxic-anoxic interfaces in general is that they are typically regimes where the reactions are compressed into small spatial scales that make detailed sampling difficult. For example, suboxic conditions exist in hemipelagic sediments deposited on the shelf, slope and rise of continental margins but these sites are more difficult to study with high resolution because of their centimeter scale distribution [14, 17]. The Black Sea is a unique place to study these reactions because the complete progression of redox reactions is well resolved on characteristic density surfaces, within a suitable depth range for resolution by pump-type sampling.

2. Scientific Background

Oxidation-reduction (redox) processes influence the distribution of many major and minor elements in the ocean. Equilibrium calculations can be used to predict concentrations but real systems are typically far from equilibrium because of slow kinetics (sometimes even when mediated by bacteria) [64]. The most common redox end-members in marine systems are oxic (0_2 present) and anoxic (H_2S present) environments. The interface between oxic and anoxic environments is a site of oxidation-reduction reactions involving oxidized and reduced forms of oxygen, nitrogen, sulfur, carbon, manganese and iron. The flux of electrons and concentrations of the different species of these elements determines the position of the oxic-anoxic interface [6].

The results from the 1988 R/V Knorr cruises resulted in dramatic new views about the circulation of the Black Sea and the distribution of properties around the oxic/anoxic interface [40]. Most previous data suggested some overlap at low concentrations in the distributions of oxygen and sulfide at the oxic-anoxic interface. Scientists at the Shirshov Institute in Moscow still view the interface this way (e.g. [70]). The coexistence of oxygen and sulfide was difficult to justify, considering the rapid kinetics of their reaction [36]. During the 1988 expedition no overlap was observed. Instead, a suboxic zone that ranged in thickness from 20 to 50m was present. In this suboxic zone oxygen varied from less than 2 μM to 10 μM (Fig. la) and sulfide was less than 5 nM (Fig. 1b) [33]. Neither oxygen nor sulfide exhibited any perceptible vertical or horizontal gradients within the suboxic zone [40]. Recently, V. Zaburdaev (MHI, personal communication, 1996) used chemical probes to confirm that there is a thick zone with no 0_2 and H_2S gradients. He also found a sharp gradient in H_2S at the top of the H_2S layer, similar to that of Luther *et al.* [33].

Nitrate was low in the euphotic zone and increased to a maximum approximately coincident with the upper boundary of the suboxic zone [12]. Nitrate then decreased to zero within the suboxic zone (Fig. le). Nitrite maxima were observed near the upper and lower boundaries of the nitrate maximum, corresponding to zones of nitrification and denitrification (Fig. 1d). Dissolved manganese (Fig. le), iron (Fig. 1f) and ammonia (Fig. lg) all started to increase at the depth where nitrate went to zero. Manganese and ammonia both increased steeply after the disappearance of nitrate. Iron began to increase at the same point but then increased more steeply after the first appearance of sulfide. It may be difficult to separate the iron and manganese redox cycles because iron and manganese are well correlated in the suboxic zone [60], but there is a clear difference in their distributions versus depth.

Features in the water column occur at different depths at different locations, but they occur on the same density surface [12, 31, 69]. The density structure is dominated by the salinity fields. The high quality of the 1988 data set allowed precise analysis of the characteristic density values of critical points in the various profiles (Table 1) [43].

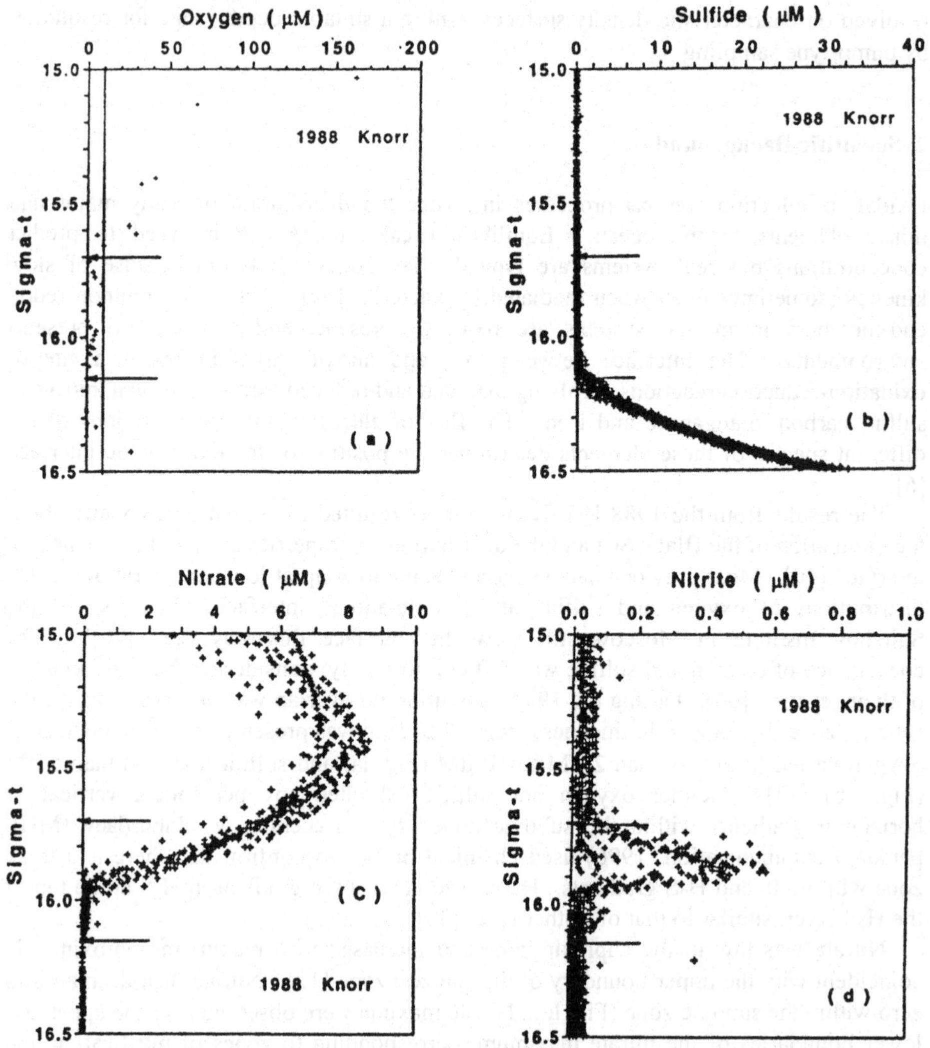

Figure 1. Pump cast profiles of a)oxygen, b)sulfide, c)nitrate, d)nitrite, e)manganese, f)iron, g)ammonia and h)phosphate as a function of density from the 1988 Knorr cruises. Arrows at density values of 15.7 and 16.15 represent the upper and lower boundaries, respectively, of the suboxic zone as described in the text.

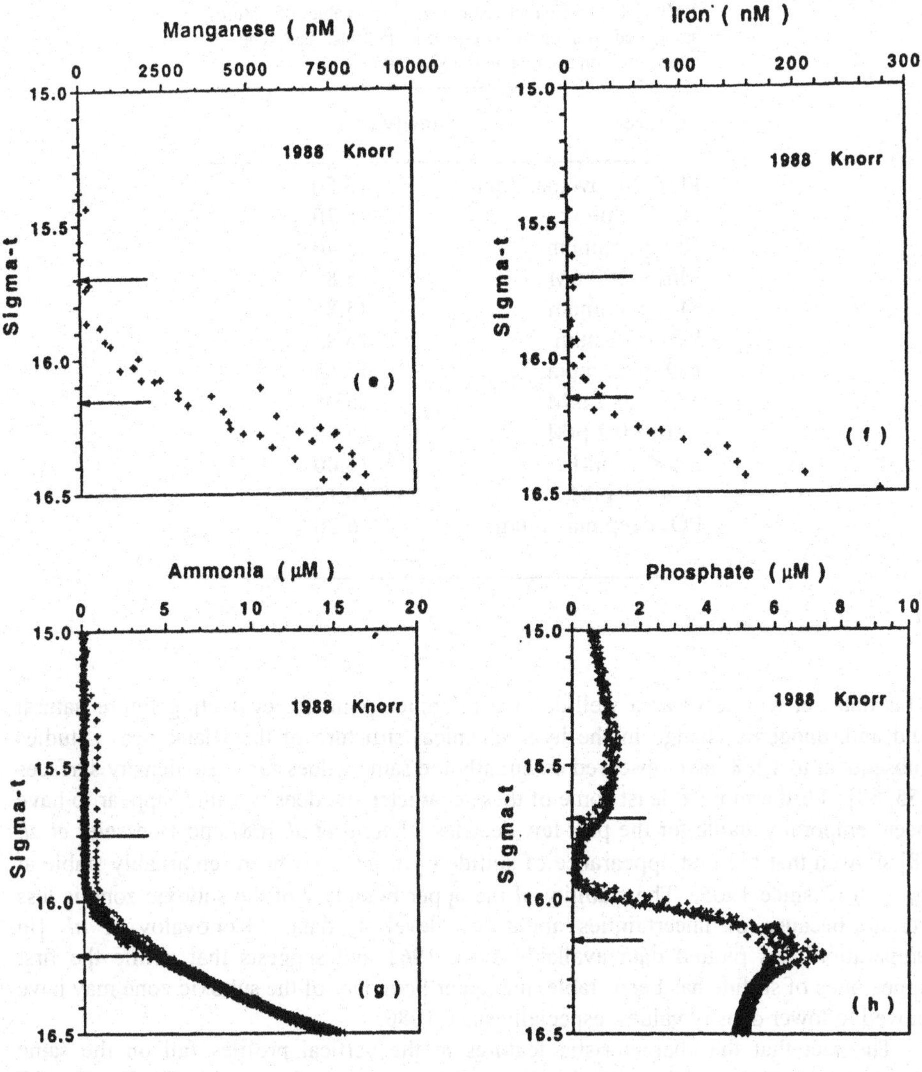

Figure 1, continued

TABLE 1 The characteristic density (σ_t) values of features associated with the suboxic zone in the Black Sea The uncertainty on each value is about 0.05

Feature	Density (σ_t)
PO_4 shallow maximum	15.50
O_2 < 10 ptm	15.70
NO_3 maximum	15.40
Mn_d < 200 nM	15.85
Mn_p maximum	15.85
PO_4 minimum	15.85
NO_2 maximum	15.85
NO_3 < 0.2 ptM	15.95
NH_4 > 0.2 ptM	15.95
Fe_d < 10 nM	16.00
H_2S > 1 piM	16.15
PO_4 deep maximum	16.20

This data set will serve as a well-defined reference point for evaluating future natural and anthropogenic change in the hydrochemical structure of the Black Sea. Studies subsequent to 1988 have observed essentially the same values for these density surfaces [56, 68]. Furthermore at least some of these characteristic density values appear to have been temporally stable for the past few decades. Tugrul *et al.* [68] and Buesseler *et al.* [8] showed that the first appearance of sulfide (>10 µM) has been remarkably stable at σ_t = 16.15 since 1965. The stability of the upper boundary of the suboxic zone is less certain because of uncertainties about low level 0_2 data. Konovalov *et al.* [in preparation] has plotted data available since 1961 and suggests that, while the first appearance of sulfide has been stable, the upper boundary of the suboxic zone may have moved to lower density values, especially since 1980.

The fact that the characteristic features in the vertical profiles fall on the same density surfaces means that distributions and reactions can be discussed in terms of a composite profile with a vertical density scale. The breaks in slope suggest narrow zones of intense biochemical reactions. The various chemical species and redox reaction zones have good vertical resolution which means they can be clearly identified and sampled on future visits.

The oxygen and sulfide distributions are of particular interest because they are used to define the oxic-anoxic interface. Dissolved oxygen decreases to concentrations less than 10 µM by σ_t = 15.65 (Fig. la) and sulfide does not begin to increase (>10 µM) until σ_t = 16.15 (Fig. lb). There is no formal or consistent definition of suboxic conditions but these are the boundaries suggested by [43] for the suboxic zone. The suboxic zone varies in thickness at different locations but is always equivalent to a density separation of $\Delta\sigma_t$ = 0.50+0.10. Previously oxygen and sulfide were observed to coexist (e.g. [62]) and it was assumed that oxygen oxidized the upward flux of sulfide at the oxic-anoxic interface. The new data sets suggest that there is a density barrier that separates oxygen and sulfide gradients and implies that sulfide is not oxidized by oxygen. Today, we have much improved analytical techniques for 0_2 and HS^- that we can apply to this problem.

3. Hypotheses Regarding the Suboxic Zone

Five main hypotheses, largely based on the 1988 Knorr results, should be addressed in future studies. None of them is easy to quantify and all have significant uncertainties.

3. 1. HYPOTHESIS 1: ANAEROBIC ABIOGENIC SULFIDE OXIDATION

In this hypothesis, Fe (III) and Mn (III, IV) species are the direct oxidants of HS-. A schematic reaction cycle shown in Fig. 2. Millero [38] proposed that sulfide is oxidized by settling particulate $MnO_2(s)$ and possibly FeOOH(s) formed in shallower oxic/suboxic waters., Luther *et al.* [33] proposed that, since particulate Mn concentrations were small, dissolved MN(III) was responsible for sulfide oxidation. Fry *et al.* [18] reached this conclusion independently using sulfur isotopes. Lewis and Landing [31] observed two maxima in particulate Mn at nM levels in the central basin but no particulate maxima for Fe above the sulfide interface. Iron may form colloidal sulfide species below the interface. Tebo [66] observed that the deeper Mn maxima contained oxidized Mn but the shallower maxima did not. The deep maximum in particulate Mn, which is probably Mn (III, IV) manganate material, appears to be much larger at the margins than in the central parts of the Black Sea [27, 31]. Tebo [66] argued that the Mn cycle may be the key process for maintaining the broad suboxic zone. Oxidized Mn may oxidize reduced sulfur, whereas reduced Mn(II) may contribute to oxygen consumption. Murray *et al.* [43] pointed out that one difficulty with the metal oxide hypothesis is that the sum of the vertical electron equivalent gradients of Mn(II) and Fe(II) is much less than that of sulfide. The problem could be resolved if Mn(III,IV) and Fe(III) are recycled in a catalytic role at the depth where HS^- first appears.

The proposed cycle is that HS^- is oxidized by Mn (IV) oxides or Mn(III) organic complexes to elemental sulfur (S^0) and Mn^{2+}. The S^0 is reduced back to HS^- by the facultative anaerobe, Shawanella putrefaciens MR-4 [53]. The Mn^{2+} reacts with $N0_3^-$ to produce N_2 and some form of oxidized Mn(III, IV). Mn(III) is potentially a powerful oxidant and can be formed by both Mn(IV) reduction and Mn(II) oxidation. It can form stable solution complexes with organic matter [29, 34]. The rate of reaction of both Mn(IV) and Mn(III) with HS^- is fast (seconds) and S^0 is the oxidized product. Yao and Millero [71, 72] have shown, in lab studies, that both MnO_2 and FEOOH can oxidize HS^- rapidly to S^0. Aller and Rude [1] proposed a similar reaction of solid phase sulfides with Mn(IV) oxides in sediments of the Amazon Fan. Their experiments suggested bacterial mediation. Experiments in the Black Sea by [31] and Luther *et al.*[33], and the stable sulfur isotope data by Fry *et al.* [18], suggested that abiogenic reaction could be important there.

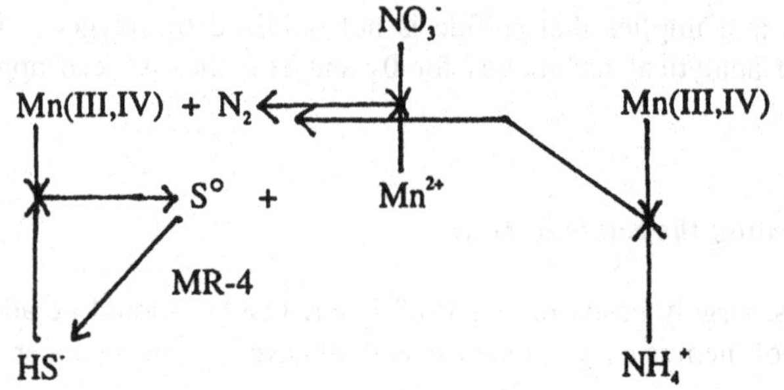

Figure 2. Schematic reactions of sulfur, nitrogen and manganese species showing how Mn can be a catalyst. Sulfide is oxidized by Mn (III, IV) species to elemental S, which is reduced in turn back to HS^- by the bacterium Shawanella putrefaciens (MR-4). Mn (II) is also formed which reduces NO_3^- to N_2 and results in more Mn (HI,IV). This Mn(III,IV) is available to react with more HS^-. Ammonia can also react with Mn (III,IV) to form MN(II) and N_2. Both NO_3 and NH_4^+ end up as N_2.

3.2. HYPOTHESIS 2: MANGANESE AS A CATALYSIS FOR NITROGEN TRANSFORMATIONS

Nitrogen transformations are always complicated at oxic-anoxic interfaces because of the large number of oxidation states available for N. Nitrate increases to a maximum at $\sigma_t = 15.40$ then decreases to zero at $\sigma_t = 15.95$. The simplest explanation for the decrease in $N0_3^-$ with depth is heterotrophic denitrification according to:

$$5\{CH_20\} + 4N0_3^- = 2N_2 + 5HCO_3^- + H^+ + 2H_20$$

There is a $N0_2^-$ maximum at $\sigma_t = 15.85$ that reflects $N0_2^-$ as an intermediate in denitrification. Ammonia increases below a density of $\sigma_t = 15.95$, which is the same density where $N0_3^-$ goes to zero and is deeper than where 0_2 decreases to <10 µM. MN(II) and Fe(II) are also high in deeper water and decrease to low values at $\sigma_t = 15.85$ and 16.00 respectively. Murray *et al.* [45] suggested that the distributions and gradients were consistent with chemo-denitrification reactions where $N0_3^-$ reacts with NH_4^+, Mn^{2+} and Fe^{2+} to form N_2 according to:

$$3N0_3^- + 5NH_4^+ = 4N_2 + 9H_20 + 2H^+$$

$$2N0_3^- + 5Mn^{2+} + 4H_20 = N_2 + 5MnO_2(s) + 8H^+$$

$$2N0_3^- + 10Fe^{2+} + 24H_20 = N_2 + 10Fe(OH)_3(s) + 18H^+$$

The energeties are favorable at in situ concentrations but the kinetics of these reactions is unknown. Regardless, the concentration of total fixed N ($NO_3^- + NO_2^- + NH_4^+$) goes through a dramatic minimum in the suboxic zone and there must be an increase in N_2. Recent pore water measurements, thermodynamic calculations and experiments by Luther *et al.* [35] support the reaction of NO_3^- with Mn^{2+} and also suggest that the upward flux of NH_4^+can be oxidized by MN(IV) or Mn(III) species to form N_2 and Mn^{2+}. These reaction pathways are also shown in Figure 2.

$$2\ NH_4^+ + 3\ MnO_2(s) + 4H^+ = N_2 + 3\ Mn^{2+} + 6\ H_20$$

Sorensen *et al.* [61] and Schulz *et al.* [58] also used deep-sea pore water distributions to suggest that NO_3^- may be reduced by reaction with NH_4^+ and Mn^{2+} . A complete description of the distribution of inorganic nitrogen species and nitrogen isotopes is one of the major omissions from our knowledge about the chemistry of the Black Sea.

3.3. HYPOTHESIS 3: VENTILATION

It is difficult to discuss the suboxic zone without reference to the circulation of the upper layers of the Black Sea. Murray *et al.* [40, 41], Buesseler *et al.* [7] and Ozsoy *et al.* [52] suggested that horizontal water exchange by ventilation is a significant process, especially in the upper 500m. According to this view the pycnocline (including the suboxic zone) originates from horizontal input of water produced by entrainment of the salinity Cold Intermediate Layer (CIL) water by warm, salty Mediterranean water flowing in through the Bosporus. Dissolved oxygen in these ventilation injections would react to oxidize sulfide (Konovalov *et al.*, [in preparation]). An analogous situation appears to occur in Saanich Inlet [2]. Such ventilation could create a suboxic zone and account for the imbalance in the gradients of oxygen and sulfide.

The Cold Intermediate Layer (CIL) ($T < 8\ ^{\circ}C$; $\sigma_t = 14.7$) appears to have two sites of formation. Some CIL is generated in the winter over the northwest shelf [4, 16, 20, 67] and is advected to the eastern and southern parts of the sea. Ovchinnikov and Popov [51] argued for generation of CIL in the cyclonic gyres in the central parts of the Black Sea. It appears that the Rim Current prevents penetration of waters generated in the central cyclonic gyres to the periphery of the basin. These different sources show up as a bi-modal distribution of T-S properties in the core of the CIL [23]. As might be expected, recent data shows significant interannual variability in the depth of the pycnocline and presumably the rate of CIL formation [23]. Though the density of key features appears constant, the depth of those horizons is sensitive to climate variability, primarily the fresh water balance.

Recent data analysis and modeling studies have greatly advanced our understanding of the upper layer circulation and its wind and thermohaline forcing [45, 46, 47, 49]. Coupled biological~physical models are also being produced [48].

Lee *et al.* [submitted] used chlorofluoromethane (CFM) distributions and a multi-layer horizontal-box model to estimate rates of ventilation of different density layers in the upper 500m. F-11 and F-12 are two CFMs used to calibrate oceanographic models.

This is because their accumulation in the atmosphere began in the 1950s (Figure 3). They increased at different rates and their ratio can serve as a clock for introducing time into box models (Figure 3). The CFMs are thought to be chemically inert and they exchange across the air-sea interface by gas exchange. The inert assumption does not hold for strongly reducing waters and F-11 appears to be non-conservative relative to F-12 in the suboxic and anoxic waters of the Black Sea [9, 30].

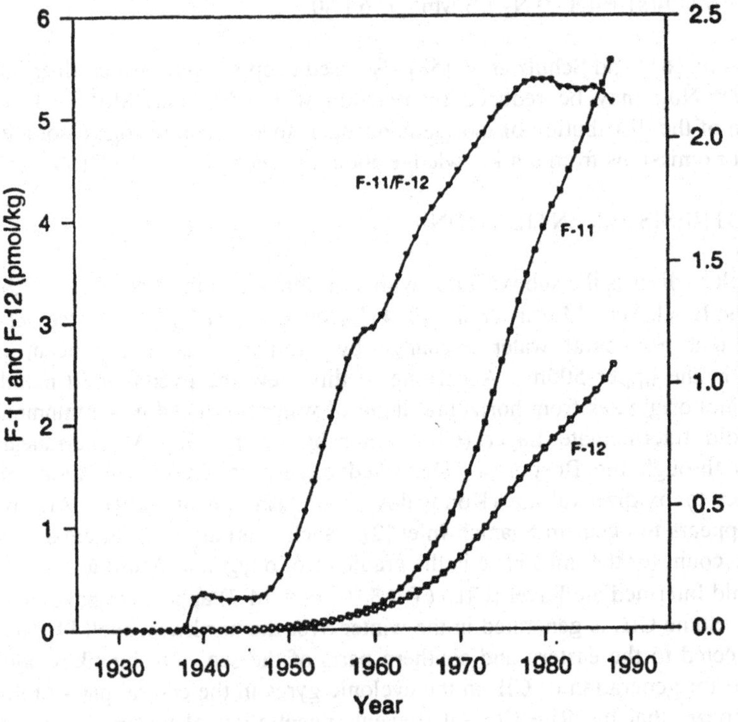

Figure 3. Atmospheric concentrations of F-11 and F-12 and the ratio of F-11 /F-12 as a function of time.

Because these tracers have only been present in the atmosphere since the 1950s they are only present in the upper layers of the Black Sea. The distribution of the saturation ratios F-11 and F-12 versus density are shown in Figure 4. The values approach zero at a density of about $\sigma_t = 17.0$, which corresponds to a depth of about 500m.

Lee *et al.* [30] recently constructed a CFM calibrated multi-box model of ventilation in the upper 500m (to $\sigma_t = 17.01$). A schematic diagram of the box model is shown in Figure 5. There are 5 zone boxes and the density boundaries and approxiamte depth intervals are shown in the figure. According to the ventilation model, each layer is formed by mixing varying proportions of Bosporus Inflow (BI) with Cold Intermediate Layer (CIL) water [42]. Ale proportions of the endmembers are calculated from the salinity. These mixtures (the Interface Boxes in Figure 5) exchange at a rate, with the respective zone boxes. [See Lee *et al.* [30] for the details.] The density boundaries and residence times of the boxes are given in Table 2. The CFM based residence times increased from 2 yr for the CIL to 575 yr for the 16.89-17.01 layer (Table 2).

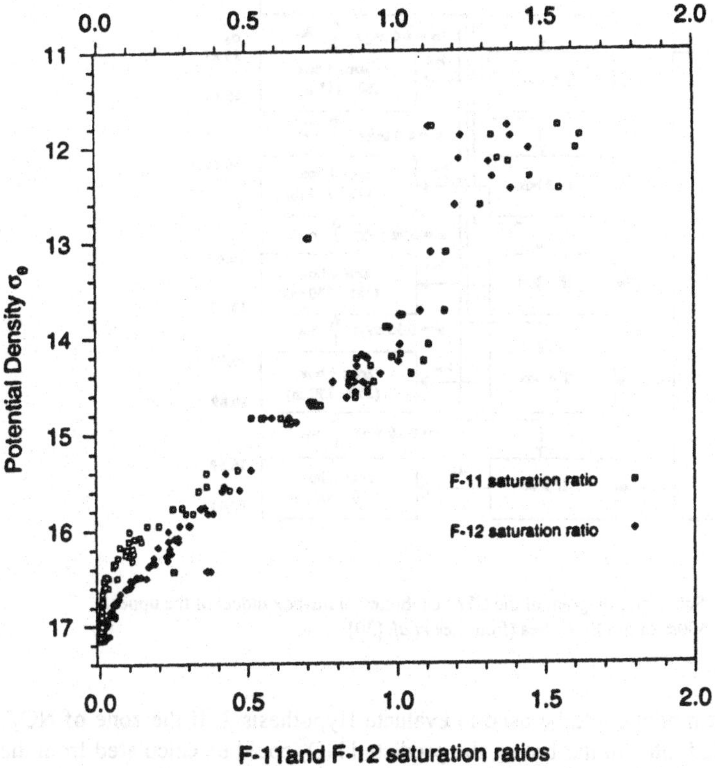

Figure 4. The distribution of the saturation ratios of F-11 and F-12 as a function of density in the Black Sea.

TABLE 2. Residence times of different density intervals determined by Lee *et al.* [30] using a ventilation box model calibrated with CFM data

Density Interval	CFM based Residence Times (yr)
CIL	2
15.45-16.16	7.6
16.16-16.40	14
16.40-16.73	115
16.73-16.89	180
16.89-17.01	575

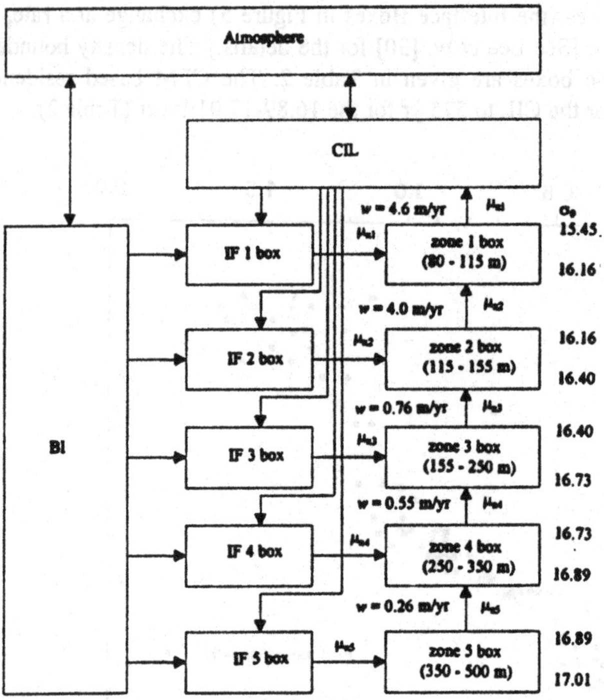

Figure 5. Schematic diagram of the CFM calibrated multi-box model of the upper 500m of the Black Sea (from Lee *et al.* [30].

This ventilation model could be used to evaluate Hypothesis 2. If the zone of NO_3^-, decrease was divided into similar boxes, the predicted NO_3^-could be calculated from the proportions of Bosporus Inflow and CIL. The difference between predicted NO_3^- and measured NO_3^- would indicate the NO_3^- lost by denitrification. Division by the CFM based residence time would give the apparent rate of denitrification.

3.4. HYPOTHESIS 4: ANAEROBIC PHOTOSYNTHESIS

Jorgensen *et al.* [26] and Jannasch [25] proposed that sulfide is oxidized anaerobically in association with phototrophic reduction Of $C0_2$ to organic carbon. This hypothesis was supported by the discovery of large quantities of bacteriochlorophyll pigments within and below the suboxic zone [54]. This pigment is characteristic of the brown sulfur bacteria, Chlorobium. This bacterium requires reduced S (H_2S or S^0) for growth and is capable of growth at very low light levels. The light levels at the depth of the bacteriochlorophyll maximum, were in fact very low (<<0.1% I_0 where I_0 is the incident radiation) and the carbon assimilation rates necessary to verify the hypothesis are difficult to calculate or measure.

If there are fluctuations in the depth of the oxiclanoxic interface, this hypothesis will be more important when the interface is shallow. Repeta *et al.* [54] stated that Chlorobium was not observed in 1969 or 1975. The first appearance of H_2S at that time was slightly deeper at $\sigma_t = 16.35$. It is possible that the suboxic zone only formed after the oxic/anoxic interface shoaled to a slightly shallower level that received more light. This may also be the explanation why a suboxic zone is not observed in the Cariaco Trench. The interface there is deeper than 200m.

3.5. HYPOTHESIS 5: ALKALINITY CONSTRAINS THE STOICHIOMETRY OF SUBOXIC REACTIONS

Total $C0_2$ and alkalinity increase with depth through the suboxic zone into the deep anoxic water. They are key parameters because they reflect the net effect of all the oxidation-reduction reactions on the carbon and proton balances. In the oxic euphotic zone, total $C0_2$ increases while alkalinity stays approximately constant. In the anoxic, sulfide containing, waters total $C0_2$ and alkalinity are dominated by sulfate reduction.

$$106\ CH_20 + 53\ SO_4^{2+} = 106\ HCO_3^- + 53\ H_2S$$

Goyet *et al.* [21] observed that dissolution of $CaCO_3$ caused both total $C0_2$ and alkalinity to increase faster than expected based on sequential oxidation of Redfield organic matter by 0_2, $N0_3^-$and $SO_4^{2=}$. Thus, total alkalinity needs to be corrected for the minor bases: borate, ammonia, phosphate, silicate and bisulfide and with Ca^{2+} representing $CaCO_3$ dissolution. Previous models of oxic-anoxic interfaces assumed that sulfide was oxidized by oxygen. Alkalinity could be a sensitive tracer. For example, the ratios of $\Delta alk/\ \Delta HS^-$ are different depending on whether bisulfide is oxidized by 0_2 to SO_4^{2-}, SO_3^{2-}, $S_20_3^{2-}$ or S^0.

	$\Delta alk/\Delta HS^-$
$20_2 + HS^- = SO_4^{2-} + H^+$	2
$1.5\ O_2 + HS^- = SO_3^{2-} + H^+$	2
$O_2 + HS^- = 0.5\ S_2O_3^{2-} + 0.5\ H_2O$	1
$0.5\ O_2 + HS^- + H^+ = S^0 + H_2O$	0

In the suboxic zone, where SO_4^{2-} reduction is less dominant, the balance of Atotal $C0_2/\Delta alk$ should be a sensitive constraint on the choice of reactions. For example, the ratios of alkalinity to $N0_3^-$ would be very different depending on whether $N0_3^-$ is reduced by the heterotrophic pathway or by reaction with NH_4^+, Mn^{2+} or Fe^{2+}. This is just one reaction, and the alkalinity is influenced by many others as well, but it does represent an important constraint on the stoichiometry of suboxic reactions.

4. Conclusions

The suboxic zone in the Black Sea is a uniquely well defined site for studying oxidation-reduction reactions that are important for suboxic reagions throught the ocean basins and sediments. Using primarily data from the 1988 R/V Knorr Expedition we present five hypotheses for further study. These hypotheses are 1) that the upward tlux of sulfide is oxidized by Mn (III,IV) and Fe(III) species, 2) that Mn species act as a catalyst in which the downward flux of nitrate is reduced by Mn^{2+} and the upward flux of NH_4^+ is oxidized by Mn (III,IV) species, 3) that ventilation of the upper 500m is rapid compared to the deep water and can be calculated using CFM distributions, 4) that sulfide is oxidized anacrobically in association with phototrophic reduction of CO, to organic carbon (this hypothesis is an alternative to hypothesis 1), and 5) that detailed alkalinity profiles can constrain the stoichiometry of suboxic reactions.

Acknowledgements

This amnuseript benefitted from stimulating discussions with S. Konovalov, T. Oguz and and E. Yakushev. This work was supported by Petroleum Research Fund Grant 29302 (JWM) and NSF OCE-9314349 (GWL). This is University of Washington Contribution Number 2188.

References

1 Aller R.C. and P.D Rude (1988) Complete oxidation of solid phase sulfides by manganese and bacteria in anoxic marine sediments. *Geochimica* et *Cosmochimica Acta* **52**, 751-765

2. Anderson J.J. and A H. Devol (1973) Deep water renewal in Saanich Inlet, an intermittently anoxic basin *Estuarine and Coastal Marine Science* **1**, 1-10

3 Berger W and P H. Roth (1975) Oceanic micropaleontology progress and prospect *Rev. Geophys Space Phys* **13**, 560-585.

4. Blatov A S., N P. Bulgakov, V A Ivanov, A.N. Kosarev and V S Tuzhilkin (1984) Variability of hydrophysical fields of the Black Sea Leningrad, Hydrometcoizat, pp. 240 (Russian)

5 Brandes J.A., S W.A. Naqvi and A.H. Devol (submitted) ^{15}N of nitrate in suboxic regions a tracer for suboxic diagenesis, mixing and nitrogen cycles *Limnology and Oceanography*

6 Brewer P G and Murray J W. (1973) Carbon, nitrogen and phosphorus in the Black Sea. *Deep-Sea Research* **20**, 803-818.

7 Buesseler K O , H. D. Livingston and S A. Casso (1991) Mixing beween oxic and anoxic waters of the Black Sea as traced by Chemobyl cesium isotopes *Deep-Sea Research* **38**, S725-S746

8. Buesseler K O , H.D Livingston, L Ivanov and A. Romanov (1994) Stability of the oxic-anoxic interface in the Black Sea. *Deep-Sea Research* **41**, 283-296

9. Bullister J L. and B. S. Lee (1995) Chloroflurorcarbon-11 removal in anoxic marine waters *Geophysical Research Letters* **22**, 1893-1896.

10. Caspers H. (1957) Black Sea and Sea of Azov In Treatise on Marine Ecology and Paleocology, J.W Hedgpeth, editor, Geological Society of America Memoire 67, 1, 801-921

11 Cline J.D and F.A Richards (1972) Oxygen deficient conditions and nitrate reduction in the eastem tropical North Pacific Ocean *Limnology and Oceanography* **17**, 885-900

12 Codıspoti L.A., G.E. Friederich, J.W. Murray and C. Sakamoto (1991) Chemical variabılity in the Black Sea: implıcatıons of data obtained with a continuous vertical profiling system that penetrated the oxic-anoxic interface. *Deep- Sea Research* **38**, S691-S710.

13. Degens E T. and Ross D A , editors (1974) The Black Sea-geology, chemistry and biology, AAPG Memoir 20.

14. Emerson S., R. Jahnke, M. Bender, P. Froelich, C. Klinkhainmer, C. Bowser and G Setlock (1980) Early diagenesis in sediments from the eastern equatorial Pacific, I. Pore water nutrıent and carbonate results *Earth and Planetary Scıence Letters* **49**, 57-80.

15. Emerson S., S. Kalborn, L. Jacobs, B. M. Tebo, K.H. Nealson and A.R Rosson (1982) Environmental oxidation rate of manganese (II): bacterial catalysis. *Geochımıca et Cosmochımıca Acta* **46**, 1073-1079

16 Filippov D.M. (1965) The cold intermediate layer in the Black Sea. *Oceanology* **5**, *47-52*

17. Froelich P.N., G.P. Klinkhammer, M.L. Bender, N.A Luedtke, G R. Heath, D. Cullen, P. Dauphın, D Hammond, B. Hartman and V. Maynard (1979) Early dıagenesis of organic matter in pclagic sediments of the eastern equatorial Atlantic suboxic diagenesıs *Geochımıca et Cosmochımıca Acta* **43**, 1075-1090

18 Fry B , H W Jannasch, S.J. Molyncaux, C 0. Wırsen, J A Muramoto and S. King (1991) Stable isotope studies of the carbon, nitrogen and sulfur cycles in the Black Sea and the Cariaco Trench. *Deep-Sea Research* **38**, 1003-1019.

19 Gagosıan R.B. (1975) Hydrographıc data from the Black Sea, R V Chain 120 Leg 1, April 15-30, 1975 WHOI Report

20 Georgıev Yu S. (1967) On dynamıcs of the cold ıntermediate layer in the Black Sea. In: Okeanograflcheskiye ıssledovaniya Chernogo Morya (Oceanographic investigations of the Black Sea), Naukova Dumka, Kiev, 105-113 (Russian).

21. Goyet C., A.L. Bradshaw and P G. Brewer (1991) The carbonate system ın the Black Sea *Deep-Sea Research* **38**, S1049-S1068.

22. Hastıngs D and S. Emerson (1988) Oxidation of manganese by spores of a marine Bacillus: kınetic and thennodynamic considerations. *Geochımıca et Cosmochımıca Acta* **50**, 1819-1824

23 Ivanov L I , S Besiktepe, E.G. Nikolaenko, E. Ozsoy, V Diakonu and E. Demirov (1997) Volumetric fine structure of the Black Sea cold intermediate layer.

24. Izdar E. and J.W. Murray, eds. (1991) Black Sea Oceanography. Kluwer Acadeniic Publishers, Dordrecht, 487pp

25. Jannasch H.W, (1991) Microbial processes in the Black Sea water column and top sedıment: An overview. In. (E Izdar and J.W. Murray, eds) Black Sea Oceanography. Kluwer Academic Publishers, Dordrecht, 271-286.

26. Jorgensen B.B., H. Fossıng, C. O Wırsen and H.W. Jannasch (1991) Sulfide oxidation in the anoxıc Black Sea chemocline. *Deep-Sea Research* **38**, 1083-1103.

27. Kempe S., G Liebezeit, A. R Diercks and V Asper (1990) Water balance in the Black Sea. *Nature* **346**, 419

28 Konovalov S.K., J.W. Murray, L I Ivanov and A.S Samodurov (ın preparation) Temporal variatıons in basic hydrochemıcal properties of the Black Sea.

29. Kostka J.E., G.W. Luther and K.H. Nealson (1995) Chemical and bıological reductıon of Mn(III)pyrophosphate complexes: Potentıal importance of dissolved Mn(III) as an environmental oxidant. *Geochımıca et Cosmochımıca Acta* **59**, 885-894

30. Lee B.S., J.L. Bullister and J.W. Murray (submitted) Anthropogenıc chlorotluoromethanes ın the Black Sea *Deep-Sea Research.*

31. Lewıs B.L. and Landing W. (1991) The bıogeochenistry of manganeseandiron in the Black Sea. *Deep-Sea Research* **38** S773-S804.

32. Luther G W., T. Ferdelman and E. Tsainakıs (1988) Evidence suggesting anaerobic oxıdation of the bısulfide ion in Chesapeake Bay *Estuarıes* **11**, 281-285.

33. Luther G W. III, T.M. Church and D Powell (1991) Sulfur speciatıon and sulfide oxıdation in the water column of the Black Sea. *Deep-Sea Research* **38**, S1121-S1138.

34 Luther G.W III, D. Nuzzio and J. Wu (1994) Speciation of manganese in Chesapeake Bay waters by voltammetric methods. *Anal. Chım. Acta* **284**, 473-480.

35. Luther G.W. III, B. Sundby, B.L Lewis, P J. Brendel and N. Silverberg (1997) The interaction of manganese wıth the nıtrogen cycle ın contınental margin sediments: alternatıve pathways for dinıtrogen formation. *Geochım. Cosmochım. Acta* **61**

36. Millero F J., S. Hubinger, M. Fernandez and S. Garnett (1987) Oxidation of H,S in seawater as a function of temperature, pH and ionic strength. *Environmental Science and Technology* **11**, 1114-1120
37. Millero F.J. (1991) The oxidation of H_2S in Black Sea waters. *Deep-Sea Research.* **38**, S1139-S1150
38. Millero F.J. (1991) The oxidation of H_2S with 0_2 in the Black Sea. In: (E. Izdar and J.W Murray, eds) Black Sea Oceanography. Kluwer Acad. Publ., Dordrecht, pp 205-227.
39. Millero F.J. (1991) The oxidation of H_2S in the Chesapeake Bay. *Estuarine and Coastal Shelf Science* **33**, 521-527.
40 Murray J.W., H.W. Jannasch, S. Honjo, R.F. Anderson, W.S. Reeburgh, Z. Top, G.E. Friederich, L.A. Codispoti and E. Izdar (1989) Unexpected changes in the oxiclanoxic interface in the Black Sea. *Nature* **338**, 411-413.
41. Murray J.W. (1991) The 1988 Black Sea Oceanographic Expedition: introduction and summary. *Deep-Sea Research* **38,** S655-S661.
42. Murray J.W. Z Top and E. Ozsoy (1991) Hydrographic properties and ventilation of the Black Sea Deep-*Sea Research* **38**, S663-S690.
43 Murray J.W. , L.A. Codispoti and G.E. Friederich (1995) Oxidation-reduction environments: The suboxic zone in the Black Sea. In: (C.P. Huang, C.R. O'Melia and J.J. Morgan, eds.) Aquatic Chemistry: Interfacial and Interspecies Processes. American Chemical Society
44 Nameroff T J , A H Devol, L. Balistrieri and J.W Murray (submitted) Suboxic trace metal geochemistry in the eastern tropical north Pacific, *Geochimica et Cosmochimica Acta*
45. Oguz T , P.E. LaViolette and U. Unluata (1992) The upper layer circulation of the Black Sea. Its variability as inferred from hydrographic and satellite observations. *Journal of Geophysical Research* **97**, 12,569-12,584.
46 Oguz T., V S Latun, M.A. Latif, V.V. Vladimirov, H I. Sur, A.A. Markov, E. Ozsoy, B.B Kotovshchikov, E.E. Eremeev and U. Unluata (1993) Circulation in the surface and intermediate layers of the Black Sea. *Deep-Sea Research* **40**, 1597-1612.
47. Oguz T., P. Malanotte-Rizzoli and D. Aubrey (1995) Wind and thermohaline circulation of the Black Sea driven by yearly mean climatological forcing. *Journal of Geophysical Research* **100**, 6845-6863
48. Oguz T., H. Ducklow, P. Malanotte-Rizzoli, S. Tugrul, N.P. Nezlin and U. Unluata (1996) Simulation of annual plankton productivity cycle in the Black Sea by a one-dimensional physical-biological model. *Journal of Geophysical Research* **101**, 16,585-16,599.
49. Oguz T. and P. Malanotte-Rizzoli (1996) Seasonal variability of wind and thermohaline-driven circulation in the Black Sea. Modeling studies. *Journal of Geophysical Research* **101**, 16,551-16,569.
50. Ostlund H.G. (1974) Expedition "Odysseus 65". Radiocarbon age of Black Sea Water. In: The Black Seageology, chemistry and biology, E.T. Degens and D.A. Ross, editors, AAPG Memoir 20, pp 127-132
51. Ovchinnikov I.M. and Yu. I. Popov (1987) Evolution of the cold intermediate layer in the Black Sea. *Oceanology* **27**, *555-560.*
52. Ozsoy E., Z. Top, G. White and J.W. Murray (1991) Double diffusion intrusions, mixing and deep sea convective processes in the Black Sea. In: (E. Izdar and J.W Muffay, eds) Black Sea Oceanography Kluwer Academic Publishers, Dordrecht, 17-42
53. Perry K.A., J.E.Kostka, G.W. Luther and K.H Nealson (1993) Mediation of sulfur speciation by a Black Sea facultative anacrobe. *Science* **259**, 801-803.
54. Repeta D.J., D.J. Simpson, B.B. Jorgenson and H.W. Jannasch (1989) Evidence for anoxygenic photosynthesis from the distribution of bacteriochlorophylls in the Black Sea. *Nature* **342**, 67-72.
55. Rue E L., G.J. Smith, G, A. Cutter and K.W. Bruland (1997) The response of trace element redox couples to suboxic conditions in the water column *Deep-Sea Research* **44**, 113-134
56. Saydam C., S. Tugruk, O. Basturk and T Oguz (1993) Identification of the oxiclanoxic interface by isopycnal surfaces in the Black Sea *Deep-Sea Research* **40**, 1405-1412.
57. Schlanger S O. and H.C. Jenkyns (1976) Cretaceous oceanic anoxic events-causes and consequences. *Geol. Mijnb.* **55**, 179-184.
58. Schultz H D., A. Dahmke, T. Schinzel, K. Wallmann and M. Zabel (1994) Early diagenetic processes, fluxes and reactions rates in sediments of the South Atlantic *Geochimica et Cosmochimica Acta* **58**, 2041-2060.
59. Sen Gupta R. (1971) Oceanography of the Black Sea: inorganic nitrogen compounds. *Deep-Sea Research* **18**, 457-475

60. Sholkowitz E.R. (1992) Comment on "Redox cycling of rare earth elements in the suboxic zone of the Black Sea" by C.R. German, B.P. Holliday and H. Elderfield. *Geochimica et Cosmochimica Acta* **56**, 4305-4307.

61 Sorensen J., K.S. Sorensen, S. Colley, D.J. Hydes, J. Thompson and T.R.S. Wilson (1987) Depth localization of denitrification in deep-sea sediment from the Madeira Abyssal Plain. *Limnology and Oceanography* **32**, 758-762.

62 Sorokin Yu. I (1983) The Black Sea. In: Ecosystems of the World 26. Estuaries and Enclosed Seas, B H Ketchum, editor, Elsevier, Amsterdam, pp. 253-292.

63. Spencer D.W. and Brewer P.C. (1971) Vertical advection diffusion and redox potential as controls on the distribution of managnese and other trace metals dissolved in waters of the Black Sea. *Journal of Geeophysical Research* **76**, 5877-5892.

64. Stumm W. and J.J. Morgan (1996) Aquatic Chemistry, 3rd Edition. John Wiley, New York

65 Tebo B. , K H. Nealson , S. Emersonand L. Jacobs (1984) Microbial mediation of Mn(II) and Co(II) precipitation at the O_2/H_2S interfaces in two anoxic fjords. *Limnology and Oceanography* **20**, 1247-1258.

66 Tebo B.M. (1991) Manganese (II) oxidation in the suboxic zone of the Black Sea. *Deep-Sea Research* **38**, S883-S906.

67. Tolmazin D. (1985) Changing coastal oceanography of the Black Sea. I: Northwestern Shelf *Progress in Oceanography* **15**, 217-276.

68 Tugrul S., O. Basturk, C Saydam and A. Yilmaz (1992) Changes in the hydrochemistry of the Black Sea inferred from water density profiles. *Nature* **359**, 137-139.

69 Vinogradov M. Ye and Yu. R Nalbandov (1990) Effect of changes in water density on the profiles of physicochemical and biological characteristics in the pclagic ecosystem of the Black Sea. *Oceanology* **30**, 567-573.

70. Yakushev E.V. and L.N. Neretin (1997) One dimensional modeling of nitrogen and sulfur cycles in the aphotic zones of the Black and Arabian Seas *Global Biogeochemical Cycles* **11**, 401-414

71. Yao W. and F.J. Millero (1993) The rate of sulfide oxidation by MnO_2 in seawater. *Geochimica et Cosmochimica Acta* **57**, 3359-3365.

72. Yao W. and F.J. Millero (1996) Oxidation of hydrogen sulfide by hydrous Fe(III) oxides in seawater. *Marine Chemistry* **52**, 1-16.

73 Zhang J-Z and F J. Millero (1993) The rate of sulfide oxidation in seawater. *Geochimica et Cosmochimica Acta* **55**, 677-685.

AN APPROACH TO MODELLING ANOXIC CONDITIONS IN THE BLACK SEA

EVGENIY V.YAKUSHEV
P.P.Shirshov Institute of Oceanology,
36 Nakhimovskiy Pr., Moscow, 117851, Russia

Abstract

In anoxic conditions the oxidation of organic matter occurs in different stoichimetric reactions. Therefore modelling of oxic/anoxic transformation requires parameterization of the cycles of several elements simultaneously, in contrast to models dealing only with nutrient cycles under oxic conditions, where it is possible to use the Redfield ratios. An O-N-S-Mn model is considered to describe the biogeochemical sources. Rates of biochemical processes mediated by bacteria are described by first-order equations using semiempirical functions of O_2 concentration.

The processes of turbulent diffusion, sedimentation, and biogeochemical transformation of compounds were parameterized in the frames of one-dimensional and two-dimensional coupled models. The model was calibrated using data observed for the vertical distribution of compounds in the upper layers of the Black Sea. The calculated spatial distributions of nitrogen compounds (total organic nitrogen, ammonium, nitrate, nitrite), inorganic reduced sulfur compounds (hydrogen sulfide, elemental sulfur, thiosulfate, sulfate), dissolved and particulate manganese, as well as dissolved oxygen agree reasonably well with the observations.

Model estimations confirm that the existence of anoxic conditions is controlled primarily by the peculiarities of organic matter decay (a consequence of oxidant consumption) in conjunction with restricted aeration. According to the model simulations, the most sensitive hydrochemical parameters from point of view of vertical advection anomalies are particulate manganese, organic matter and elemental sulphur. The results of work undertaken so far suggests that future development of the model should concentrate on the improvement of parameterization of peculiarities of sedimentation connected with manganese cycle, organic matter balance, and the use of hydrophysical model results for the description of advection and diffusion.

1. Introduction

The study of anoxia conditions formation is important because they can be formed in both natural and anthropogenic ways. Such a study is particularly important in the case of the Black Sea, the largest natural water body with extensive anoxia, where about 80% of waters are anoxic and localised areas of hypoxia and anoxia occur in polluted shelf

S. Beşiktepe et al. (eds.),
***Environmental Degradation of the Black Sea: Challenges and Remedies*, 93–108.**

regions. A representative case study of the oxic/anoxic transformation is the Black Sea redox layer. This is the zone of contact between oxic and anoxic waters, which can be named also [22] Nepheloid Redox Layer (NRL), because of its low transparency.

Modelling of the oxic/anoxic contact zone can provide, in addition to quantitative results, insight into the qualitative nature of the zone, about which relatively little is currently known. Hence, the main objective was to compare model results based on our current knowledge of events, processes and systems with the situation observed in nature. Improvements to the model can then be made in order to successively reduce discrepancies between modelled output and reality.

Modelling the process of anoxia formation is also important, because these processes need to be parameterized in models with different assumptions explaining the development of the ecological state of the Black Sea. [12, 20].

Modelling of biogeochemical processes in sea water requires parameterization of the chemical elements cycling. The first stage of the construction of each model is the choice of the relevant variables. Two basic approaches can be used: the ratio of concentrations of certain forms of chemical elements present in the sea water and/or the time scales over which the process is studied.

The current data for nutrient (N, P, Si, C, O, S) distribution confirm that even in the euphotic zone the majority of these elements are contained in the form of dissolved inorganic material, while nutrients in the cells of living organisms account for only a very small percentage. Below the euphotic zone, the share of inorganic matter increases to more than 95-99 %.

The formation and maintenance of the chemical structure occurs over time scales varying from one week to decades. The time required exceeds the time scales of biological processes.

Therefore if the goal of the study is the formation of hydrochemical structures and their variability, we should describe the inorganic forms in detail, while the biological parameters can remain generalised.

If we compare the characteristics of the biogeochemical cycles in oxic and anoxic conditions we find that there is a principal difference: in oxic conditions the cycles of different nutrients are parallel. It is therefore possible to use the stoichiometric law (Redfield ratios) and to describe the cycle of only one element and to extend the results to other elements. For example, it is possible to use either phosphorus or nitrogen as the photosynthesis limitation, and to then calculate the oxygen changes on the base of the Redfield ratio.

In anoxic conditions, every element cycle plays its own geochemical role because the oxidation of organic matter occurs in different stoichimetric reactions. Modelling of oxic/anoxic transformation therefore requires the simultaneous parameterization of cycles of several elements.

This work began with a model of coupled nitrogen and oxygen cycles [24], which demonstrated that the existence of anoxic conditions is controlled predominantly by the characteristics of organic matter decay under conditions of restricted aeration. Then the sulphur cycle was added, and we now consider the O-N-S-Mn model (Figure 1):

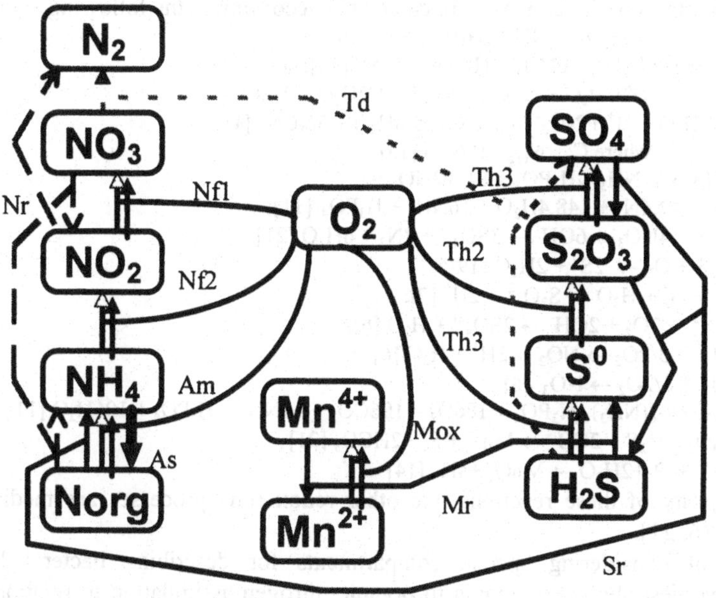

Figure 1 Diagrammatic representation of the coupled model of sulfur, nitrogen, manganese and oxygen cycling showing the compartments and the modeled rates of transformation processes

2. Chemical-biological sources parameterization

The model (Figure 1) computes the content of hydrogen sulfide (H_2S), total elemental sulfur (S^0+S_n^{2-}), thiosulfate (and sulfites) ($S_2O_3^{2-}$+SO_3^{2-}), sulfate (SO_4^{2-}), total organic nitrogen (N_{org}), ammonium (NH_4^+), nitrite (NO_2^-), nitrate (NO_3^-), dissolved manganese (Mn^{2+}), particulate manganese (Mn^{4+}) and dissolved oxygen (O_2).

The transformation of sulfur occurs as a result of hydrogen sulfide oxidation and sulfate reduction. Nitrogen transformation occurs as a result of ammonification, nitrification, nitrate reduction (denitrification), thiodenitrification and ammonium assimilation. In addition, the processes of manganese oxidation and reduction are also considered (Figure 1).

2.1. CHEMICAL-BIOLOGICAL SOURCES

To describe the transformation of matter between the model compartments carried out by micro-biological or chemical means, first order equations were used.

For the dependence of processes considered on oxygen content, the following linear functions were proposed (Table 1).

Stoichiometric coefficients m_i were calculated according to the following equations:

m_1, m_5: $(CH_2O)_{106}(NH_3)_{16}H_3PO_4 + 53SO_4^{2-} =$
$= 106CO_2 + 106H_2O + 16NH_3 + H_3PO_4 + 53S^{2-}$ [13];

m_2, m_3: $1/2CH_2O + NO_3^- \rightarrow NO_2^- + 1/2H_2O + 1/2CO_2$
$3/4CH_2O + H^+ + NO_2^- \rightarrow 1/2N_2 + 5/4H_2O + 3/4CO_2$ [1]
where $C_{org}:N_{org} = 106:16$ [13];

m_4: $(CH_2O)_{106}(NH_3)_{16}H_3PO_4 + 84.8HNO_3 =$
$= 106CO_2 + 42.4N_2 + 148.4H_2O + 16NH_3 + H_3PO_4$ [11];

m_6: $3H_2S + 4NO_3^- + 6OH^- \rightarrow 3SO_4^{2-} + 2N_2 + 6H_2O$ [21];

m_7: $2H_2S + O_2 \rightarrow 2S^0 + 2H_2O$ [19];

m_8: $2S^0 + O_2 + H_2O \rightarrow S_2O_3^{2-} + 2H^+$ [7];

m_9: $S_2O_3^{2-} + 2O_2 + 2OH^- \rightarrow 2SO_4^{2-} + H_2O$ [6];

m_{10}: $NH_4^+ + 3/2O_2 \rightarrow NO_2^- + 2H^+ + H_2O$ [6];

m_{11}: $NO_2^- + 1/2O_2 \rightarrow NO_3^-$ [6];

m_{12}: $(CH_2O)_{106}(NH_3)_{16}H_3PO_4 + 106O_2 = 106CO_2 + 16NH_3 + H_3PO_4 + 106H_2O$ [11].

m_{13}: $MnO_2 + H_2S + 2CO_2 \rightarrow Mn^{2+} + S^0 + 2HCO_3$ [21]

m_{14}: $Mn^{2+} + O_2 + 2H_2O \rightarrow MnO_2 + 4H^+$ [14]

The majority of these reactions like other redox-zone processes are mediated by bacteria [7, 16, 21].

Instead of considering special compartments for describing bacteria biomass dynamics, we describe the ammonia to organic nitrogen assimilation in relation to the intensity of the microbiological processes (chemosynthesis in nitrogen units).

$As = p_1\,Th1 + p_1\,Th2 + p_1\,Th3 + p_2\,Nf1 + p_2\,Nf2 + p_3\,Td + p_4\,Nr1 + p_4\,Nr2$ - where p_i - the quantity of consumed nitrogen of NH_4^+ in units of consumed substrate (S^{2-}, S^0, $S_2O_3^{2-}$, NH_4^+, NO_2^-). This assumption can be made because the main objective of this study is to describe the chemical structure of the redox layer, and not detailed modelling of its bacterial ecosystem.

Therefore the chemical sources of the model are the following:

$$R_{Norg} = -Am - m_2\,Nr1 - m_3\,Nr2 - m_5\,Sr1 + As$$
$$R_{NH4+} = Am + m_4\,Nr1 + m_5\,Sr1 - Nf1 - As$$
$$R_{NO2-} = Nf1 - Nf2 + Nr1 - Nr2$$
$$R_{NO3-} = Nf2 - Nr1 - m_6\,Td$$
$$R_{H2S} = -Th1 + Sr2 - Td - m_{13}\,Mr$$
$$R_{S0} = Th1 - Th2$$
$$R_{S2O32-} = Th2 - Th3 + Sr1 - Sr2$$
$$R_{SO42-} = Th3 - Sr1 + Td$$
$$R_{O2} = -m_7\,Th1 - m_8\,Th2 - m_9\,Th3 - m_{10}\,Nf1 - m_{11}\,Nf2 - m_{12}\,Am - m_{14}\,Mox,$$
$$R_{Mn2+} = Mr - Mox;$$
$$R_{Mn4+} = Mox - Mr;$$

where m_i (i = 2-12) - stoichiometric coefficients of the model.

The values of the coefficients used are the values presented in Table 2. The detailed description of the model is in[25].

TABLE 1. Formulations, names of processes, dependence on oxygen content

Formula	**Process**	**Dependence on oxygen**
$Am = K_{Am}\,[N_{org}]\,F_{Am}(O_2)$	Ammonification	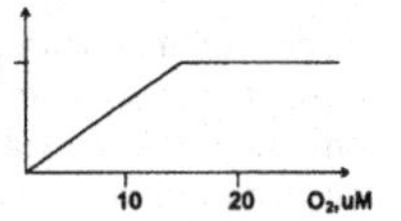
$Nf1 = K_{Nf1}\,[NH_4^+]\,F_{Nf}(O_2)$ $Nf2 = K_{Nf2}\,[NO_2^-]\,F_{Nf}(O_2)$	1 stage of nitrification 2 stage of nitrification	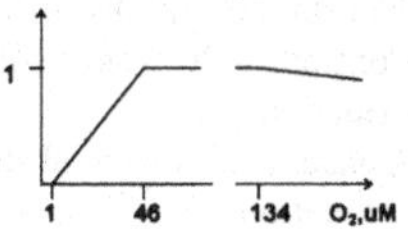
$Nr1 = K_{Nr1}\,[NO_3^-]\,F_{Nr}(O_2)$ $Nr2 = K_{Nr2}\,[NO_2^-]\,F_{Nr}(O_2)$	nitrate reduction denitrification	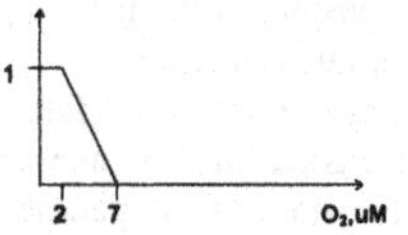
$Th1 = K_{Th1}\,[H_2S]\,F_{Th}(O_2)$ $Th2 = K_{Th2}\,[S^0]\,F_{Th}(O_2)$ $Th3 = K_{Th3}\,[S_2O_3^{2-}]F_{Th}(O_2)$	1 stage of H_2S oxidation 2 stage of H_2S oxidation 3 stage of H_2S oxidation	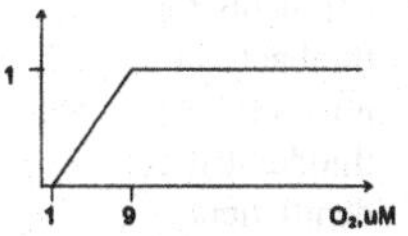
$Sr1 = m_1\,K_{Sr1}\,[N_{org}]\,F_{Sr}(O_2)$ $Sr2 = K_{Sr2}\,[S_2O_3^{2-}]\,F_{Sr}(O_2)$	1 stage of sulphate reduction 2 stage of sulphate reduction	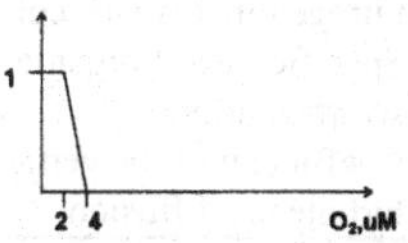
$Td = K_{Td}\,[H_2S]\,F_{Td}(O_2)\,F_i\,(NO_3)$	thiodenitrification	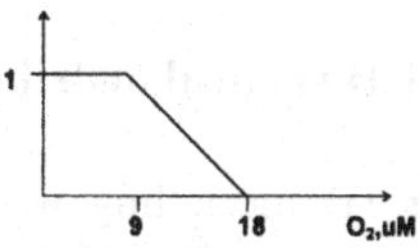
$Mox = K_{Mox}\,[Mn^{2+}]\,F_{mox}(O_2)$	manganese oxidation	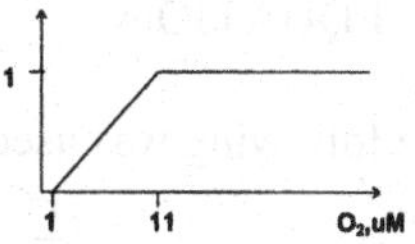
$Mr = K_{Mr}\,[Mn^{4+}]\,F_{mox}(O_2)$	manganese reduction	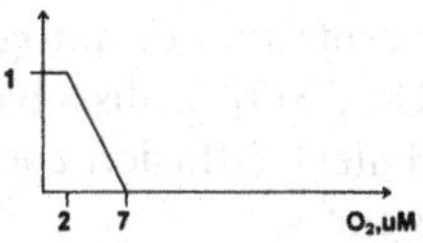

Where K_i are constants of the corresponding reactions, m_i are the stoicheometric coefficients and $F_i(O_2)$, $F_i(NO_3^-)$ are the dependencies of reaction rates on oxygen and nitrate content respectively.

TABLE 2. Model coefficients

Name	Symbol, units	Value	Literature estimation	Source
Constant of ammonification	K_{Am}, day^{-1}	0.01	0.01-0.1	[15,21]
Constant of ammonium oxidation	K_{Nf1}, day^{-1}	0.1	0.01-0.13	[15]
Constant of nitrite oxidation	K_{Nf2}, day^{-1}	0.3	0.01-0.63	[15]
Constant of nitrate reduction	K_{Nr1}, day^{-1}	0 16	-	-
Constant of denitrification	K_{Nr2}, day^{-1}	0.22	-	-
Constant of thiodenitrification	K_{Td}, day^{-1}	0.006	-	-
Constant of hydrogen sulfide oxidation	K_{Th1}, day^{-1}	0.45	0.01-0.3	[17]
Constant of elemental sulfur oxidation	K_{Th2}, day^{-1}	0.7	0.01-0.7	[18]
Constant of thiosulfate oxidation	K_{Th3}, day^{-1}	0.4	0.01-0 5	[17]
Constant of the first stage of sulfate reduction	K_{Sr1}, day^{-1}	0.001	-	-
Constant of thiosulfate reduction	K_{Sr2}, day^{-1}	0.004	-	-
Constant of manganese oxidation	K_{Mox}, day^{-1}	0 10	-	-
Constant of manganese reduction	K_{Mr}, day^{-1}	0.75	-	-
Ammonium assimilation coefficients for:			-	-
thiobacteria	p_1, μM(N)/μM(S)	1.0	-	-
nitrifiers	P_2	0 2	-	-
thiodenitrifiers	p_3	0.2	-	-
denitrifiers	p_4	0.2	-	-
Percentage of total organic nitrogen that is particulate	a, %	10	6-14	[16]
Specific rate of organic matter sedimentation	W, m day^{-1}	1-50	30-400	[15]
Coefficient of the vertical turbulence diffusion	A_z, cm^2 s^{-1}	0 4	0 05-10	[5, 8]

3. 1-D vertical distribution modelling

For 1D modelling we considered a column of water with a depth range from 50 m (lower boundary of the euphotic zone) to 150 m.

3.1. EQUATION

The following was used as the basic equation for dissolved components:

$$\frac{\partial C_i}{\partial t} = \frac{\partial}{\partial z} Az \frac{\partial C_i}{\partial z} + R_{C_i}. \qquad (1)$$

where R_{C_i} - sources and sinks of a substance (rates of transformation), C_i - concentration of nitrogen compounds (NH_4^+, NO_2^-, NO_3^-), sulfur compounds (H_2S, S_0, $S_2O_3^{2-}$, SO_4^{2-}), dissolved manganese (Mn^{2+}), and dissolved oxygen (O_2). A_z - vertical turbulent diffusion coefficient. We assume A_z to be a model constant with the value of 0.3 cm^2/s.

The total organic nitrogen and particulate manganese (Mn^{4+}) concentrations were calculated according to (1) supplemented by the introduction of a particulate matter sinking rate term:

$$\frac{\partial C_i}{\partial t} + W\frac{\partial (a_i C_i)}{\partial z} = \frac{\partial}{\partial z} Az \frac{\partial C_i}{\partial z} + Rc_i \qquad (2)$$

Where a = the fraction of organic matter that is particulate and where W = sinking rate of the particulate matter (Table 1).

3.2. BOUNDARY CONDITIONS

It was assumed that at the upper boundary of the water columns being studied here, the chemical, biological and physical processes are balanced and maintain constant concentrations of most chemical elements throughout the year. Therefore, we specify constant values of all model components at the upper boundary.

The values of all the compounds in the model were set to zero, except for the following: N_{org}=3.8 μM/l, NO_3^-=1.0 μM /l, O_2=300 μM /l, SO_4=15 M/l.

The pronounced halocline (pycnocline) in the Black Sea restricts vertical motion and hence the contact of deep waters with the surface. Because of this weak mixing, the existence of the deep Bosphorus current has little effect on the renewal of the intermediate horizons. Thus, the Black Sea system may be considered to be closed to the input of oxygen-rich water from the deeper levels. Therefore, at the lower boundary in the Black Sea model, a flux boundary condition (radiation condition) was implemented for all compounds (except the Mn^{2+} and sulfate ion):

$$\frac{\partial C_i^*}{dt} + Cf\frac{\partial C_i}{dz}|_\sigma = 0, \qquad (3)$$

where the "σ" symbol represents the boundary, C_i^* is the next time step concentration at the boundary (node outside the integration area), $\frac{\partial C_i}{dz}$ is the gradient at the boundary. The phase velocity C_f was calculated on the basis of the Orlanski approximation:

$$Cf = -\frac{\partial C_i}{\partial t} / \frac{\partial C_i^*}{\partial z}, \qquad (4)$$

where $\frac{\partial C_i}{\partial t}$ is the change on the adjacent-to-boundary node, C_i^* is the previous step concentration at the boundary (the node outside the zone of integration).

The implication of this radiation condition is that the value at the lower boundary is determined by parameter concentration changes within the integrated area. Therefore concentrations at the lower boundary in the Black Sea model are consequently formed only by the processes that occur within the zone of integration.

For sulfate at the lower boundary, a boundary condition of the first order was given, because model processes do not essentially change the content of this compartment. The constant value was accepted also for the Mn^{2+}(Mn^{2+}=5.3 μM)

For all sulfur compounds, with the exception of sulfates, zero concentrations were given as initial conditions; for nitrogen compounds and oxygen, constant values at all depths corresponding to the upper boundary values were given.

Therefore, the calculations were started with initial conditions characteristic of an "oxic" ocean, without any anoxia even at the lower boundary.

3.3. APPLICATION

The model was calculated for the layer 50-150 m with vertical resolution of 2 m. The model equations (1) were integrated using a first-order Eurelian method with time step of 0.1 day for at least 1 year after the stable solution was reached (Figure 2).

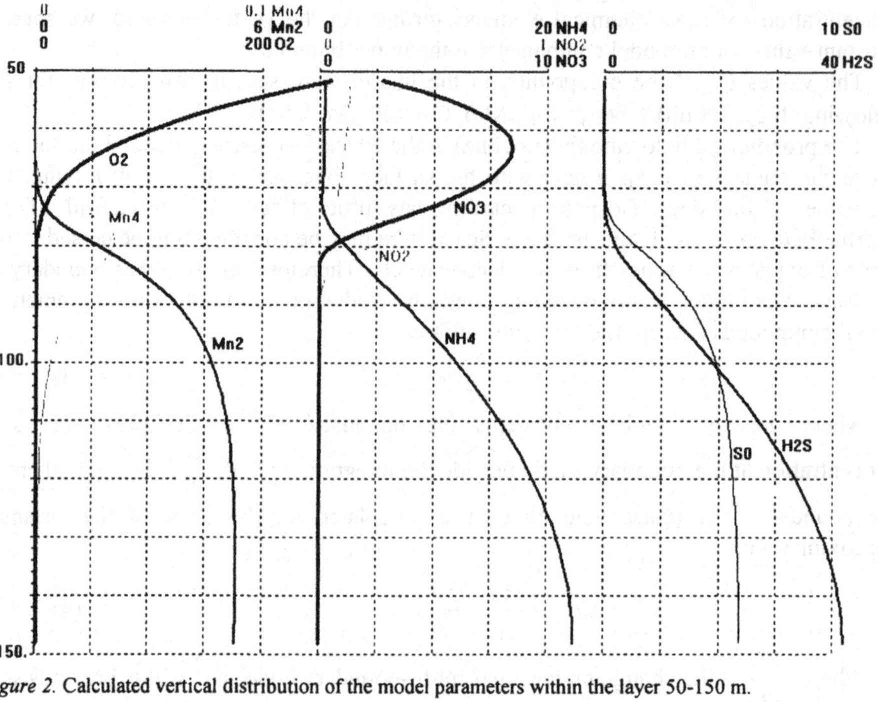

Figure 2. Calculated vertical distribution of the model parameters within the layer 50-150 m.

4. 2D section distribution modeling

4.1. EQUATION

For the numerical experiments, a vertical section (200 m X 100 km) was considered. For the hydrophysical scenario description a 2D equation, was used.

$$\frac{\partial C_i}{\partial t} + u\frac{\partial C_i}{\partial z} + w\frac{\partial C_i}{\partial z} + W_s\frac{\partial C_i}{\partial z} = \frac{\partial}{\partial x}A_l\frac{\partial C_i}{\partial x} + \frac{\partial}{\partial z}A_z\frac{\partial C_i}{\partial z} + R_{C_i} \quad (5)$$

where u,w - advective velocity components and A_l - horizontal turbulent diffusion coefficient.

The advection components were calculated on the base of a simplified flow function field (Figure 3) which permits parameterization of a horizontal flow with a zero vertical velocity and then a region of local upwelling with an intensity of 3 10^{-4} cm/s with a corresponding downwelling region with the same intensity.

4.2. BOUNDARY CONDITIONS

At the upper and lower boundaries the conditions described in 3.2 were accepted. At the left and right boundaries the condition of an absence of flux was assumed.

4.3. APPLICATION

The model equations (1) were integrated using a first-order Eulerian method with time step of 0.1 day for at least 1 year beyond the point at which the stable solution was reached.

5. Results

5.1. 1D MODELING

The results of vertical distribution 1D modelling are presented Figure 2. They reflect the main features of the vertical structure of compounds in the aphotic layer. One can see the decrease of oxygen concentrations from about 300 μM at the upper boundary to zero at a depth about 100 m.

The vertical profiles of nitrogen compounds (N_{org}, NH_4^+, NO_2^-, NO_3^-) calculated in the model reflect the main features of the distributions of these compounds observed in nature [16].

In upper level oxic conditions, where ammonification is the dominant process, one can see the maximum nitrate concentration. Below this gives way to a small nitrite maximum with increased ammonia concentrations. Concentrations of ammonia in deep waters correspond to those actually observed [16].

H_2S profile also reflects the observed situation of an increasing concentration of this parameter with depth [23]. Elemental sulphur values increase near the depth at which hydrogen sulphide appears. Then the distribution of this parameter is uniform. This is a consequence of the model assumption that S^0 forms as an intermediate compound in sulphide oxidation processes which take place near the upper boundary of the anoxic zone. In the layers below the are no sources or sinks for this parameter. Thiosulfates are intermediate products of both the sulfate reduction and sulfide oxidation processes.

Manganese compound profiles derived in the model also reflect the main features of this parameter's vertical distribution in nature. Dissolved manganese is oxidised by oxygen and the particulate manganese forms. The particulate manganese reacts with hydrogen sulphide and is transformed into dissolved manganese. A dissolved manganese maximum appeared in the model experiments when the oxygen supply increased and both processes intensified.

A significant feature of the model is the increase of organic matter in the contact zone (Figure 2) resulting from the assimilation of ammonia during bacteria mediated processes of matter transformation. This corresponds to empirical data [18] and confirms the view that the nature of the NRL is connected primarily with intensive bacteria activity at this depths [23].

Compared with the previous version of this model [25], which didn't consider manganese, this version illustrates significant differences of H_2S depletion and NH^{4+} depletion with depth. This relates to the fact that H_2S can be oxidised by oxygen and by particulate manganese, while ammonia can only be oxidised by oxygen.

Nevertheless within the framework of the parameterization of processes used in this model Mn can not be the sole or even the dominant oxidiser of H_2S.

According to model, the H_2S depletion point is characterised by the following concentrations: H_2S = 1.89 μM, O_2= 1.84 μM, Mn4=0.035 μM, NO_3=0.78 μM. Hence Mn can oxidise only a small fraction of the hydrogen sulphide. It can therefore be concluded that the model supports the idea of Tebo [19] that Mn cycle processes "help to maintain a broad suboxic zone", because this metal is being used as an electron-transfer mediator between oxygen and sulfide [11]. Model concentrations of oxygen are formally in agreement with the opinion that in the suboxic zone, oxygen concentrations can vary from 2 to 10 μM [10]. But nevertheless the present model can not explain H_2S oxidation in the absence of oxygen.

The main difference between the model results and the empirical observations appears to be the distance between the oxycline depletion point and the H_2S depletion point (10 m in the model and about 30-40 m in nature). During iterations of the quantitative experiments it was possible to reduce this divergence, but in these cases the gradient of all the parameters in the redox layer was significantly smoother than that observed in nature. Similar smoothed curves or an absence of distance between O_2 and H_2S depletion points can be found in the results of other models [2, 4, 11] describing the contact between oxic and anoxic layers.

5.2. 2D MODELING

One of the objectives was to test the hypothesis that horizontal turbulence and advection influence the structure of the redox layer.

The 2D model presented is devoted to estimating the role of hydrophysical process oscillations on the hydrochemical structure of the zone of contact between the oxic and anoxic layers.

The goal of this study was also to assist in the analysis of data obtained in the Sakaria canyon region during the r/v "Bilim" cruise in July 1997 [3]. During this expedition a well defined NRL was observed at the stations with chemical structure anomalies: appearance of hydrogen sulphide at the isopycnal surface of 16.40 (deeper by up to 20 m compared with other regions) and high (up to 50 μM) concentrations of oxygen at the surface 16.20.

The results of the 2D model calculations are presented in Figure 3a, 3b.

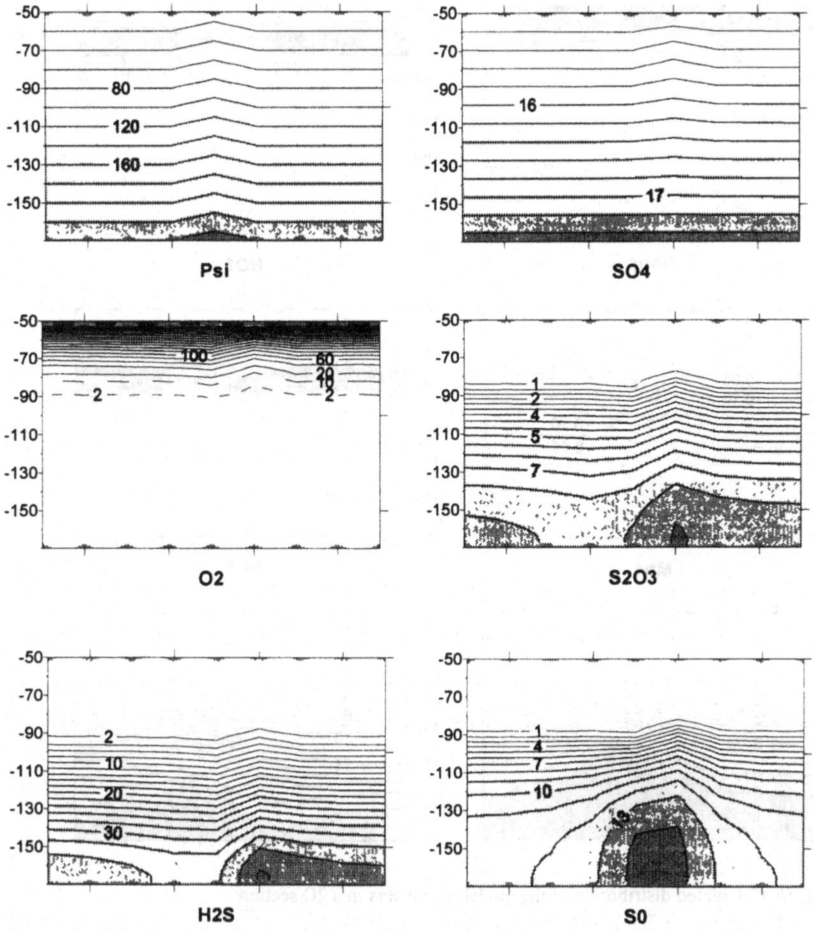

Figure 3a Calculated distribution of the model parameters in a section. Psi is the flow function (cm^2/s)

In Figure 3a the flow function field shows the existence of horizontal advection component with a velocity of 2 cm/s travelling from right to left and the presence of a local upwelling region followed by downwelling with a velocity of 3 10^{-4} cm/s.

In the model we obtained oxidation processes intensification in the upwelling region, where the particulate manganese formation took place. In the subsequent downwelling region this enriched particulate manganese is used together with oxygen for hydrogen sulphide oxidation. This continues to the depth of the hydrogen sulphide depletion point where elemental sulphur concentrations increase and where, according to the model assumption, the total organic nitrogen content also increases.

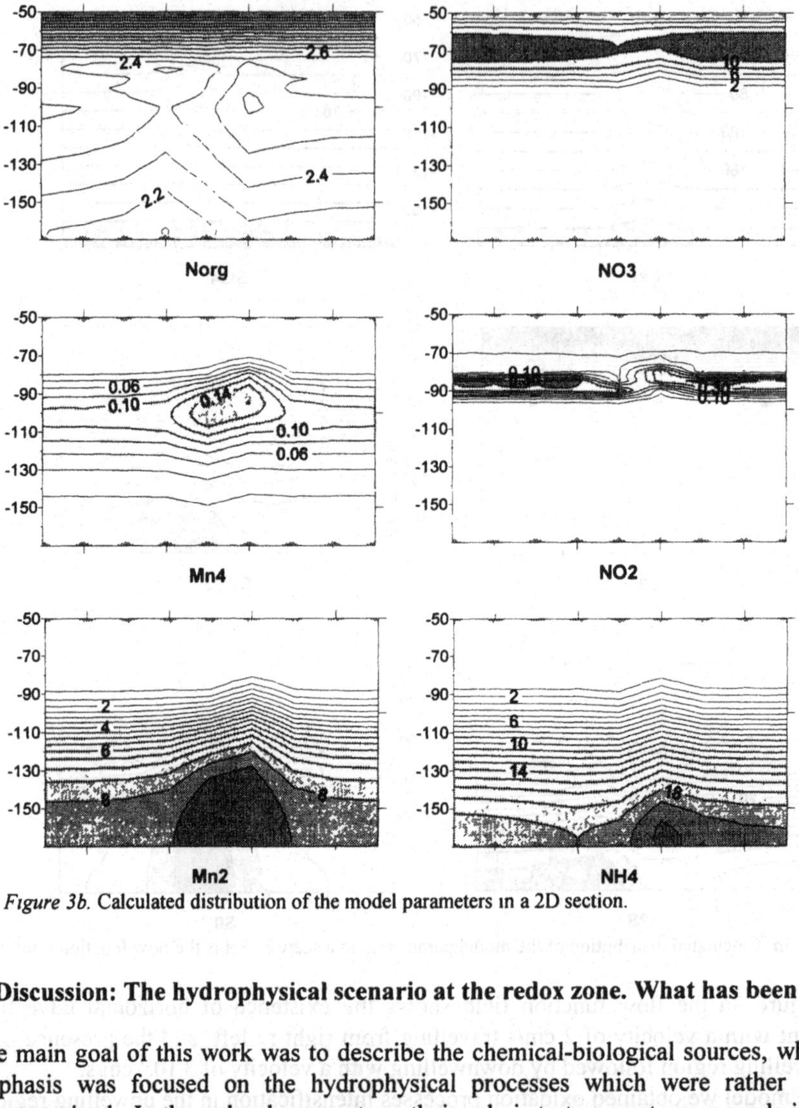

Figure 3b. Calculated distribution of the model parameters in a 2D section.

6. Discussion: The hydrophysical scenario at the redox zone. What has been lost?

The main goal of this work was to describe the chemical-biological sources, while less emphasis was focused on the hydrophysical processes which were rather roughly parameterized. In the redox layer system, their role is to transport compounds involved in the reactions. According to general opinion, the vertical exchange (primarily turbulent) is of major significance, and dominates in comparison with horizontal exchange. For this reason an attempt was made to analyse the vertical distribution of the second derivative as a basis for flux estimates.

Three surface layers were chosen, characterised by depletion layers of maximum gradients for the main compounds involved in redox cycles in the zone [10, 21], and hence the peaks of the second derivatives for compounds. In Table 3 the results are presented.

TABLE 3. Estimates of sinks/sources of the main oxidants and reductants connected with the vertical turbulent fluxes ($Az\ \partial^2 C_i/\partial z^2$, where Az=0.3cm^2/s) calculated for different layers of redox zone:

Level	Oxidants			Reductants		
	NO3	O2	Mn4	H2S	NH4	Mn2
15 70-15.85 Oxycline depletion Upper boundary of NRL	0	+11.6[1] +11.0	-0.001[2]	0	0	0
	+11 0			0		
16.05-16.10 Mn2 and NH4 depletion. Middle part of NRL	+0.7	0[1] 0	0	0	+0.49	+0.41
	+0.7			+0.9		
16 15-16.20 H2S depletion Lower part of NRL	0	0[1] +0.58	+0.001[2]	+1.84	0	0
	0[1] - +0.58			+1 84		

[1] - According to data from J Murray [9]
[2] - Data from B.L Lewis and V.M.Landing [7].
The latter fluxes were calculated on the base of I. Volkov's vertical gradients estimates [22]and our own data, obtained during the r/v Akvanavt cruise in November 1997.

As follows from Table 3 the imbalance between oxidant and reductant fluxes occurs at every depth. But nevertheless we face not only a deficit of oxidants in the lower part of the NRL (which is well known) but also a deficit of reductants in the upper part. Organic matter also can play the role of a "reductant", but there are no data indicating that it is characterised by significant gradients in this layer.

To explain this phenomenon a joint mechanism of chemical, biological and physical processes should be elaborated.

Up to now the main attention has been concentrated on the chemical mechanism. The possible chemical reactions which can take place in NRL have been described by a number of authors [9, 13, 21, 22].

The role of biological processes is not restricted to mediating redox chemical reactions. Intensive organic matter formation connected with bacteria biomass synthesis should affect the observed distribution of inorganic nitrogen compounds (first of all ammonia) and phosphates.

Both biological processes and chemical processes can be reflected in models as sources which balance the fluxes of compounds connected with hydrophysics.

The role of hydrophysical processes have not yet been seriously investigated. In models and the field data analysis [3, 7, 11, 25] investigators usually assume the vertical turbulent coefficient *Az* constant or describe its linear dependence on depth with the minimum at the pycnocline. It is assumed that there are no changes of *Az* in NRL.

Another process which can be included in the hydrophysical list is sedimentation of particulate matter. In previous models the rate of sedimentation is assumed to be constant with depth. This assumption implies that the density of particles remains constant with depth. However, the formation of particulate manganese concentrations at particular depths should affect density. According to observations, the particulate matter is represented mainly by organic detrital particles. In the layers of particulate manganese formation the share of particulate manganese can reach 11% [13] or 40% [7] of dry weight. Morphologically, one can see a thin film of particulate manganese which covers the detrital particles of different origin (Ivanov, 1997, personal communication). This indicates an increasing of density of particles and hence a corresponding increase in their sedimentation rate. After detrital particles descend to the lower part of NRL, the particulate manganese dissolves, the particles become lighter, and the change in density will cause a decrease in the rate of sedimentation and suspension or even floatation to higher levels. We can therefore suggest a mechanism to explain the additional mixing connected with manganese reduction/oxidation. The intensity of this mixing should be connected with the size of particles, concentration of the particulate matter, and the rate at which the manganese cycle processes occur. More detailed investigation of this process might help explain the different depths of layers with dominating oxidants and reductants shown in Table 3. The further development of the redox layer models requires parameterization of this process.

7. Conclusion

The main goal of this model was to study the role of chemical-biological sources affecting the profiles of the compounds investigated.

Although this is a simplistic model, it is capable of reproducing the basic features of the redox zone structure. This correlation suggests that the theoretical knowledge generated by the approach taken in this model can explain the main features of the phenomena observed.

Modelling of oxic/anoxic transformation requires parameterization of the cycles of several elements simultaneously, in contrast to models dealing only with nutrient cycles under oxic conditions.

Model estimations confirm that the existence of anoxic conditions is controlled primarily by the peculiarities of organic matter decay (a consequence of oxidant consumption) in conjunction with restricted aeration.

According to the model simulations, the most sensitive hydrochemical parameters from point of view of vertical advection anomalies are particulate manganese, organic matter and elemental sulphur. All these parameters influence transparency, and the transparency layer anomalies can result from the hydrophysical process peculiarities.

The model was calibrated using data observed for the vertical distribution of compounds in the upper layers of the Black Sea. The results obtained can be used to describe the nitrogen, sulphur and manganese cycles in other natural aquatic ecosystems where anoxic environments are present or possible.

The results of work undertaken so far suggests that future development of the model should concentrate on the improvement of parameterization of peculiarities of sedimentation connected with manganese cycle, organic matter balance, and the use of hydrophysical model results for the description of advection and diffusion.

Acknowledgements

This work was conducted with the financial support of the Russian Foundation for Basic Research, grants 96-05-66169 and 96-05-65134. The author would like to thank James Murray and Igor Volkov for thought-provoking discussions on the nature of the oxic/anoxic interface and acknowledge the assistance of Andrew Reed in improving the clarity of the paper.

References

1. Anderson, J.J., Okubo, A., Robbins, A S , and Richards, F A , (1982) A model for nitrite and nitrate distributions in oceanic oxygen minimum zones, *Deep Sea Res* **29**, 1113-1140.
2. Ayzatullin, T.A , and Leonov, A V. (1975) Kinetics and mechanism of oxidative transformation of inorganic sulfur compounds in sea water, *Okeanologiya* **15**, 1026-1033, (Russian).
3 Basturk, O., Volkov, I.I , Gokmen, S., Gungor, H , Romanov, A.S, and Yakushev, E V (1998, in press) International expedition on r/v "Bilim" in July 1997 in the Black Sea, *Oceanology 2*.
4. Belyaev V.I., E.E Sovga, and Lyubartseva S P (1997) Modeling the hydrogen sulfide zone of the Black Sea, *Ecological Modelling* **96**, 51-59.
5. Bezborodov, A.A., and Eremeev V.N (1993), *Chernoe more Zona vzaimodeystviya aerobnych i anaerobnych vod [Black Sea. The oxic/anoxic interface]*, AS of the Ukraine, the Marine Hydrophysical Institute, Sevastopol, 299,(Russian, English summary)
6 Jorgensen, B B (1989) Biogeochemistry of chemoautotrophic bacteria, in H G Shlegel and B. Bowien B. (eds.), *Autotrophic Bacteria*, Sci. Tech. Publ & Springer-Verlag, Madison, pp 117-146.
7. Kondrat'eva, E N. (1983) *Chemolitotrofy i metilotrofy (Chemolithotrophs and methilotrophs)*, Moscow State University, Moscow, 172 pp.
8 Lewis, B.L., and Landing, W.M.(1991) The biogeochemistry of manganese and iron in the Black Sea, *Deep Sea Res* **38(2A)**, S773-S803
9 Murray, J W., Codispoti, L.A , and Friederich, G E. (1995) The suboxic zone in the Black Sea, in C.P.Huang, R O'Melia and J J.Morgan (eds) *Aquatic chemistry interfacial and interspecies processes*, American Chemical Society, pp 157-176
10. Murray, J.W., Jannasch, H W , Honjo, S., et al., (1989) Unexpected changes in the oxic/anoxic interface in the Black Sea, *Nature* **338**, 411-413
11 Nealson, K.H , Myers, C R., and Wimpee, B B. (1991) Isolation and identification of manganese-reducing bacteria and estimates of microbial Mn(IV)-reducing potential in the Black Sea, *Deep Sea Res* **38**, S907-S920
12 Oguz, T , Ducklow, H , Malanotte-Rizzoli, P., Tugrul, S , Nezlin, N.P., and Unluata, U. (1996) Simulation of annual plankton productivity cycle in the Black sea by a one -dimensional physical-biological model. *J. Geophys. Res* **101**, 16585-16599
13 Richards, F.A., (1965), Anoxic basins and fjords, in J P Riley and G.Skirrow (eds), *Chemical Oceanography, Vol.1*, Acad Press, New York, pp.611-645
14. Rozanov A G. (1995) Redox stratification of the Black Sea water, *Oceanology* **35**, *4*, 544-549 (Russian)
15 Sergeev, Yu N. (ed.), (1979), *Modelirovaniye perenosa i transformacii veschestv v more [Modeling of transport and transformation of substances in sea]*, LSU, Leningrad, 296 pp., (Russian).
16 Skopintsev, B A , (1975), *Formirovaniye sovremennogo chimicheskogo sostava Chernogo morya [Formation of the recent chemical composition of the Black Sea]*, Gidrometeoizdat, Leningrad, 336 pp.(Russian).
17 Sorokin, Yu.I., Sorokin, D.Yu , and Avdeev, V A (1991) Aktivnost' microflory i okislitel'nye processy sernogo cycla v tolsche vody Chernogo morya [Microbial activity and sulfur cycle oxidation processes in the Black Sea water column], in M E Vinogradov (ed), *Izmenchivost' ecosystemy Chernogo morya (estesstvennye i antropogennyye faktory)*, Nauka, Moscow, pp.173-188 (Russian).
18. Sorokin, Yu.I., Sorokin, P Yu , and Sorokina O V (1992) Raspredeleniye i funkcional'naya aktivnost' microflory v tolsche vody Chernogo morya zimoy i v nachale vesny 1991 g [Distribution and functional microflora activity in the Black Sea water column during winter and beginning spring 1991], in

M.E.Vinogradov (ed.) *Zimnee sostoyaniye ecosystemy otkrytoy chasti Chernogo morya,* IO RAS, Moscow, pp. 89-102,(Russian).

19. Tebo, B M. (1991) Manganese (II) oxidation in the suboxic zone of the Black Sea, *Deep Sea Res.* **38**, S883-S906.
20. Van Eeckhout D., and Lancelot C (in press) Modeling of the functioning of the North-Western Black Sea ecosystem from 1960 to present, in *NATO Advanced Research Workshop on "Sensitivity of North Sea, Baltic Sea and Black Sea to anthropogenic and climatic changes" (14-18 November 1995)* NATO-ASI Series.
21 Volkov, Igor I, (1984) *Geochimiya sery v osadkach okeana [Sulfur geochemistry in ocean sediments],* Nauka, Moscow, 272 pp (Russian)
22 Volkov, I I., Kontar, E.A., Lukashev, Yu.F, Neretin, L.N, Nyffeler, F., Rozanov, A.G. (1997) Upper boundary of hydrogen sulfide: Implications for the nephrloid redox layer in waters of Caucasian Slope of the Black Sea, *Geochemistry International* **35(6)**, 540-550.
23. Volkov, I.I., Rozanov, A.G., Demidova, T.P. (1992) Soedininiya neorganicheskoy vosstanovlennoy sery i rasvorennyy marganets v vode Chernogo morya [Inorganic reduced sulfur compounds and dissolved manganese in the Black Sea water column], in M.E.Vinogradov (ed.) *Zimnee sostoyaniye ecosystemy otkrytoy chasti Chernogo morya,* IO RAS, Moscow, pp. 38-50. (Russian).
24 Yakushev, E.V (1992) Numerical modeling of transformation of nitrogen compounds in the redox zone of the Black Sea, *Oceanology* **32(2)**, 257-263.
25. Yakushev, E.V. and Neretin, L.N (1997) One-Dimensional Modeling of Nitrogen and Sulfur Cycles in the Aphotic Zones of the Black and Arabian Seas, *Global Biogeochemical Cycles* **11(3)**, 401-414

TEMPORAL (SEASONAL AND INTERANNUAL) CHANGES OF ECOSYSTEM OF THE OPEN WATERS OF THE BLACK SEA

M. E. VINOGRADOV, E. A. SHUSHKINA, A. S. MIKAELYAN,
N. P. NEZLIN
P. P. Shirshov Institute of Oceanology, Russian Academy of Sciences
36 Nakhimovskiy Avenue, Moscow, 117851, Russia

Abstract. The pelagic ecosystem of the open part of the Black Sea was analyzed from the point of view of its temporal changes on interannual and seasonal basis. The material was collected during interdisciplinary expeditions to the Black Sea, between 1978 - 1996. The observed interannual variability is discussed for all plankton groups except protozooplankton. During 1980 - 1993 a gradual decrease of mean air temperature in winter and an increase in phytoplankton biomass in summer were observed. The lowest and the highest phytoplankton biomasses corresponded to high and low temperatures in 1980 and 1992 respectively. The climatic quasi-periodic 20-years oscillations of winter air temperature determine the general intensity of Black Sea current system and, as a result, favorable conditions for growth of phytoplankton. The analysis of historical phytoplankton data corroborates this hypothesis. According to surface chlorophyll "a" satellite measurements the interannual variations were seen in winter and spring during the period from 1978 to 1986.

The intrusion of *Mnemiopsis leidyi* to the Black Sea in 1989 led to radical changes in the structure and functioning of the ecosystem. After its outbreak, the biomass of phytoplankton increased in summer-autumn; the abundance of bacterioplankton was higher in spring; protozoan biomass did not changed. The fodder mesozooplankton as well as jelly-fishes biomasses sharply decreased. Within the jellies group, a significant decrease in the *Aurelia* biomass was compensated partly by developing of *Mnemiopsis*. The role of different groups of plankton in the community also changed significantly. As an example, the percentage of gelatinous macroplankton increased from 10-20% to 72-78% of the total zooplankton biomass. The decrease of biomass of mesozooplankton resulted in sharply diminishing of catches of pelagic fish.

The characteristics of the plankton community in temperate regions exhibit regular seasonal oscillations of a relatively high range, the biomass values varying by factors of 5-10 during the year. Two different seasonal scenarios of phytoplankton succession (with winter or spring blooms) for open waters of the Black Sea were considered. The spring phytoplankton mass development starts in the upper mixed layer after the seasonal pycnocline appearance, while the winter bloom ends with the formation of the seasonal pycnocline.

S. Beşiktepe et al. (eds.),
***Environmental Degradation of the Black Sea: Challenges and Remedies*, 109–129.**

Due to the radical changes in the zooplankton communities, associated with the intrusion of *Mnemiopsis leidyi,* the seasonal pattern has changed. The seasonal variations of zooplankton indicate that the grazing pressure of *Mnemiopsis leidyi* is the principle factor that determines these changes. The variations in the biomasses of its potential preys were the most pronounced. Influence of *Mnemiopsis* towards *Aurelia aurita* was the competitive one. It appears that the typical pattern of *Aurelia* seasonal development has not changed, however, the absolute values of its biomass has decreased.

1. Introduction

The earlier investigations of the seasonal dynamics of the plankton communities in the Black Sea were all carried out in coastal regions [1, 9, 13, 35]. The seasonal changes of the different groups of organisms illustrate the typical pattern for temperate and subtropical areas [18]. The formation of seasonal thermocline was followed by the development of the spring bloom of phytoplankton. The subsequent steps of ecological succession are also typical for all pelagic communities: the increase of zooplankton biomass and decrease of the phytoplankton, till its rise again in late autumn. The mechanism of this pattern seems to be clear and has been well studied [8, 16]. In contrast to the shelf areas, data on seasonal changes of plankton communities in the deep sea waters are virtually absent. The few available observations show that due to the characteristics of the current system, the seasonal pattern of ecosystem in the deep areas could well be different from the coastal ones. For example, unusual blooms of phytoplankton have been observed during the winter in the temperature homogenous upper mixed layer [10, 23].

The interannual changes of the Black Sea ecosystem have received special attention during the last few years. These changes are associated with the drastic changes in the main ecosystem parameters, which occurred in the early 90s [34]. It is assumed that the main reason of the changes was the anthropogenic impact.

The anthropogenic impact on marine ecosystems is manifested in various forms: toxic contamination, controlled river flows which decrease inflows affecting the salinity and water stratification, eutrophication, i.e. overfertilization, of the sea with nutrients and dissolved organic substance, fisheries, and finally, accidental or deliberate introduction of new species. The anthropogenic impact is often distorted by natural fluctuations of environmental parameters such as water supply and intensity of circulation, with mixing and temperature contrasts influencing water stratification. However, it is obvious that during the recent 15-25 years anthropogenic effects have resulted in profound alterations of the Black Sea ecosystem in the north-western part of the Sea [34]. In the deep areas of the basin, the observed changes in the productivity level of plankton community which in turn led to changes in the optical and chemical properties, were recorded by many authors in the late 80s - early 90s [25, 33]. At the same time, the mass development of a newcomer ctenophore *Mnemiopsis leidyi* in the last few years has been followed by drastic drops in stocks of small herbivorous fish (particularly, brisling and anchovy) [15, 19, 28, 35]. Nevertheless, the role of the anthropogenic factor in these changes is not evident for the deep part of the Black Sea.

In the present paper we describe the patterns of seasonal and the interannual changes of ecosystem of the deep waters of the Black Sea.

2. Material and methods

The material was collected during interdisciplinary expeditions to the Black Sea. It covers the period from 1978 to 1996. For the present analysis, only stations with depths greater than 200 m were used. The distribution of stations is shown in Figure 1.

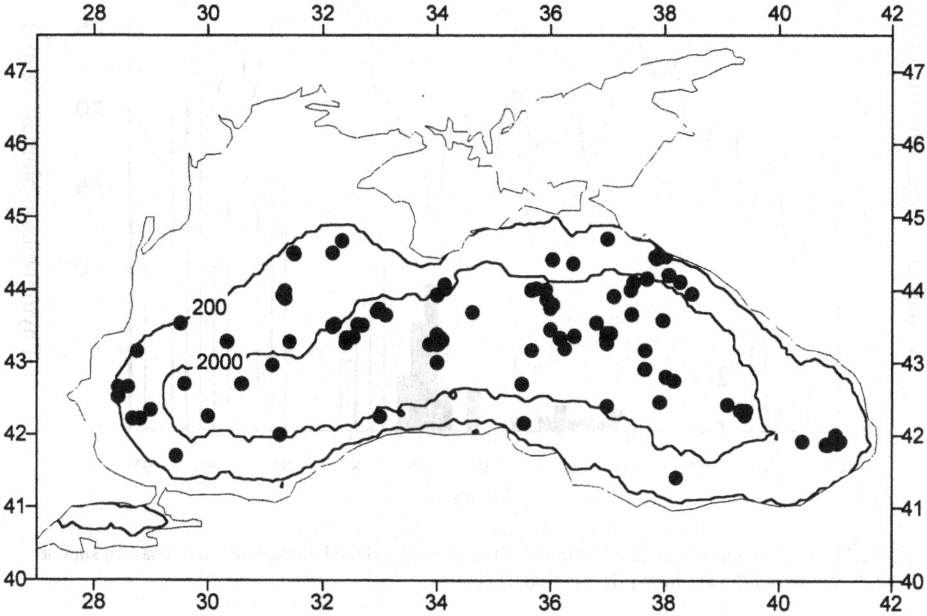

Figure 1. Map of stations of the cruises of P. P. Shirshov Institute of Oceanology RAS in the open part of the Black Sea (depth > 200 m).

At each station the following parameters were measured: vertical hydrological profiles (CTD soundings), nutrient contents, light penetration, biomass and the species composition of phytoplankton, bacterioplankton, ciliates, zooflagellates, mesozooplankton, macrozooplankton, fish eggs and larvae, the chlorophyll "a" concentration and the primary production. The detail description of methods used may be found in the literature cited in the following.

3. Results and discussion

3.1. INTERANNUAL VARIATIONS

The clearest indication of interannual changes is the pattern of fluctuations of phytoplankton biomass. The mean value of phytoplankton biomass in summer-autumn (June - October) changed several times during the period from 1978 to 1995 (Figure 2). These changes are inverse to oscillations in air temperature in winter (Figure 2, curve 1).

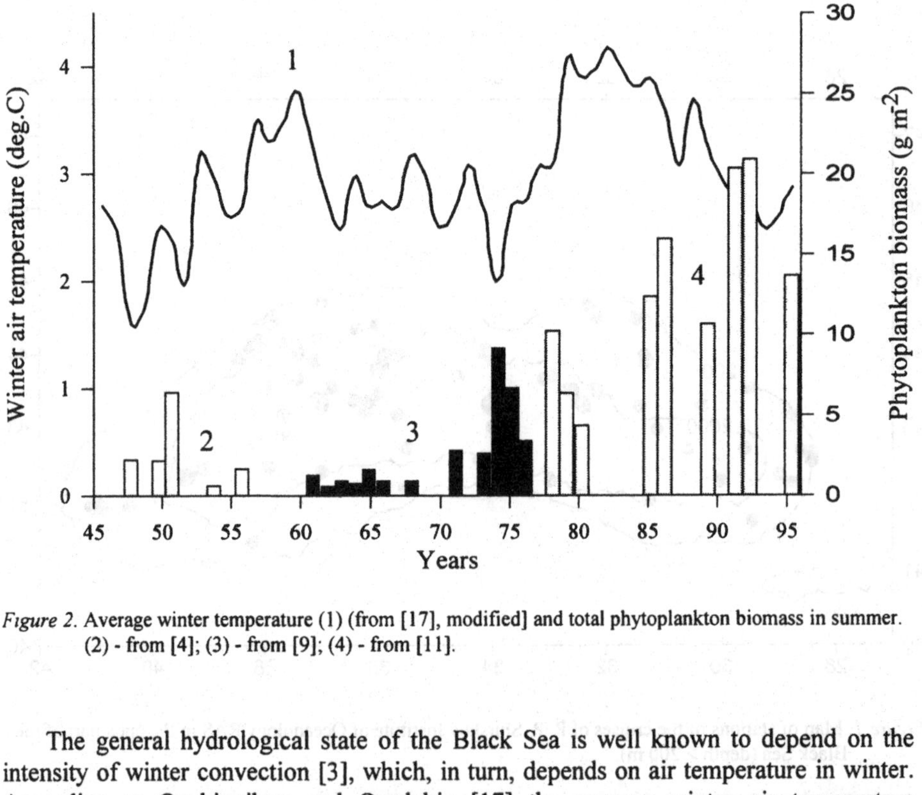

Figure 2. Average winter temperature (1) (from [17], modified] and total phytoplankton biomass in summer. (2) - from [4]; (3) - from [9]; (4) - from [11].

The general hydrological state of the Black Sea is well known to depend on the intensity of winter convection [3], which, in turn, depends on air temperature in winter. According to Ovchinnikov and Osadchiy [17] the average winter air temperature undergoes a quasi-periodic 20-years oscillation. During the coldest years of the observed period the low winter temperature causes intense water uplift in the centers of both gyres ("Black-Sea upwelling") and intensifies the circulation of the Rim Current. Such an intensification takes place both in winter and in summer. In summer this leads to more intensive penetration of the nutrients to the upper mixed layer in the center of the Sea, and, hence, creates favorable conditions for phytoplankton growth.

The variations in the phytoplankton biomass seem to corroborate this hypothesis. In the period from 1980 to 1993 a gradual decrease of mean air temperature in winter and an increase in phytoplankton biomass in summer were observed (Figure 2). The lowest and the highest phytoplankton biomasses corresponded to high and low temperatures in 1980 and 1992 respectively. The decrease in biomass from 1978 to 1980 and from 1992 to 1995 corresponded to inverse changes in air temperature. While the earlier data (2 and 3 in Figure 2) can not be compared directly with the more recent data, due to differences in methods of collection and treatment of phytoplankton samples, the position of peaks in the biomass data does indicate the inverse relationship between the phytoplankton biomass and air temperature. Data on phytoplankton biomass obtained in the eastern part of the sea from 1961 to 1976 [9] confirm the correspondence of low winter air temperatures and maximum of algae abundance in summer (Figure 2, curve 3). A similar pattern is also seen in the data for the period 1948 to 1956 [4]. Cold winter periods from 1948 to 1952 corresponded to highest summer-autumn phytoplankton biomasses (Figure 2, curve 2).

The next warm and cold extremes are expected in 1998-2000 and 2010-2012 respectively. Hence, the level of productivity of the sea observed in 1980 should be compared with 2000, and the same should be done for the 1990 and 2010 data.

The changes between the years could be expected also during the winter and spring seasons. Due to lack of the field observations during these periods, the data on surface chlorophyll "a" from the CZCS radiometer of the satellite "Nimbus-7" were used [2]. During 1978 to 1986 large variations in surface chlorophyll concentration were seen in winter: October-December (10 times), January - February (5 times) (Figure 3, curves 1 and 2). The spring bloom period (March-April) also shows the wide range in surface pigment content (Figure 3, curves 1 and 2). It is interesting that in contrast to summer phytoplankton biomass, the highest chlorophyll values corresponded to the most warm winters.

The interannual changes are seen not only in biomass but in size and taxonomic structure of phytoplankton (Figure 4). These changes are illustrated by the percentage of picophytoplankton (small algae with cell size under 2 μm), nanophytoplankton (algae with cell size from 2 to 15 μm), and microphytoplankton (over 15 μm) in the summer phytoplankton biomass. The taxonomic composition is presented by the percentage of two main taxonomic groups of algae: diatoms and dinoflagellates. It is evident that after 1991 the structure of phytoplankton community of the Black Sea essentially differs from that observed earlier. The role of large cells of diatoms significantly increased. It could be supposed that before 1991 these large cells of diatoms were grazed by mesozooplankton, but after 1991 the biomass of mesozooplankton decreased due to the influence of *Mnemiopsis leidyi* and the abundance of large diatoms increased. It will be recalled, that the total value of phytoplankton biomass varied under the influence of other factors (Figure 2), zooplankton grazing being not crucial.

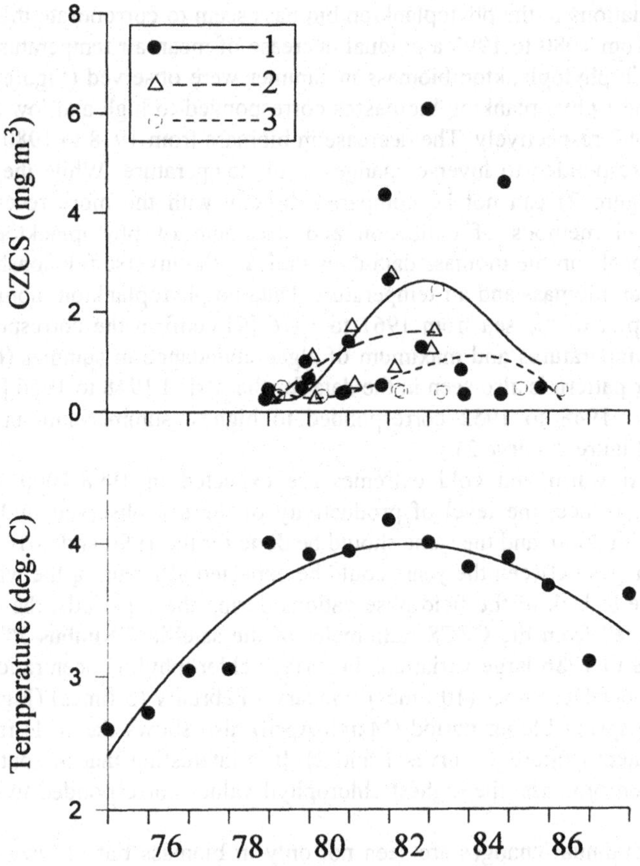

Figure 3. Interannual variation of CZCS-measured surface chlorophyll *a* concentration (Y-axis, mg m^{-3}) during cold season (October-April) in the open part of the Black Sea (depth >200 m) during 1978-1986. (1) - October-December, (2) - January-February, (3) - March-April The values of mean winter air temperature (degrees C) are also given.

Variations of summer chlorophyll "a" and in the level of primary production were observed from the 1960's to early 1990's [25] (Figure 5). During this period both parameters varied *ca* 3 times. The significant increase in chlorophyll "a" concentration in the early 90's is evident and corresponds well with the changes in phytoplankton biomass (Figure 2). The sharp peak in primary production observed in 1986 (Figure 5) is not in coincidence with phytoplankton biomass and chlorophyll "a" dynamics as well as with the changes in optical water transparency. The latter showed the sharp decrease in disk Secchi transparency since 1990 [33]. Unfortunately, it is impossible to compare earlier peaks of phytoplankton biomass with chlorophyll "a" concentration and with primary production changes. These parameters were not measured during the phytoplankton peaks of 1951 and 1974.

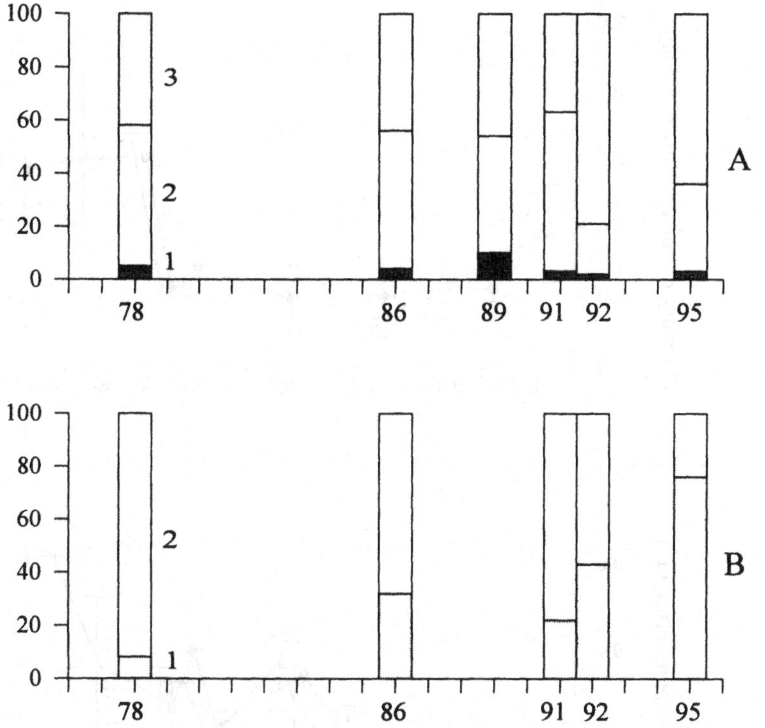

Figure 4. Changes in (A) percentage (%) of 3 size groups ((1) - pico-, (2) - nano-, and (3) - microplankton) in the total phytoplankyon biomass, (B) percentage (%) of biomass of (1) diatoms and (2) dinoflagellates

However the recent changes in phytoplankton community are not as drastic as in the zooplankton community. The latter resulted from the development of newcomer ctenophore *Mnemiopsis leidyi* in late 1980s. This species is a voracious predator [24]. It is believed that *Mnemiopsis leidyi* reached the Black Sea in the ballast water of ships cruising between the western North Atlantic and the Black Sea.

The mass development of *Mnemiopsis leidyi* in the Black Sea started in 1987 (Figure 6) and, at first, covered the bays, gulfs and coastal waters. Since the spring of 1988, its juveniles were encountered in all open sea areas and in the autumn of that year its biomass reached 1.5 kg m^{-2} [31]. During the summer of 1989, the biomass of *Mnemiopsis* considerably grew and its total value for the whole sea attained 1 Gt [21]. In 1990, the development of *Mnemiopsis* remained at the same rate (Figure6). Its abundance reached 10-12 kg m^{-2} at several coastal areas (Anapa, the south-western Bulgarian coast), but did not exceed 1.5-3 kg m^{-2} in the open sea.

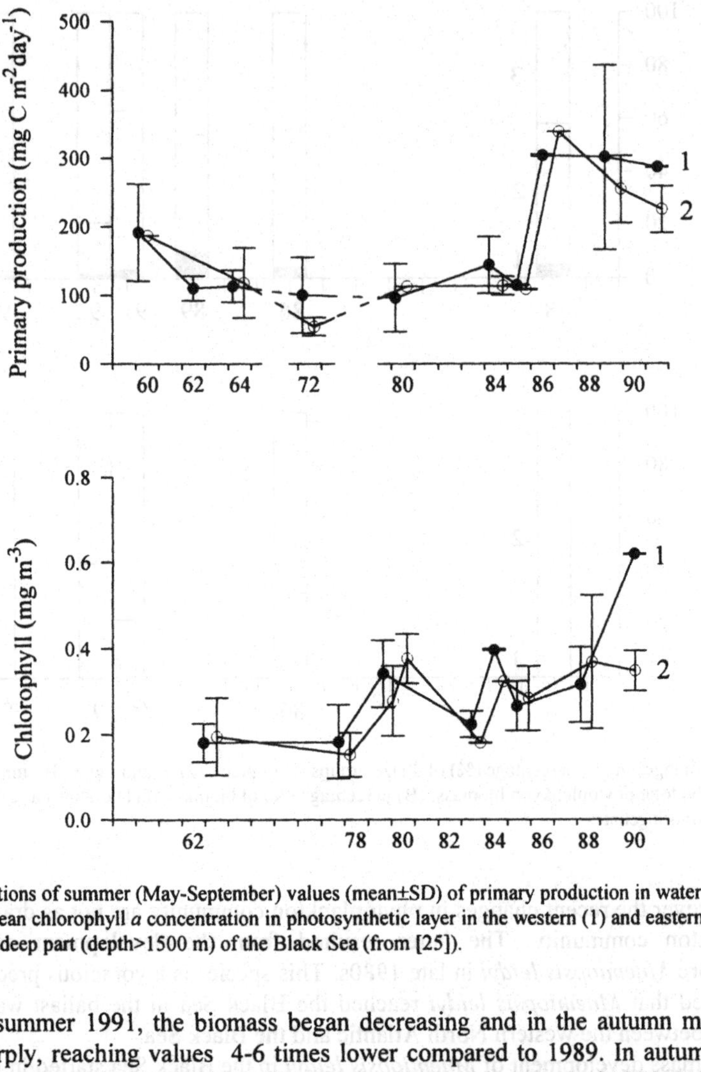

Figure 5. Variations of summer (May-September) values (mean±SD) of primary production in water column and mean chlorophyll *a* concentration in photosynthetic layer in the western (1) and eastern (2) parts of the deep part (depth>1500 m) of the Black Sea (from [25]).

During summer 1991, the biomass began decreasing and in the autumn months it dropped sharply, reaching values 4-6 times lower compared to 1989. In autumn 1992 (September-October) the *Mnemiopsis* biomass remained at the level of 1991. It is likely that the ctenophore development passed its peak at the end of 1989 and in 1990, and its place in community will be more modest in the future.

Another species which determined the level of community biomass before the *Mnemiopsis* appearance, was jellyfish *Aurelia aurita.* Its abundance sharply increased in the 1970s and exceeded the estimates of the 1950s - early 1960s by more than two orders of magnitude. The jellyfish probably occupied the ecological-trophic niche that became vacant as a result of abrupt decrease in the stock of plankton-feeding fish. It can be considered as an indicator of profound changes in the Black Sea ecosystem in the 1970s. Jellyfish biomass varied slightly since the late 1970s and its autumn peak was on average *ca* 1 kg m^{-2} in the open sea. A sharp drop in its abundance followed the mass development of *M. leidyi*. Changes in the biomass of *A. aurita* from 1978 to 1991 are shown in Figure 6. Against the background of seasonal variations, a sharp decrease in their abundance can be clearly seen after summer of 1988. While the wet biomass of the jellyfish, averaged over the period of 1978-1988, was about 1 kg m^{-2} (*ca* 400 Mt for the whole sea area), in the summer of 1989, it decreased to 0.14 kg m^{-2} i.e. to about one-seventh of its former value and was as low as 60 Mt for the whole sea area. At the same time, changes in the size structure of *A. aurita* population were observed. The mean size of organisms decreased. It seems that the jellyfish were incapable of attaining larger sizes because of sharp food competition with *M. leidyi*. It is interesting also that the biomasses of *M. leidyi* and *A. aurita* during peaks of their outbreak, were approximately the same in terms of organic carbon (Figure 6), although their wet weights differed by several orders of magnitude

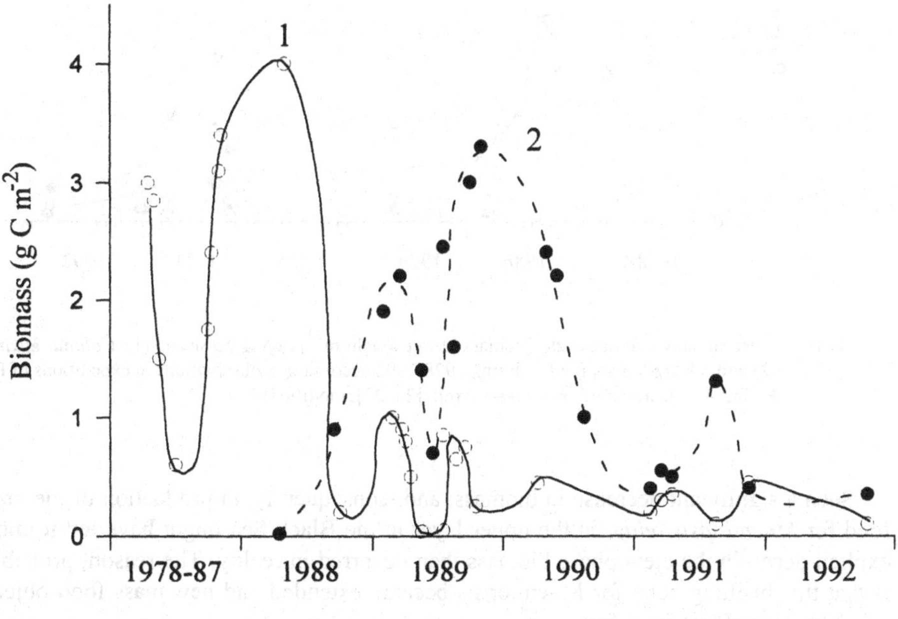

Figure 6. **Interannual variations of the biomass (gC m^{-2}) of *Aurelia aurita* (1) and *Mnemiopsis leydyi* (2) during 1978-1992 according to observations in expeditions of P. P. Shirshov Institute of Oceanology (from [22, 27], modified).**

The taxonomic structure of zooplankton community has been deformed. The species which are permanent inhabitants of the upper mixed layer, the initial basic biotope of *M. leidyi*, and those regularly moving up there prove to be easily accessible, whereas the species remaining in the deep water layers (*C. euxinus, P. elongatus*) were biotopically isolated from *M. leidyi* and, therefore, were less exposed to grazing pressure.

Due to this reason, the changes in zooplankton communities, first of all, concerned the mesozooplankton inhabiting the layer over intermediate cold waters [20, 22]. In

1989 its abundance declined in 2-2.5 times on average, as compared to the previous period. Biomass of some species and groups decreased 3-10 times or even more (Figure 7).

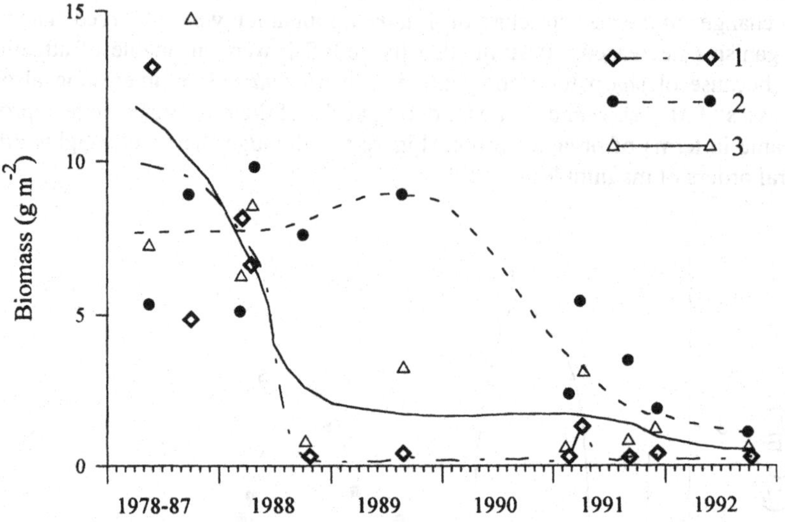

Figure 7 Interannual variations of the biomass (g wet weight m^{-2}) of *Sagitta setosa* (1), *Calanus euxinus* (2) and other zooplankton (3) during 1978-1992 according to observations in expeditions of P. P Shirshov Institute of Oceanology (from [22, 27], modified).

Such a significant decrease in biomass, and, consequently, in production of the main food for *Mnemiopsis leidyi* in the upper layer in the Black Sea might have led to more explicit decay in the ctenophore biomass than occurred in reality. The reason, probably, is that the dwelling zone for Mnemiopsis became extended and new mass food objects were introduced into its ration.

During the first period of development *Mnemiopsis leidyi* was strictly limited to the upper mixed layer over the thermocline and its assemblages often occurred just under the surface. However, in autumn 1989, we observed from the submersible "Osmotr" large-size ctenophore individuals in the under-thermocline layer as well. During the observations from submersible "Argus" in spring 1991 the penetration of the bulk of ctenophore population below the upper boundary of pycnocline was recorded. During the summer periods of 1991 and 1992, unlike 1989, only a few individuals were encountered in the surface layer, while the whole population inhabited the thermocline layer and underneath. In the under-thermocline layer the ctenophore, probably, began to feed on the population of *Calanus euxinus*, which lifts up from the deep waters at night time. During this period, the biomass of *C. euxinus* exceeded that of other zooplankton which were less than in summer 1989 by factors of 2.5 - 3.5. At all stations in 1991 and 1992 the biomass of *C. euxinus* was less by 1.7-8.3 times as compared with the values observed at the same sampling sites in 1989.

The intrusion of *Mnemiopsis leidyi* led to radical changes in the structure and functioning of the ecosystem. Such a strong carnivore pressure caused the decrease in the population density of *C. euxinus*. The data obtained in summer and autumn 1991 and in autumn 1992 evidently corroborated this fact. The biomass of *C. euxinus* in 1991 and 1992 changed significantly. Evident differences could be seen between two periods: before and after the *Mnemiopsis* appearance (Table 1). After the outburst of Mnemiopsis the biomass of phytoplankton increased in summer-autumn; the abundance of bacterioplankton was higher in spring; protozoan biomass did not change. The fodder mesozooplankton as well as jelly-fishes biomasses sharply decreased. Within the jellies group the significant decay of *Aurelia* biomass was partly compensated by development of *Mnemiopsis.*

TABLE 1. Averaged biomasses (B±S.E., g C· m^{-2}) and percentages of total community biomass of the main elements of planktonic communities in the open part of the Black Sea (depth >1000 m) during different seasons before the bloom of newcomer ctenophore *Mnemiopsis leidyi* (1978-1988) and after the bloom (1989-1992).

	Winter (February)		Spring (March-April)		Summer-autumn (May-November)	
	before	after	before	after	before	after
Phytoplankton	-	5.06±1 36 (64%)	3.31±0.54 (36%)	4.02±0 54 (48%)	0.87±0.10 (12%)	1.72±0.15 (31%)
Bacteria	-	0.68±0.07 (9%)	0.54±0 04 (7%)	0.89±0.07 (11%)	0.60±0.07 (8%)	0.71±0 04 (13%)
Protozoa	-	0.22±0.04 (3%)	0.21±0.04 (3%)	0.22±0.03 (3%)	0.20±0.03 (3%)	0.23±0.03 (4%)
Mesozooplankton	-	0.25±0.04 (3%)	1.48±0.11 (16%)	0.56±0.18 (7%)	1.64±0.46 (22%)	0.68±0.10 (12%)
Jelly-fishes (*Pleurobrachia pileus* + *Aurelia aurita* + *Mnemiopsis leidyi*)	-	1 71±0 31 (22%)	3.46±0 60 (38%)	2 61±0.51 (31%)	4.04±0.28 (55%)	2.24±0.19 (40%)
Aurelia aurita	-	1 01±0.29 (13%)	2.98±0.54 (32%)	1 57±0.32 (19%)	3.67±0.22 (50%)	0.98±0.18 (18%)
Mnemiopsis leidyi	-	0.43±0.07 (5%)	0.00±0.00 (0%)	0.50±0.15 (6%)	0 00±0 00 (0%)	0 78±0.08 (14%)

The role of different groups of plankton in the community also significantly changed (Figures 8 and 9). Before the introduction of Mnemiopsis bacterioplankton contributed from 5 to 10% to the total zooplankton biomass. This rate was similar to this ratio in the ocean ecosystems. After outburst of Mnemiopsis the role of bacteria sharply increased (Figure 8). The role of gelatinous macroplankton increased as well. Earlier, they comprised from 10 to 20% of the total zooplankton biomass. After the development of *Mnemiopsis leidyi*, the portion of jelly organisms in the Black Sea zooplankton (including macroplankton) amounted to 72-78% in carbon content and more than 99% in the wet weight units. Thus, the thin aerobic layer of the Black Sea, the only dwelling place for all its inhabitants (except bacteria), turned to be conqured by mucilaginous zooplankton - the trophic deadlock in the food chains of the sea. It should be mentioned that in undisturbed marine ecosystems the percentage of jellies usually does not exceed 10%.

The elimination of fodder mesoplankton, which consumes microplankton (phytoplankton, bacteria, protozoa) removed its pressure from microplankton and its role in total plankton biomass. The changes of production of the main groups of the community were also important. After the outburst of Mnemiopsis the production of total mesoplankton decreased about two-fold, especially the fodder zooplankton (Figure 9).

The decrease of biomass of mesozooplankton resulted in sharp decline of catches of pelagic fish in the Black Sea (Figure 10). The dominant fish (about 70% of total catch) was anchovy (*Engraulis encrasiccolus*), the fish with the shortest life cycle. Immediately before *Mnemiopsis leidyi* mass development (1985-88) the total (for all countries of the Black Sea region) catch values varied from 360 to 530 thousand tons per year (average 435). After the bloom of *Mnemiopsis* the catch dropped to $161 \cdot 10^3$ tons. During 1990-1991 the abundance of ctenophore decreased, but the catches continued to decrease as well.

The life duration of the horse mackerel (*Trachurus mediterraneus*) is longer. Thus the generation of 1985-1988 continued to share in the catch during 1989. The slump in catches occurred in 1990-1991 [29]. In the northern part of the sea the decrease of catches occurred in 1986 (Figure 10) and was not connected with *Mnemiopsis*, but in 1988-1989 the catches of horse mackerel decreased 10 times (from 2.5 to 0.3-0.4 thousand tons). In 1990-1991 this species disappeared from the official statistical reports, and appeared again only in 1992, when the biomass of *Mnemiopsis* started to decrease.

At the beginning of mass development population of *Mnemiopsis* was rather strictly confined to warm upper mixed layer, but as early as 1990 its main aggregations were concentrated in the thermocline while some specimens penetrated even deeper. Hence the *Mnemiopsis* population came in contact with the population of *Calanus euxinus*, which resulted in sharp decrease in its abundance and biomass [32]. Hence the food resources of sprat (*Sprattus sprattus*) were undermined. The catches of this fish started to decrease in 1990. The total catch in the basin was maximum in 1987-1989 (66-105 thousand tons). In 1991 it was as low as 16.9 thousand tons. In 1986-1988 the catch of USSR was 43-54 thousand tons. In 1989 the fleet tried to compensate the fall of anchovy catch by the intensified fishery of sprat. That year its catch rose to 89 thousand tons. However since 1990 the catch dropped; in 1991-1992 the catch values were as low as 15.0-14.8 thousand tons.

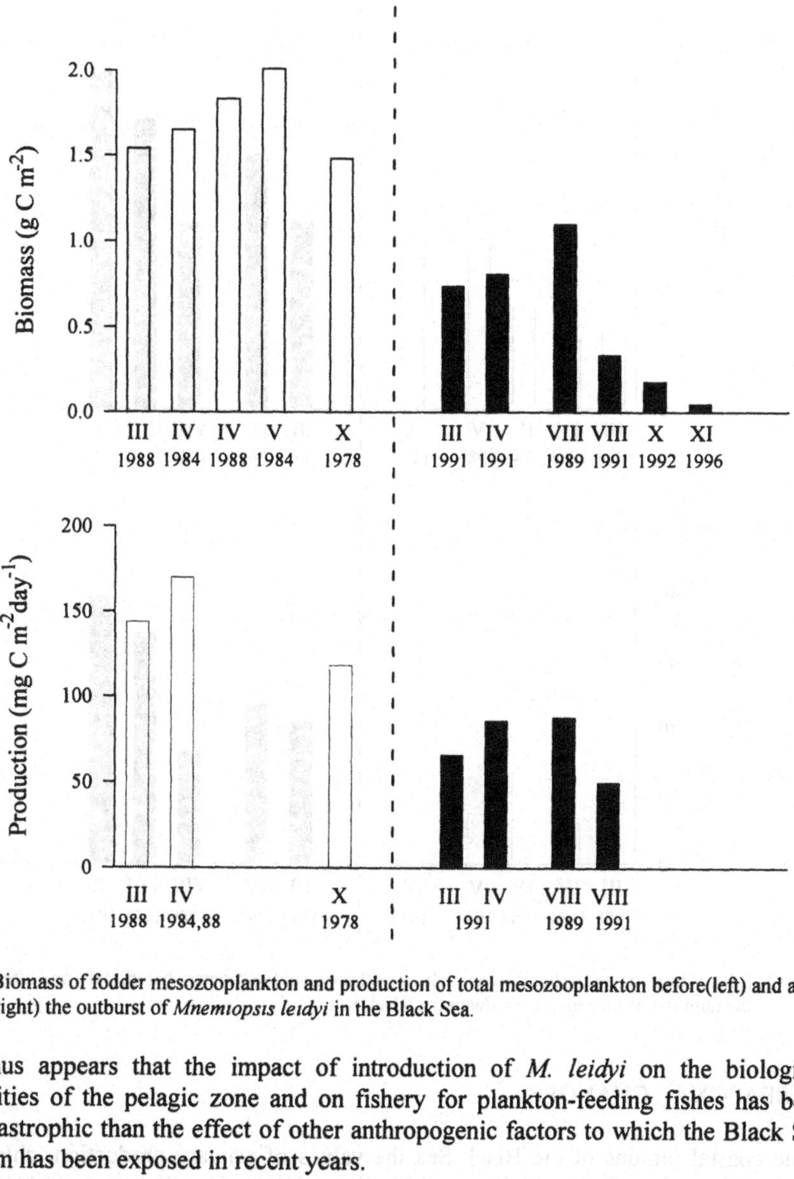

Figure 8 Biomass of fodder mesozooplankton and production of total mesozooplankton before(left) and after (right) the outburst of *Mnemiopsis leidyi* in the Black Sea.

It thus appears that the impact of introduction of *M. leidyi* on the biological communities of the pelagic zone and on fishery for plankton-feeding fishes has been more catastrophic than the effect of other anthropogenic factors to which the Black Sea ecosystem has been exposed in recent years.

Nevertheless, it seems impossible to separate the influence of climatic and anthropogenic factors on the Black Sea ecosystem in the early 1990's. At least, for low trophic level (phytoplankton, bacterioplankton and mesozooplankton) the influence of both climatic oscillations and *Mnemiopsis* development can be said to be equally important.

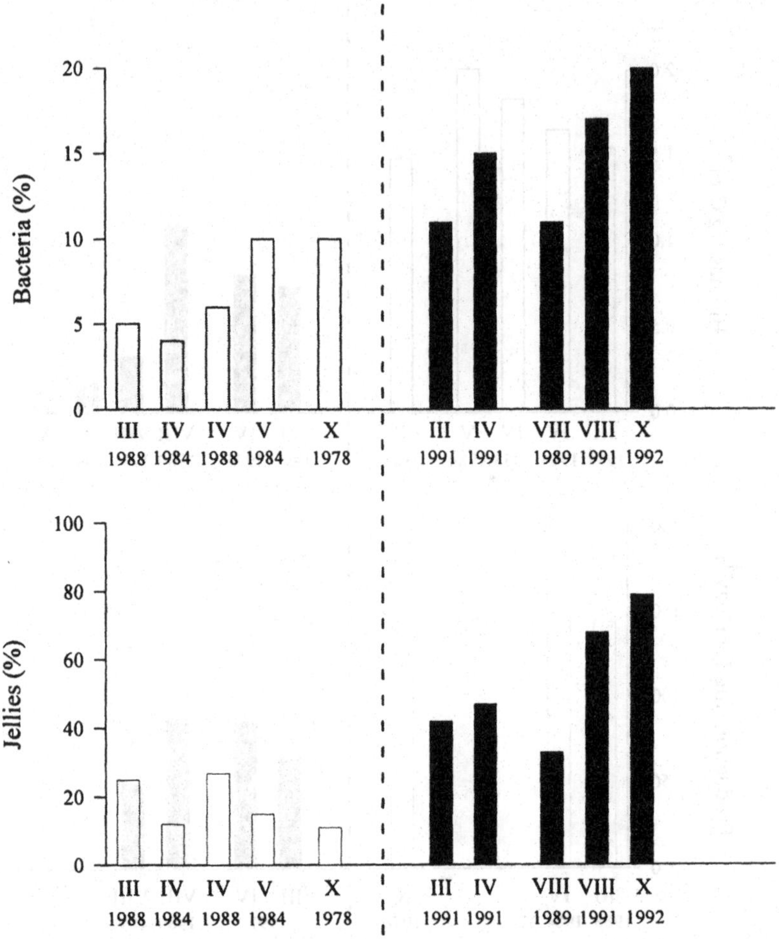

Figure 9. Role of bacteria and gelatinous animals in total zooplankton biomass before (left) and after (right) the outburst of *Mnemiopsis leidyi* in the Black Sea.

3.2. SEASONAL CHANGES

In the coastal regions of the Black Sea the values of primary production, chlorophyll concentration, phyto- and zooplankton biomasses change according to typical pattern of seasonal variations of temperate and subtropical areas [18]: spring bloom (March) of phytoplankton is followed by its decrease in summer due to depletion of nutrients. At the same time in the open regions the bloom of diatom phytoplankton occurs in winter (January-February). The reason is that during cold winter time with the absence of seasonal thermocline the main pycnocline in the centers of the cyclonic gyres rises as close to surface as 25-30 m. This phenomenon is called "Black-Sea upwelling" [27]. As a result the whole upper mixed layer apparently becomes located within the euphotic zone that favors primary production. This leads to the phytoplankton bloom. As an example, in 1988 and 1991 the biomass of diatom *Nitzschia delicatula* was observed as high as 120 g m^{-2} [10, 12].

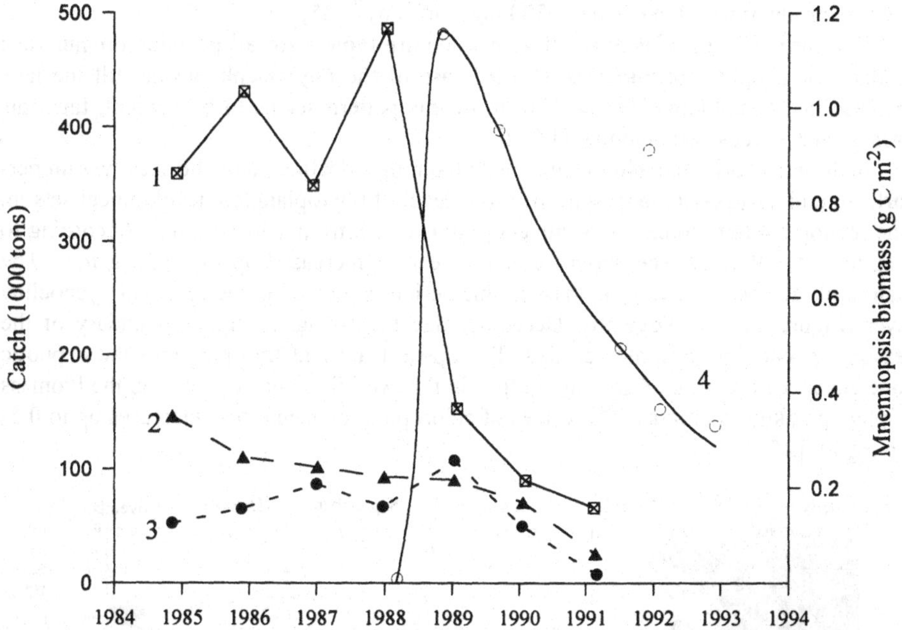

Figure 10 Total catchs of anchovy (1), sprat (2) and horse-mackerel (3) in the Black Sea, and mean biomass of *Mnemiopsis leidyi* during summer months (4) (from [29]).

It should be noted, that the lifting up of pycnocline does not take place every year. In normal winters the pycnocline is located at depths of 50-60 m, which prevents the development of phytoplankton. In these years, the seasonal pattern in the open waters with spring bloom in March is similar to that in the coastal areas. The spring phytoplankton mass development starts in the upper mixed layer after the seasonal pycnocline appearance. The winter bloom, on the other hand, ends when the seasonal pycnocline is formed. This processes is clearly seen in changes of vertical distribution of phytoplankton biomass (Figure 11). At the end of March during a few days cells of diatoms sink down to seasonal pycnocline, cleaning up the upper mixed layer [23].

Thus, two different seasonal scenarios (with winter or spring blooms) for open waters of the sea, should be considered. Unfortunately, the very scarce data for winter season do not permit to construct these patterns separately. Usually, the data have been averaged over many years, which results in mixing the two types of blooms .For example, the primary production percentage averaged for many years for the whole open regions of the Black Sea is distributed as follows: 35% - winter, 30% - spring, 16% - summer and 19% - autumn, the total winter-spring period being two-fold productive than the summer-autumn one [25]. During the year with the winter bloom, over 60% of total year primary production is created in winter [6].

During the winter or spring bloom, the areas of high average chlorophyll concentrations (0.4-2.7 mg m^{-3}) were spread over a large part of the open sea. The primary production value reached 1300 mg C m^{-2} day^{-1} [25].

The further steps of ecological succession are typical for all pelagic communities: increase of zooplankton biomass and decrease of the phytoplankton one, till the new increase in late autumn. The mechanism of this pattern seems to be evident, thus this process was successfully modeled [8, 16].

In March-April, the phytoplankton "blooming" declines and the steady summer-early autumn (April-October) state with low level of phytoplankton development sets in. The phytoplankton biomass in summer-autumn varies from 4 to 10 g m^{-2}. According to long-term oscillations (see above) in some years it increased up to 20-30 g m^{-2} . The main bulk of phytoplankton biomass in this period is located in the seasonal pycnocline layer (Figure 11). In November-December due to mixing the upper boundary of the seasonal pycnocline becomes eroded. The upward nutrient transport into the euphotic zone enhances favoring of growth of algae in this well illuminated zone and the biomass of phytoplankton increases. The values of chlorophyll concentration increases up to 0.5-1.6 mg m^{-3}[25].

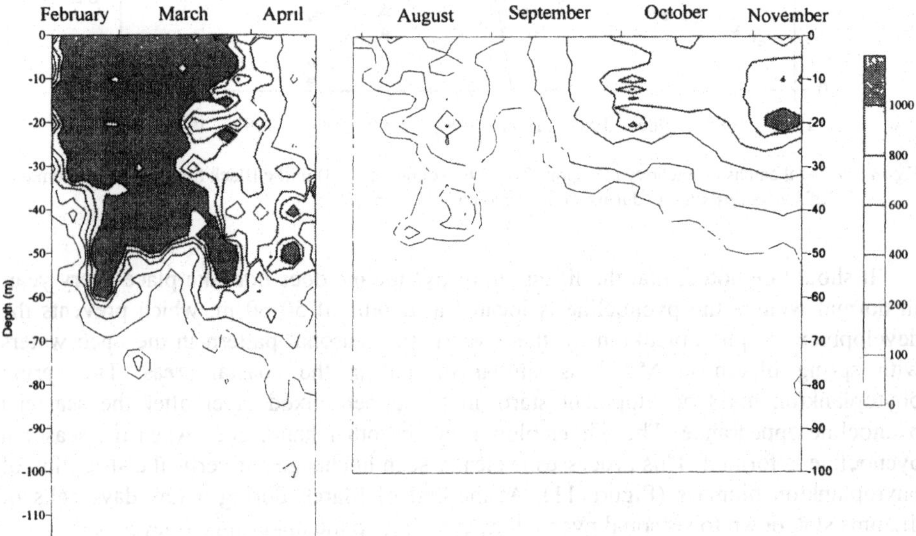

Figure 11. Seasonal variation of phytoplankton biomass(mg m^{-3}) in the open part of the Black Sea (depth >1000 m)

Due to the radical changes in the zooplankton communities, connected with the intrusion of *Mnemiopsis leidyi* (see above), the seasonal pattern today is different today than in the past. The averaging of long-term data for construction of seasonal zooplankton changes makes no sense. Instead, here we present the seasonal variability of zoocene structure that has been monitored during 1991 at five representative points in the open sea [28].

The seasonal trend in the biomass of the ctenophore *Mnemiopsis leidyi*, edificator of the open sea communities, is presented in Figure 12A. In winter (February) its quantity was minimum over the entire area and did not exceed 200-800 g m^{-2}. Most of its population was composed of medium-sized individuals. In spring (March-April), the biomass rapidly increased at the periphery of the gyres while it remained at the same level.in the centers. In summer, abundance of the population along with juvenile individuals number increased, which resulted in high total biomass. In autumn (November) the biomass of the population decreased to values typical for winter.

Fluctuations of biomass of jellyfish *Aurelia aurita* were synchronous over the entire sea area and had two peaks in spring (March-April) and autumn and two minima in winter and summer (Figure 12B). These changes were determined by biological cycles of this jellyfish. It has been analyzed and successfully modeled [7].

The seasonal changes of *Pleurobrachia pileus* (Figure 12C) were similar to that of *Mnemiopsis*. Biomass of *P.pileus* significantly increased from winter to summer and sharply decreased by late autumn. The increase of the biomass was the result of growth of animals. During winter the specimens of 5-10 mm size predominated, during summer the average diameter was twice higher (10-20 mm).

The seasonal trends in biomass of the species consumed by jelly carnivores seem to be determined by the trophic pressure of the latter. This is evident from the pattern of seasonal variations of total biomass of *Mnemiopsis* prey (Figure 12D). This group consists of small crustacean plankton (*Paracalanus, Acartia, Oithona, Pseudocalanus, Cladocera*), larvae of benthic animals and appendicularia (*Sagitta setosa* and *Calanus euxinus* which are also the potential food for ctenophores will be considered separately). The bulk of these animals consists of small-sized species-opportunists with rather short life cycle. These species are capable of increasing their biomass when the limiting factors slacken. Biomass of these species in winter was minimum over all the sea surface. They began reproducing, growing and increasing their biomass only during spring, after phytoplankton bloom, when the seasonal thermocline was established. At the same time a lot of larvae of benthic animals appeared in plankton. The total biomass of zooplankton that seems to be potential prey of *Mnemiopsis* grew rapidly reaching in some areas 9 g m^{-2}. The amounts of *Mnemiopsis, Aurelia* and *Pleurobrachia* increased concurrently, and the strong pressure of these carnivores led to decay of the biomass of their potential food by end of summer: in August it was as low as in winter. It is likely that this substantial decrease in food resources caused decay of *Mnemiopsis* biomass in the autumn. The autumn weakening of the carnivores' pressure caused the counter-phase increase in biomass of this group.

The affect of carnivores (*Mnemiopsis*) on seasonal variations of biomass of *Sagitta setosa* was also evident (Figure 12F). The early spring rise in *Sagitta* biomass was related to rapid growth of population consisting of large, mature specimens. Death of the large-size animals and consumption of juveniles by *Mnemiopsis* led to the sharp reduction of the population. Its biomass in summer declined to the values about 10 mg

m^{-2}. In contrast to small opportunist species, the population of *Sagitta* has slow growth rate; hence the abundance of its population failed to grow during autumn weakening of the pressure of carnivores, therefore the biomass increase of *Sagitta* was very low.

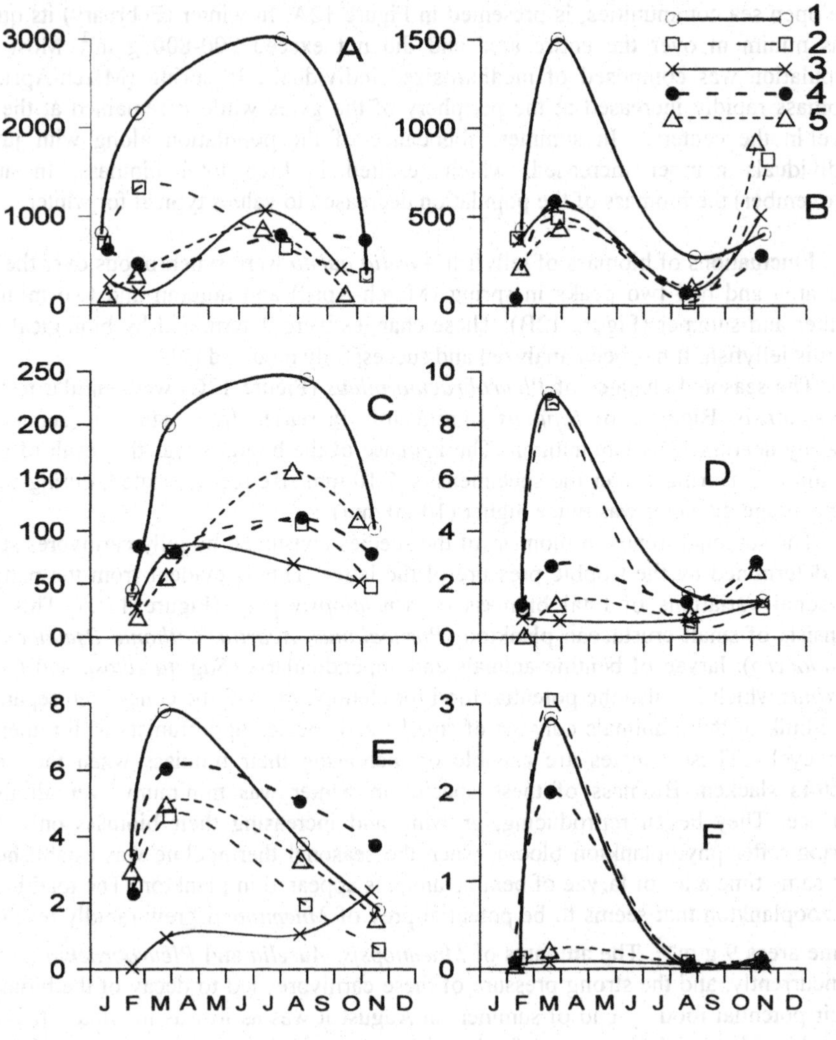

Figure 12 Seasonal variation of the biomasses (g m^{-2}) of main compounds of zooplankton during 1991 (from [28]) (A) - *Mnemiopsis leydyi*; (B) - *Aurelia aurita*; (C) - *Pleurobrachia pileus*; (D) - small zooplankton (prey of *Mnemiopsis*), (E) - *Calanus euxinus*; (F) - *Sagitta setosa*. (1) - south-eastern region; (2) - outer periphery of the Rim Current near Gelendzhik; (3) - eastern cyclonic gyre; (4) - convergence south of Yalta, (5) - western cyclonic gyre

It is interesting to note that in the 1960-80s before the introduction of *Mnemiopsis* the mesoplankton biomass reached maximum in summer months - July-August [5].

The seasonal variations in biomass of *Calanus euxinus* should be considered separately (Figure 12 E). The bulk of its population is associated with the under-thermocline layer and the upper main pycnocline waters [26]. Existence of the population in the open sea is considerably determined by replenishment from the shoals, primarily from the north-western shelf where mass reproduction occurs during late winter [30]. Thus, the seasonal cycle of this species in the Black Sea is determined by both temporal and spatial redistribution of the population.

The seasonal changes in biomass of *Calanus euxinus* in the open sea region, 90-95% of which consisted of V and VI, and less by IV copepodits were synchronous during the whole period of observations over the entire sea area. In winter the biomass of *C. euxinus* was low. In the open sea the amount of nauplii and early (I-II) copepodits was not high in March or during other seasons. It increased later in spring, obviously, due to migration of quickly growing juveniles from the shoals. Maximum in *C.euxinus* concentration was observed in March-April. Then it gradually decreased till November to the values lower than that at the beginning of the year. The reason of this decrease seems to be primarily the feeding off by *Mnemiopsis, Pleurobrachia*, and some pelagic fishes.

The described pattern of seasonal variations of zooplankton illustrates that the grazing pressure of *Mnemiopsis leidyi* is the principle factor that determines these changes. Thus, the variations of the biomasses of its potential preys were the most pronounced. Influence of *Mnemiopsis* towards *Aurelia aurita* was the competitive one. It apparently did not change the typical pattern of *Aurelia* seasonal development, but decreased its absolute biomass values. *Noctiluca miliaris* was not influenced by *Mnemiopsis*. *Pleurobrachia pileus* has the similar pattern of seasonal variations as *Mnemiopsis*. It seems to be strange because these two species are the competing ones. However the cores of the populations of these species are separated vertically: *Mnemiopsis* inhabits the upper mixed layer, and the *Pleurobrachia* occurs in the cold intermediate waters below the thermocline.

The seasonal variations in the deep part of the sea are the least known part of the temporal ecosystem variations. As was shown before, significant interannual variability occurs and the seasonal patterns differ between the years. Undoubtedly, it should be taken into account during the elaboration of the scheme of studying of the seasonal changes in the open waters of the Black Sea. Evidently, this study should be primarily based on the monitoring of the ecosystem within the same year.

References

1 Bodeanu N (1995) Algal blooms in Mammaia Bay (Romanian Black Sea coast). *in Harmful Marine Algal Blooms*, Lassus, P., Arzul, G. Denn, E., Gentien, P., Marcaillou-leBaut C. (eds), Lavoiseir Publ. Inc p. 127-132.

2. Feldman, G C., Curing, N., Ng, C , Esaia, W , McClain, C R., Elrod, J , Maynard, N., Edres, D., Evans, R., Brown, J., Walsh, S., Carle, M. and Podesta, G. (1989) Ocean colour availability of the global data set, *Eos Trans. Amer. Geophys. Union.* **70**, 634-641.

3. Filipov, D.M. (1968) *Circulation and structure of the Black Sea waters*, Nauka, Moscow. (In Russian)

4. Georgieva, L.V (1993) Species composition of phytoplankton communities, in A.V.Kovalev and Z.Z Fınenko (eds.), *Plankton of the Black Sea*, Naukova Dumka, Kiev, pp. 33-55. (In Russian)
5. Kovalev, A. V. (1991) *Structure of zooplankton communıtıes of Atlantıc and Mediterranean basın*, Naukova Dumka, Kiev, 140 p.
6. Krupatkina D.K., Finenko Z.Z., Shalapyonok A.A. (1991) Primary production and size-fractionated structure of the Black Sea phytoplankton in the winter-spring period. *Mar.Ecol. Progr. Ser.* 73, 25-31
7. Lebedeva, L. P. and Shushkina, E. A. (1991) Evaluation of population characteristics of the medusa *Aurelia aurıta* in the Black Sea, *Oceanology, English Translation*, **31** (3), 314-319.
8. Lebedeva, L. P and Shushkina, E. A. (1994) The model investigation of the Black Sea communıty changes caused by Mnemiopsis, *Oceanology, English Translatıon*, **34** (1), 79-87.
9. Mashtakova, G.P. and Roukhiyainen, M.P. (1979) Seasonal dynamics of phytoplankton, in V.N.Greese (ed.), *Productivıty of the Black Sea*, Naukova Dumka, Kiev, pp. 79-85. (In Russian)
10. Mikaelyan, A. S. (1995) Winter bloom of the diatom Nitzschia delicatula in the open waters of the Black sea, *Marıne Ecology - Progress Series*, **129**, 241-251.
11. Mikaelyan, A. S. (1996) Longtime variability in phytoplankton communities in the open waters of the Black Sea related to environmental changes, in E.Ozsoy and A.Mikaelyan (eds.), *Sensivıty of North Sea, Baltıc Sea and Black Sea to Anthropogenıc and Clımatıc Changes*, Kluwer Academic Publishers, Dordrecht.
12. Mikaelyan, A. S , Nesterova, D. A and Georgieva, L. V. (1992) Winter blooming of the algae Nitzschia delicatula in the open regions of the Black Sea, *Ecosystem of open Black Sea ın wınter*, Moscow, IO RAN, 58-72. (In Russian)
13. Moncheva S., Petrova-Karadjova V. and A.Palasov (1995) Harmful Algal blooms along the Bulgarian Black Sea coast and possible patterns of fish and zoobenthic mortalities, *ın Harmful Marıne Algal Blooms*, Lassus, P., Arzul, G. Denn, E., Gentien, P., Marcaıllou-leBaut C. (eds), Lavoiseir Publ. Inc. p. 193-198
14 Nezlin N.P (1997) Seasonal variation of surface pigment distribution in the Black Sea on CZCS data, in E.Ozsoy and A.Mikaelyan (eds.), *Sensıtıvity to Change: Black Sea, Baltic Sea and North Sea*, Kluwer Academic Publishers, Dordrecht, pp. 131-138.
15. Niermann U. and W.Greve (1997) Distrubution and fluctuation of dominant zooplankton species ın the southern Black Sea in comparison to the North and Baltic Sea, in E.Ozsoy and A.Mikaelyan (eds.), *Sensıtıvity to Change: Black Sea, Baltıc Sea and North Sea*, Kluwer Academic Publishers, Dordrecht, p. 65-78.
16. Oguz, T , Ducklow, H., Malanotte-Rizzoli, P , Tugrul, S., Nezlin, N. P. and Unluata, U. (1996) Simulation of annual plankton productivity cycle in the Black Sea by a one-dimensional physical-biological model, *Journal of Geophysıcal Research*, **101**, C7, 16585-16599.
17. Ovchinnikov, I.M. and Osadchy, A.S. (1991) Secular variability of winter climatic conditions influencıng pecularities of hydrological conditions in the Black Sea, *Variabılıty of the Black Sea ecosystem*, Nauka , Moscow, pp. 85-89. (In Russian).
18. Raymont, J. E. G. (1980) *Plankton and productıvıty ın the oceans*. Vol. 1 *Phytoplankton*. 2nd ed., Pergamon Press, New York.
19. Shiganova T.A. (1997) *Mnemiopsıs leıdyı* abundance in the Black Sea and its impact on the pelagic community, in E.Ozsoy and A.Mikaelyan (eds.), *Sensıtıvıty to Change: Black Sea, Baltic Sea and North Sea*, Kluwer Academic Publishers, Dordrecht, p. 117-131.
20. Shushkina, E. A. and Musaeva, E. I. (1990) Structure of planktonic communıty of the Black Sea epipelagic zone and its variation caused by invasion of new ctenophore species, *Oceanology, Englısh Translatıon*, **30** (2), 225-228.
21. Shushkina, E. A., Nikolaeva, G G. and Lukasheva, T. A. (1990) Changes of structure of plankton community of the Black Sea during mass development of the ctenophore Mnemiopsis leidyi (Agassiz), *Zhurnal Obshchey Bıologıı*, **51** (1), 54-60 (ın Russian).
22. Shushkina, E. A. and Vinogradov, M. E. (1991) Long-term changes in the biomass of plankton in open areas of the Black Sea, *Oceanology, English Translation*, **31** (6), 716-721.
23. Sukhanova I.N., Pogosyan S.I and V.S.Vshivtsev (1991) Temporal variations of the populations structure of the mass species of phytoplankton in spring bloomıng, in M.E.Vinogradov (ed.) *Changes in ecosystem of the Black Sea*. Nauka, Moscow, p. 117-127 (In Russian)

24. Tsikhon-Lukanina, E. A., Reznichenko, O. G. and Lukasheva, T. A. (1991) Quantitative patterns of feeding in the Black Sea ctenophore Mnemiopsis leidyi, *Oceanology, English Translation*, **31** (2), 196-199.
25. Vedernikov, V. I. and Demidov, A. B. (1993) Primary production and chlorophyll in deep regions of the Black Sea, *Oceanology. English Translation*, **33** (2), 193-199.
26. Vinogradov, M. E., Musaeva, E. I. and Semenova, T. N. (1990) Factors determining the position of the lower layer of mesoplankton concentration in the Black Sea, *Okeanologiya*, **30** (5), 295-305. (in Russian, English summary)
27 Vinogradov, M. E., Sapozhnikov, V. V. and Shushkina, E. A. (1992) *The Black Sea ecosystem*, Nauka, Moscow, pp. 1-112. (In Russian)
28 Vinogradov, M. E. and Shushkina, E. A. (1992) Temporal changes in community structure in the open Black Sea, *Oceanology, English Translation*, **32** (4), 485-491
29. Vinogradov, M. E., Shushkina, E. A , Bulgakova, Yu. V. and Seobaba, I. I. (1995) Consumption of zooplankton by the comb-jelly Mnemiopsis leidyi and pelagic fishes in the Black Sea, *Oceanology, English Translation*, **35** (4), 523-527.
30 Vinogradov, M. E., Shushkina, E. A , Musaeva, E. I. and Nikolaeva, G. G. (1992) Vertical distribution of the Black Sea mesozooplankton in winter, 1991, in M. E. Vinogradov (ed), *Ecosystem of open Black Sea in winter*, Moscow, IO RAN, 103-119. (In Russian)
31 Vinogradov, M. E., Shushkina, E. A., Musaeva, E. I. and Sorokin, P Yu. (1989) A newly acclimated species settlers in the Black Sea: the Ctenophore Mnemiopsis leidyi (Ctenophora: Lobata), *Oceanology, English Translation*, **29** (2), 220-225.
32. Vinogradov, M. E., Shushkina, E. A. and Nikolaeva, G. G. (1993) The state of zoocoene in open areas of the Black Sea during late summer, 1992, *Oceanology, English Translation*, **33** (3), 329-333
33. Vladimirov V.L., Mankovsky M.V , Solov'ev M.V. and A.V.Mishonov (1997) Seasonal and long-term variability of the Black Sea Optical parameters in E.Ozsoy and A.Mikaelyan (eds.), *Sensitivity to Change: Black Sea, Baltic Sea and North Sea*, Kluwer Academic Publishers, Dordrecht, p 33-48
34. Zaitsev Yu.P. and Alexandrov B.G. (1997) Resent man-made changes in the Black Sea ecosystem, in E.Ozsoy and A.Mikaelyan (eds), *Sensitivity to Change· Black Sea, Baltic Sea and North Sea*, Kluwer Academic Publishers, Dordrecht, p. 25-32.
35. Zernova V V. and N P. Nezlin (1983) Seasonal changes of phytoplankton in the western part of the Black Sea in 1978, Yu. I. Sorokin and M.E.Vinogradov (eds.), *Seasonal changes of plankton on the Black Sea*, Nauka, Moscow, p. 12-33 (In Russian)

DISTRIBUTION OF PLANKTONIC PRIMARY PRODUCTION IN THE BLACK SEA

ALEXANDRU S. BOLOGA[1], PETRE T. FRANGOPOL[2], VLADIMIR I. VEDERNIKOV[3], LUDMILA V. STELMAKH[4], OLEG A. YUNEV[4], AYSEN YILMAZ[5], TEMEL OGUZ[5]
[1]*Romanian Marine Research Institute, RO-8700 Constanta, Romania*
[2]*A. I. Cuza University, RO-6600 Iassy, Romania*
[3]*P.P. Shirshov Institute of Oceanology, RAS, RU-117851 Moscow, Russian Federation*
[4]*Institute of Biology of Southern Seas, UA-335000 Sevastopol, Ukraine*
[5]*Institute of Marine Sciences, TR-33731 Erdemli-Icel, Turkey*

Abstract

Planktonic primary production data from the entire Black Sea are reviewed for the last two decades. Surface and vertical profile of data are spatially and seasonally compared for different significant areas (north-western shelf, western and southern coast, eastearn and western halistatic zones). High production rates, especially in the coastal waters of the NW and W Black Sea exhibit large inter- and intraseasonal variations. Such values clearly reflect meso- and eutrophic feature of these waters. There are major annual spring (diatoms) and autumn (coccolithophorids) blooms, followed in recent years by additional summer (dinoflagellates, coccolithophorids) blooms. The major primary producers are usually *Skeletonema costatum, Chaetoceros curvisetus, Peridinium trochoideum, Exuviaella cordata and Prorocentrum micans*. Factorial analyses reveal very high correlation coefficients between chlorophyll *a* concentration and salinity, primary production and salinity, and between chlorophyll *a* concentration and primary production. The annual cycle of plankton dynamics in the central Black Sea is being studied by means of an one-dimensional vertically resolved physical-biological upper ocean model, involving interactions between inorganic N (NO_3, NH_4), phytoplankton and herbivorous zooplankton, and detritus.

1. Introduction

Historically, before the last decade maxima in primary production of the Black Sea occurred twice a year, with a major bloom, mainly diatoms, in early spring, followed by a secondary bloom, mainly coccolithophorids, in autumn [26]. Such blooms occurred mainly in coastal areas. Recently, additional summer blooms composed predominantly of dinoflagellates and coccolithophorids (*Emiliana huxleyi*) have increasingly been observed

S. Beşiktepe et al. (eds.),
Environmental Degradation of the Black Sea: Challenges and Remedies, 131–145.

[1, 7, 15, 16, 17, 26, 32]. Massive red tides along the Romanian [2, 3, 4, 5, 6] and Bulgarian [19, 20, 21, 24, 25] coasts have been reported. Intense blooms of certain plankton species, have also been cited during winter along the western Anatolian coast [29].

2. Methodology

Primary production (PP) was measured using two varieties of the ^{14}C technique [27]. The first modification was an *in situ* determination which was used during Russian and Romanian cruises [8, 9, 10, 33, 34]. According to this method two light and one dark (275 ml (Russia) or 250 ml (Romania)) bottles containing phytoplankton from different depths were each inoculated with 3-12 (Russia) or 5-25 (Romania) μCi aqueuos $NaH^{14}CO_3$ solution and incubated *in situ* (at the depths from where they were collected) during half of the light day (usually from noon to sunset). After recovery the bottles ware filtered under vacuum through 0.3-0.6 μm membrane filters. After filtration the filters were rinsed five times with 30 ml of filtered seawater. Then they were dried in open Petri dishes overnight at room temperature and placed into dark vials with 10 ml of liquid scintillator per vial. Activity of the filters was counted with a radiometer RZhS-05 (Russia) and Nuclear Enterprises LSCI manual scintillation counter or a Packard Tri-Carb 3385 automatic scintillation spectrometer (Romania).

The production in the sample per period of day light (12hrs) was calculated using the common formula with correction for dark fixation of carbon dioxide. The value for total carbon content in these samples were given by the hydrochemists or were set equal to 36000 mg C m^{-3}. Results were expressed as mg C m^{-3} d^{-1} for each investigated depth or as mg C m^{-2} d^{-1} for integrated primary production in the water column (PPI).

The detection limit for ^{14}C determination of PP at a given depth was approximately 0.5 mg C m^{-3} d^{-1}. Relative errors are very much dependent on the value determined and usually extended from 5 to 30% for most depths in the euphotic zone.

The second technique was a simulated *in situ* measurement of PP using a flow-through incubation box in which light conditions were simulated to be like those in the natural marine environment [29, 30, 38]. Two light and one dark glass or plastic bottles (50-250 ml, Ukrainian cruises and 25-50 ml, Turkish cruises) containing samples from different depths were inoculated with 5-20 μCi aqueuos $NaH^{14}CO_3$ solution and exposed to light for 4-8 hrs. After exposure the filtering and counting procedures were the same as in the first modification.

Absolute values of primary production obtained by these ^{14}C techniques should be compared only with much caution. The comparison is fraught with difficulties as intercalibration was not performed between the two techniques. Nevertheless, the estimates obtained are applicable for description of the distribution and seasonal dynamics of primary production.

3. Results and discussion

3.1. NORTHERN AND WESTERN SHELF: SURFACE, VERTICAL AND SEASONAL DISTRIBUTION

Surface primary production (PPS) in Sevastopol Bay is subject to great variation, both within each season and from season to season [28]. Usually, in such bays as Sevastopol and Gelendzhik, the annual cycle of PPS is characterized by three maxima in spring, late summer and autumn [29]. In the early months of spring (February-March) PPS ranged from 11-35 mg C m^{-3} d^{-1}. Toward the end of this period (April-May) one could observe intensive blooms of *Skeletonema costatum* and *Chaetoceros curvisetus* which were accompanied by PPS of 240-400 mg C m^{-3} d^{-1}. In summer (June-August), when phytoplankton was dominated by the pyrrhophyte algae *Peridinium trochoideum, Exuviaella cordata and Prorocentrum micans*, PPS varied from 14 to 280 mg C m^{-3} d^{-1}. In autumn (October-November), when plankton was again dominated by diatoms, PPS varied from 20 to 80 mg C m^{-3} d^{-1}. In winter (December-January) the potential PPS constituted only 7-8 mg C m^{-3} d^{-1}. According to these results this part of the Black Sea can be regarded as an eutrophic region.

Maximum productivity during spring - autumn was found on the NW shelf. High productive areas were also observed on the NE shelf. The next highest values were reported along the Romanian (western) and the Anatolian (south - western) coasts, and extended into the central region separating the eastern and western gyres [7, 32]. Limited observations suggest that the western Anatolian coast is also a region of relatively high productivity [31].

Eutrophication started in the NW shelf area influenced by high nutrient outflow from the Danube and Dnestr rivers, and progressed southward along the western shelf [7, 13].

Early results obtained off Constanta during summer 1977 gave PP values ranging between 297-825 mg C m^{-3} d^{-1} in the upper water layer (0 m) and between 238-243 at 5 to 10 m.

Other results obtained in the same shelf zone off Constanta (0 m) during 1978 ranged between 4.7 to 67 mg C m^{-3} d^{-1} in December and 71.3 to 1,162.8 in July [8]. Thus, the development of photosynthetic assimilation and primary productivity, which is maximum in summer (July) and minimum in winter (December), closely follows the evolution of environmental factors (e.g. light and temperature: July = 62,163 lx, 24^0 C; December = 2,123 lx, 11^0 C). Monthly data for primary production in surface (0m) waters were obtained in the nearshore and offshore up to 30 nautical miles zones of Constanta during 1979 [9]. Microscopic analysis (taxonomic composition, weight of the major groups) of phytoplankton was performed at the same time. The results gave mean PPS values that ranged from 47.2 (January) to 475.4 (March) mg C m^{-3} d^{-1} and a mean annual PPS of 54.8 g C m^{-3} y^{-1}. The variability in these values was usually correlated with phytoplankton density and biomass. The highest PPS values in March were correlated with maximum phytoplankton biomass reaching 13.9 to 32.2 g m^{-3}. This production was primarily due to the diatoms *Skeletonema costatum, Thalassiosira subsalina and Detonula confervacea.* In April, diatoms (almost exclusively *Skeletonema costatum*) had a density of 33,180 cells l^{-1} and a biomass of 9,978 mg m^{-3}. In May, maximum densities and biomasses were also recorded nearshore and decreased seaward. These data indicate the spatial and seasonal variability of PPS in Constanta shelf waters. Maximum level during spring was due to abundant development of diatoms, whereas high values during early summer were a consequence of the massive occurrence of dinoflagellates. In this area higher values were found nearshore compared with those in the offshore zone. The results in the offshore zone of Constanta ranged between 1.7 to 1,938 mg C m^{-3} d^{-1} in the surface layer and the integrated values (0-50m) ranged between 0.2 to 27.4 g C m^{-2} d^{-1} [10]. Total annual PP was estimated to be 7,625,000 t C y^{-1} in this sector of the Romanian coast. The highest PP occurred in the 0-10 m layer (except for June, when the maximum value was found at 25 m).

Other results were obtained in 1982 at a distance of 1.2 n.m. off Constanta from depths of 0,5 and 10 m [11]; analyses of chlorophyll *a* and microscopic counting of phytoplankton (density and biomass) were performed simultaneously. Mean annual PP values at these depths were 5.2, 1.4 and 1.0 mg C m^{-3} h^{-1}, respectively. The annual average integrated primary productivity was 2.3 mg C m^{-2} h^{-1}. Maximum productive rates occurred in late spring and summer during this particularly year (Fig.1). The vertical distribution of PP showed higher values at 0 m (except in May at 5 and 10 m, respectively). The maximum annual mean value also occurred at 0 m. The higher PP level in the surface layers (Fig.1), was consistent with 1980 data from the same sector collected down to 50 m depth [10, 11].
The maximum PP in July was due to an intense bloom of the dinoflagellate *Exuviella cordata.*

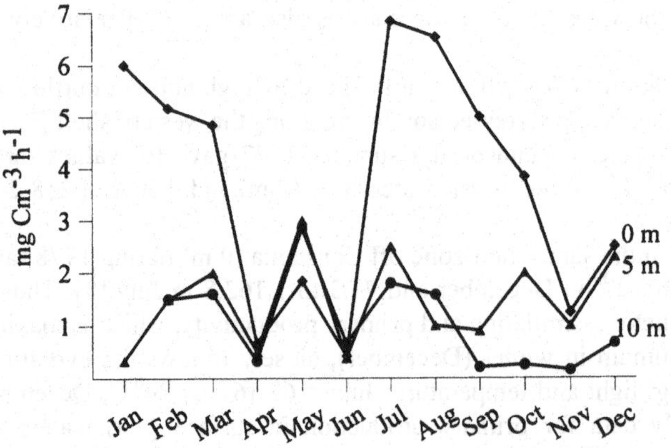

Figure 1. Annual cycle of planktonic primary productivity at 0, 5 and 10 m in the Constanta nearshore sector in 1982 (cf. [11])

Comparisons between primary productivities in different areas along the whole western Black Sea coast from the region north of the Danube to north of the Bosporus can be made with the 1982 data set [12]. The very high values offshore north of the Danube (much higher than those during the last two decades) were due to the intense eutrophication of the north-western part of the Black Sea as a consequence of the nutrients carried by the Danube and implicitly by the frequent occurrence of intense phytoplankton blooms.

The highest PP values compared to the other sectors occurred offshore of the Danube in May and reached 1,530 mg C m^{-3} d^{-1} and 1.6 g C m^{-2} d^{-1}. In this sector, in contrast to the two southern ones, maximum PP was determined in the upper layer between 0 to 5 m (Fig.2). Factorial analysis gave very high correlation coefficients between chlorophyll *a* concentration and salinity (-0.91), between primary production and salinity (-0.94) and between chlorophyll *a* and primary production (0.91), suggesting that the waters with the most river influence had the most biological activities.

All PP data collected in the offshore zone of Constanta between 1977 and 1982 by means of the ^{14}C method showed evidence of the highly eutrophic feature of the Romanian Black Sea coastal waters [7]. Estimates of primary production and phytoplankton biomass along the Romanian coast in the 1980's were much higher than in the previous two decades.This was due to a general increase of eutrophication and more frequent and intense blooms.

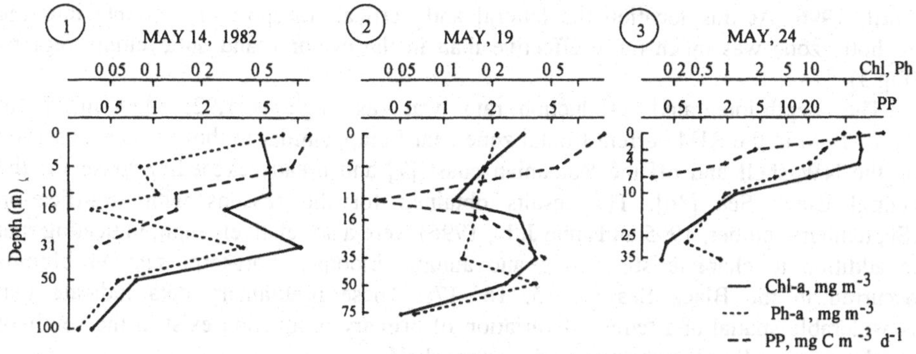

Figure 2. Vertical distribution of primary production (PP), chlorophyll-a (Chl-a) and phaeopigments-a (Ph-a) off the Bosporus (Turkey) (1), off Caliacra (Bulgaria) (2) and off the Danube (Romania) (3) in May, 1982 (cf. [12])

3.2. SOUTHERN OFFSHORE WATERS

The spring-time surface productivity in the southern Black Sea is modulated by transient dynamics [32]. Species differentiation and competition are evident along the boundary current systems. Early summer blooms coincide with peak flood discharges from major rivers, and influence the spread of eutrophication in the basin.

Until recently, data on optical transparency, phytoplankton biomass and PP were very limited for the southern Black Sea [36]. Light penetration is generally limited to upper 15-40 m, with the downward attenuation coefficient varying between 0.1 and 0.3 m^{-1}. Coastal waters of the region are fed by the riverine input whereas the cyclonically dominated open ecosystem is mainly controlled by influx from the oxygenated lower layers by vertical mixing that is much more effective in winter. However, the input from the anoxic layer is limited due to the presence of a strong permanent halocline in the basin. Halocline coincides with the suboxic zone which is in between the oxic and anoxic waters, where intense denitrification and redox-dependent processes also limit nitrogen and phosphorus input to the productive layer [22]. Thus, surface waters are always poor in inorganic nutrients during the stratified seasons because the photosynthetic uptake rate exceeds the nutrient input rate in the spring.

The photosynthetic carbon production rate range from 247-1,925 and 405-687 mg C m^{-2} d^{-1} in the spring and summer-autumn periods, respectively.

Relatively high PP rates were observed in the upper euphotic zone to 10-20 m and these values always showed similar vertical profiles (Fig.3). Below this layer, PP decreased markedly due to light limitation. This was confirmed by the maximum production P(M) rates, which are the production rates when deep samples are exposed to full light conditions. P(M) rates were similar to the near surface production rates down to the base of the euphotic zone.

Production rates P (during true noon) and daily production rates P(D) ranged between 1-10 mg C m^{-3} h^{-1} and 2-200, respectively, for the top 10 to 20 m. The highest rates were observed at a frontal station (off Sinop) in the RIM current region during April, 1996. At this location the latteral and vertical transport of nutrients into the euphotic zone was much more effective than in the cyclonic and anticyclonic regions (Fig.3).

The depth-integrated production rate was as high as 1,925 mg C m^{-2} d^{-1} in April, 1996 in the RIM current frontal zone near Sinop, similar to those values observed on the NW shelf and off the Romanian coast [7] and greatly exceeding those for the central Black Sea [36]. The results obtained for the seasons with stratification (September-October, 1995 and June-July, 1996) were also relatively high, indicating that in addition to characteristic spring and autumn blooms, short-term summer blooms occurred in the Black Sea [7, 15, 16, 17]. These preliminary data indicate that considerable spatial and temporal variation of primary production exist in the southern Black Sea as well as the northern and western shelf.

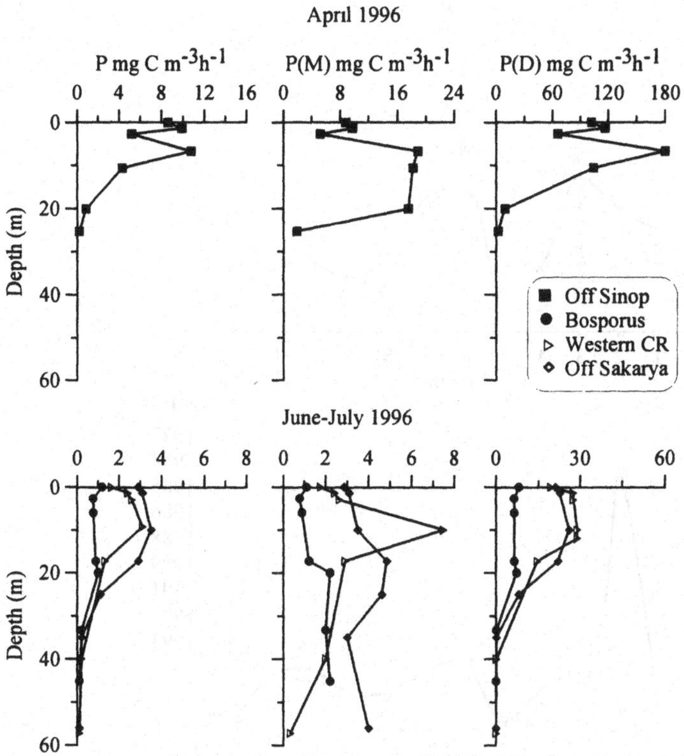

Figure 3. Primary production (P), maximum rates (M) and daily rates (D) off Bosporus, Sinop, Sakarya and in the western cyclonic region during April and June-July 1996 ([36])

3.3. OPEN REGIONS (EASTERN AND WESTERN HALISTATIC ZONES)

Despite the deficiency of year-long seasonal data for the open sea, it is possible to use the historical data sets [28, 34] to draw conclusion about seasonal changes of PPI within the western and eastern cyclonic gyres.

Enhanced mixing of the cyclonic gyres in winter was responsible for estimates of PPI that amounted to 300-700 mg C m^{-2} d^{-1} with a maximum in the centre of the gyre (Fig. 4).

PPI reached a maximum in March with an average of 640 mg C m^{-2} d^{-1}. Production was also the largest (1200 mg C m^{-2} d^{-1}) in the centre of the gyres, decreasing to 200 mg C m^{-2} d^{-1} towards the periphery. The community *Nitzschia delicatissima* and *Rhizosolenia calcar-avis* reached maximum population during this period. In different years the winter-spring peak of phytoplankton growth in the central area took place either in February or March.

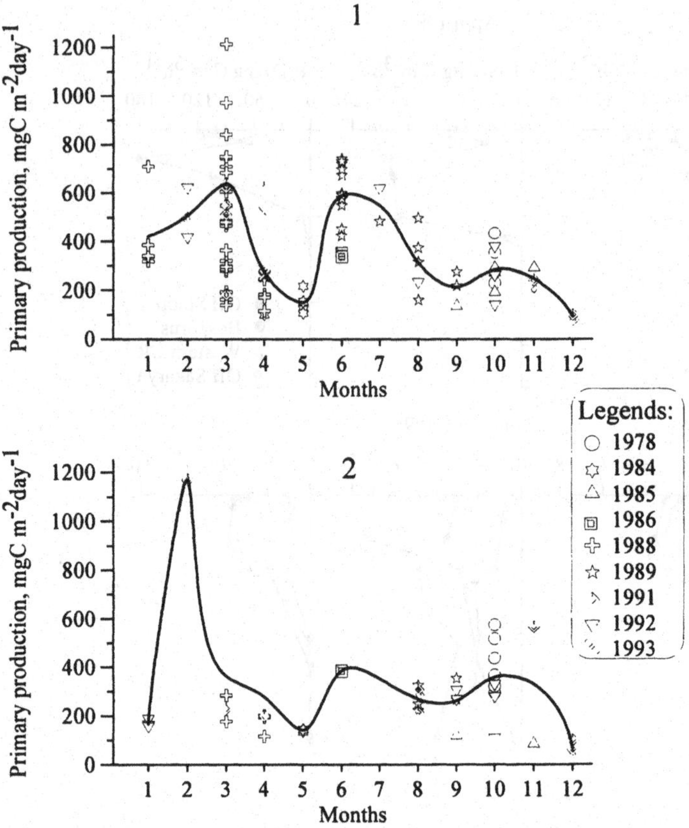

Figure 4. Seasonal variability of the primary production in the central part (depths > 1500 m) of the Black Sea [29] 1 - western cyclonic gyre (WCG) & convergence zone, 2 - eastern cyclonic gyre (ECG)

For example, in 1988, in the western cyclonic gyre, the maximum was observed in the beginning of March, and in 1991, in the eastern cyclonic gyre it was in February (Fig. 4). Western spring values of PPI for open waters of the Black Sea are 5 to 10 times higher than those for coastal waters where the spring bloom of phytoplankton takes place usually 2-3 months later.

In late March-early April, diatoms die out rapidly in central areas of the sea, and are replaced by the pyrrophytae *Ceratium furca*, *Ceratium fusus*, *Exuviaella cordata* and small coccolithophores [17]. At the same time PPI decreases to 200-400 mg C m^{-2} d^{-1}, decreasing to a spring minimum (150 mg C m^{-2} d^{-1}) in May.

Phytoplankton production of the central area increased to a second maximum of 600 mg C m^{-2} d^{-1} in June (Fig. 4). The summer peak is comparable with the spring one and is induced by the growth of small varieties of pyrrophytae and coccolithophores capable of thriving at lower concentrations of nutrients compared with diatom algae.

In August - September PPI declines and but then it increases again to 300 mg C m^{-2} d^{-1} in October. The lowest production values were measured in late November - early December.

The total annual gross PPI in the middle of the Black Sea is 130 g C m^{-2}. 50% of the total occurs from January to May, 36 % in the summer, and 14% during autumn and winter.

Analysis of long-time changes of summer mean primary production in the the photosynthetic layer in the western (1) and eastern (2) deep water regions of the Black Sea (cf. [36]) indicate that PPI has increased by approximately a factor of two during the last 20 years (e.g. Tab.1, Fig.5). Such changes reflect the progressive eutrophication of the deep sea regions [14].

Table 1 Summer (May-September) primary production (P) in the photosynthetic layer of deep water regions of the Black Sea (N = number of measurements, M = mean value, σ = standard deviation) in different years (cf [36])

Region		Primary production (mg C m^{-2} d^{-1})		
		1960-1972	1980-1985	1986-1991
Western	N	17	12	8
	M	138	128	347
	σ	58	58	135
Eastern	N	16	18	11
	M	150	128	298
	σ	123	47	59
All deepwater regions	N	42	36	23
	M	170	139	321
	σ	120	61	100

The vertical distribution of PP on different transects occupied during the summer (Fig.6) showed that the layer of intensive photosynthesis (> 4-5 mg C m^{-3} d^{-1}) extended to depths of 10 m on the NW shelf and 20-40 m in the central regions of the sea (Fig.7). The changes of the thickness of the photosynhetic layer in deep offshore regions from 50 to 65 m characterized the Black Sea in other seasons as well [33, 35]. The depth of the photosynthetic layer decreased to 25-45 m in the region of continental slope near Burgas and Gelendzhik and to minimum values between 15-26 m in the highly productive NW shelf region [34].

Analyses of PPI data for the period of 1960-1991 shows the following variability in the water column: between 570-1,200 mg C m^{-2} d^{-1} on the north-western shelf, between 320-500 in the regions of continental slope, and between 100-370 in the central deep sea

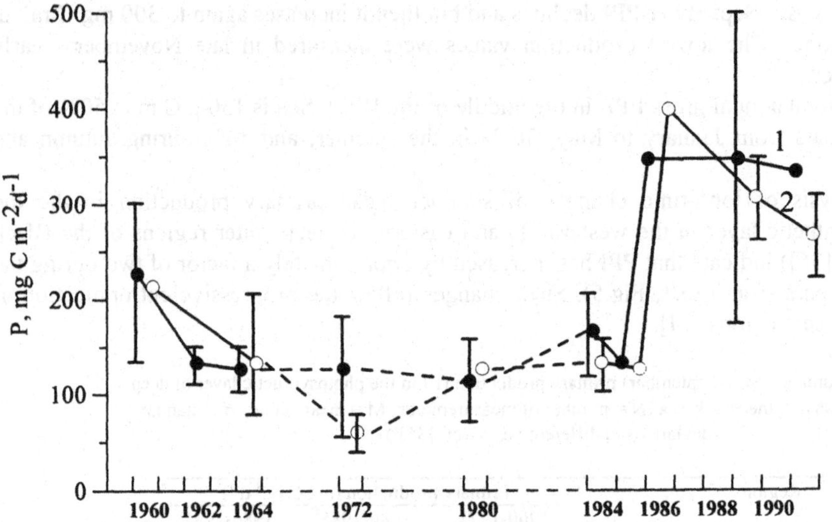

Figure 5. Historical change of summer (May-September) mean primary production in the western (1) and eastern (2) deep water (>1000 m) regions of the Black Sea (cf. [36])

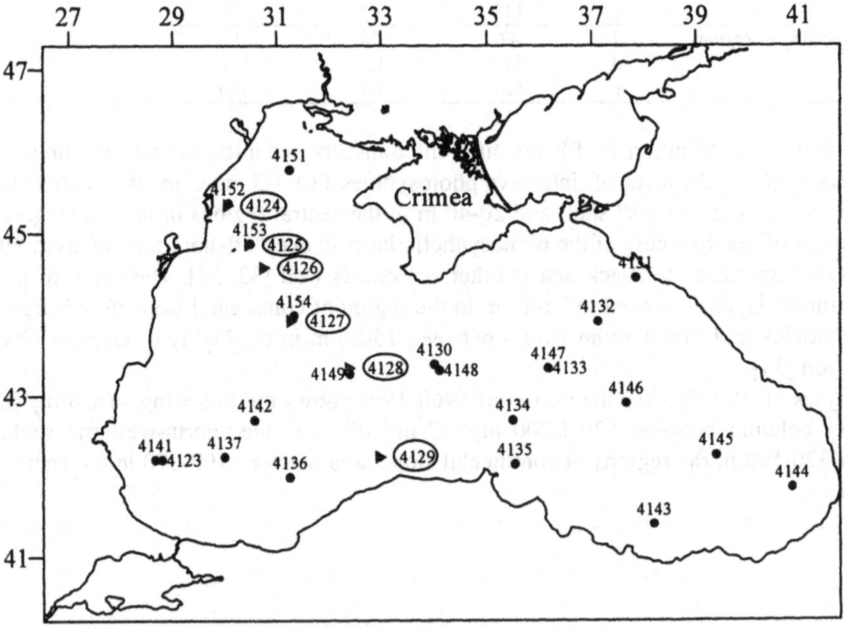

Figure 6. Location of primary production sampling stations during the 44th cruise of R/V "Dmitri Mendeleev" between July-September 1989 (Data from the NATO TU-Black Sea joint data bank (METU IMS, Erdemli-Icel, Turkey)

regions [37]. The autumn changes in the distribution of vertical phytoplankton started in the western parts of the sea and gradually extended to the east.

3.4. MODELLING

The annual cycle of plankton dynamics in the central Black Sea was recently studied using an one-dimensional vertically resolved coupled physical - biological upper ocean model [23]. This model involved interactions between inorganic N (NO_3 , NH_4), phytoplankton and herbivorous zooplankton biomasses, and detritus. It simulated

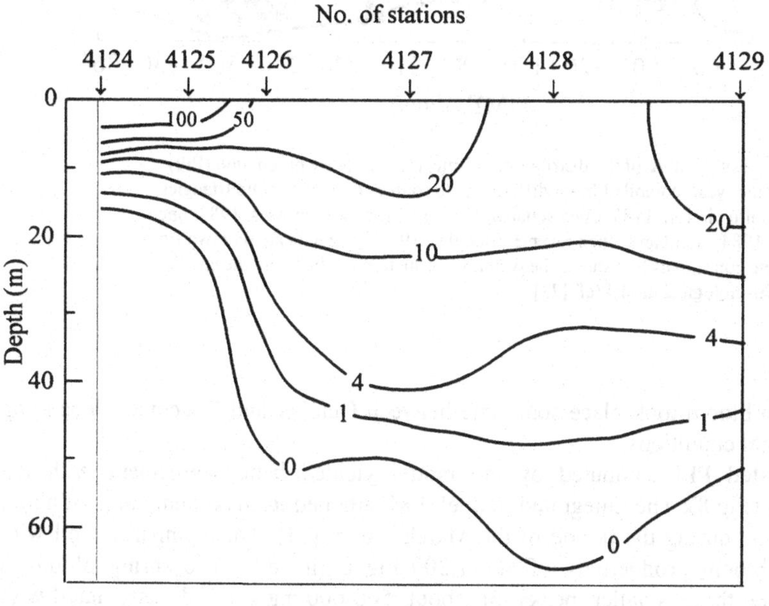

Figure 7. Vertical distribution of primary production (mg C m^{-3} d^{-1}) along the western transect shown in Fig. 6 (cf. [34]): the section was constructed using the stations circled in Fig. 6

seasonal and vertical characteristic features, in particular, formation of the cold intermediate water mass and yearly evolution of the upper layer stratification, annual cycle of production with the spring and fall blooms, and the subsurface phytoplankton maximum layer in summer. The computed seasonal cycles of the PPI distribution compared reasonably well with the data. The annual cycle of PP distribution was constructed by combining the field observations within the central basin during the last decade [36].The continuous line gives the corresponding model predictions (Fig.8). Year-to-year variability in the productive cycle was observed. Initiation of the spring bloom was shown to be critically dependent on the stability of the water column. It was followed by lower PP at the time of establishment of the seasonal thermocline in April.

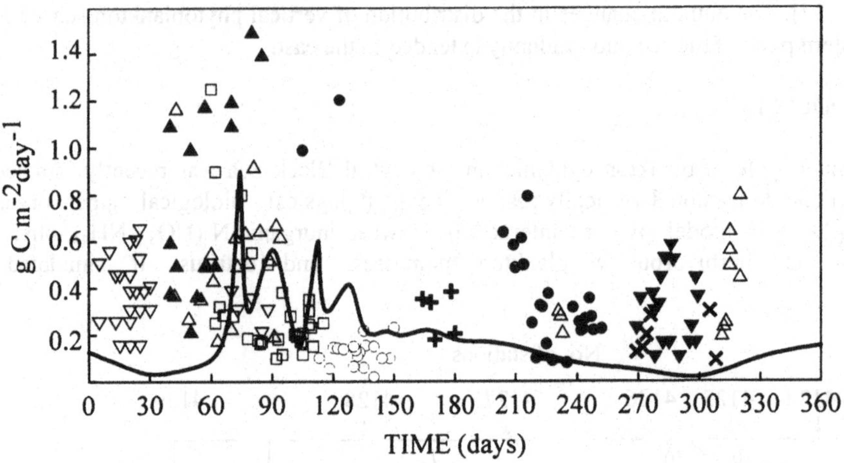

Figure 8. A composite picture of the distribution of integrated primary production (PPI) during the year compiled from different data sources (open and solid triangles, 1991; solid circles, 1989, open squares, 1988; plusses 1986; crosses, 1985; open circles,1984; solid and open reverse triangles, 1978) The data are redrawn from [36].The continuous line shows the primary production predicted by the coupled physical-biological model (cf. [23]

The autumn bloom took place sometime between October and December depending on environmental conditions.

The integrated PPI computed by the model yielded better agreement with the composite data (Fig.8). The integrated model PPI attained its maximum value of about 900 mg C m^{-2} d^{-1} during the period of the March bloom [23]. This compared well with the measured bloom productivity of 400-1,300 mg C m^{-2} d^{-1} . The spring bloom is followed by the three smaller peaks of about 500-600 mg C m^{-2} d^{-1} associated with the regenerated PPI in the spring. PPI estimates of about 300-500 mg C m^{-2} d^{-1} during May were also comparable with the May, 1988 measurements of 339 and 573 mg C m^{-2} d^{-1} at two stations within the Central Black Sea [17]. August to October was shown to be the least productive period characterized by model values of less than 200 mg C m^{-2} d^{-1}, whereas observed values varied from 200 to 500 mg C m^{-2} d^{-1} in the same period.

This model predicts PPI values reasonably well for most parts of the year except during the autumn bloom season. The maximum model estimate of autumn PP of 200 mg C m^{-3} d^{-1} was somewhat less than the observations.

4. Conclusions:

1. A general increase of eutrophication is reflected by more intense and frequent phytoplankton blooms that have characterized the Black Sea during the last two decades.
2. In addition to the usual early spring and autumn blooms, additional summer and sometimes even winter blooms have occurred in more recent times.
3. High primary production rates have been registered especially in the north-western and western coastal waters.
4. In the NW area, primary production varies greatly seasonally as well as within each season.
5. In the highly eutrophic western coast, e.g. off Constanta, the highest production values observed in shallow waters reached up to 1,162 mg C m^{-3} d^{-1} in July; offshore values reached up to 1,938 mg C m^{-3} d^{-1} or 27.4 g C m^{-2} d^{-1} in August.
6. Factor analysis revealed high correlation coefficients between chlorophyll *a* and salinity, primary production and salinity, and chlorophyll *a* and primary production along the western coast.
7. Primary production in the centers of the eastern and western halistatic zones were typically low during late spring, summer and autumn.
8. Average monthly production values in both gyres varied from 200 to 745 mg C m^{-2} d^{-1} indicating an increase in this parameter by approximately a factor of two during the last 20 years.
9. Primary production ranged from 570 to 1,200 mg C m^{-2} d^{-1} on the NW shelf, between 320 to 500 in the regions of continental slope, and between 100 to 370 in the central deep sea regions.
10. A recently developed one-dimensional vertically resolved coupled physical - biological upper ocean model predicted primary production values reasonably well for most periods of the year.

Acknowledgements

The authors acknowledge the preparation of this joint paper under the fruitful NATO TU - Black Sea Program headed by Prof. U. Unluata and express hereby their gratitude for his hospitality during all working meetings.

References

1 Benli H. (1987) Investigation of plankton distribution ın the southern Black Sea and its effects on particle flux, ın E T Degens, E. Izdar and S Honjo (Eds.), *Partıcle Flux ın the Ocean,* Mitt. Geol.-Paleontol. Inst., Univ. Hamburg **62**, 77-87.

2 Bodeanu N (1984) Donnees nouvelles concernant les devellopements massifs des especes phytoplanctonıques et deroulement des phenomenes de floraison de l'eau du littoral Roumain de la mer Noıre, *Cercetarı marıne* **17**, 63-68

3. Bodeanu N. (1992 a) Algal blooms and development of the main planktonic species at the Romanian Black Sea littoral in conditions of intensification of the eutrophication process, in R.A.Vollenweider, R. Marchetti and R.V. Viviani (Eds.), *Marine Coastal Eutrophication*, Elsevier Sci. Publ., Amsterdam, 891-906.
4. Bodeanu N. (1992 b) Microalgal blooms in the Romanian area of the Black Sea and contemporary eutrophic conditions, in D. L. Taylor, H. H. Seliger (eds.), *Toxic Phytoplankton Blooms in the Sea*, Elsevier Sci. Publ., Amsterdam, 203-209.
5. Bodeanu N., Usurelu M. (1979) Dinoflagellate blooms in Romanian Black Sea coastal waters. in D. L. Taylor, H. H. Seliger (Eds.), *Toxic Dinoflagellate Blooms*, Elsevier Sci. Publ., Amsterdam, 151-154.
6. Bodeanu N., Cociasu A., Tiganus V (in press) Eutrophication, plankton blooms, hypoxia at the Romanian Black Sea littoral, *Deep-Sea Res* , **2** (Black Sea special issue).
7. Bologa A S. (1985/1986) Planktonic primary productivity of the Black Sea: A review, *Thalassia Jugoslavica*, **21/22,** 1/2, 1-22.
8. Bologa A. S., Frangopol P. T., Frangopol M., Stanef I. (1980) Marine phytoplankton photosynthesis in the offshore zone of Constantza (Black Sea) during June-December, 1978, *Rev. Roum. Biol.-Biol. veget.*, **25**, 2, 129-133.
9 Bologa A. S., Usurelu M., Frangopol P. T. (1981) Planktonic primary productivity of the Romanian surface coastal waters (Black Sea) in 1979, *Oceanologica Acta,* **4**, 3, 343-349.
10. Bologa A. S., Frangopol P. T. (1982) Data on the vertical distribution of planktonic primary productivity in the offshore zone of Constantza (the Black Sea), *Rev. Roum. Biol.-Biol. veget* , **25**, 27, 2, 141-146
11 Bologa A. S., Skolka H. V , Frangopol P. T. (1984) Annual cycle of planktonic primary productivity off the Romanian Black Sea coast, *Marine Ecology - Progr. Ser.*, **19**, 25-32.
12. Bologa A. S., Burlakova Z. P., Chmyr V. D., Kholodov V. I. (1985) Distribution of chlorophyll *a*, phaeophytin *a* and primary production in the western Black Sea (May, 1982), *Cercetari marine,* **18**, 97-115.
13. Bologa A. S , Bodeanu N., Petran A., Tiganus V., Zaitsev Yu. P. (1995) Major modifications of the Black Sea benthic and planktonic biota in the last three decades, in F. Briand (Ed.), *Les mers tributaires de la Mediterranee*, Bull. Inst. Oceanogr. Monaco, Num. spec. 15, p. 85-110.
14. Codispoti L. A., Friederich G. E., Murray J W., Sakamoto C. M. (1991) Chemical variability in the Black Sea: implications of continuous vertical profiles that penetrated the oxic/anoxic interface, *Deep-Sea Research,* **38,** Suppl. 2, S691-S710.
15. Hay B. J , Honjo S. (1989) Particle deposition in the present and holocene Black Sea, *Oceanography*, **2**, 26-31.
16 Hay B. J., Honjo S., Kempe S , Itekkot V. A , Degens E. T , Konuk T , Izdar E. (1990) Interannual variability in particle flux in the southwestern Black Sea, *Deep Sea Res.*, **37**, 911-928.
17. Hay B J., Arthur M. A., Dean W. E., Neff E. D. (1991) Sediment deposition in the late holocene abyssal Black Sea· Terrigenous and biogenic matter, *Deep-Sea Res* , **38** (suppl.), 5711-5723.
18. Karl D. M., Knauer G. A. (1991) Microbial production and particle flux in the upper 350 m of the Black Sea, *Deep-Sea Res.*, Part. A, **38**, suppl.2, 5655-5661.
19. Moncheva S. (1991) Ecology of some common phytoplankton species from the Black Sea under the condition of eutrophication, Doctorate Thesis, I0-BAS, 197 pp.
20 Moncheva S. (1992) Cysts of blooming dinoflagellates from the Black Sea coast, *Rapp. Comm. int mer Medit* , **33**, 261.
21. Moncheva S., Petrova-Karadjova V , Palasov A (1993) Harmful algal blooms along the Bulgarian Black Sea coast and possible patterns of fish and zoobenthic mortalities, 6th Int. Conf "Toxic Marine Phytoplankton", 18-22 Oct., Nantes, France.
22. Murray J. W , Codispoti L. A., Friederich G. E , (1995) Oxidation-Reduction Environments. The Suboxic Zone in the Black Sea, *ACS Advances in Chemistry Series* No. 244, 157-176
23 Oguz T., Ducklow H., Malanotte-Rizzoli P , Tugrul S., Nezlin N. P., Unluata U (1996) Simulation of annual plankton productivity cycle in the Black Sea by a one-dimensional physical-biological model, *J Geophys. Res.*, **101**, C7, 16585-16599.
24. Petrova-Karadjova V. (1985) Cherveniyat priliv ot *Prorocentrum micans* i *Exuviaella cordata* vav Varnenski zaliv i po krajbrezhieto prez noemvri 1984 g. (Red-tide of *Prorocentrum micans* (Ehr) and *Exuviaella cordata* (Ost.) in Varna Bay and along the coast in November, 1984), *Hydrobiology,* BAS, **26**, 70-78.

25. Petrova-Karadjova V. (1990) Monitoring of the blooms along the Bulgarian Black Sea coast, *Rapp. Comm. int. mer Medit.*, **31**, 1, 209.
26 Sorokin Yu. I. (1983) The Black Sea, in B H. Ketchum (Ed.), *Estuaries and Enclosed Seas. Ecosystems of the World*, Elsevier, Amsterdam, pp. 253-292.
27 Steeman Nelsen, E. (1952) The use of radioactive carbon (c^{14}) for measuring organic production in the sea, *J Cons. Int. Explor. Mer.* **18**, 117-140.
28. Stelmakh L V. (1988) The contribution of picoplankton to primary production and the content of chlorophyll "a" in eutrophic waters as exemplified by Sevastopol Bay, *Okeanologya*, **28**, 1, 95-99
29. Stelmakh L. V. (1995) Sezonnye izmenenia pervichnoy produktsii Cernogo morya (Seasonal changes of primary production in the Black Sea) in *Kompleksnye ekologicheskie issledovania Chernogo morya (Complex ecological research of the Black Sea)*, Sevastopol, 111-120.
30. Stelmakh L. V., Yunev O. A., Finenko Z. Z., Churilova T Ja , Bologa A. S. (1997) Peculiarities of seasonal changes of primary production in various bioprovinces of the Black Sea, in *NATO TU-Black Sea Project: Symposium on scientific results*, Crimea, Ukraine, June 15-19, Abstracts, 91-92.
31 Sur H. I., Ozsoy E., Unluata U. (1994) Boundary current instabilities, upwelling, shelf mixing and eutrophication processes in the Black Sea, *Progr. Oceanogr.*, **33**, 249-302.
32. Sur H. I., Ozsoy E., Ilyin Y. P., Unluata U. (1995) Coastal/deep interactions in the Black Sea and their ecological / environmental impacts, *J. Mar Systems*, **7**, 293-320
33. Vedernikov V I. (1987) Pervichnaya produktsya v Chernom more vesnoy 1984 g. (Primary production of the Black Sea during spring 1984), in *Sovremennoe sostoianie ekosistemy Chernogo morya (Contemporaneous State of the Black Sea Ecosystem)*, Nauka, Moscow, 105-118.
34. Vedernikov V. I. (1991) Osobennosti raspredeleniya pervichnoy produktsii i khlorofilla v Chernom more v vesenniy i letniy periody (Pecularities of primary production and chlorophyll distribution in the Black Sea in spring and summer), in M. V. Vinogradov (Ed.), *Izmenchivnosti ekosistemy Chernogo morya: estestvennye i antropogennye faktory (Variability of the Black Sea Ecosystem :Natural and Anthropogenic Factors)*, Nauka, Moscow, 128-147
35. Vedernikov V. I., Konovalov B. V., Koblents-Mishke O. I. (1980) Osobennosti raspredelenya pervichnoy produktsii i khlorofilla v Chernom more osenyu 1978 g. (Some pecularities of primary production and chlorophyll distribution in the Black Sea in autumn 1978). in *Ekosistemy pelagiali Chernogo morya (Pelagic Ecosystems of the Black Sea)*, Nauka, Moscow, 105-117.
36. Vedernikov V. I., Demidov A. B. (1993) Pervichnaya produktsya i khlorofill v glubokovodnykh raionakh Chernogo morya (Primary production and chlorophyll in the deep regions of the Black Sea), *Okeanologya*, **33**, 2, 229-235.
37. Vedernikov V. I., Demidov A. B., Korneeva G. A. (1996) Osobennosti raspredelenya produktsionnykh kharakteristikikh fitoplanktona i skorosti gidroliza prirodnykh polimerov v Chernom more osennii period (Patterns of phytoplankton production and hydrolysis rate of natural polymers in the Black Sea in autumn), *Okeanologya*, **36**, 2, 250-259.
38. Yilmaz A., Tugrul S , Polat C., Ediger D., Coban Y., Morkoc E. (in press) On the production, elemental composition (C,N,P) and distribution of photosynthetic organic matter in the southern Black Sea, *Hydrobiologia*.

FLUCTUATIONS OF PELAGIC SPECIES OF THE OPEN BLACK SEA DURING 1980-1995 AND POSSIBLE TELECONNECTIONS

U. NIERMANN, A. E. KIDEYS
Institute of Marine Sciences,
Middle East Technical University,
P.O. Box 28, 33731 Erdemli, Turkey,

A.V. KOVALEV, V. MELNIKOV
Institute of Biology of the Southern Seas,
Nachimov Ave. 2, Sevastopol, Crimea, Ukraine,

V. BELOKOPYTOV
Marine Department of Ukrainian Research
Hydrometeorological Institute,
61, Sovetskaya av., 335011,
Sevastopol, Crimea, Ukraine

Abstract

The drastic changes in the Black Sea ecosystem, i.e. the harsh decline of the Black Sea fishery in 1989 and the dramatic changes in the zooplankton were often related to the outburst of the accidentally introduced ctenophore *Mnemiopsis leidyi* and to other man made events, such as pollution, eutrophication, regulation of river outflows (irrigation, damming) and overfishing. Beginning with the question as to why such changes in the Black Sea ecosystem occurred specifically at the end of the 1980s, the fluctuation of zooplankton stocks in other regions of the world are reviewed and compared with the changes in the Black Sea ecosystem. It transpires that changes in the zooplankton community and in small pelagic fish stocks in the second half of the 1980's until the beginning of the 1990s were evident in all seas under consideration. These changes were discussed in connection with changes in the climatic regime. Striking changes were observed in the NAO (North Atlantic Oscillation), SO (Southern Oscillation), ENSO (Southern Oscillation (El Niño Index), and ALPI (Aleutian Low Pressure Index) in the second half of the 1980s resulting in changes of the hydrological and meteorological regime (river run off, salinity, sea- and air temperature, atmospheric pressure, precipitation and strength of westerly winds) in the northern hemisphere. It is concluded (hypothetically), that possibly, changes in the weather regime during the 1980s could have triggered the changes in the phyto- and mesozooplankton communities of the Black Sea, which caused the conditions for the outburst of *M. leidyi* and the decline of the anchovy stock.

S. Beşiktepe et al. (eds.),
Environmental Degradation of the Black Sea: Challenges and Remedies, 147–173.

1. Introduction

The ecosystem of the Black Sea has undergone dramatic changes, starting in the 1960s. These changes became very obvious due to the spectacular mass development of the accidentally introduced ctenophore *Mnemiopsis leidyi* during 1988-1990 and the collapse of the anchovy fishery in 1989 [1, 2, 3, 4, 5, 6, 7, 8]. During the same period, at the end of the 1980s and beginning of the 1990s, severe changes in the phytoplankton and mesozooplankton composition and a sharp decrease in the total zooplankton biomass became obvious [9, 10, 11, 12]. Gradual changes in the biomass and species composition of the phyto- and zooplankton had already been observed during the 1970s. Blooms of small algal species became very frequent and the former dominant valuable fodder zooplankton species were replaced for the most part by small species, which did not provide good quality fodder for higher trophic levels [13, 14]. Since 1992 however, the Black Sea ecosystem has shown positive signs of recovery [15]. The zooplankton biomass has increased slightly whilst the number of *Mnemiopsis leidyi* has decreased to a moderate level and the total Black Sea anchovy catch in 1995 was 400,639 tonnes, nearly at the same level as it was in the good period of the mid 1980s (449581 tonnes) [16].

The decrease in biomass and changes in the species composition of the plankton and fishes, especially in coastal areas since the 1960s, were related to the increasing chemical and oil pollution, eutrophication, and other anthropogenic impacts such as dumping, dredging, and damming of the rivers [4, 14, 17, 18]. In particular the eutrophication of the northern shelf areas had far reaching effects on the benthic and pelagic communities and fish stocks [6, 14]. The northwestern shelf was the most productive area of the Black Sea and the most important feeding and nursery ground of the commercially exploited Black Sea fish stocks [19]. This means that fish larvae in their most sensitive stages, the, have to grow up in the most polluted and highly eutrophic area of the Black Sea, leading to high mortality of the young fishes. Since most of the commercially exploited Black Sea fish stocks are migrating species with feeding and spawning areas in the northern Black Sea and overwintering at the southern Turkish coasts, negative recruitment affects the entire Black Sea fishery.

The sudden and drastic changes in the mesozooplankton community and the decline of the anchovy stock during 1989/1990 were related, besides the ongoing anthropogenic impacts, to two main reasons; the impact of the mass development of *Mnemiopsis leidyi*, and the overfishing. One of the most striking causes for the drastic changes in the zooplankton composition may be the mass development of the new invader *Mnemiopsis leidyi* during 1988-90. The sharp decrease in the zooplankton biomass, and the sharp decline in anchovy eggs and larvae, reflecting the collapse in anchovy catches for the entire Black Sea in 1989, coincided with the mass development of the new invader *Mnemiopsis* which appeared in 1988. Laboratory experiments and in situ mesocosm studies on the predation and relative predation potential of *Mnemiopsis* showed that *Mnemiopsis* is an important predator of zooplankton, anchovy eggs and larvae (especially of yolk-sac larvae) [20, 21, 22, 23, 24, 25, 26]. *Mnemiopsis* could, therefore, be a threat to fishery year-class recruitment [27]. The high fecundity (an average of 8000 eggs within 23 days) and the huge growth rates (up to daily doubling of the individual biomass) observed in this ctenophore could only be sustained by high feeding rates [28]. Since the seasonal bloom of *Mnemiopsis* starts as well in the mean spawning season of the anchovy in July/August and continues until late autumn [29], *M. leidyi* could decimate a high amount of fish larvae either by predation or due to food competition.

An argument for this theory is that the anchovy stock recovered after 1992, when the biomass of *Mnemiopsis* declined to moderate levels (Fig.1). However, in contradiction, during the mass development of *Aurelia aurita* during the 1970s (feeding on the similar prey organisms), which reached about the same biomass in terms of carbon as the *M. leidyi* bloom in 1989 [15], the impact to the anchovy stock was very little. This suggests that other factors played a role in the decline of the anchovy.

Overfishing could be seen as another reason for the decline of the Black Sea fish stock and changes in the entire ecosystem of the Black Sea [8, 30, 31, 32, 33]. Until now there has been no international management of the fish stocks of the Black Sea, of which anchovy is the most important. During winter the anchovy migrate from the northern Black Sea towards the warmer waters of the Turkish coast and form very dense schools in a narrow coastal band. Due to this schooling behavior the anchovy become extremely vulnerable to the Turkish fishing vessels, which since the mid 1980s have become very well equipped with sonar and purse seiners. Due to the intense fishing of the Turkish fishing fleet the size of the spawning stock was reduced to the extent, that not enough young fish were produced to ensure the survival of the future stock [31]. Gücü [31] concluded that the effects of over-fishing, especially that of Turkey at the end of the 1980s, should not be dismissed as a minor reason for the collapse of the anchovy fishery.

The question remains, why did the outburst of *Mnemiopsis leidyi* and the subsequent collapse of the anchovy fishery occur specifically at the end of the 1980s and not before or after? It could be assumed, that due to the man-made and natural environmental changes occurring since the end of the 1960s, the pelagic ecosystem was driven to an evermore unstable state with a changed prey-predator relationship, and that it required only a certain trigger for the outburst of *Mnemiopsis leidyi*, which was then favoured by the altered trophic structure. This trigger could be assumed to be a certain climatic signal, which occurred either during or at the end of the 1980s. In order to detect this signal we compared the long-term fluctuations of the plankton and anchovy of the Black Sea with fluctuations in different regions of the world, to discuss to which extent it is possible to link the fluctuation of species in the Black Sea and other seas to large-scale meteorological and hydrographical events.

2. Fluctuation patterns in the Black Sea and other seas

In order to detect similarities in the fluctuation patterns of different taxonomic groups in the Black Sea and in other seas, striking fluctuation patterns are listed and described. We focus mainly on events, which happened in the Black Sea at the end of the 1980s and the beginning of the 1990s. These patterns are compared with findings in other seas (Baltic Sea, North Sea, Atlantic, waters off California) and in European fresh water lakes (Lake Windermere; Great Britain, Bodensee; Germany,) and related to possible hydrographical and meteorological events.

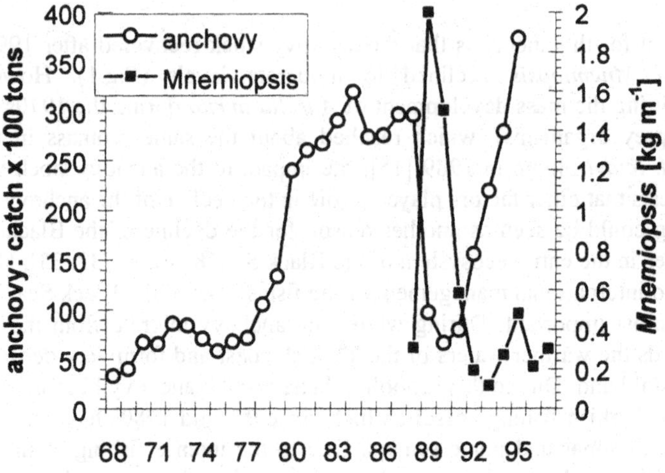

Figure 1: Black Sea 1968 - 1997: Turkish anchovy catch and *Mnemiopsis* biomass (wet weight). (Data on anchovy catch: State Statistical Institute, Turkey; data on *Mnemiopsis* (1988: Institute of Biology of the Southern Seas, IBSS, Ukraine; 1989-1990: Shirshov Institute, Russia; 1991-1997: Institute of Marine Sciences, Turkey)

2.1. OBSERVATIONS IN THE BLACK SEA

2.1.1. *Phytoplankton*

An increasing trend in the phytoplankton biomass was obvious since the beginning of the 1970s in the offshore regions of the Black Sea (see Mikaelyan, [34]). Especially after 1985, the phytoplankton biomass fluctuated, but at a steady high level. In 1989 however, the biomass was very low, but during the summer of 1990/91 very high biomasses were recorded in the open area of the Black Sea. The author analysed the fluctuation of the phytoplankton biomass in the open Black Sea in relation to the concentration of nutrients, (PO_4, NO_3, NO_2), to the fluctuation of zooplankton, to hydrodynamic features of the Black Sea and the winter air temperature. He concluded, that interannual oscillations of the average air temperature in winter are inversely related to the mean phytoplankton biomass in summer, indicating that the biomasses of the bulk of phytoplankton species are linked to the hydrodynamic processes in the Black Sea driven by climatic variations.

Changes in the species composition of the phytoplankton off the Bulgarian coast, were also obvious since 1970, when the biomass ratio of the two dominant phytoplankton groups (Bacillariophyta: Dinophyta) had undergone a dramatic change [35]. Up to 1970 the Bacillariophyta (diatoms) were dominant in terms of biomass (86

% of the total phytoplankton biomass). After 1970 the biomass of both groups displayed high oscillations indicating a period of instability. In subsequent years the biomass of the Dinophyta increased rapidly and became the dominant group in the Black Sea (1973-1974; 1978-1980; 1985-1988). Then from 1989-1993 the Bacillariophyta became dominant again (Fig. 2).

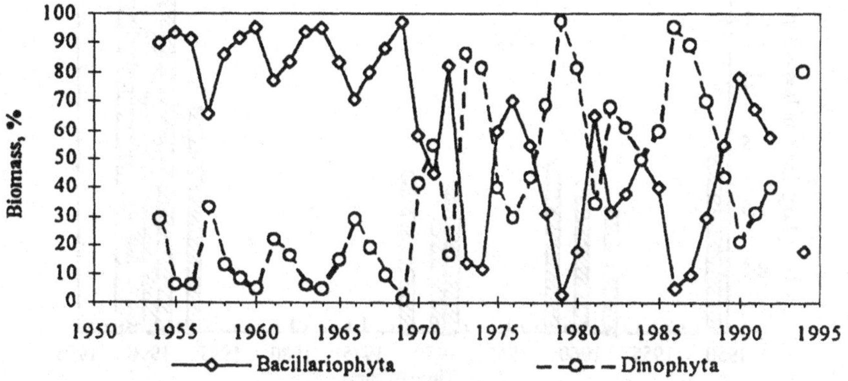

Figure 2. Black Sea 1954 - 1990. Long-term trend in Bacillariophyta : Dinophyta biomass ratio (redrawn from [32] and [36]).

An enhanced bloom of coccolithophores (mainly *Emiliana huxleyi*), with a biomass 1.5-2 orders of magnitude larger than in previous years, was observed in the Black Sea during 1989-1992 (Fig. 3; [37]. The mass occurrence of this species reduced the transparency of the sea water more than 3 fold from 21.3 to 6.2 m.

2.1.2. *Zooplankton*

The wet weight of the total zooplankton increased from the 1960s until 1990 due to mass developments of gelatinous organisms such as *Aurelia aurita* (1960s; 1970s, 1980s), *Noctiluca scintilans* (1980s, 1990s) and the new invader *Mnemiopsis leidyi* (since 1988). When the biomass of *M. leidyi* reached its climax during the end of the 1980s the biomass of *A. aurita* decreased. In the Black Sea *Mnemiopsis* showed the typical pattern of a new coloniser: After mass development in the years 1989 and 1990 with a biomass of 2 kg m^{-2} in the open sea and 4.7 kg m^{-2} at the north-western shelf [29], the numbers and biomass had dropped to a moderate level by 1991 and fluctuated till 1996 in the range of 0.2 - 0.5 kg m^{-2} [7, 10, 33]. After the decrease of *Mnemiopsis* in the summer of 1990 the biomass of *Aurelia* increased again and since the summer 1991 until 1994 the biomass of both species remained more or less at the same level [7].

Long-term series of the mesozooplankton biomass in coastal and offshore areas of the Black Sea, established by different Institutes displayed similar patterns. The composition of the zooplankton communities was different during the period before 1986 and after 1989-1992. One obvious point was the drastic decline of the mesozooplankton biomass by the end of the 1980s and beginning of the 1990s [9, 11, 12, 29, 33, 38].

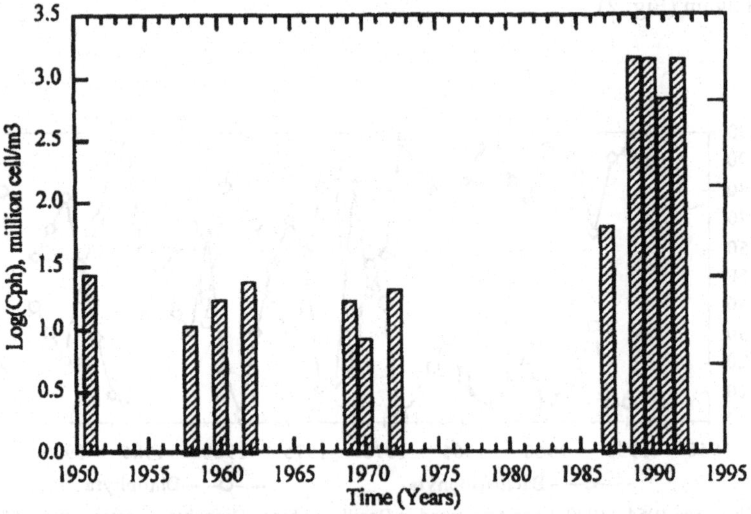

Figure 3 Phytoplankton concentration in the photic zone of the western deep part of the Black Sea during 1950 - 1992 (redrawn from [37]). The phytoplankton blooms consisted mainly of coccolithophorides (*Emiliana huxleyi*). Years without bars. no measurements

Changes in the dominant species were already observed in Sevastopol Bay at the beginning of the 1980s. During 1960-1964 the percentage of *Acartia clausi* of the total copepod community was 17 %. This increased to over 30 % during 1981-1985, and during 1991-1994 rose to 75 %. Copepod species such as *Oithona nana, O. similis* and the cold water species *Pseudocalanus elongatus* and *Calanus euxinus*, which where dominant before the 1980s, decreased during the 1990s [11, 39]. The species *O. nana* was not observed from 1986-1997. In offshore areas the dominant species *Pseudocalanus elongatus, Calanus euxinus* and *Sagitta setosa* decreased about 10 fold and 100 fold respectively [15, 29]. Since 1993/1994 the biomass of the mesozooplankton has been increasing in all areas of the Black Sea [9, 11, 29, 33, 38]. And in 1995 the species *Oithona nana* appeared again in the zooplankton samples [39].

2.1.3. *Ichthyoplankton and anchovy*

In the early 1960s the number of anchovy eggs varied between 120-390 eggs m^{-2} in the Sevastopol area (From 1963 to 1986 no sampling was carried out.). In the 1980s egg and larval numbers were low compared to the 1960s. In 1989 a significant decline in the number of anchovy eggs and larvae occurred in inshore and offshore waters [6]. Since 1992 the number of eggs increased again and in July 1995 the average number of anchovy eggs in Sevastopol Bay were in the same range (180 m^{-2}) as the numbers found during the early 1960s (Fig. 4). Similarly, the highest egg and larval abundances for the anchovy of the southern Black Sea were obtained in the summer of 1996 [40]. Despite the high egg number, the number of larvae in the northern Black Sea was very poor during 1995/96 compared to the 1960s. The amount of larvae observed, compared to the egg numbers of *Engraulis encrasicolus* and *Trachurus mediterraneus* was 0.1-7% in 1995/96 whereas during the early 1960s this proportion was 30 to 50% [41].

The sharp decline in anchovy eggs and larvae at the end of the 1980s reflected the collapse in anchovy catches for the whole of the Black Sea which fell from 550,000 tonnes in 1988 to 180,000 tonnes in 1989. Since 1992 the anchovy stock has recovered and during 1995 the catch of anchovy was nearly as high as in the best period during the 1980s [16].

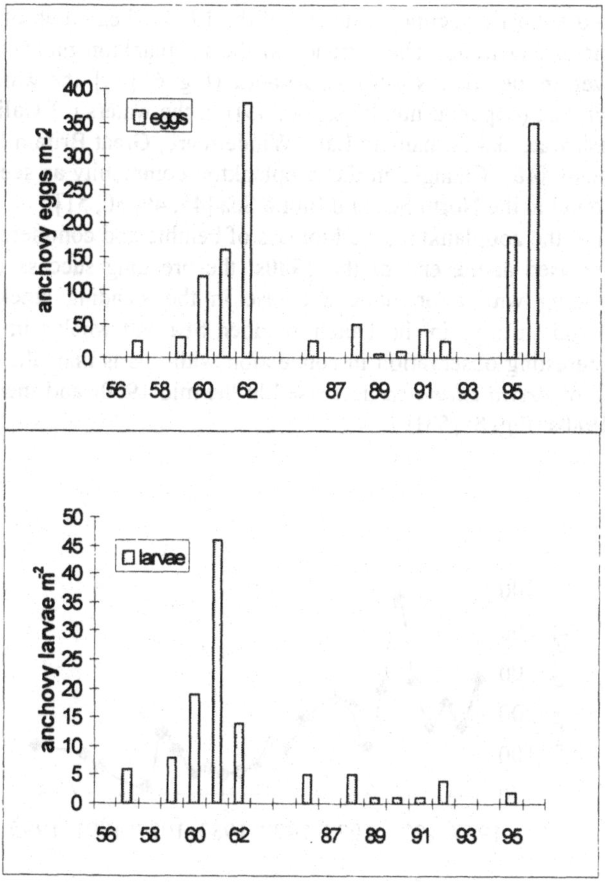

Figure 4. Black Sea: Fluctuations in the abundance of anchovy egg and larva (numbers m^{-2}) along the Crimean peninsula (1957-1996). ([6], and unpublished data of A. Gordina, IBSS, Sevastopol, Ukraine)

In summary, it is obvious that long-term data series established by different institutes in different areas of the Black Sea displayed a similar overall fluctuation pattern: high biomass in the 1970s, a decline in the biomass at the beginning of the 1980s, a rise in the biomass during the mid 1980s and a drastic decline at the end of the 1980s / beginning of the 1990s (Fig. 5). After 1993, the mesozooplankton biomass recovered and is continuing to display an increasing trend [9, 11, 12, 29, 33, 38].

2.2. FLUCTUATION PATTERNS IN OTHER SEAS

A comparison with long-term data series of other regions of the world showed similar fluctuation patterns in the zooplankton as seen in the Black Sea: high biomass in the 1970s, a decline in the biomass at the beginning of the 1980s, a rise in biomass during the mid 1980s and a drastic decline at the end of the 1980s / beginning of the 1990s and after 1993 an increasing trend. These trends in the zooplankton biomass or numbers were found as well in the open North East Atlantic (Fig. 6; [42], the waters off Iceland [43], in the North Sea (copepod numbers, [44, 45]) in the waters off California (Fig. 7; [46]), and in fresh water lakes, namely Lake Windermere; Great Britain [47] and in the Bodensee; Germany [48]. Changes in the zooplankton community as seen in the Black Sea, were also found in the North Sea and Baltic Sea [45, 49, 50, 51].

In addition to the zooplankton, the biomass of benthic and commercially exploited species also decreased at the end of the 1980s: the breeding success of eider ducks collapsed, coinciding with a dramatic decrease in the standing stocks of mussels, cockles, whelks and shrimps in the Dutch Wadden Sea and whelks in the North Sea [52]. Another interesting observation in connection with this is that the annual growth rings in the shell of *Arctia islandica* decreased in the mid 1980s and increased again at the end of the 1980s (Fig. 8; [53])

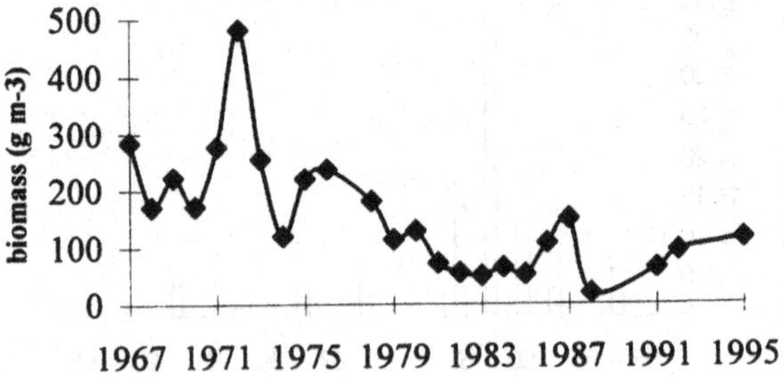

Figure 5. Zooplankton biomass off the Bulgarian Black Sea coast during summer 1967-1995. (without *Noctiluca scintilans)* [9]. This figure reflects the same trends, as were observed by Romanian, Ukrainean, Russian and Turkish Institutes [11, 12, 29, 33, 38].

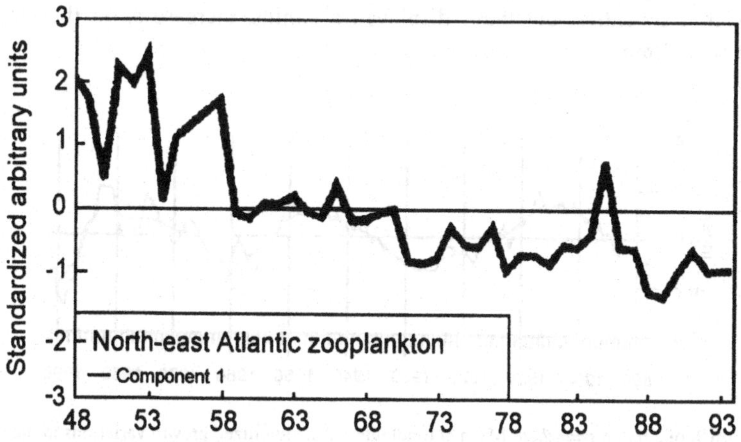

Figure 6. Trends in zooplankton of the northeastern Atlantic from 1948-1993 according to the Continuous Plankton Recorder (redrawn from [42].

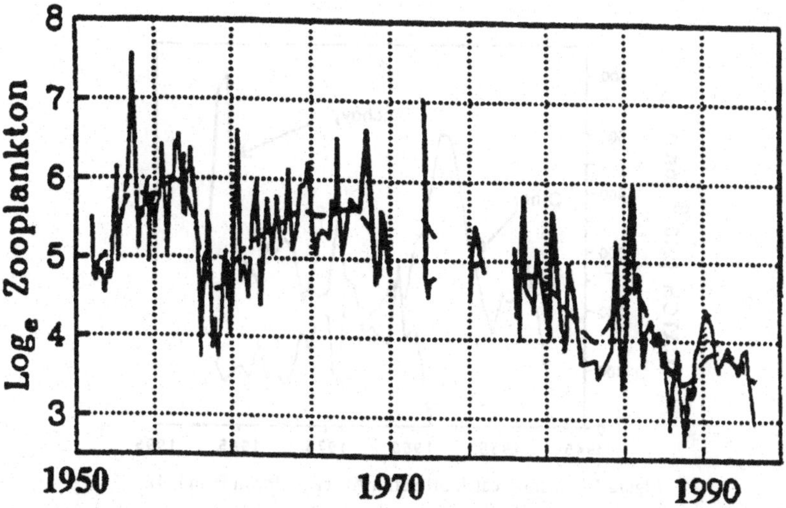

Figure 7 Fluctuation in zooplankton biomass off California from 1951-1993 (redrawn from [46]).

Comparison of the anchovy catches in different upwelling systems of the world showed that the anchovy stock of South Africa (Fig. 9) Benguela, and California (Fig. 10) collapsed as well in the same period as the anchovy stock of the Black Sea during the second half of the 1980s despite no impact of *Mnemiopsis leidyi* and in spite of different stock management regimes and fishing efforts [54, 55]. The anchovy stock in the Kuroshio-Ohashio upwelling system and the Humbolt area had already decreased during the mid seventies, but they displayed a similar increase as in the other areas during the mid 1980s.

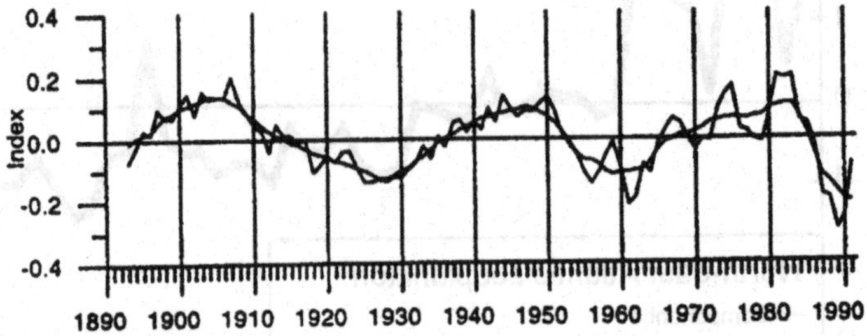

Figure 8. Growth of *Arctica islandica*: Mean chronologies of standardized growth variations in old shells from the Fladen Ground, North Sea. Shown are the unsmoothed mean chronologies with a 3-year adjacent averages. The increase at the end of the 1990s could be related to an Atlantic inflow event (redrawn from [53]).

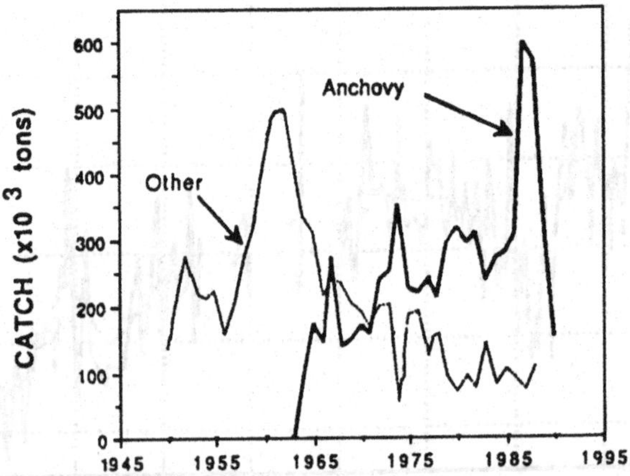

Figure 9 Anchovy catch off South Africa (redrawn from [54].

3. Meteorological and hydrological forcing and species densities

The findings of so many similarities in the fluctuation patterns of different species at the end of the 1980s leads to the question: what causes this behaviour in the different marine ecosystems mentioned above? Anthropogenic impacts, which have increased since the 1960s undoubtedly have an effect on marine ecosystems, especially in enclosed and semi-enclosed seas such as Black Sea, Baltic Sea and North Sea. Since this topic has already been discussed by many authors (reviewed by Zaitsev [14], we want to focus solely on the influence of large scale weather patterns on the fluctuation of marine species.

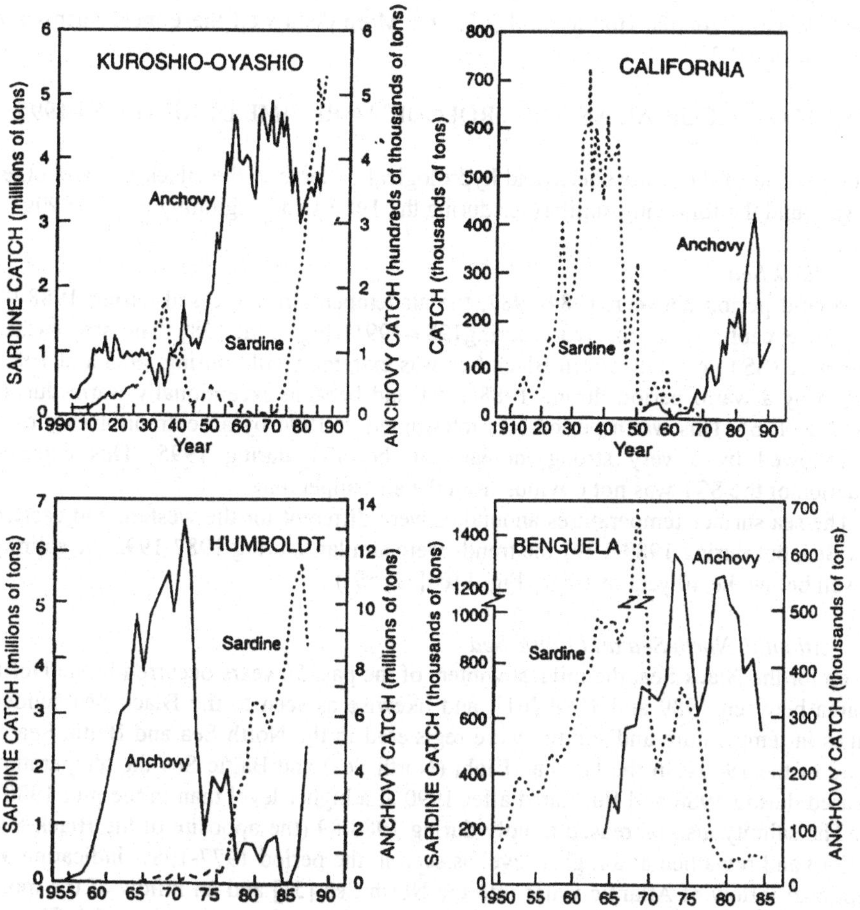

Figure 10 Annual catch of sardine and anchovy in four major current systems during the 20th century (redrawn from [55])

Links with short-term or large-scale weather patterns have been suggested by many authors. Wind, extreme winter and summer temperatures, rainfall, high river discharges, and salinity anomalies are mentioned as possible causes in the literature [56]. Examples are: the mean phytoplankton biomass in summer is inversely related to the average air temperature in winter (Black Sea; [34]. The average zooplankton abundance in May-August is dependent on the sea temperature in March (the eastern Black Sea, in 0-100 m depth; [57]. The abundance of copepod species is dependent on the salinity: *Calanus finmarchicus* is associated with salinity in deep waters (North Sea; [44], high densities of *Pseudocalanus elongatus* correspond to years with low salinity at 5 m to 20 m depth before and during the reproduction season (Baltic Sea; [50]). Thus a shift in storm frequencies or wind directions might cause changes in the sediment water exchange in shallow areas or influence the depth of stratified waters in deeper areas, which affects the abundance of zooplankton as well [45, 56]. More details on these mechanisms are given in section 3.3.

3.1. METEOROLOGICAL AND HYDROLOGICAL REGIME DURING 1980-1995

By comparison of the atmospheric and hydrological patterns in the Black Sea and other seas we found the following similarities during the 1980's and beginning of the 1990s:

3.1.1. Black Sea

After a cold period between 1984-1987, the air temperature increased during 1988-93 [12] with exceptional warm years during1989-1991 (Fig. 11a; [58]. The sea surface temperature (SST) of the western Black Sea was extremely cold during 1985 and 1987, followed by a warm period during 1988-1991 and became exceptionally warm during 1989 (Fig. 11b). This warm period was interrupted in 1994 by an extraordinarily cold SST followed by a very strong increase of the SST during 1995. This extreme fluctuation of the SST was not obvious from the air temperature.

The sea surface temperatures anomalies were different for the western and eastern Black Sea, but during 1989-1991 the trends were similar. During 1987-1992 the salinity was well below the 40 year average (Fig. 11c; [12, 59]).

3.1.2. Atlantic, North Sea and Baltic Sea

As seen for the Black Sea, the mildest winters of the past 50 years occurred in the North Sea area between 1989 and 1994 [61], and likewise as seen in the Black Sea sudden changes in temperature and salinity were registered in the North Sea and Baltic Sea at the end of the 1980s. In the German Bight (North Sea) and Baltic Sea the temperature increased during 1988 and fluctuated after 1990 at a higher level than in the mid 1980's [51]. The salinity also increased rapidly during 1988/89 (the opposite of the Black Sea situation) and remained at a higher level as seen in the period 1977-1987, indicating an exceptional inflow of Atlantic water into the North Sea [51] and an inflow of the more saline North Sea water into the Baltic Sea [50]. Evidence for an exceptional inflow of Atlantic water into the North Sea having taken place during 1989 is the presence of Lusitanian fish species like *Trachinus vipera, Zeus faber, Mullus surmuletus* [62, 63] and other warm water species such as the siphonophore *Muggiaea atlantica* [89], the tunicate *Doliolum nationalis* [65], the cladoceran *Pelina* [64] and four phytoplankton species [66] since then. The hydrological changes during the end of the 1980s were reflected in the Baltic Sea [49] and Kiel Bight [50] as a change in the species composition and high interannual fluctuations of some species, indicating a period of instability, which often is induced by a change of environmental parameter [67].

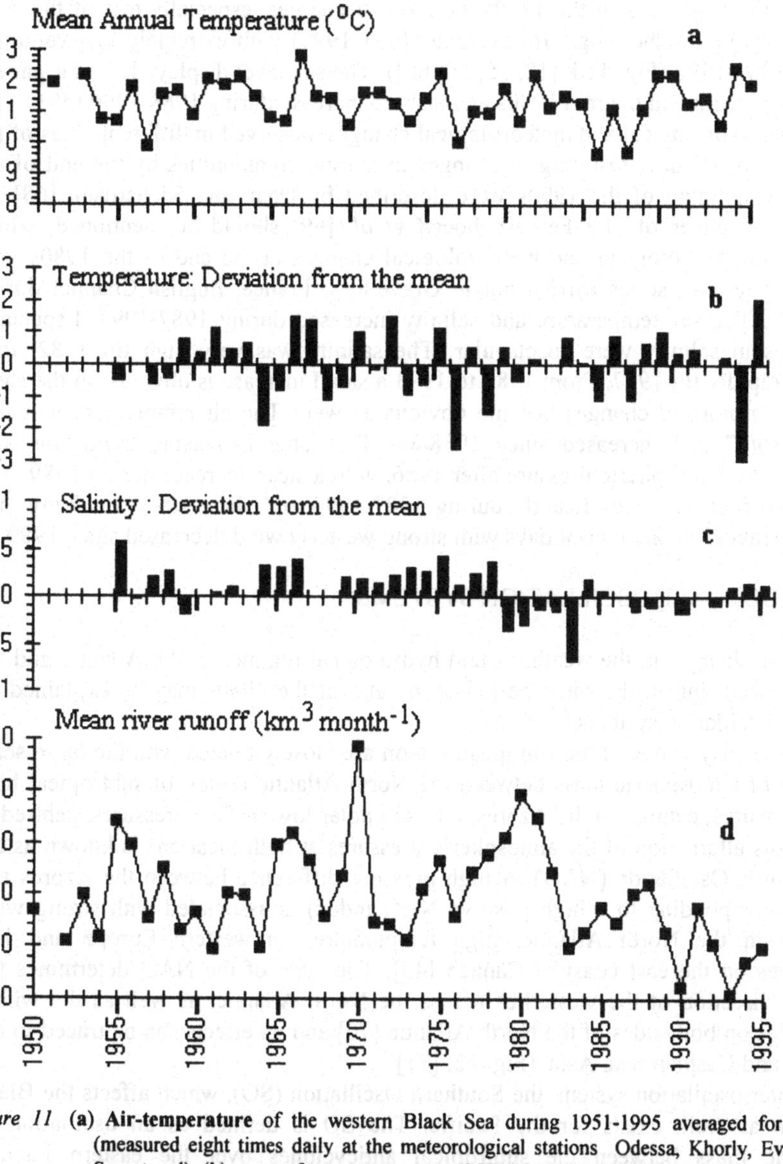

Figure 11 (a) **Air-temperature** of the western Black Sea during 1951-1995 averaged for March (measured eight times daily at the meteorological stations Odessa, Khorly, Evpatoria, Sevastopol). (b) **sea surface temperature** (SST) 1951-1994, March; (c) **surface salinity (SSS)** for March, inshore, 1951-1994. SST and SSS were registered eight times daily in the north-western area (44^0 -44^0 40'N and 31^0-33^0 E) and averaged for February (d) **annual discharges of the Black Sea rivers** of the former Soviet Union were calculated as a sum of the monthly run-off of the Danube, Dniepr, Dniestr, Yuzniy Bug, Rioni, Ingouri, Bzyb, Mzymta, Chorokh, Kodory (redrawn from [12])

Since the beginning of the 1980s the river discharges, especially that of the river Danube, were below the long-term average (1931-1995) with extremely low values in 1982 and 1989-1990 (Fig. 11d, [12, 58, 59, 60]). The sea level displayed similar pattern with a strong decline in the mid 1980s and a sharp increase during 1988-1990 [59].

Similar hydrological and meteorological changes, observed in different areas of the North Sea and Atlantic, resulting in changes in marine communities by the end of the 1980s and beginning of the 1990s were described by Daan and Richardson [68]. In particular, the paper of Le Fevre-Lehoerff *et al.* [69] should be mentioned, which documents the hydrological and meteorological changes at the end of the 1980s very clearly (a long-term series carried out in Gravelines; France, English Channel during 1974-1993). The sea temperature and salinity increased during 1987-1993. Especially the changes in salinity were spectacular. The salinity was very high till 1982, then decreased rapidly till 1987. From 1988 to 1993 a small increase is obvious. In the same period meteorological changes became obvious as well: The air temperature was low during 1986/87 and increased since 1988/89. The same increasing trend was also obvious for the atmospheric pressure after 1986, with a steep increase during 1989. The precipitation increased significantly during 1987-1990 and since 1991 was below the 19-year average. The number of days with strong westerly wind decreased since 1989.

3.2. LARGE-SCALE OSCILLATION SYSTEMS

These major changes in the weather - and hydrological regimes of the Atlantic and the Black Sea occurring in the same period at the end of the 1980s may be explained by large-scale oscillation systems:

The westerly winds of the European region are closely related with the large-scale alternation of atmospheric mass between the North Atlantic region of subtropical high surface pressures, centred in the Azores, and sub polar low surface pressures, centred in Iceland. This alternation of the atmospheric pressures at both locations is known as the North Atlantic Oscillation (NAO). A high pressure difference between the Azores and Iceland (corresponding to a high positive NAO index) is associated with strong wind circulation in the North Atlantic, high temperatures in western Europe and low temperatures on the east coast of Canada [45]. The state of the NAO determines the speed and direction of the westerlies across the North Atlantic, as well as the winter temperatures on both sides of the North Atlantic [70] and its effects can be traced to the Black Sea and Caspian Sea, Asia (Fig. 12; [71]).

Another oscillation system, the Southern Oscillation (SO), which affects the Black Sea region as well, exists for the Pacific. The SO is defined as an oscillation of atmospheric mass between the subtropical anticyclones over the eastern Pacific, particularly the Eastern Island anticyclone, and the Indonesian low-pressure area. The SO is associated with the El Niño phenomenon, resulting in the term: El Niño/Southern Oscillation (ENSO).

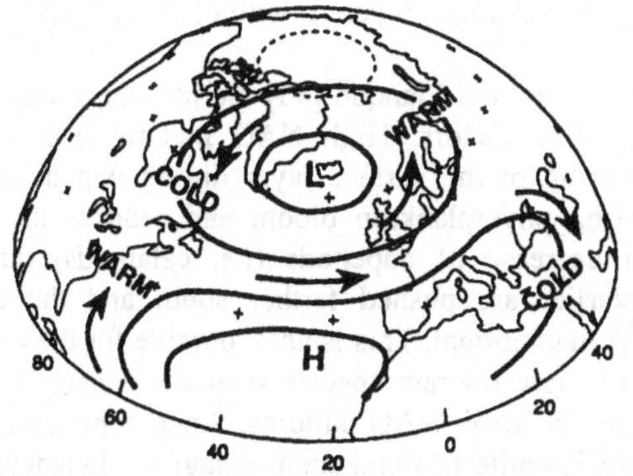

Figure 12. Idealised scheme of the NAO. If the NAO index is high, the westerly winds reach the Black Sea (redrawn from [71]).

Although the short and long term atmospheric variabilities in the eastern Mediterranean and Black Sea regions are well known, hypotheses on teleconnections with global atmospheric events have recently been put forward [58]. Polonsky *et al.* [59] have shown good correlation between the Black Sea hydrology, the North Atlantic sea surface temperature (SST) variability and the ENSO-type variability of the tropical Pacific Ocean, which indicates a global ocean-atmosphere coupling. Both types of variability lead to changes of the cyclone trajectories, precipitation over the river drainage basins and changes in the river discharges, resulting in changes in the north-western Black Sea hydrography [59], which may finally lead to changes in the Black Sea plankton community. (The effects of the recent El-Nino event on the pelagic communities are not mentioned in this paper, since the data are still being processed).

It is known, that global atmospheric changes, such as the Southern Oscillation (SO), El Niño Southern Oscillation (ENSO), and North Atlantic Oscillation (NAO) influence pelagic communities [70, 72, 73, 74, 75]. The 1982-83 ENSO event has been associated with a reduced phytoplankton biomass in the western and eastern Pacific. It has also been linked with a rise in temperature and a reduced flow of the Californian current, that induced radical changes in zooplankton biomass (Mann, 1993). It is known, that the ENSO and the Aleutian low affect the fishery in the Pacific as well [73].

3.3. HOW DO WEATHER PATTERNS AFFECT THE ZOOPLANKTON COMMUNITY?

One hypothesis as to how weather patterns may affect the zooplankton community is given by [45]. In the North Atlantic, the fluctuations of the *Calanus finmarchicus* (cold water species) and *C. helgolandicus* are closely linked with the state of the NAO. In years with high NAO a significant decline in total *Calanus* abundances is obvious [45]. Due to differences in the biology of both species (different seasonal cycles, opposite temperature affinities and different geographical locations) the species react differently. In years with high NAO the cold water species *C. finmarchicus* displays lower abundances than in high NAO years, whereas with *C. helgolandicus* the opposite is true [45].

According to Fromentin and Planque [45] the mechanism of these changes could be explained roughly by 2 factors: High NAO pattern leads to high wind stress, generating a strong mixing of the surface layer during winter and spring. Enhanced mixing delays the spring phytoplankton bloom and reduces the primary production leading to a general decrease of copepods (i.e. calanoids) due to lack of food. Additionally the westerlies are pushed farther south and the air and sea surface temperatures are higher than normal. This is unfavourable for the cold water species like *C. finmarchicus* and so more tolerant species such as *C. helgolandicus* are favoured. The situation is reversed by weak NAO patterns. Since copepod species constitute the main food resources for juvenile fish and small pelagics, like anchovy and sprat, it can be assumed, that a change in the copepod stock in terms of species composition and biomass has consequences for the pelagic fish stocks. Additionally wind-induced turbulence and temperature influence the capture efficiency, growth and development of fish larvae [76].

But not only the strength of the NAO and the westerlies influence the fluctuations of the zooplankton in the Atlantic, the position of the Gulf stream could be related with the abundance of the zooplankton in the European area as well. Larger zooplankton abundances occur around the British Isles during years with northward displacements of the northern wall than during those with southward displacements [47]. An interpretation of this association 'Gulf stream : Zooplankton biomass' is, that displacements of the northern wall may result in weak local perturbations of the atmospheric circulation, that are felt downstream on the European Continental shelf. Evidence to support this interpretation is given by the good relationship between the zooplankton fluctuation in Lake Windermere (a fresh water lake in England) located at the same latitude as the northern wall. These observations confirm, that the coupling is atmospheric. Associated air temperature and water temperature data from Lake Windermere indicate that the coupling operates through the seasonal thermal stratification [47].

3.4 CLIMATIC SITUATION AT THE END OF THE 1980S AND BEGINNING OF THE 1990s

The most important changes in the zooplankton biomass and species composition have taken place during the end of the 1980s and the beginning of the 1990s. What type of large weather patterns could trigger the hydrological and meteorological changes during this period?

According to Rodionov [71] the 1980s differed from the previous decades by the following peculiarities: The 1983-1990 period was dominated by a positive phase of the NAO (Fig. 13). The winter SST averaged over the North Atlantic was particularly low in the mid 1980s; in 1986 it reached the lowest recorded value for the entire period of observation since the late 19th century [77]. Enhanced climatic variability during the 1980s was also observed in the Atlantic European sector. The magnitude of the NAO index (computed as the difference between normalised mean winter surface air temperature (SAT) at 60°N, 10°E, Jakobshaven, Greenland and 70°N, 50°W, Oslo,

Norway) increased since 1970 and its variance was about 5 fold higher during the 1970-1989 period than in the previous 2 decades (Fig. 14). Most striking was the fact that the index jumped from a record low value in 1986 to a record high value in 1989. The extremely low value in 1986 resulted in severe winters in northern Europe [77]. The high-pressure anomaly over Greenland and frequent northerly winds along its eastern flank were responsible for the negative temperature anomalies in the northeastern Atlantic. A contrasting situation occurred in winter 1989, when intensive advection of warm and moist air from the north Atlantic led to a warm winter in western and northern Europe influencing the Black Sea region as well.

The same strikingly high climatic variability as seen in the North Atlantic was also obvious in the Pacific. The decade from 1980-1990 displayed the largest global scale year to year variability of the 20th century [78]. The Southern Oscillation experienced the strongest warm (El Niño) episode of the century (1982/83) and the strongest cold (La Niña) episode in 50 years (1988). According to Trenberth and Hurrell [79], the very strong La Niña event of 1988 apparently terminated the climate regime, that was established in 1977.

Long-term data series suggest that strong climatic changes in the Black Sea surface waters was in synchronism with the adjacent seas [58]: For example, in the Black Sea an extreme event of cooling evidently took place in 1987, when similar effects were noted in the surrounding seas, *e.g.* dense water intrusion into the Marmara Sea from the Aegean [80] and deep water formation in the Rhodes Gyre region in the Mediterranean [81]. That the period at the end of the 1980s was extraordinary could be seen as well in the ice coverage in the Golf of Gdansk (Poland), which was very strong during the middle of the 1980s (1984-87) but was negligible after 1987, when the warm period started [82].

To summarise, the following hydrological and meteorological weather pattern changed over the northern hemisphere at the end of the 1980s:

- In the open Black Sea the air and the sea surface temperatures decreased after 1986, were extremely low in 1987, rose again in 1988 until 1991 and decreased again until 1994 [11, 12, 58, 83].
- Changes in hydrological and meteorological parameters as in air temperature, atmospheric pressure, precipitation and changes in the strength of westerly winds were recorded at end of the eighties (Baltic Sea; [49]; North Sea; English Channel: [51, 52, 68, 69]) and in other regions of the northern hemisphere like the Barents Sea [71].
- Distinct changes were measured in the NAO (North Atlantic Oscillation; mean sea level pressure difference between Azores and Iceland, Fig. 13), SOI (Southern Oscillation Index; mean sea level pressure difference between Darwin and Tahiti), ENSO (Southern Oscillation (El Niño) Index), and ALPI (Aleutian Low pressure Index, Fig. 15) at the end of the 1980s [46, 75]. Both the Iceland low and the Aleutian low were stronger than normal. Additionally the Iceland low shifted westward and the Aleutian low shifted south eastward from their long-term mean positions [71].

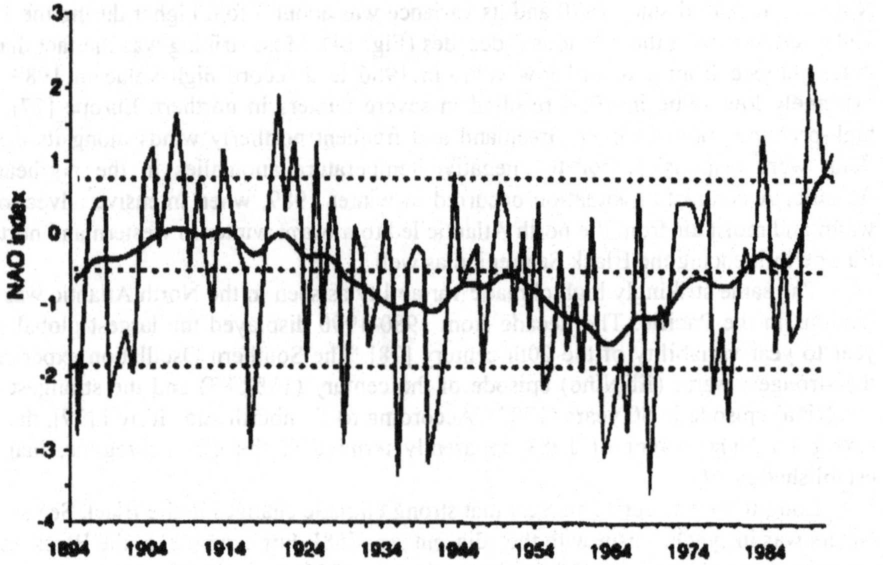

Figure 13. North Atlantic Oscillation (NAO) index based on the average pressure difference between the Azores and Iceland for the period December - April between 1985 and 1992. Dashed lines represents mean values (middle line) and standard deviations (upper and lower lines) (redrawn from [45]).

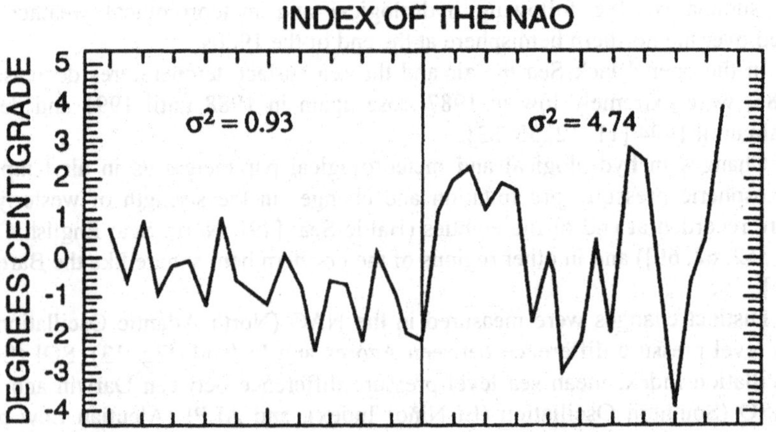

Figure 14. Index of the North Atlantic Oscillation computed as the difference between normalised mean winter surface air temperatures(SAT) at 60°N, 10°E Jakobshaven, Greenland, and 70°N, 50°W Oslo,Norway (redrawn from[71])

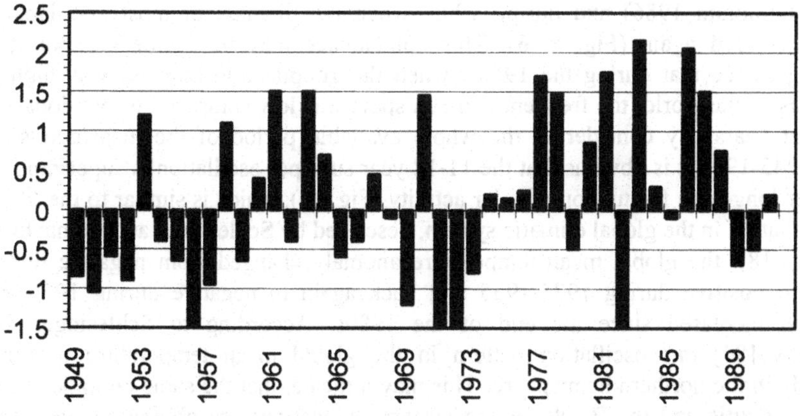

Figure 15 Fluctuation of Aleutian Low Pressure Index (ALPI) from 1949-1991 (redrawn from [46]).

- The shift of the northern wall of the Gulf Stream was far southwards during 1986-1988, compared to the previous and subsequent years [47].

In this connection it may be interesting to note, that :

- The oscillation of the global air temperature, which has a period of 65-70 years had its lowest amplitude during the mid 1980s [84].
- The sun activity, measured in terms of sun spot numbers, which has a period of 11-12 years, was very low during the mid 1980s and rose back to high values during 1988 [71].
- The moon's orbit crossed the ecliptic in 1988 (lunar tide, with a period of 18. 6 years), which causes an extreme peak in tidal forces (and has well documented effects on the sedimentation rates in the Wadden Sea) [52].

By mentioning the sun activity the question occurs, to which extent it is possible to connect the oscillation of the global coupled ocean-atmosphere system with the sun activity? Storm patterns in temperate zones may be influenced by the occurrence of sun spots [85]. Cycles of 11 to 22 years have been observed in several sedimentary records which hint at a possible influence of the 11 year sunspots cycle or the 22 year Hale cycle on marine systems [56, 86]. It was suggested that in the Black Sea, the water transparency [37] and the ratio Bacillariophyta : Dinophyta [35] could be associated with solar activity.

Comparing the fluctuation of the Black Sea communities with the sun spot cycles (Fig. 16) it becomes obvious, that the number of sun spots increased after 1987/88, when the biomass of the zooplankton decreased and were minimal during the mid eighties (minimum 1986) and during 1995, when the biomass of most zooplankton species increased again (Fig. 5; 6). This coincidence may be accidental, but it is interesting to see that during the 1970s, when the zooplankton biomass was high in many areas of the world, the frequency of sun spots was low compared to previous and subsequent years. By considering the whole available period of the measured solar activity 1745-1995 it is obvious that the 11-12 year sun spot oscillation is superimposed by another long-term fluctuation of solar activity (Fig. 17), which is similar to the 65-70 years oscillation in the global climatic system, described by Schlesinger and Ramankutty ([84]; Fig. 18): the global mean temperature anomaly changed from negative during 1900-20 to positive during 1935-1955 and back again to negative during 1970-mid 1980s and increased since the end of the 1980s. According to Schlesinger and Ramankutty [84] this oscillation pattern in the global mean temperature occurred exclusively in the northern hemisphere. This may indicate, that the same climatic forces drive the Atlantic and the Black Sea ecosystem. Indications are obvious as well that similar biological amplification may occur in terrestrial populations [87]. That climatic cycles affect vegetation and fish stocks in the same area was already shown by Ottestad in 1942 by a most interesting comparison: the yield of the cod in Lofoten fishery correlates well with periodicities in the width of annual rings in the pine trees [90].

4. Limitations

The mention of teleconnections and simultaneous meteorological events in different parts of the world, as outlined in this paper is not sufficient to explain the Black Sea events. Our ideas should be taken as stimulation for the evaluation of the existing long-term data of the Black Sea in relation to global weather patterns.

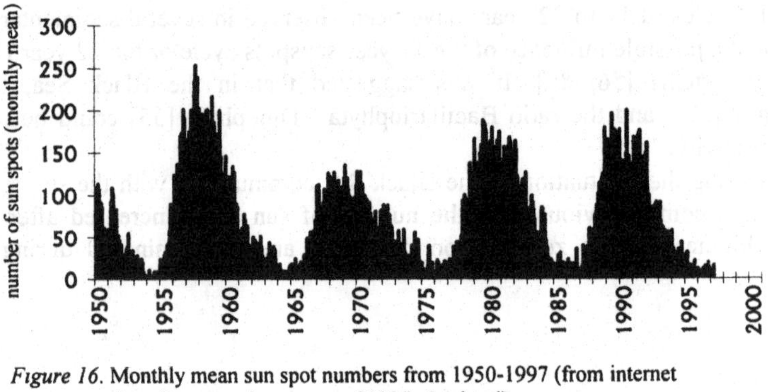

Figure 16. Monthly mean sun spot numbers from 1950-1997 (from internet http //www.ngdc.noaa.gov:8080/index.html).

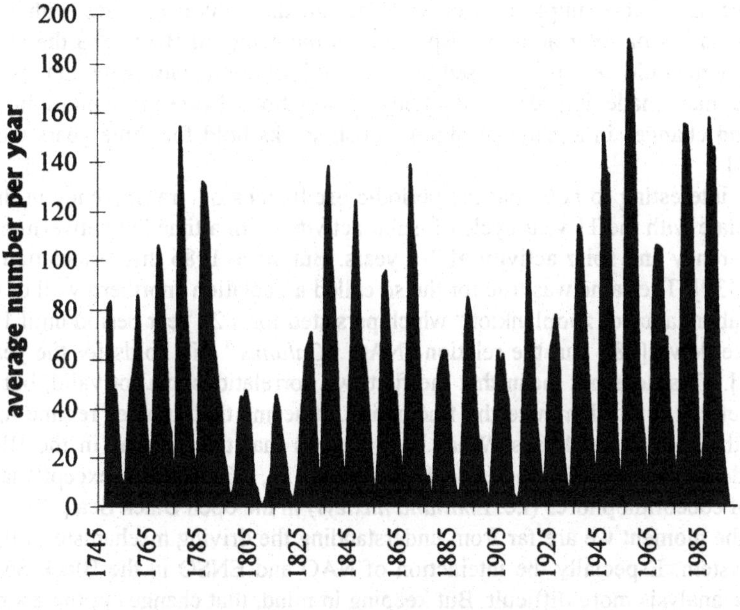

Figure 17. Average annual sun spot numbers from 1745-1997 (from internet http://www.ngdc.noaa.gov:8080/index.html)

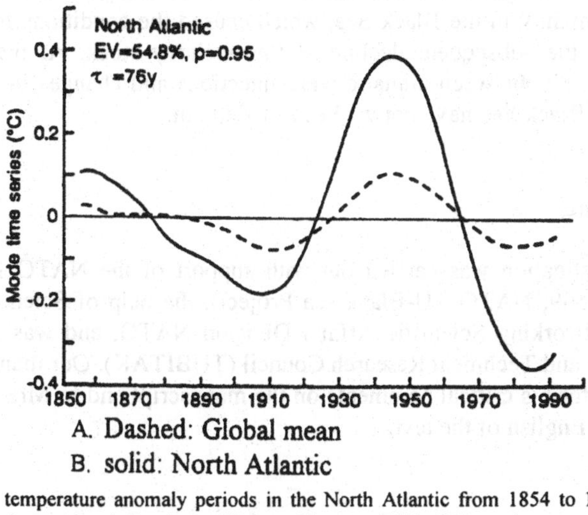

A. Dashed: Global mean

B. solid: North Atlantic

Figure 18 Global temperature anomaly periods in the North Atlantic from 1854 to 1991 (extracted from IPCC -International Panel on Climate Change, [84]) EV = eigen vector, P = correlation coefficient between regional and global means.

It would be too simple as well to relate all the natural changes with long-term climatic changes or solar activity. Especially in the enclosed Black Sea the changes in the ecosystem could be superimposed by effects of pollution, eutrophication, overfishing and other man made impacts. Commonly, correlations between changes in physical factors and changes in aquatic plankton or fish stocks hold for some years then break down [73].

It is interesting to note that the periodic oscillations of transparency in the Black Sea correlate with the 11 year cycle of solar activity, with a time lag between maximum of transparency and solar activity of 2-3 years. But since 1986 this association has not held true [37]. The same was true for the so called association "northern wall of the Gulf Stream : abundance of zooplankton" which persisted for a 22 year period until 1987 and then broke down [88]. But the relation "NAO : *Calanus*" still holds for the years after 1987 [45]. This does not mean that the first two correlations are not valid, but another stronger effect could dominate the fluctuation. It seems that another regime has taken over by the end of the 1980s. What is striking is that the diatoms in the Black Sea became dominant again by the end of the 1980s [32, 36], and the exceptionally high blooms of coccolithophores (i.e. *Emiliana huxleyi*) in the open Black Sea [37].

At the moment we are far from understanding the driving mechanism of the Black Sea ecosystem. Especially the interaction of NAO and ENSO in the Black Sea region makes the analysis more difficult. But keeping in mind, that changes in the zooplankton community and in small pelagic fish stocks in the late 1980's were evident in all seas under consideration, a climatic impact (or superimposition of different cycles or events; hydrological, meteorological and biological) could have triggered the changes in the zooplankton community in the Black Sea, which caused the conditions for the outburst of *M. leidyi* and the subsequent decline of the anchovy stock. At the present time investigations on links between climatic teleconnections and changes in the planktonic community in the Black Sea have not yet been carried out.

Acknowledgements

The present investigation was carried out with support of the NATO Linkage Grant (ENVIR.LG. 951569; NATO TU-Black sea Project), the help of Science for Stability and computer networking Scientific Affairs Division NATO, and was funded by the Turkish Scientific and Technical Research Council (TUBITAK). Our thanks go to Dr H. v. Westernhagen for his critical comments on the manuscript and to Mrs. A. M. Kideys for correcting the English of the text.

References

1. Shushkina, E A., and Musayeva, E. I (1983) The role of jellyfish in the energy system of Black Sea communities. *Oceanology* 23, 92-96.
2. Vinogradov, M. Ye., Shushkina, E A., Musayeva, E. I. and Sorokin, P.Yu. (1989) A newly acclimated species in the Black Sea: The ctenophore *Mnemiopsis leidyi* (Ctenophora: Lobata). *Oceanology* 29(2), 220-224.

3. Zaika, V. E., and Sergeeva, N. G (1991) Diurnal dynamics of population structure and vertical distribution of ctenophore *Mnemiopsis maccradyi* Mayer in the Black Sea. *Zhurnal Obshchbiologii, Kiev*, 27(2), 15-19 (in Russian).
4. Mee, L. D. (1992) The Black Sea in crisis: The need for concerted international action. *Ambio 21* (3), 278-286.
5. Caddy, J. F. (1993) Towards a cooperative evaluation of human impacts on fishery ecosystems of enclosed and semi-enclosed seas. *Fishery Science* 1(1), 57-95.
6. Niermann U., Bingel, F., Gorban, A., Gordina, A. D., Gücü, A. C., Kideys, A. E., Konsulov, A., Radu, G., Subbotin, A. A. and Zaika, V. E. (1994) Distribution of anchovy eggs and larvae (*Engraulis encrasicolus* Cuv.) in the Black Sea in 1991 and 1992 in comparison to former surveys. *ICES Journal of Marine Sciences* 51, 395-406.
7. Mutlu E., Bingel, F., Gücü, A. C, Melnikov, V. V., Niermann U., Ostr, N. A., Zaika, V. E. (1994) Distribution of the new invader *Mnemiopsis sp.* and the resident *Aurelia aurita* and *Pleurobrachia pileus* populations in the Black Sea in the years 1991-1993. *ICES Journal of Marine Sciences* 51, 407-421.
8 Kideys, A. E. (1994) Recent dramatic changes in the Black Sea ecosystem: The reason for the sharp decline in Turkish anchovy fisheries, *Journal of Marine Systems* 5, 171-181.
9. Konsulov, A. and Kamburska, L. (1998) Black Sea zooplankton structural dynamic and variability off the Bulgarian Black Sea coast during 1991-1995, in L. Ivanov and T. Oguz (eds.), *NATO TU Black Sea Assessment Workshop: NATO TU-Black Sea Project: Symposium on Scientific Results*, Vol. 1, Kluwer Academic Publishers, pp. 281-292.
10. Vinogradov, M. Ye., Sapozhnikov, V. V., and Shushkina, E. A. (1992) *The Black Sea ecosystem.* Moskva, Russia, Nauka: 112 pp.
11. Kovalev, A. V, Gubanova, A. E Kideys, Melnikov, V. V. Niermann, U., Ostrovskaya, N. A., Prusova, I. Yu., Skryabin, V. A., Uysal, Z., and Zagarodnyaya, Ju. A. (1998) . Long-term changes in the biomass and composition of fodder zooplankton in coastal regions of the Black Sea during the period 1957-1996, in L. Ivanov and T Oguz (eds.), *NATO TU Black Sea Assessment Workshop: NATO TU-Black Sea Project: Symposium on Scientific Results*, Vol. 1, Kluwer Academic Publishers, pp. 209-220.
12. Kovalev A V, Niermann, U , Melnikov, V. V., Belokopytov, V., Uysal, Z., Kideys, A. E., Ünsal, M., and Altukhov, D. (1998) Long-term changes in the Black Sea zooplankton: the role of natural and anthropogenic factors, in L. Ivanov and T. Oguz (eds.), *NATO TU Black Sea Assessment Workshop: NATO TU-Black Sea Project: Symposium on Scientific Results*, Vol. 1, Kluwer Academic Publishers, pp. 221-234.
13. Bodeanu, N., (in press). Long-term evolution of phytoplankton blooms in the Romanian Black Sea area, in Ü Ünlüata and A. Bologa (eds.) *NATO Advanced Research Workshop (ARW). Environmental Degradation of the Black Sea : Challenges and Remedies*, Kluwer Academic Publishers.
14 Zaitsev, Yu. P. (1993) Impacts of eutrophication on the Black Sea fauna, *General Fisheries Council for the Mediterranean, Studies and Reviews, Rome, FAO*, 64 (2): 63-86.
15. Shiganova T. A., Kideys, A E., Gucu, A. C., Niermann, U., and Khoroshilov, V. S. (1998) Changes of species diversity and abundance in the main components of the Black Sea pelagic community during last decade, in L. Ivanov and T Oguz (eds.), *NATO TU Black Sea Assessment Workshop: NATO TU-Black Sea Project: Symposium on Scientific Results*, Vol. 1, Kluwer Academic Publishers, pp. 171-188.
16 MacLennan, D. N., Yasuda, T , and Mee, L. (1997) Analyses of the Black Sea fishery fleet and landings, BSEP-PCU, Dolmabahce Sarayi, II Harekat Kosku, 80680 Besiktas, Istanbul.
17. Balkas, T , Dechev, G , Mihnea, R., Serbanescu, O., and Ünlüata, U. (1990) State of the marine environment in the Black Sea region. *UNEP Regional Seas Reports and Studies* 124, 41 pp.
18. Volovik, S. P., Dubinina, V. G., and Semenov, A. D. (1993) Fisheries and environment studies in the Black Sea system. Part 1· Hydrobiology and dynamics of fishing in the Sea of Azov. *Studies and Reviews, General Fisheries Council for the Mediterranean, Rome, FAO*, 64, 1-58.
19. Ivanov, L., and Beverton, R J. H. (1985) The fisheries resources of the Mediterranean. Part two: The Black Sea, *Studies and Reviews, General Fisheries Council for the Mediterranean*, 60, 135 pp.
20 Burrell, V. G., and Van Engel, W. A. (1976) Predation by and distribution of a ctenophore, *Mnemiopsis leidyi* A. Agassiz, in the York River estuary, *Estuarine and Coastal Marine Science*, 4, 235-242.

21. Mountford, K. (1980). Occurrence and predatıon by *Mnemıopsis leidyi* in Barnegat Bay, New Jersey, *Estuarine and Coastal Marine Science* 10, 393-402.
22. Cowan, Jr. V. G. and Houde, E.D. (1993) Relative predation potentials of scyphomedusae, ctenophores and planktivorous fish on ichthyoplankton in Chesapeake Bay, *Marıne Ecology Progress Serıes* 95, 55-65.
23. Purcell, J. E., Nemazie D. A., Dorsey S , Gamble J. C., and Houde E. D. (1993) In situ predation rates on bay anchovy eggs and larvae by scyphomedusae and ctenophores in Chesapeake Bay, USA, ICES CM 1993/L:42, 22 pp.
24. Sergeeva, N. G., Zaika, V. E., and Mikhailova, T. V. (1990) Nutrition of ctenophore *Mnemıopsıs mccradyi* under conditions of the Black Sea, *Ekologya Morya (Kiev)* 35, 18-22, (in Russian).
25. Tsikhon-Lukanina, Ye. A., Reznıtchenko, O. G., and Lukasheva, T. A. (1991) Quantitative patterns of feeding of the Black Sea ctenophore *Mnemiopsıs leıdyı, Oceanology* 31, 196-199.
26. Tsikhon-Lukanina, Ye. A. and Reznıtchenko, O. G. (1991) Diet of the ctenophore *Mnemiopsis* in the Black Sea as a function of size, *Oceanology* 31, 320-323.
27 Monteleone, D. M. and Duguay, L E. (1988) Laboratory studies of the predation by the ctenophore *Mnemiopsıs leıdyı* on early stages in the life history of the bay anchovy, *Anchoa mitchıllı, Journal of Plankton Research* 10 (3), 359-372.
28. Reeve, M. R., Walter, M. A., and Ikeda, T. (1978) Laboratory studies of ingestion and food utilization in lobate and tentaculate ctenophores, *Lımnology and Oceanography* 23(4), 740-751.
29. Shushkina, E. A. and Vinogradov, M. Ye. (1991) Long-term changes in the biomass of plankton ın open areas of the Black Sea, *Oceanology* 31 (6), 716-721.
30. Caddy, J. F. (1992) Update of the fishery situation in the Black Sea and revision of the conclusions of the 1990 GFCM Studies and Reviews No 63, by Caddy and Griffiths: "A perspective on recent fishery-related events in the Black Sea". Unpublished manuscript.
31. Gücü, A. (1997) Role of fishing ın the Black Sea ecosystem, in: E. Ozsoy and A. Mikaelyan (eds.), *Sensıtivity to change: Black Sea, Baltıc Sea and North Sea, NATO ASI Series,* Kluwer Academıc Publishers, pp. 149-162.
32. Prodanov, K., Mikhailov, K., Daskalov, G., Maxim, K., Chashchin, A., Arkhipov, A., Shlyakhov, V., and Ozdamar, E. (1997) Environmental impact on fish resources in the Black Sea, ın: E. Ozsoy and A. Mikaelyan (eds.), *Sensıtıvıty to change: Black Sea, Baltıc Sea and North Sea. NATO ASI Series,* Kluwer Academic Publishers, pp. 163-181.
33. Shiganova, T. A. (1997) *Mnemıopsıs leıdyı* abundance in the Black Sea and ıts impact to the pelagic community, ın: E. Ozsoy and A. Mikaelyan (eds.), *Sensıtıvıty to change: Black Sea, Baltıc Sea and North Sea. NATO ASI Serıes*, Kluwer Academic Publıshers, pp. 163-181
34. Mikaelyan, A.S. (1997). Long-term variabilıty of phytoplankton communities in open Black Sea in relation to environmental changes, ın: E. Ozsoy and A. Mikaelyan (eds.), *Sensıtivıty to change: Black Sea, Baltıc Sea and North Sea. NATO ASI Serıes,* Kluwer Academic Publishers, pp. 105-116.
35 Petrova-Karadjova, V. (1992) Solar control upon the phytoplankton in the Black Sea, *Rapp. Com. Int. Mer. Medıt*, 33
36 Moncheva, S. and Krastev, A (1997) Some aspects of phytoplankton long-term alterations of Bulgarian sea shelf, in: E. Ozsoy and A. Mıkaelyan (eds.), *Sensıtıvıty to change: Black Sea, Baltıc Sea and North Sea. NATO ASI Serıes,* Kluwer Academic Publishers, pp. 79-93.
37. Mankovsky, V. I., Vladimirov, V. L., Afonin, E. I., Mishonov, A V., Solovev, M. V., Anninsky, B E., Georgieva, L. V , and Yunev, O. A. (1996) Long-term varıabılity of the Black Sea water transparency and reasons for its strong decrease in the late 1980s and early 1990s, MHI & IBSS, Sevastopol, 32 pp. (ın Russian).
38. Petran, A. and Moldoveanu, M. (1997) Post-invasion ecological impact of the Atlantic ctenophore *Mnemıopsıs leıdyı* Agassiz, 1865 on the zooplankton from the Romanian Black Sea waters, *Cercetarı marıne* 27-28, 135-157.
39. Gubanova, A. D., Prusova, I. Yu., Shadrın, N. V., Polikarpov, I. G , (in press) Dramatical changes ın the copepod community in Sevastopol Bay (the Black Sea) durıng the last decades (1976-1996), *Journal of Marine Systems*
40. Kideys, A. E., Gordına, A. D., Niermann, U., Uysal, Z., Shiganova, T. A., and Bingel, F.(1998) Distributıon of eggs and larvae of anchovy with respect to ambıent condıtions ın the southern Black Sea durıng 1993 and 1996, ın L. Ivanov and T. Oguz (eds.), *NATO TU Black Sea Assessment Workshop: NATO TU-Black Sea Project. Symposium on Scıentific Results,* Vol. 1, Kluwer Academıc Publishers, pp.189-198.

41. Gordina, A. D., Niermann, U., Kideys, A. E., Subbotin, A. A., and Artyomov, Yu.G. (1998) State of summer ichthyoplankton in the Black Sea, in L. Ivanov and T. Oguz (eds.), *NATO TU Black Sea Assessment Workshop: NATO TU-Black Sea Project: Symposium on Scientific Results*, Vol. 1, Kluwer Academic Publishers, pp.367-379.
42. Harris, R. (1996) Global ocean ecosystem dynamics, *Global Change Newsletter* 28, 2-5.
43 Astthorsson O.S. and Gilason, A. (1995) Long-term changes in zooplankton biomass in Icelandic waters in spring, *ICES J Mar. Sci* 52 657-668
44. Broekhuizen, N. and McKenzie, E. (1995) Patterns of abundance for *Calanus* and smaller copepods in the North Sea: time series decomposition of two CPR data sets, *Mar. Ecol. Prog. Ser.* 118, 103-120
45. Fromentin, J. M.and Planque, B. (1996) *Calanus* and environment in the eastern North Atlantic II. Influence of the North Atlantic Oscillation on *C finmarchicus* and *C. helgolandicus*, *Mar. Ecol. Prog Ser.* 134, 111-118
46. Roemmich, D. and McGowan, J. (1995). Climatic warming and the decline of zooplankton in the California current, *Science* 267, 1324-1326.
47. Taylor, A (1995) North- South shifts of the Gulf Stream and their climatic connection with the abundance of zooplankton in the UK and its surrounding seas, *ICES J. mar. Sci.* 52, 711-721.
48. Straile, D. (in press) Impacts of eutrophication and climate variability on zooplankton population dynamics in a large and deep lake.- ICES Symposium on The variability of plankton and their physico-chemical environment; Kiel, Germany· 12-21 March 1997.
49. Schulz, S., Ertebjerg, G, Behrends, G, Breuel, G., Ciszewski, P., Horstmann, U., Konogonen, K., Kostrichkina, E., Leppanen, J.M, Mohlenberg, F., Sandsröm, O., Viitasalo, M., and Willen, T. (1990) Baltic Marine Environment Protection Commission -Helsinki Commission- Second periodic assessment of the state of the marine environment of the Baltic Sea, 1984-1988; Background document. Baltic Sea Environment Proceedings 35B, 210 pp.
50. Behrends, G. (1996) Long-term investigation of seasonal mesoplankton dynamics in Kiel Bight, Germany, in: A. Andrushaitis (ed.) *Proceedings of the 13th BMB Symposium, Riga (1-4 September, 1993)*, Institute of Aquatic Ecology, University of Latvia, pp. 93-98.
51. Niermann, U. and Greve, W. (1997) Distribution and fluctuation of dominant zooplankton species in the southern Black Sea in comparison to the North Sea and Baltic Sea, in: Ozsoy and Mikaelean (eds.), *Sensitivity to change: Black Sea, Baltic Sea and North Sea NATO ASI Series*, Kluwer Academic Publishers, pp. 65-77
52. Lindeboom, H. J., van Raaphorst, W, Beukema, J J., Cadee, G. C., and Swennen, C. (1994) Sudden changes in the biota of the North Sea Oceanic influences underestimated, *ICES C M. 1994/L:27*, 1-16.
53. Witbaard, R., Duineveld, G. C. A., and de Wilde, P. A. W. J. (1997). A long-term growth record derived from *Arctica islandica* (mollusca, bivalvia) from the Fladen Ground (northern North Sea).- *J. mar. biol. Ass U. K.* 77, 801-816
54. Shelton, P. A., Armstrong, M. J, and Roel, B. A. (1993) An overview of the application of the daily egg production method in the assessment and management of anchovy in the southeast Atlantic, *Bulletin of Marine Science* 53, 778-794
55. Hunter, J. R. and Alheit, J. (1995) International GLOBEC small pelagic fishes and climate change program. Report of the First Planning Meeting, La Paz, Mexico, June 20-24, 1994 GLOBEC Report No. 8, 72 pp.
56 Lindeboom, H. J. (in press) Variability of the marine ecosystem and possible causes, ICES Symposium on the variability of plankton and their physico-chemical environment, Kiel, Germany: 12-21 March 1997.
57. Fedorina, A. J (1987) Dynamic development of the Black Sea zooplankton and its reasons, Thesis, Moskow State University (in Russian)
58. Ozsoy, E. and Ünlüata, Ü. (1997) Oceanography of the Black Sea· a review of some recent results, *Earth-Science Reviews* 42, 231-272.
59. Polonsky, A., Voskresenskaya, E., and Belokopytov, V. (1997) Variability of northwestern Black Sea hydrography and the river discharges as part of global ocean-atmosphere fluctuations, in: E. Ozsoy and A Mikaelyan (eds), *Sensitivity to change: Black Sea, Baltic Sea and North Sea. NATO ASI Series*, Kluwer Academic Publishers, pp 11-24.
60. Cociasu, A., Diaconu, V., Popa, L, Nae, I., Dorogan, L, and Malciu, V. (1997). The nutrient stock of the Romanian shelf of the Black Sea during the last three decades, in: E. Ozsoy and A. Mikaelyan (eds.), *Sensitivity to change. Black Sea, Baltic Sea and North Sea. NATO ASI Series*, Kluwer Academic Publishers, pp. 49-63.

61. Becker, G. A. and Pauly, M. (1996) Sea surface temperature changes in the North Sea and their causes *ICES Journal of Marine Science*, 53, 887-898.
62. Corten, A. and van de Kamp, G. (1996) Variation in the abundance of the southern fish species in the southern North Sea in relation to hydrography and wind, *ICES Journal of Marine Science* 53, 1113-1119
63. Heessen, H. J. L. and Daan, N. (1995) Long term trends in fish species caught during the international bottom trawl survey, *ICES Journal of Marine Science* 53, 1063-1078.
64. Greve, W., Reiners, F, and Kleinfeld, R. (1991). Systemökologie I Langzeituntersuchungen des Zooplanktons. *Jahresbericht der Biologischen Anstalt Helgoland*, 1990, 50-51.
65. Lindley, J. A., Gamble, J. C., and Hunt, H. G (1995) A change in the zooplankton of the central North Sea 55°N to 58°N: a possible consequence of changes in the benthos, *Mar. Ecol Prog. Ser* 119, 299-303.
66. Nehring, S. (in press) Establishing of southern plankton algae in the North Sea - biological indicators for long-term climatic changes in the pelagic ecosystem?- ICES Symposium on The variability of plankton and their physico-chemical environment; Kiel, Germany: 12-21 March 1997
67. Pearson, T. H. and Rosenberg, R. (1978) Macrobenthic succession in relation to organic enrichment and pollution of the marine environment, *Oceanogr. Mar. Biol. ann. Rev.* 16, 229-311.
68. Daan, N. and Richardson, K. (1996) Changes in the North Sea ecosystem and their causes: Århus 1975 revisited, *ICES Journal of Marine Science* 53, 875-1226.
69. Le Fevre-Lehoerff, G,. Ibanez, F., Poniz, P., and Fromentin, J. M. (1995) Hydroclimatic relationships with planktonic time series from 1975 to 1992 in the North Sea off Gravelines, France, *Mar. Ecol Prog. Ser.* 129, 269-281
70. Mann, K H and Lazier, J R. N. (1991) *Dynamics of marine ecosystems,* Blackwell, Oxford.
71. Rodionov, S. N. (1994) *Global and regional climate interaction. the Caspian Sea experience*,-Kluwer Academic Publishers, The Netherlands· 241pp.
72. Childers, D. L., Day Jr., J. W., and Muller, R. A. (1990) Relating climatological forcing to coastal water levels in Louisiana estuaries and the potential importance of El Nino-southern oscillation events, *Climate Research* 1, 31-42.
73 Mann, K. H (1993) Physical oceanography, food chains, and fish stocks: a review, *ICES J mar. Sci.* 50, 105-119
74. Cushing, D H. (1995) The long-term relationship between zooplankton and fish, *ICES J mar Sci* 52, 611-626
75. Alheit, J and Hagen, E (1997) Long-term climate forcing of European herring and sardine populations, *Fisheries Oceanography* 6, 130-139.
76. Conover, R J., Wilson, S., Harding, G. C. H, and Vass, W P. (1995) Climate, copepods and cods· some thought on the long-range prospects for a sustainable northern cod fishery, *Climate Research* 5, 69-82.
77. Rodionov, S. N. and Korvnin, A. S. (1991) Influence of thermal condition on abundance of pollack stock in the eastern Bering Sea, *Rybnoye Khozyaistvo (Fisheries)* 3, 26-31 (in Russian).
78. Halpert, M. S. and Ropelewski, C. F. (1991) Climate assessment. A decadal review, 1981-1990.- Camp Springs, MD: US. Department of Commerce, NWS/NMC/CAC.
79. Trenberth, K. E, and Hurrell, J. W. (1994) Decadal atmosphere-ocean variations in the North Pacific, *Climate Dynamics* 9, 303-319.
80 Besiktepe, S T., Sur, H. I, Özsoy, E, Latif, M. A., Oguz, T, and Ü. Ünlüata (1994) The circulation and hydrography of the Marmara Sea, *Prog. Oceanogr* 34, 285-334.
81. Gertman, I. F., Ovchinnikov, I. M., and Popov, Y. I. (1990) Deep convection in the Levantine Sea, *Rapp. Comm. Mer Medit.*, 32, 172
82. Sztobrryn, M., Kalas, M., and Staskiewicz, A (1997) Changes of mean sea level and ice conditions in Gdynia as indicators of climate changes in the gulf of Gdansk, in: E. Ozsoy and A. Mikaelyan (eds.), *Sensitivity to change: Black Sea, Baltic Sea and North Sea. NATO ASI Series*, Kluwer Academic Publishers, pp. 1-9
83 Ovchinnikov, I. M. and Osadchy, A. S. (1991) Secular variability of winter climatic conditions influencing peculiarities of hydrological conditions in the Black Sea, in M.E Vinogradov (ed.), *Variability of the Black Sea ecosystem*, Nauka, Moscow, pp. 85-89, (in Russian)
84. Schlesinger, M. E. and Ramankutty, N. (1994) An oscillation in the global climate system of period 65-70 years, *Nature* 367, 723-726
85. Haigh, J. D (1996) The impact of solar variability on climate, *Science* 272, 981-984

86. Burroughs, W. J. (1992) *Weather cycles real or ımagınary?* Cambridge Universıty Press, Cambridge, UK, 207pp.
87. Willis, A. J., Dunett, N P., Hunt, R., and Grimme, J P. (in press) Does Gulf Stream position affect vegetation dynamıcs in Western Europe? *Oıkos*.
88 Hays, G C., Carr, M. R., and Taylor, A. H (1993) The relationshıp between Gulf Stream position and copepod abundance derived from the continuous plankton recorder survey. separating biologıcal signal from sampling noise, *Journal of Plankton Research* 15, 1359-1373
89. Greve, W (1994) The 1989 German Bight invasion of *Muggıaea atlantıca, ICES Journal of Marıne Scıences* 51, 355-358.
90. Ottestad, P., (1942) On perıodıcal varıations in the yield of the great fisheries and the possibility of establishing yield prognoses, *Fısk Dır Skr. (Ser Havunders)* 7, 1-11.

STATUS AND EVOLUTION OF THE ROMANIAN BLACK SEA COASTAL ECOSYSTEM

ADRIANA PETRANU[(1)], MUKADER APAS[(1)], NICOLAE BODEANU[(1)], ALEXANDRU S. BOLOGA[(1)], CAMELIA DUMITRACHE[(1)], MARIA MOLDOVEANU[(1)], GHEORGHE RADU[(1)] and VICTORIA TIGANUS[(2)]
[(1)] Romanian Marine Research Institute, RO-8700 Constantza, Romania
[(2)] University "Ovidius" Constantza, RO-8700 Constantza, Romania

Abstract

During the past three decades the Romanian coastal zone of the Black Sea has been subject to severe ecological disturbance, as a result of complex and multiple anthropogenic pressures, leading to changes in biotopes and in the biological components of the ecosystem. The consequences of this influence can be evaluated as the ecosystem degradation. Until 1965-70, investigations showed the existence of an ecosystem with a rich variety of marine life and high productivity. Research undertaken after 1970 indicated structural and functional changes, especially in the coastal ecosystem where anthropogenic influences act more directly. The major consequence of these disturbances was a reduction of biodiversity, both for plant and animal species. Since 1970, monitoring has documented the changes in the main biotic components of the pelagic and benthic systems and the effect of eutrophication and pollution on the floral and faunal communities. The ecosystem is now dominated by eutrophication/pollution with their associated unfavorable consequences. The evolution, current state and trends in pelagic plant and animal populations and changes in benthic communities are discussed in this paper. The main problems associated with the introduction of exotic species are also analyzed in order to assess their impact on the ecosystem. The onset of eutrophication has led to profound transformation throughout the northwestern shelf area of the Black Sea, including the Romanian coastal zone. The main ecological consequence of increasing eutrophication has been an increase of planktonic primary production. The algal blooms, a rare phenomena until the 1970's, increased in frequency and magnitude (for the decade 1981-1990, 47 monospecific blooms). Several species of mycoflora have developed excessively during the last ten years, generating fungal blooms, especially since 1987, and sometimes containing pathogenic forms. During 1976-1986, zooplankton populations produced higher seasonal peak abundance, compared to the previous period; the total biomass increased, but the species diversity fell. An abundance of gelatinous organisms (*Aurelia aurita, Noctiluca scintillans*), producing spectacular blooms, is a characteristic of this period. After 1987, a degradation of the zooplanktonic communities was noted; a severe reduction of populations of fodder zooplankton occurred, especially

S. Beşiktepe et al. (eds.),
***Environmental Degradation of the Black Sea: Challenges and Remedies,* 175–195.**

during summer, concurrently with a spectacular reduction in catches of planktivorous fish. Profound modifications have occurred in the benthic communities, involving both macrophytobenthos and zoobenthos, and resulting in qualitative and quantitative impoverishment. Of note is the almost complete disappearance of the formerly important *Cystoseira* belts. In conditions of hypoxia, created by the already mentioned blooms, the most affected species over large areas were the zoobenthos. Populations of the more sensitive species have diminished their frequency in biocoenoses, while those resistant to the alteration of the environment have proliferated, becoming dominant in some communities. In a period when the eutrophication was less advanced (1976-1986) the macrobenthos densities and general biomasses increased owing to the proliferation of opportunistic species, while the zoobenthos reduced constantly especially in 1989-1992. The predatory ctenophore *Mnemiopsis leidyi* had a significant impact in the pelagic environment. After its invasion of Romanian waters, a reduced level of the zooplanktonic species with trophic value was recorded with a major impact on the Romanian Black Sea fisheries. The environmental monitoring over the past 4-5 years reflects perceptible improvements in the state of some components of the ecosystem. However, the state of the whole ecosystem continues to be damaged and threatened by the consequences of eutrophication.

1. Introduction

Over the last three decades, in addition to atypical natural factors, e.g. exceptionally high fresh water inflow from tributary rivers and decreases in water temperature, the Romanian coastal zone has encountered severe ecological disturbances as a result of multiple, complex, anthropogenic pressure; the man-made factors interact with the natural ones, amplifying their effects in varying degrees of severity.

Anthropogenic factors that have caused ecological disturbance along the Romanian Black Sea coastal zone are:

- Change in chemical composition of fresh water flowing into the sea. The largest rivers flow through regions of intensive agriculture, developed industries and densely populated cities, contributing to the discharges of numerous allochthonous substances, toxic or nontoxic, resulting from human activity.
- Discharges of untreated waste water, both urban and industrial. Whereas the northern coast between Sulina and Constantza is mostly uninhabited, the southern coast from Constantza to Mangalia is heavily populated and industrially developed. Thus the southern zone produces a large volume of waste water for discharge to sea.
- Increase in the volume of maritime traffic, number of harbors and coastal resorts and the development of oil drilling on the shelf.
- Coastal engineering constructions during the last few decades, in order to consolidate and protect the shores (dams, barrages), or to exploit the sand resources from the beaches.

All these various human impacts have led to changes in the biotopes and in the biological components of the ecosystem. The effect of these influences can be evaluated in terms of ecosystem degradation.

In Romania, during the post-war period, certain biological oceanographic laboratories performed floristic and faunistic inventories, producing remarkable results concerning the identification of taxonomic groups, the dynamics of plankton and the major benthic associations. Thus, until 1965-1970, an inventory of species in pelagic and benthic systems supplied information of the status of almost all inhabitants, as well as an assessment of natural productivity. All this knowledge has indicated a rich variety of marine life and a high productivity at all trophic levels.

The research undertaken after 1970 demonstrated structural and functional changes, especially of coastal ecosystem where the anthropogenic influences acted more directly. The major consequence of these disturbances was a reduction of biota, both plant and animal, including fish stocks and fish catches.

Since 1970, the research has shown the effect of eutrophication and pollution on the faunal and floral communities. This has involved the transformation from a diverse ecosystem supporting a rich variety of marine life to one that is dominated by eutrophication with its unfavorable consequences.

There is extensive literature regarding the development of marine life, the phenomena and processes which have modified the species composition and quantitative indices of different components of the ecosystem [e.g.13].

The extensive data obtained by Romanian scientists in fields of various specialties, over the last three decades, has produced a framework for this multidisciplinary study of the status of the main biotic components of the ecosystem, and their evolution under increasing anthropogenic influences.

The most important event in the past three decades has been the increase in eutrophication, the levels of nitrogen, phosphorus and organic matter in the marine environment increasing as a result of inflows from the Danube and other rivers in the north-western part of the Black Sea and from local discharges of waste water [17,18,41]. For example, during the 1986-1988 period the mean quantities of organic matter were 1.2 times higher than in 1960-1970, the nitrate almost five fold and phosphate 25-fold. The oligo-mesotrophic coastal waters in 1959-1969 gradually became eutrophic in 1960-1970 before becoming polytrophic as the nutrients ceased to be growth limiting factors.

A significant rise in nitrates and phosphates created a completely new environment for planktonic algae. Intense nutrification resulted in an increase in the frequency and amplitude of the algal blooms. The increase in algal cell number during blooms has had a negative effect on all biotic components of the ecosystem. The resulting lack of oxygen combined with an increase in organic matter has frequently produced mass mortalities of whole groups of benthic organisms and severe impoverishment of the marine fauna.

Thus, the marine ecosystem has undergone a series of severe disturbances affecting the plankton, nekton and benthos.

2. The impact of eutrophication on pelagic communities

2.1. MICROALGAL BLOOMS

The main ecological consequences resulting from intensification of the eutrophication process has been an increase of planktonic primary production and an increase in the frequency, magnitude, and spatial extension of algal blooms. If the large algal blooms were an exceptional phenomenon until 1970, they are now the norm occurring every year [7,8].

With a progressive increase since the 1970's, the red tides reached maximum magnitude during the 1990's when 49 such events were recorded in the Romanian sector of the Black Sea, 15 of them having an exceptional intensity, with numerical values ranging from 50 million to 1 billion cells l^{-1}.

The intensification of the frequency and magnitude of blooms, the increase of the overall species number which produces algal developments and the increase of cells abundance of majority algal species induced a continuous increase in total phytoplankton. As a result the mean biomass of Romanian coastal waters in the period 1983-1990 was 4,105 $mg.m^{-3}$ (Table 1) - a value almost 10 times greater than the value of 495 $mg.m^{-3}$ recorded for the period 1959-1963 [9]. The growth of the algae in inshore waters are also remarkable. For example, the phytoplankton numerical densities close to the Mamaia shore increased almost eight times and the biomasses - almost four times compared with the values obtained for the years 1962-1965 (Table 2) [8].

TABLE 1. Mean phytoplankton quantities in the Romanian offshore waters of the Black Sea

Period	Density ($cells.l^{-1}$)	Biomass ($mg\ m^{-3}$)
1971-1975	259,600	719
1976-1980	789,489	2,244
1983-1990	2,235,577	4,105

TABLE 2. Mean phytoplankton quantities at Mamaia in the periods 1983-1988 and 1962 -1965

Parameter	1983-1988	1962-1965
Density ($cells.l^{-1}$)	6,519,686	887,067
Biomass ($mg.m^{-3}$)	7,135. 29	2,004.13

The increase in organic matter is favored during the warm part of the year and is associated with massive development of algae with mixotrophic affinity such as dinoflagellates and euglenids. Thus, the proportion of diatoms in the total community progressively decreased, while the proportion of dinoflagellates and other non-diatom groups increased. One of the direct effects of anthropogenic eutrophication on the phytoplankton included the stimulation of the growth of dinoflagellates at the expense of diatoms, whose proportion in total phytoplankton has been reduced. Thus, the phytoplankton community structure in the 80's was very different from that during 1960-1970, when the community was dominated by the diatoms.

The non-diatom proportion increased especially in the summer, due to the warm water affinities of the main species of this group (Fig.1). The increased contribution of the non-diatom groups reflects the enhancement in the magnitude and frequency of the blooms produced by the species from these groups (the dinoflagellates *Prorocentrum cordatum* and *Heterocapsa triquetra,* the euglenid *Eutreptia lanowii*, the chrysophite *Emiliana huxleyi*) (Table 3) [9].

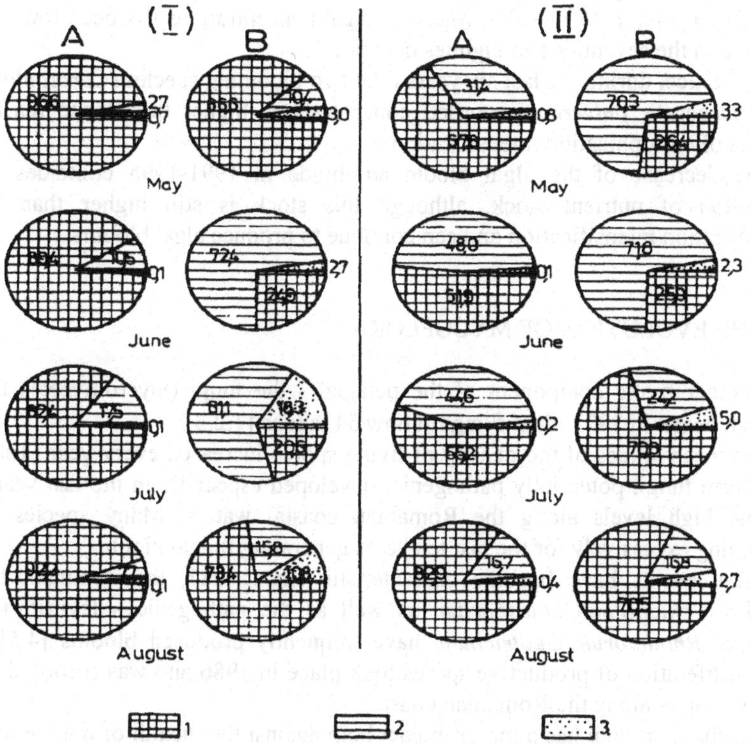

Figure 1 Proportions (%) of the main algal groups in the long-term monthly means of phytoplankton Numerical densities (I) and biomass (II) between May and August, 1960-1970 (A) and 1983-1988 (B).

1.Diatoms 2. Dinoflagellates 3. Other algal groups

TABLE 3. Proportional (%) representation of the main algal groups in the phytoplankton density and biomass levels during the periods 1960-1970, 1971-1980 and 1983-1990

Parameter	Algal group	Proportion (%)		
		1960-1970	1971-1980	1983-1990
	Diatoms	92.3	84.1	38.3
Numerical density	Dinoflagellates	7 6	11.8	20.0
(cells.l^{-1})	Other groups	0.1	4 1	41.7
	Diatoms	84.5	60.9	69 2
Biomass	Dinoflagellates	15 1	37 7	26 4
(mg.m^{-3})	Other group	0.4	1 4	4.4

Many of the species producing blooms in the seventies and eighties did not cause such phenomena in 1991-1996, when the algal concentrations has been lower than those achieved in the seventies and eighties decades.

In practice, during the last six years only two blooming species (*Cerataulina pelagica* and *Microcystis pulverea)* recorded concentrations higher than those reached in the decades of eutrophication intensification.

The decrease of the algal bloom amplitude in 1991-1996 coincides with some diminution of nutrient stock, although this stock is still higher than before the eutrophication intensification and can continue to produce algal blooms.

2.2. THE EVOLUTION OF MYCOFLORA

For another biotic component of the pelagial - the fungi (mycromycetes flora) - the research began in 1971 identifying till now 54 taxons [1,2,3].

The mean values of the density of living spores increased every year (Table 4). The levuriform fungi, potentially pathogenic, developed especially in the last years reaching extreme high levels along the Romanian coastal waters. Many species have been developing excessively for the last ten years generating fungal blooms.

These events have become characteristic since 1987, the dominant filamentous complex *Penicillium/Cladosporium* as well as the pathogenic - levuriform complex *Candida/ Rhodotorula/Geotrichum,* have frequently produced blooms [4,5]. The first such proliferation of productive spores took place in 1986 and was recorded throughout shallow waters along the Romanian coast.

The fungic bloom has a major impact both against the health of marine environment and the human population. The frequency of pathogenic genera *Candida/Rhodotorula* ranged between 7-80% of the total annual production. As a result of this high pathogenity and production of large quantities in 1990-1994 in particular, the littoral waters are becoming increasingly infested.

TABLE 4. Evolution of number of viable fungus spores in the area of the Danube months between 1987 and 1996.

Year/Station	Sulina	Mila 9	Sf. Gheorghe
1987	9 700	33 360	13 320
1988	6 820	7 350	6 790
1990	13 640	29 300	12 520
1991	8 160	11 740	16 480
1992	3 620	8 220	8 460
1993	16 540	13 680	26 760
1994	33 870	51 240	42 940
1995	72 620	31 620	41 840
1996	46 500	14 780	49 080

The proliferation of the fungal blooms represents another aspect of the deteriorated conditions in the coastal waters, a consequence and at the same time an indicator of eutrophic - polluted environment.

2.3. CHANGES IN ZOOPLANKTONIC COMMUNITIES - SOME PECULIARITIES IN ZOOPLANKTON DYNAMICS

Anthropogenic eutrophication has had an indirect effect on the zooplankton through its impact on the phytoplankton. The ecological impact produced by abundance of the phytoplankton food against the zooplankters was evidenced by a quantitative, more intensive development, especially organisms feeding on them (e.g. *Noctiluca scintillans*), as compared to previous periods.

The most notable feature of the zooplankton communities dynamics has been the changes in the relative abundance of groups or species. Thus, as eutrophication intensified, some species disappeared, and others were considerably reduced, while others thrived, increasing markedly.

The superficial and ultra-superficial biotope, inhabited by the hyponeustonic species has been most exposed to pollution. Here, the species of copepods belonging to the family Pontellidae - *Anomalocera patersoni*, *Pontella mediterranea*, *Labidocera brunescens*, once forming large concentrations [29,30], suffered a considerable reduction, only isolated individuals being found in the offshore waters in 1983 and also 2 ind.m^{-3} in 1992.

Other zooplanktonic species which have not been found for the last 25 years are the copepods monstrilloid - *Monstrilla grandis*, *M. helgolandica* and *M. longiremis,* constantly occurring in large numbers in the Romanian Black Sea waters in the 1960's [42].

The reduction in the populations of some more sensitive holoplanktonic species - the copepod *Centropages ponticus*, the cladocer *Penilia avirostris*, which were dominant especially in the area of the Danube mouths, represents another modification in the structure of the zooplanktonic communities.

Many studies have emphasized the importance of the larval component in the Romanian waters, who formed a substantial part of the fodder zooplankton biomass until the 1970's, [31,32,33,36,38,43]. The size of meroplanktonic populations mirrored the changes in the benthic communities which suffered important modifications in the 1990's, especially the larval stages of molluscs which diminished very much, as a consequence of mass mortality of the benthic fauna in 1986-1990 period.

On the other hand, the success of a small number of species has resulted in high values of density and biomass. Opportunistic species, both herbivorous and detritivorous (feeding on detritus), have become dominant leading to biomass values higher than those recorded two decades ago.

During the period 1980-1987 the mean annual biomass values in the Romanian waters were continuously increased mainly as a result of high summer abundance by the copepod *Acartia clausi* and the cladoceran *Pleopis polyphemoides*. Mean annual density of copepods exceeded 7,100 ind.m^{-3} in the 1980-1986 period, while in previous years it represented only 4,300 ind.m^{-3} [44]. The mean annual values of biomass were six time higher than the mean for the decade 1960-1970 (Table 5).

TABLE 5. Copepod contribution (%) to the total zooplankton density (D=ind m^{-3}) and biomass (B=mg.m^{-3}) (cf. Porumb,1989)

	Winter		Spring		Summer		Autumn		Annual mean	
Period	D	B	D	B	D	B	D	B	D	B
1970-1979	72.0	79 6	16.3	19.9	16.3	58.9	58.8	48.5	19.0	50.6
1980-1986	90.3	57.1	7 6	33.6	39.8	66.2	57.6	70.9	26.4	55.9

Copepods were considerably more important in the quantitative development of the pelagic ecosystem in the period 1980-1986, than any other category of organisms. The number of copepods reached a peak in warm seasons (spring, summer and autumn) and the increase was preceded by blooms of some small sized dynoflagellates and diatoms. These were of importance as a food resource for planktivorous fish such as *Sprattus sprattus* whose production increased during this period .

In addition to the trophic species already mentioned, the cystoflagellate *Noctiluca scintillans* underwent the most spectacular quantitative development during the last decade. As eutrophication has increased, so the biomass of *Noctiluca* in the overall biomass of zooplankton has risen steadily. Intensive blooms of *Noctiluca* were recorded in the summers of 1986-1989, when the total quantities of zooplankton were at their highest (Table 6).

There was an explosion in the population of scyphozoan jellyfish *Aurelia aurita* also. It is known that in general the mass development of so-called gelatinous planktonic organisms, whose bodies consist of up to 98-99% water, is a characteristic feature of marine zooplankton under eutrophication [51].

Thus, in the last decade the total zooplanktonic biomass (including *Noctiluca*) has increased, but the species diversity has fallen. The tendency of simplification in the qualitative structure and the reduction in the biodiversity was more obvious in the near shore areas which are closer to sources of pollution and constitute additional stress to the more sensitive species.

TABLE 6. The numerical and weight maximum dominant (D%) of *Noctiluca scintillans* in zooplanktonic populations in Constantza area

Years	Month	D% in total density	D% in total biomass
1982	July	91.5	95.8
1983	June	94 7	96.7
1984	August	95.6	99.1
1985	July	41.0	34.3
1986	July	91 5	98.3
1987	July	92.3	98.5
1988	August	91.5	98.5
1989	July	97 9	99.9
1990	August	97 5	99.3
1991	July	99 2	99.8

Long term studies in such areas have delineated communities that consist of only two or three species. These species are capable of utilizing large quantities of available organic matter (e.g. ciliates, rotiferes, *Noctiluca*) [34,35].

Since 1987, a degradation of zooplanktonic communities has been noted. Their status has become very precarious particularly during summer, a severe reduction of populations being registered for most fodder species. The quantitative indices for the period 1989-1994 demonstrate this decrease. In 1990 and 1991, the mean densities were six to seven times lower than in 1989, because dominant species such as *Acartia clausi* and *Pleopis polyphemoide*, were less abundant [40].

The structural degradation and decline of fodder biomass must be attributed to the impact of the predatory ctenophore *Mnemiopsis leidyi.* The *Aurelia* populations collapsed almost immediately and zooplankton and ichthyoplankton dropped sharply.

After 1987, the seasonal patterns of zooplankton showed that, for all trophic groups, summer biomass values had declined to levels never previously recorded for the Romanian littoral [39] (Fig. 2,3).

The decrease in the quantity of zooplankton, mostly filter feeding herbivorous species, caused reduction in the number of consumers of phytoplankton which is usually present in excess, contributing to the increase of primary production.

The reduction in the quantity of zooplankton has been correlated with a spectacular reduction of the catches of planktivorous fish, their stocks drastically decreasing after 1988.

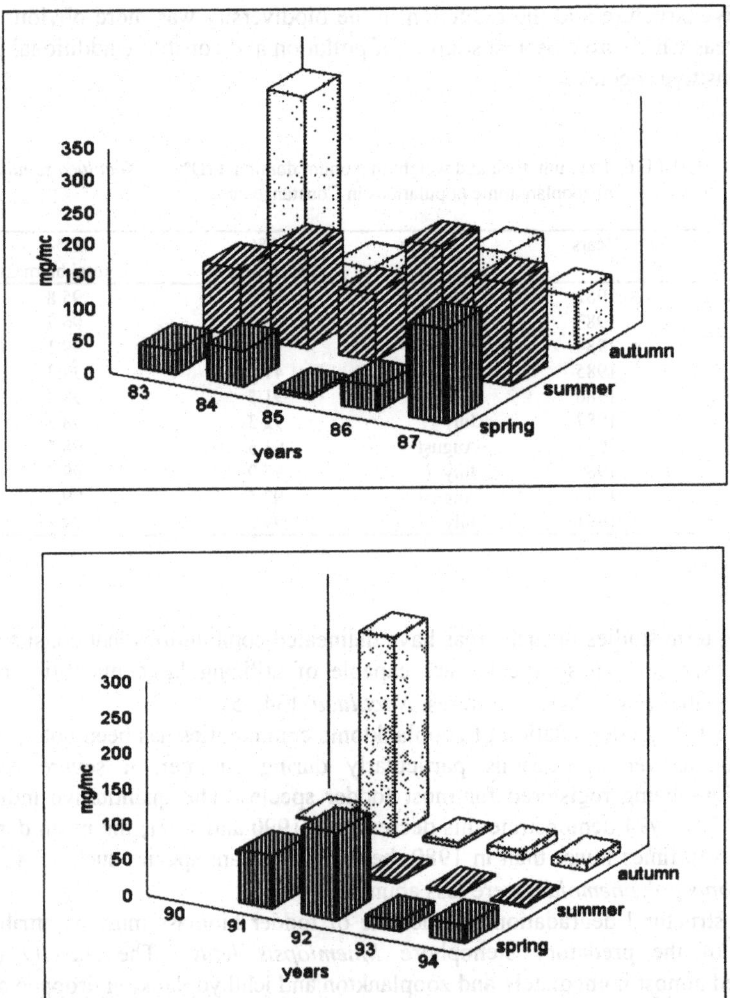

Figures 2 -3 - **Seasonally evolution of the fodder zooplanktonic biomasses in the shallow waters during 1983-1987 and 1990-1994**

3. Current state of bottom communities

3.1. ALGAL MACROFLORA

The Romanian research concerning the Black Sea macrophytobenthos covers a period of over 100 years [11] and the inventory of benthic marine algae lists 187 species [6]. On the total of 36 Cyanophytes, 45 Chlorophytes, 2 Xantophytes, 30 Phaeophytes and 74 Rhodophytes mentioned along the Romanian shore beginning with 1935, only 105 were still found after 1960. Simultaneously with an increasing eutrophication and influence of anthropic factors, the decline of the macrophytobenthos became evident both quantitatively and qualitatively. The most obvious consequences were owing to the landslides provoked by some construction works (dams, ways) when the algal carpet was covered by sand or mud, worsening transparency.

Investigation in the southern sector of the Romanian shore in 1982 showed only 24 taxa: 1 Cyanophyta, 11 Chlorophyta, 2 Phaeophyta and 10 Rhodophyta [16].

Along the Romanian shore *Cystoseira barbata* was the dominant perennial brown alga with high productivity and an important ecological role [10]. The largest and most compact field was the Vama Veche site. Between 0.5 and 2m depth this alga formed a belt of 150-200m in width and 3 km length. The biomass of *Cystoseira* has decreased from 5,400 t fresh weight in 1971 to 755 t in 1973 and 120 t in 1979. At present the almost complete disappearance of the former important *Cystoseira* belts has to be mentioned [12].

The fields of *Cystoseira barbata* constituted a very important ecological niche for the development of marine life in the rocky infralittoral zone especially for the associated fauna inhabiting that field [49].

The free substrata, previously populated by *Cystoseira* are now covered by opportunistic species with a short life cycle, having a rapid growth. The most frequent species belong to the genera *Enteromorpha* and *Ceramium* and to a lesser extent to *Ulva*, *Cladophora*, *Porphyra* and *Callithamnion* their diversity being a biological indicator of saprobic sea water. The present vegetation is characterized by a very uniform aspect due to a small number of component species.

A recent paper presented a Red List of the benthic macrophytes from the Romanian Black Sea sector, where 20 extinct and endangered species, 41 rare species and 4 insufficiently known species are listed [14].

The phanerogams *Zostera marina* and *Z. nana* have also declined considerably. The impoverishment of algal communities that were preferable biotopes for many invertebrate and fish species caused a decline of animal populations and even the disappearance of some faunistic components of the ecosystem.

3.2. BOTTOM FAUNA IN HYPOXIC CONDITION

During 1960-1970, considerable effort was directed to the study of the taxonomy of benthic invertebrates almost 800 taxa being identified. The great majority of these were small - sized forms (meiobenthos). The study of benthic ecology up to 1970 provided an insight into faunal associations (biocoenoses). Trends in the benthic ecosystem can therefore be documented over a period of three decades. The enhanced growth of phytoplankton during the last few decades, culminating in frequent and extensive blooms, has had serious consequences for the zoobenthic communities. Algal blooms have resulted in severe oxygen depletion leading to hypoxia and causing the death of the benthic biota especially.

This phenomenon also adds to the increasing quantity of dead matter, and thus to an increased oxygen shortage perpetuated over time and culminating in mass of the benthic fauna. As a consequence, over the last two decades there has been a precarious survival of benthic populations with chronic mortality in the warmer part of each year.

In comparison to the data obtained in 1960-1970 period, the zoobenthos recorded in the last two decades was greatly diminished both qualitatively and quantitatively.

Profound modifications have occurred in the zoobenthic community structure and these changes have been reported in several papers [20,21,22].

There has been a notable fall in the number and abundance of species, especially those that are members of more vulnerable biocoenoses, which tend to be rich in filter feeders.

Thus, in the decade 1960-1970, in the most widespread biocoenosis of the northern infralittoral zone- the *Corbula mediterranea* biocoenosis, more than 100 taxa were identified, the molluscs representing the most important group. Its present status reflects depletion, both qualitatively and quantitatively. From 100 species recorded in 1965 only 12 were to be found in 1982. New opportunistic species have appeared in these zones, mainly in the highly eutrophised marine area- *Mya arenaria, Cunearca cornea, Neanthes succinea* becoming dominant [21,48] (Table 7).

The biocoenoses of coarser sands from the midlittoral of the southern zone was characterized in the 1960-1970 period by the association of the bivalve *Donacilla cornea* and the polychaete *Ophelia bicornis*, both of which have now disappeared .

Rocky substrata occupy only a small proportion of the sea floor (0.3% of the total area of the Romanian shelf) but the associated benthic communities are ecologically important due to the high abundance of organisms, the biocoenosis of *Mytilus galloprovincialis* being the most important.

TABLE 7. Frequency, density and biomass of zoobenthic species identified in the *Corbula* biocoenosis in 1982

Organisms	Frequency (%)	5 m		10 m	
		ind.m^{-2}	g.m^{-2}	ind. m^{-2}	g.m^{-2}
Nematoda	100	52,850	0.09	2,860	0.005
Nereis succinea	80	780	0.47	260	0.150
Spio filicornis	75	1,170	0.70	140	0.800
Cardium edule	10	10	6.00	0	0
Telinna tenuis	10	15	6.27	0	0
Mya arenaria	90	90	18.48	180	4 90
Corbula mediterranea	90	1,850	27.45	680	6 69
Copepoda	20	20	0.0004	180	0 003
Balanus improvisus	5	10	0.0002	0	0
Iphinoe maeotica	5	10	0.005	0	0
Ampelisca diadema	80	305	1.83	40	0 240
Bathyporea guilliamsoniana	5	5	0.002	0	0

A number of species in these communities have disappeared or become rarer through degradation of the biotope caused by terrigenous pollution, or indirectly by the reduction in macrophytes with their associated diverse fauna.

In 1993, the research on benthic fauna from rocky zones showed marked changes [50]. The number of macrobenthic species and the frequency of crustaceans had fallen: in previous years, 28 crustaceans were registered at 3m depth, but only 14 were found in 1993. The decline in the abundance of vagile forms of crustaceans and polychaetes, consequently induced a general reduction of total density.

The benthic communities of muddy bottoms have been influenced by a number of factors, including increased turbulence and sedimentation. The Danube carries a high load of alluvial deposits, which continually modify the substrata, inducing instability.

In the two subcoenoses in front of the Danube mouths, the subcoenosis *Spisula subtruncata/Syndesmia fragilis/ Cardium paucicostatum* (at 10m up to 50m depth) and the subcoenosis *Spisula/Corbula* (at 5-8m depth), the consequences of hypoxia induced mass mortality determining a decline in biological diversity (Fig.4,5) [22,23].

For the first subcoenosis, from 32 species of animals identified during 1976-1977, there were 27 species in 1978, 22 in 1979, and only 14 in 1980. In 1987, the fauna was poor: only 15 macrobenthic species, occurred with the opportunistic polychaetes - *Neanthes succinea* and *Polydora limicola* dominant [46,47].

During 1960-1965 the muddy biocoenosis with *Mytilus* (at 30-45m depth), was considered to be the richest biocoenosis in the entire sector (50 species), with *Mytilus* as dominant. After 1977, the impoverishment was progressive, the species most affected being molluscs and crustaceans. In 1980, the general biomass had dropped, to 30 times smaller than in 1977. The main cause of this decline was hypoxia conditions following the phytoplankton blooms.

Muddy bottoms supporting *Modiolus phaseolinus* biocoenosis were highly developed off Sulina, Sf. Gheorghe and Tuzla-Mangalia before 1970. Initially, the eutrophication had only a limited impact on this offshore biocoenosis but surveys at 60-120m in 1994 indicated that the fauna was much poorer (Fig.6).

In a period when eutrophication was less developed (1976-1986) the macrobenthos density and biomass increased owing to the proliferation of the opportunistic species, but recently, especially between 1989-1992, the zoobenthos has declined constantly. Eventually, even tolerant species have been affected and reduced by the progressively deteriorating conditions.

At present, the benthic communities are characterized by great instability, every year the hypoxia or even anoxia causing mortalities over large areas, their location depending on the specific hydrological conditions.

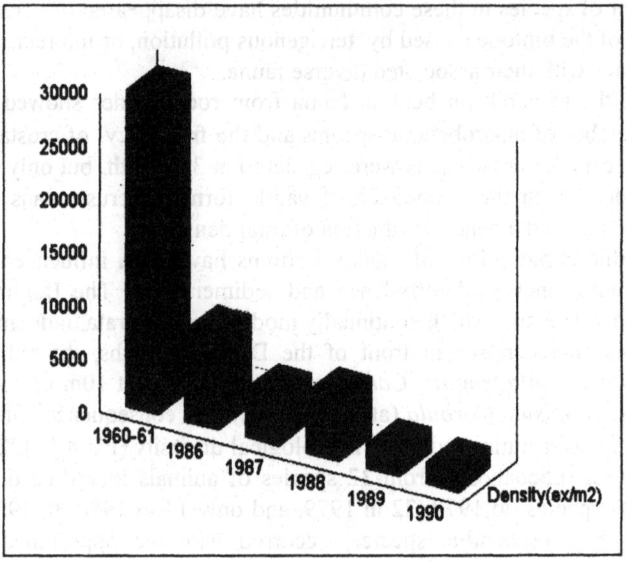

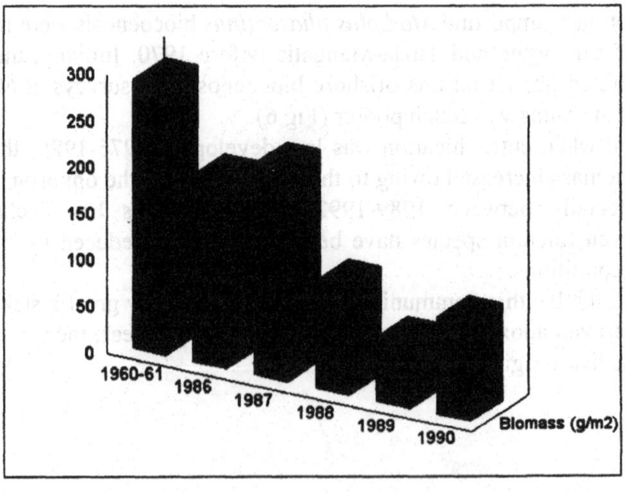

Figure 4-5 **The densities (ex/m2) and biomass (g/m2) of the macrozoobenthos from the Danube mouth zone, during 1960-61 and 1986-1990 period.**

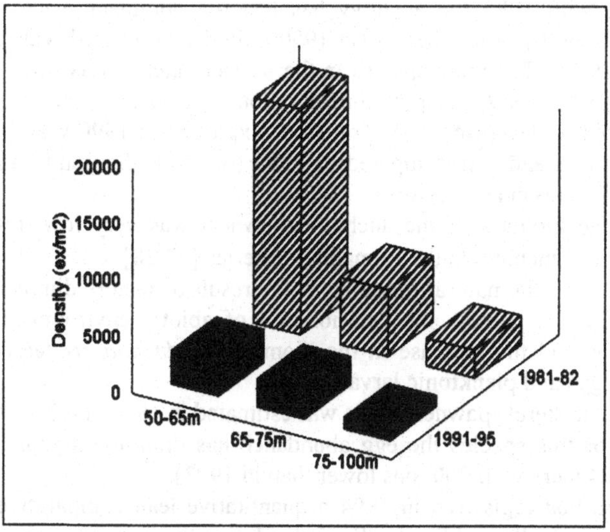

Figure 6 The mean densities of the macrozoobenthic species found in the offshore benthos, during 1981-1982 and 1991-1995.

4. Decline in the biological diversity of ichtyofauna and strong fluctuations in stocks of exploitable fish

Changes in the Black Sea ecosystem in the Romanian littoral were reflected also in the qualitative composition of the ichtyofauna, as shown by the taxonomic composition of non-commercial species, or in commercial catches. Before the 1960's over 20 commercially valuable fish species were caught, in the 1980's this amount declined to only five - *Sprattus sprattus phalericus, Alosa kessleri pontica, Engraulis eucrasicholus ponticus, Odontogadus merlangus euxinus* and *Trachurus mediterraneus ponticus* [15]. Some of these changes had an impact on coastal and shelf waters, others on the pelagic zone, affecting common and rare species.

The bottom fish inhabiting the shallow coastal waters sharply declined owing to hypoxia. This began with mass mortalities of fish as well as other benthic organisms, after 1970. For example, the turbot (*Psetta maeotica*) catches, on the Romanian shelf totaled 354.4 tons in 1950-1954, 124.4 tons in 1965-1969 and 70.2 tons in 1970-1974.

In the open sea, the large predators, such as common tunafish, mackerel and bluefish have been caught in smaller and smaller quantities almost to their complete disappearance.

With the disappearance of predatory fish, the number of small pelagic species with shorter life cycles such as anchovy, sprat and horse mackerel have increased until 1985-1986 [45].

Beginning with 1985 the anchovy has lost the dominant place in catches, being gradually replaced by other species (in 1990 only 5 t, compared with 110-3230 t in the period 1950-1979). The main species in the trawler catches was sprat (more than 80%) but beginning with 1989 the sprat concentrations tended to be more offshore and their densities tended to decrease.. The sprat stock exploited in 1990 was lower every month by about 3-4 thousand t as compared to 1989 (6.5-14.6 thousand t in 1990 compared with 12.6-26.7 thousand t in 1989).

In 1992, the biomass of the anchovy spawners was very low, reflecting the poor occurrence of the anchovy in the Romanian fisheries [27,28] .

The increase of the natural mortality is the result of factors diminishing the survival of eggs and larvae, namely the deterioration of abiotic environmental factors in the spawning areas and the increase in the biomass of *Mnemiopsis leidyi,* as a voracious consumer of eggs and planktonic larvae.

The horse mackerel spawner stocks was estimated in 1990 at 4.2 $x10^3$ t and in 1991 at $2.57x10^3$ t. For this species the egg abundance has diminished year by year also (the catches in 1994 were with 900 tons lower than in 1993).

The genera that registered in 1994 a quantitative leap comparatively with previous years were *Clupeonella* and *Sardinella*. Considering the last two years, the sizes of 1994 captures are similar to those of 1992 and 1993, 86% from the total production realized by the active fishery are for sprat, the rest being covered by whiting (13,7%), horse mackerel (0,2%) and sharks.

5. State of marine mammals population

With regards to the status of mammals, the three species of dolphins living in the Black Sea (the bottlenose dolphin *Tursiops truncatus*, the common dolphin *Delphinus truncatus delphis* and harbour porpoise *Phocaena phocaena*) are curently in a critical state in the Romanian Black Sea waters. Even though the capture of dolphins was prohibited in Romania in 1966, their populations have continued to decline.

6. Main problems related to new introduced species - their impact on the ecosystem

Among the anthropogenic factors influencing the evolution of the ecosystem, the development of sea navigation has also created a mechanism for the import of new species, especially as larvae, in the ballast water of ships.

The new species that have penetrated into the Romanian Black Sea waters are the benthic molluscs *Mya arenaria*, *Cunearca cornea* and *Rapana thomasiana,* and more recently the planktonic ctenophore *Mnemiopsis leidyi* which has had the most spectacular evolution with remarkable ecological consequences.

In the sedimentary beds of the Romanian littoral, *Mya arenaria* has become dominant in the 1970-1980 period. During a very short period, from the first record in 1969, the species settled spectacularly in sandy littoral shallow water in 1970-1972 [25,37]. For the

1973-1981 interval there were annual evaluation of the *Mya* stocks in its main area of spreading (Sulina - Mangalia at 0-30m depth, an area totaling 1,765.11 km^2). Throughout these years the population of *M. arenaria* did not fluctuate greatly remaining at significant densities.

With the increase in eutrophication and hypoxia events, the populations of *M. arenaria* have been strongly affected, mass mortalities being recorded, and the biomass being reduced even further during the decade 1980-1990.

In 1990 the *Mya* stocks for whole 0-30m zone were estimated at 417,000 tons, a considerable decline compared with the 1976-1981 period.

The introduction of the soft clam *M. arenaria*, has not represented a totally negative ecological event, since this species has contributed to the increase in the biological productivity of the sandy infralittoral zones in which it settled, although the area of distribution of aboriginal dominant species *Corbula mediterranea* has consequently decreased and the population declined..

By contrast, the introduction of the predatory gastropod *Rapana thomasiana,* originating from the Sea of Japan, represented a negative ecological event, because after destroying completely the oyster beds on the Caucasian coast, it began to extend its range rapidity along the coast, reaching the Romanian littoral over the period 1947 to 1963. The species first appeared at the Danube mouth and rapidly spread southward becoming a common element of the shallow water fauna, both on sandy and rocky bottoms [19]. On rocky bottoms it occurred most abundantly and frequently in the zones delimited by 4m and 10m exhibiting a maximum density between 8-10 m depth. An evaluation of *Rapana* stock has yet to be made but an estimation in 1976 averaged 15 $ind.m^{-2}$ at 10-15 m depth, on the rocky substratum. It is possible that with continued degradation of the coastal zone in recent years, *Rapana* too has declined.

Knowing that it feeds on bivalves, especially *Mytilus* and *Mya,* which have an essential role as natural biofilters, *Rapana* may not only damage populations of these species, but may affect the filtering capacity of such communities.

The advance of the exotic predatory ctenophore *Mnemiopsis leidyi*, along the Romanian littoral since 1988, has had a significant impact on the pelagic communities. In the summer of 1989, a reduction in zooplanktonic quantities, especially for the fodder species, was recorded, as for the whole summer period 1990-1993 [39].

Romanian research on the distribution of the ctenophore *M. leidyi* began in 1991 when a biomass of 76.000 t was estimated over an area of 4,000 sq.N.m. After 1993 the biomass had decreased and the distribution area had become smaller.

In 1993 the mean density of the ctenophore was 10 t/sq.nautical miles (n.m.), with the most dense aggregations in the southern zone of the littoral, beyond the 50m depth. In the spring of 1994 its maximum waş 7.9 t/sq.n.m., while in the autumn of the same year the biomass ranged between 11.1 and 15.1 t/sq.n.m. In the autumn of 1995, a mean density of only 0.74 ind/sq n.m. was registered in the south of the coastal zone [45].

The *M. leidyi* invasion has led to a sharp decline of the Romanian Black Sea fisheries for small fish, their stocks drastically decreasing after 1988. Although biomass of *Mnemiopsis* has decreased after 1993, the zooplankton biomass in the summer months has remained low [28].

The heavy grazing of the ctenophore on zooplanktonic organisms, which are mainly phytophagous, means not only a diminution of the fodder zooplankton but also an enormously reduction of herbivorous feeders, and thus an increasing of planktonic primary production, which is present already in excess.

7. Present status of the Romanian Black Sea ecosystem - goals for the rehabilitation

Environmental monitoring over the past four-five years suggests a perceptible and continued improvement in the state of some biotic components of the ecosystem.

Essential nutrient concentrations have continued to decline (silicate reached an extremely low value in 1994 and the concentration of phosphate continues to fall). However, in comparison with the 1960-1970 decade (i.e. prior to eutrophication), nitrate especially, and to a lesser extent phosphate, are still at high concentrations in Romanian coastal water, giving rise to phytoplankton blooms. However, the frequency of these phenomena has decreased and the algal concentrations produced during blooms have diminished also. There were 16 bloom events in 1991-1996 in comparison with 49 in the 1980's but only three blooms had high amplitudes in 1991-1996, while in the 1980's there were 15 blooms of very high amplitude.

The decrease of the algal bloom magnitude in 1991-1996 suggested a phytoplankton tendencies towards its "normal' status, before the eutrophication intensification .

However, the improvement is still fragile, despite the reduction of blooms, the nutrient stocks being still higher than before the eutrophication intensification, constituting a potential which can produce algal blooms under favourable meteorological conditions.

In the pelagic environment, also, during the last two years (1995 and 1996) in spite of rather high *Mnemiopsis* abundance, the zooplanktonic quantities were a little higher than in previous years, indicating a gradual improvement, occurring probably in the periods with decreasing abundance of the predator.

Even if in the recent period the algal blooms have not had the frequency and amplitude of those in the 1970's-80's, the synergic effects of the repeated massive mortalities of the marine animals are still maintained in the benthos. Thus, the quantitative and qualitative state of benthic communities on the Romanian continental shelf, which were the most affected, are still greatly disturbed, the situation being more obvious in the northern area of the littoral where the reduction of the specific diversity is very evident. In the last three years there were some trends of restoration for this component of the ecosystem also. For instance, at depths of 35-50m in the area of the Danube mouths, higher benthic biomass was registered due to the opportunistic species of polychaetes.

The consolidation of the gradual improvement in environmental conditions, with fewer red-tides and reduced animal mortality, is dependent on a decrease in nutrient input to the sea and on a fall in current concentrations. Nutrient input has been implicated as the main factor in ecological degradation of the Black Sea, its reduction remaining a major requirement for the continual limitation of algal blooms and mitigation of their negative effects.

These major ecological changes have not been restricted only to the Romanian littoral, they have included the whole north-western part of the Black Sea, where the effects of eutrophication/pollution process have been accumulated in time. A significant reduction of eutrophication is not possible without joint efforts of all countries of the Black Sea basin, also with the participation of danubian countries involved in the contemporary clean up effort.

Thus, a sustainable development of the Black Sea requires continued and enhanced international co-operation, through programmes designed to develop a scientifically based strategy for restoring the natural regime of this unique ecosystem.

References

1 Apas M (1985) Micromycetes du littoral roumain de la mer Noire, *Cercetari marine -Recherches marines*,**18**, 157-190.

2 Apas M. (1987-88) Structure et evolution des populations de micromycetes de la zone Constanta pendant l'annee 1987, *Cercetari marine -Recherches marines*, **20/21**, 275-284.

3. Apas M., Hulea A. (1984) Indicateurs de pollution marine. Phycomycetes *Cercetari marine - Recherches marines*, **17**, 251-266.

4. Apas M. (1990) Sur la dynamique et la structure du mycoplankton du plateau continental roumain de la mer Noire, *Cercetari marine -Recherches marines*, **23**, 49-71

5. Apas M. (1992) Mycromycetes de la zone littorale polluee de la mer Noire, *Rapp. Comm. int. Mer Medit.*, Monaco, **33** (2), 166.

6 Bavaru A., Bologa A.S., Skolka H.V (1991) A checklist of the benthic marine algae (except Diatoms) along the Romanian shore of the Black Sea, *Rev. Roum. Biol. -biol. veget.* ,**36**, 1-2, 7-22.

7 Bodeanu N. (1989) Algal blooms and development of main phytoplanktonic species at the Romanian Black Sea littoral under eutrophication conditions, *Cercetari marine -Recherches marines*, **22**,107-125.

8 Bodeanu N. (1992) Algal blooms and development of the main phytoplanktonic species at the Romanian Black Sea littoral in conditions of eutrophication process, *Science of the Total Environment*, suppl. 1992 891-906.

9. Bodeanu N. (1993) Microalgal blooms in the Romanian area of the Black Sea and contemporary eutrophication conditions, *Toxic Phytoplankton Blooms in the Sea*, 203-209.

10. Bologa A.S. (1986) Implicatii ecologice ale algoflorei bentale de la litoralul romanesc al Marii Negre, *Hidrobiologia*, **19**, 75-82.

11. Bologa A.S. (1987-88) Annotated bibliography on the macrophytobenthos along the Romanian Black Sea coast (1881-1986), *Cercetari marine - Recherches marines*, **20/21**, 353-384.

12. Bologa A.S. (1989) Present state of seaweed production along the Romanian Black Sea shore, *Vie Milieu*, **39**, 2, 105-109.

13. Bologa A.S , Bodeanu N., Petran A., Tiganus V.,and Zaitsev Yu P. (1995) Major modifications of the Black Sea benthic and planktonic biota in the last three decades, *Bulletin de l'Institut oceanographique*, Monaco, **15**, 85-110

14. Bologa A.S., Bavaru A. (in press) Lista rosie a algelor macrofite bentale disparute si pe cale de extinctie, rare si insuficient cunoscute din sectorul romanesc al Marii Negre, *Ocrot. nat. med. inconj*

15. Cautis I., Verioti F. (1976) Modificari in capturile romanesti la Marea Neagra si perspectiva exploatarii - (Changes in the Romanian catches in the Black Sea and the exploitation prospects) , *Cercetari marine - Recherches marine*, **9** suppl , 159-176

16 Celan M., Bologa A S (1983) Notice sur la flore marine du secteur sud du littoral roumain de la mer Noire, *Rapp. Comm. int. Mer Medit.*, **28**, 3, 215-217.

17 Cociasu, A (1991-1992) Niveau des principaux parametres chimiques des eaux cotieres roumaines de la mer Noire, *Cercetari marine -Recherches marine*, **24/25**, 11-24

18 Cociasu A., Popa L (1991-92) Observations sur l'evolution des principaux parametres physico-chimiques de l'eau marine de la zone Constanta, *Cercetari marine -Recherches marines*, **13**, 51-61.

19. Gomoiu M.-T (1972) Some ecological data on the gastropod *Rapana thomasiana* Crosse along the Romanian Black Sea shore, *Cercetari marine -Recherches marines*, **4**, 169-181.

20. Gomoiu M.-T. (1975) Some zoogeographical and ecological problems concerning the benthic invertebrates of the Black Sea, *Cercetari marine -Recherches marines*, **8**, 105-119.

21 Gomoiu M.-T. (1976) Modificari in structura biocenozelor bentale de la litoralul romanesc al Marii Negre (Changes in the structure of benthic biocoenoses of the Romanian littoral of the Black Sea), *Cercetari marine Recherches marines*, **9**, 119-142.

22. Gomoiu M.-T. (1981) Some problems concerning actual ecological changes in the Black Sea, *Cercetari marine -Recherches marines*, **14**, 109-127.

23 Gomoiu M-T (1983) Sur la mortalite en masse des organismes benthiques du littoral roumain de la mer Noire, *Rapp. Proc.-Verb. Reun.*, **28**, (3), 23-204.

24 Gomoiu M.-T (1984) *Scapharca inaequivalvis* (Bruguiere) - a new species in the Black Sea, *Cercetari marine -Recherches marines*, **17**, 143-165.

25 Gomoiu M.-T , Petran A., (1973) Dynamics of the settlement of the bivalve *Mya arenaria* L. on the Romanian shore of the Black Sea , *Cercetari marine - Recherches marines*, **5/6**, 263-289.

26 Mihnea P-E., (1981) Modifications des communautes phytoplanctoniques littorale sous l'influence du phenomen de pollution. *Ves Journ. Etud Poll. marines Medit.*, 869-877.

27. Nicolaev S., Radu G., Butoi G., Anton E. (1994) Structura pescariilor romanesti la Marea Neagra, evolutia capturilor si mutatiile structurale produse in ultimii 10 ani (Structure of the Black Sea Romanian fisheries, catch evolution and structural changes occurred during the last ten years) *Romanian National Report*, GEF-BSEP, Working Party on Fisheries, Constantza (11-13 April).

28 Nicolaev S., Radu G (1995) Evolution of abundance of eggs and juveniles of anchovy (*Engraulis enchrasicholus*) and horse mackerel (*Trachurus mediterraneus ponticus*) in 1990-1994, Romanian Marine Zone , National Report, GESAMP (Geneva, 20-24 March).

29. Petran A. (1962) Consideratii asupra compozitiei si variatiilor calitative ale zooplanctonului marin din dreptul litoralului romanesc al Marii Negre, *Com. Acad RPR.*, **12** (1),71-77

30. Petran A. (1976) Sur la dynamique du zooplankton des cotes roumaines de la mer Noire, pendant les annees 1974-1975, *Cercetari marine -Recherches marines*, **9**, 101-125

31. Petran A (1977) Donnees concernant le zooplancton de la zone predeltaique de la mer Noire, *Cercetari marine -Recherches marines*, **10**, 117-125

32. Petran A (1980) Sur la faune meroplanctonique du littoral roumain de la mer Noire (donnees quantitatives pour les annees 1972-1979), *Journ. Etud System Biogeogr. Medit.*, CIESM, Cagliari, 89-90.

33 Petran A (1981) Evolution du meroplancton dans le milieu portuaire de Constanta (mer Noire). *Rapp. Comm Int. mer Medit.*, Monaco, **27** (7), 121-122.

34 Petran A. (1986) Remarques sur la structuration des population zooplanctonique dans les zones des emissaires d'eaux usees du littoral roumain de la mer Noire, *Cercetari marine -Recherches marines*, **19**, 55-72

35. Petran A (1988) Upon the impact of population on the structure of zooplanktonic communities at the Romanian littoral of the Black Sea, Ziridava, *XVII-IIIth Conf. ecol*, 392-395

36. Petran A., Gomoiu M.-T. (1965) Donnees quantitatives sur le meroplancton de la region des sables a *Aloidis maeotica* de la mer Noire, *Rapp. Proc. verb. CIESM*, **18**, 2, 467-469

37 Petran A , Gomoiu M.-T (1972) The distribution of the bivalve *Mya arenaria* L on the Romanian shore of the Black Sea, *Cercetari marine -Recherches marines*, **3**, 53-67

38 Petran A., Onciu T. (1979) Donnees quantitatives concernant le meroplancton de la zone predeltaique de la mer Noire, *Rapp. Comm. int. Mer Medit.*, **25/26**, 8, 143-144.

39. Petran A., Moldoveanu M. (1994-95) Post-invasion ecological impact of the Atlantic ctenophore *Mnemiopsis leidyi* Agassiz, 1865 on the zooplankton from the Romanian Black Sea waters, *Cercetari marine -Recherches marines*, **27/28**, 135-137.

40. Petran A., Moldoveanu M (in press) Characteristics of the structure and quantitative development of zooplankton from the Black Sea shallow water during 1990-1994,*Cercetari marine - Recherches marines*

41. Popa L., Dorogan L (1991-92) Donnees nouvelles sur l 'apport de nutrilites du Danube et leur niveau de concentration dans la zone du littoral roumain, *Cercetari marine -Recherches marines*, **24/25**, 25-40.

42 Porumb F. (1961) Contributii la cunoasterea fam Monstrilidae din dreptul litoralului romanesc al Marii Negre, (Contribution a la connaissance de la fam. Monstrillidae du littoral roumain de la mer Noire), *Com Acad RPR*, **11,10**, 1223-1251

43. Porumb F. (1969) Contribution a la connaissance de la dynamique du meroplancton de la zone sud du littoral roumain de la mer Noire. *Lucr. Stat. Cerc. Mar. Agıgea*, **3**, 35-46.
44. Porumb F., (1989) On the development of *Noctıluca scıntıllans* (Macartney) Kofoıd and Swezy under the eutrophication of the Romanian Black Sea water, *Cercetarı marıne -Recherches marınes*, **22**, 247-262.
45. Radu G., Nicolaev S., Radu E. (in press) Dıstrıbutıa spatiala si evaluarea biomasei ctenoforuluı *Mnemıopsıs leıdyı* si a meduzei *Aurelıa aurıta* la litoralul romanesc al Marii Negre in perioada 1991-1995. *Cercetarı marıne -Recherches marınes.*
46. Tiganus V (1977) Observations sur la faune sessile des champs a *Cystoseıra barbata* (A.G.) de la mer Noire, *Cercetarı marıne -Recherches marınes*, **10,** 127-141
47. Tiganus V. (1982) Evolution des principales communautes benthiques du secteur marin situe devant les embouchures du Danube pendant la perıod 1977-1980, *Cercetarı marıne -Recherches marınes*, **15,** 89-106.
48. Tiganus V (1983) Modıfications dans la structure de la bıocenose des sables a *Corbula medıterranea* (Costa) du littoral roumaın, *Rapp. Comm. ınt. Mer Medıt*, **28,**3, 205-206.
49. Tiganus V. (1991/92) Fauna associated with the maın macrophyte algae from the Romanian Black Sea coast, *Cercetarı marıne - Recherches marınes*, **24/25**, 41-123.
50. Tiganus V., Dumitrache C (1991-92) Structure actuelle du zoobenthos de la zone de faible profondeur devant les embouchures du Danube, *Cercetarı marıne -Recherches marınes*, **24/25**, 125-132.
51 Zaitsev Yu. P., Polıschuk L.N. (1984) V spiska chislenosti meduzy *Aurelıa aurıta* (l) v Chernom more *Ekologya morya*, **17**, 35-46.

MODELING THE BLACK SEA PELAGIC ECOSYSTEM AND BIOGEOCHEMICAL STRUCTURE: A SYNTHESIS OF RECENT ACTIVITIES

TEMEL OGUZ, UMIT UNLUATA
Middle East Technical University, Institute of Marine Sciences, Erdemli, Icel, TURKEY
HUGH W. DUCKLOW
Virginia Institute of Marine Sciences, The College of William and Mary, Gloucester Point, VA, USA
PAOLA MALANOTTE-RIZZOLI
Massachusettes Institute of Technology, Department of Earth, Atmospheric and Planetary Sciences, Cambridge, MA, USA

Abstract: Recent modeling studies on the structure and functioning of the plankton productivity and various other features of the vertical biogeochemical structure of the Black Sea are reviewed in this study. Major findings from the available pelagic ecosystem models, catagorized as the mixed layer based physically simplified models and the vertically resolved coupled physical-biochemical models, are described first. Capability of present models in describing nutrient cycling, oxygen dynamics and suboxic-anoxic layer interactions is then assessed.

1. Introduction

The structure and functioning of plankton community in the Black Sea have been modelled using two different approaches. One of them was to employ biologically-based, physically simplified mixed layer averaged models, as given by Lebedeva and Shushkina (1994), Cokasar and Ozsoy (1998) and Eeckhout and Lancelot (1997). They considered evaluation of the mixed layer averaged properties of the ecosystem by specifying yearly cycles of the mixed layer depth

S. Beşiktepe et al. (eds.),
Environmental Degradation of the Black Sea: Challenges and Remedies, 197–223.

and temperature diagnostically from available climatological data, without invoking a mixed layer dynamics and a parameterization scheme for the interfacial transports. The second approach was to utilize z-dependent models which represent the vertical biological processes in a more detailed way and involve upper layer physical dynamics. In this type of models, the internal structure of the physical-biological system is evolved solely in response to external forcings applied as the boundary conditions at the free surface and bottom of the model ocean. A principal advantage of z-dependent models is their ability to include plankton productivity and nutrient recycling processes below the mixed layer. This is particularly important during summer when most of the production takes place below the seasonal thermocline. Its applications were given by Oguz et al. (1996, 1998a,b) and Staneva et al. (1998).

In addition to simulation of plankton productivity, Oguz et al. (1998b) incorporated various other features of the upper layer biogeochemical structure. It thereby allowed a dynamical coupling between the euphotic zone, the oxycline/nitracline, and the suboxic layers. Redox processes of the anoxic interface zone were studied to some extent by Yakushev and Neretin (1997) and Lyubartseva and Lyubartsev (1997). These models were, however, based on the "coexistence layer" assumption, thus the dissolved oxygen was accepted as a main oxidizing agent for the hydrogen sulphide. But, in the presence of an oxygen and sulphide depleted zone between the anoxic pool and the oxygenated surface layer, validity of this approach is questionable. Recently, several alternative hypotheses were proposed to explain the origin of the suboxic layer (Murray et al., 1998).

A succint review of the basic findings from these models is provided in this paper. First, those concerning with the annual plankton productivity cycle are presented in Section 2 in two groups as the "mixed layer based biological models" and "vertically-resolved coupled physical-biochemical models". Section 3 deals with the models describing the general biogeochemical characteristics of the Black Sea. A discusssion of the model results are provided in Section 4.

2. Modeling Annual Plankton Productivity Cycle

2.1. MIXED-LAYER BASED BIOLOGICAL MODELS

2.1.1. Lebedeva and Shuskina Model

Lebedeva and Shushkina (1994) explored the central Black Sea ecosystem characteristics before and after the introduction of the Mnemiopsis using a relatively simple pelagic lower trophic food web structure. The plankton community involves phytoplankton, bacteria, protozoa, mesozooplankton, medusae and mnemiopsis. They are complemented by the particulate and dissolved organic matter and nutrient compartments. Phosphate was used as the limiting nutrient whereas most of the other models were based on the nitrogen limitation which seems to be more appropriate for the central Black Sea.

In the absence of Mnemiopsis, the model simulates a major phytoplankton biomass increase up to 7 gC/m^2 in late February-early March, and a fairly uniform phytoplankton biomass level of about 1 gC/m2 throughout the summer and autumn periods (Fig. 1a). A small increase signifying the autumn bloom occurs during early-October. Mesozooplankton generally follows the evolution of phytoplankton with two maxima of 2.3 gC/m^2 in late March and 1.2 gC/m^2 in October. The medusae biomass reveals peak concentrations of 3.7 gC/m2 and 2.5 gC/m^2 in April and early-November, respectively. The reason for simulation of medusae biomass higher than those of mesozooplankton is the presence of a local source representing the advective transport of juveniles from the coastal regions at certain times of the year. Bacteria and protozoa, on the other hand, have much lower biomass values throughout the year. Bacteria exhibits spring and autumn maxima of 1.0 gC/m^2 and 0.5 gC/m^2, respectively, and a rather uniform summer concentrations of 0.2 gC/m^2. Protozoa biomass attain even smaller concentrations of about 0.1 gC/m^2 except a maximum of 0.5 gC/m^2 during late March.

Once the influence of mnemiopsis is included explicitely into the model, it predicts a five-fold reduction in the mesozooplankton biomass due to an extra grazing pressure introduced by mnemiopsis (Fig. 1b). Accordingly, the medusae concentration in the upper mixed layer is decreased twice whereas phytoplankton biomass increases gradually during the late autumn and winter months since, under these conditions, mesozooplankton experience less graz-

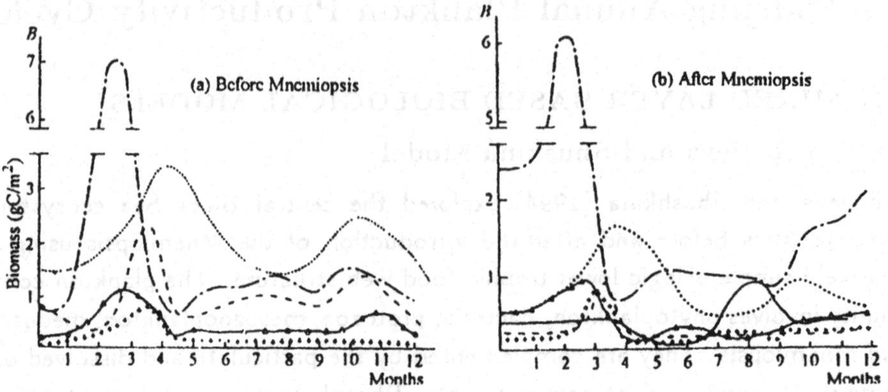

Figure 1. Annual distributions of the plankton community elements for two simulations; (a) before mnemiopsis case, (b) after mnemiopsis case computed by Lebedeva and Shushkina (1994).

ing pressure on phytoplankton. Such high winter phytoplankton biomass is followed by the early spring bloom in late February. It was noted that simulation of the high winter phytoplankton biomass agree with the observed winter bloom of 1991 within the interior part of the Black Sea. Mnemiopsis biomass stays typically at about 0.4 gC/m^2 during the autumn and winter months, followed by a slight decrease in spring. But, a two-fold increase occurs towards the end of summer.

2.1.2. Cokasar and Ozsoy Model

Cokasar and Ozsoy (1998) presented a series of simulations of the mixed layer planktonic structure using different variants of the Fasham model, ranging from a four compartment PZND model to a 9 compartment size-fractionated model for ten dynamically different regions of the Black Sea. These regions, identified according to their circulation and water mass characteristics, were the interior part of the basin, a narrow slope zone confined between the deep interior and shelf, and eight coastal regions dominated typically by anticyclonic circulation. One of these regions was chosen at the mouth of the river Danube, to observe the effects of the river supplied nutrients.

The modeling was guided by the analyses of the observed seasonal changes of mixed layer depth, nutrients and chlorophyll-a in the model regions. Nutrients content remain at a uniform level for the whole year in the regions directly influenced by river inputs. Mixed layer nutrients generally tend to increase in the other regions during the winter season. Large uncertainties however exist in estimating nutrient concentrations supplied to the coastal waters and their effects in sustaining primary production. Despite the scarcity of data, it appears that the maximum chlorophyll-a concentration occurs in February-March in the central Black Sea, and in April-May in the peripheral regions, where the level is also an order of magnitude higher, as a result of riverine and coastal sources.

The model successfully reproduced basic features of seasonal plankton and nutrient changes. The simple four compartment model was shown to provide realistic simulations of the seasonal production cycle. The PZND model had the advantage of simplicity but better results were obtained when more complexity was invoked. For example, better representation of the seasonal cycles including spring and autumn blooms were obtained with a nine compartment size fractionated model.

The model was also used to interpret the factors responsible for the observed regional differences in productivity. The computed seasonal cycle of the chlorophyll-a compared well with the chlorophyll measurements in the central Black Sea. On the other hand, advection of nutrients was found to be important along the western and southern Black Sea coastal areas downstream of the river sources. Near the Bosphorus, reasonable agreement of model results and observations could only be ensured when the seasonal pattern of advection of river nutrients were taken into account.

In the interior Black Sea, upwelling divergence resisting mixed layer deepening (greater nutricline gradient at higher density) in the cyclonic region seems to lead to a delayed bloom that is also depleted earlier as compared to the other regions, when the effects of riverine advection (whose seasonal supply has a delaying effect in coastal areas) is not included.

Simulations of the model with mnemiopsis included was, however, not satisfactory. Contrary to observed summer time increase in the mnemiopsis biomass within the interior of the basin, the model provided a late spring - early summer increase in response to the increasing zooplankton biomass, and a delayed phytoplankton bloom.

2.1.3. Eeckhout and Lancelot Model

Eeckhout and Lancelot (1997) studied the role of nutrient enrichment on destabilization of the northwestern shelf ecosystem within the last three decades. The simulation for the reference, non-perturbed coastal ecosystem of 1960's takes into account carbon, nitrogen, phosphorous and silicate cycles together with diatoms, nanophytoflagellates, bacteria, microzooplankton, copepods, as well as dissolved and particulate organic matters. This standard model structure was then extended to study the response of increased antropogenic nutrient load by including additional role of autotrophic opportunistic species on the primary production, and of gelatinous organisms called Noctiluca, Aurelia aurita and Mnemiopsis leidyi. The model also includes a benthic module.

When the model is initialized by the 1960's nutrient concentrations, it predicts an early spring diatom bloom with a maximum concentration of about 1 gC/m^2, followed by the development of a nanophytoflagellates bloom of the same size at the end of May. The zooplankton is composed of copepods and to a lesser extent of micro zooplankton. The biomass has a maximum of 0.6 gC/m^2 in April, which shows a gradual decrease in summer. When the experiment is repeated using the initial field of late 1970's nutrient concentrations, but in the absence of Mnemiopsis leidyi, the model predicts biomass increases at all trophic levels in response to this nutrient enrichment of the ecosystem. In particular, approximately 6-fold and 3-fold increases occur in the early spring diatom population and late spring nanophytoflagellates populations, respectively. The algal biomass explosion has a major impact on the copepod population which reveals a major peak of about 4 gC/m^2 at the end of March. The medusae population also shows an abrupt increase to a value of 1.2 gC/m^2 at the begining of April. This biomass concentration decreases only to about 0.8 gC/m^2 level during at the of autumn, followed by a more pronounced decrease in winter months.

Introducing Mnemiopsis leidyi into the model under the same nutrient conditions simulates the "after Mnemiopsis" scenario of the ecosystem. In this case, the model describes how Mnemiopsis takes over the control of the ecosystem as Aurelia population decreases to very low concentrations whereas Mnemiopsis exhibits similar but somewhat stronger biomass distribution over the year. The introduction of mnemiopsis, however, does not seem to introduce changes in the phytoplankton population.

2.2. VERTICALLY-RESOLVED COUPLED PHYSICAL-BIOCHEMICAL MODELS

2.2.1. Oguz et al. 1st generation Model

It was biologically simplest version involving only single phytoplankton and zooplankton groups, detritus, dissolved inorganic nitrate and ammonium (Oguz et al. 1996). It was applied to a 150 m thick upper layer water column in the central Black Sea, resolved using approximately a 3 m grid spacing. The vertical mixing is parameterized by the order 2.5 Mellor-Yamada turbulence closure parameterization. Given a knowledge of physical forcing, the model simulated main observed seasonal and vertical characteristic features; in particular, yearly evolution of the upper layer stratification, the annual cycle of production with the fall and the spring blooms, the subsurface phytoplankton maximum layer in summer, as well as realistic patterns of particulate organic nitrogen. It is found that initiation of the spring bloom depends crucially on the local mixing conditions and follows weakening of the convective overturning mechanism. As soon as the surface layer of the water column gains a slight stability, the bloom commences before the formation of the seasonal thermocline. This suggests timing of the bloom is governed by the year-to-year and/or local variabilities in the physical processes, in addition to the biological processes. The spring bloom may thus take place at an earlier period, say in February, during mild winters as pointed out by observations (Vinogradov, 1992).

Following the spring bloom, the model predicts a weaker and shorter phytoplankton growth event within April as the water column begins to stratify and the seasonal thermocline begins to form in the near-surface levels. The formation of this bloom is caused by the ammonium, generated as a by-product of the spring bloom, and trapped in the mixed layer. A period of very low primary productivity prevails throughout the summer as a consequence of severe nitrogen limitation in the surface mixed layer. However, some phytoplankton production goes on beneath the seasonal thermocline as long as this zone has sufficient light to support the phytoplankton growth. Towards the end of autumn, rapid destratification of the water column and subsequent intensification of the vertical mixing enhances the nutrient flux to the surface waters, and causes a phytoplankton bloom development of two-three weeks during the October-November-December period, depending on the local conditions.

The numerical experiments implicate the presence of a delicate balance between the growth and grazing processes in the phytoplankton dynamics. In order to get a phytoplankton distribution with two distinct blooms during the late autumn and the early spring, the grazing rate should be a certain fraction of the growth rate. If the grazing pressure is exerted too early and too strong, there will not be sufficient time for the development of sufficiently strong phytoplankton blooms. On the contrary, if it is too weak to be able to control the phytoplankton growth, one long-term bloom event occurs during the December-March period. Once the late autumn bloom is initiated, it persists whole winter since sufficient nutrient is always entrained into the surface layer to maintain the production during this period. Hence, the autumn bloom does not appear to be a robust feature of this model, contrary to the March bloom and summer subsurface chlorophyll-a maximum layer. Such modifications on the standard case of the two-bloom phytoplankton structure may also be traced in the data as a part of the year-to-year variabilities of the biological system.

2.2.2. Staneva et al. Model

Staneva et al. (1998) implemented the Oguz et al. (1996) five compartment model to different locations in cyclonic, anticyclonic and slope regions using the daily atmospheric forcing data for the period of 1980-1987. Significance of daily-to-interannual variabilities in the mixed layer thickness, the temperature and biological structures is noted in this study. The model simulates different phytoplankton structures for the cyclonic and anticyclonic regions depending on the local conditions. The anticyclones were shown to exhibit a standard two-bloom structure similar to the one presented above in Oguz et al. (1996), whereas the cyclones attain a one-bloom structure during the whole winter period. The strength of these blooms were almost half of the phytoplankton concentrations suggested by observations, implying that the model does not support a realistic nutrient cycling.

2.2.3. Oguz et al. 2nd generation Model

One drawback of the simplified five compartment ecosystem model given in Oguz et al. (1996) was underestimation of the summer production. The limited capability of such simple models in predicting summer chlorophyll values has in fact been noted by other studies (Sarmiento et al., 1993). It was also pointed out that multiple prey-multiple predator models can alleviate the limitations imposed by such simplified approaches and may generate increased chlorophyll concentrations comparable with observations (Armstrong, 1994). Oguz (1998a) therefore extended their previous "1st generation model" by introducing two phytoplankton species groups, typifying diatoms and flagellates, and two zooplankton groups (micro- and mesozooplankton). Microzooplankton (nominally $< 200\ \mu m$) consist of heterotrophic flagellates, ciliates and juvenile copepods, whereas mesozooplankton (0.2-2 mm) are formed essentially by copepods. Both of them feed on two types of prey with different prey capture efficiencies. Microzooplankton are considered to be more efficient at capturing flagellates, whereas diatoms are consumed predominantly by mesozooplankton.

Such a simple fractionation of the biogenic community structure was shown to yield increased primary production and development of more pronounced subsurface chlorophyll maximum layer during the summer period. Diatom-based early spring (March) bloom is followed by summer and autumn blooms of flagellates. They were either absent or had only a weak signature in the previous model of Oguz et al. (1996). The reason for the presence of stronger summer phytoplankton growth in the multi-species/multi-group pelagic food web model may be explained as follows. In the case of single phytoplankton and zooplankton groups, the enhanced grazing pressure exerted on phytoplankton following the March bloom prohibits noticeable phytoplankton development near the base of the euphotic zone during spring and summer months. In the presence of two phytoplankton and two zooplankton groups, the situation is somewhat different. Diatoms are responsible from March bloom, and support increased mesozooplankton activity later in spring and summer months. As a result of predator control by mesozooplankton on their grazers, flagellates do not experience any grazing pressure from the microzooplankton group, and may therefore provide a stronger subsurface production during the summer. This result implies that the choice of five compartment model may not be entirely adequate for representation of all bloom events within a year.

The second major feature of the model was its ability to reproduce a fairly realistic nutrient recycling mechanism. Dead cells and fecal matter sinking from the euphotic zone are continually remineralized to ammonium which is subsequently oxidized to nitrate. These conversion processes are accompanied at the same time with upward transport of both nitrate and ammonium to supply them back to the surface waters. The model simulations indicate that a major part of this recycling process takes place within the upper 50 m of the water column. Nearly 90 % of the primary production is recycled there. The annual nitrogen budget for the euphotic zone shows that 60% of the primary production is supported by the ammonium resources recycled within the euphotic zone. About 15% of the nitrate-based production constitutes new production whereas the rest originates from recycled nitrate within the euphotic layer as a result of the remineralization-ammonification-nitrification pathway. The remaining remineralization-nitrification occurs within the oxycline-nutricline zone confined between the euphotic zone and the suboxic layer (50-75 m). This gives rise to gradually increasing nitrate concentrations from the near-surface to about 75 m depth where the nitrate maximum occurs with concentrations of about 7.0 mmol N m^{-3}.

One interesting finding from the Oguz et al. (1998a) is the possibility of having oscillatory solutions for the linear food chain in which flagellates are consumed by microzooplankton whereas mesozooplankton are fed on diatoms and microzooplankton. In this case, all plankton community exhibit oscillatory character following the March bloom. The system is however stabilized by including some consumption of flagellates by mesozooplankton and diatoms by microzooplankton. The stability of the system was not found to depend on the type (i.e. linear, parabolic, hyperbolic, etc.) of mortality function.

Using the model, feasibility of using two different types of food preferences formulation was also explored. Assigning constant values for the coefficients of food capturing efficiencies turned out to be more practical for obtaining realistic model simulations in the vertically resolved models. On the other hand, since the weighted preferences formulation introduces depth and time dependences on these coefficients they generate an additional complexity on the model simulations.

2.2.4. Oguz et al. 3rd generation Model

The vertically resolved model of Oguz et al. (1998a) was further modified by adding carnivorous macrozooplankton, bacterioplankton and dissolved organic nitrogen compartments (Oguz et al. 1998b). The macrozooplankton compartment represents the medusae "Aurelia aurita" group dominated the Black Sea ecosystem before the invasion of the ctenophore "Mnemiopsis leidyi" (Shushkina and Musayeva, 1990; Vinogradov and Shushkina, 1992). The model simulations indicate that peaks of phytoplankton (diatoms and flagellates) and zooplankton (mesozoo and macrozoo) biomasses march sequentially one after the other as a result of their prey-predator interactions (Fig's. 2,3). Three diatom blooms occur during March (with maximum biomass of ~5 gC/m^2), late May-early June and December (~1.5 gC/m^2), whereas flagellates dominate the system during most of the summer and the entire autumn (~1-2 gC/m^2) starting from the end of June to the begining of November (Fig. 2a). This phytoplankton structure possesses more pronounced bloom characteristics during summer months as compared with the previous model described in Oguz et al. (1998a), and may point out the role of gelatinous carnivores started dominating the ecosystem during 1980's. This is due to a new "top-down" control mechanism in the food web structure in which increasing the gelatinous carnivore population puts a stronger control on the mesozooplankton community which subsequently weakens their grazing pressure on the phytoplankton structure. The most pronounced signature of this effect is observed towards the end of May and September which coincide with the periods of major increase in the Medusae population (Fig. 3a).

It may be inferred from the vertical structure of the total phytoplankton distribution (Fig. 2b) that the March and December blooms are surface-intensified events. They extend to the depth of 40-50 m coinciding approximately with the depths of winter convective overturning and the 1 % light level. Their formation is the result of entrainment of subsurface nitrates by the convective overturning and therefore is related to the new production. The late spring diatom bloom, on the other hand, is a subsurface event between the seasonal thermocline and the base of the euphotic zone and is essentially originated by the regenerated production.

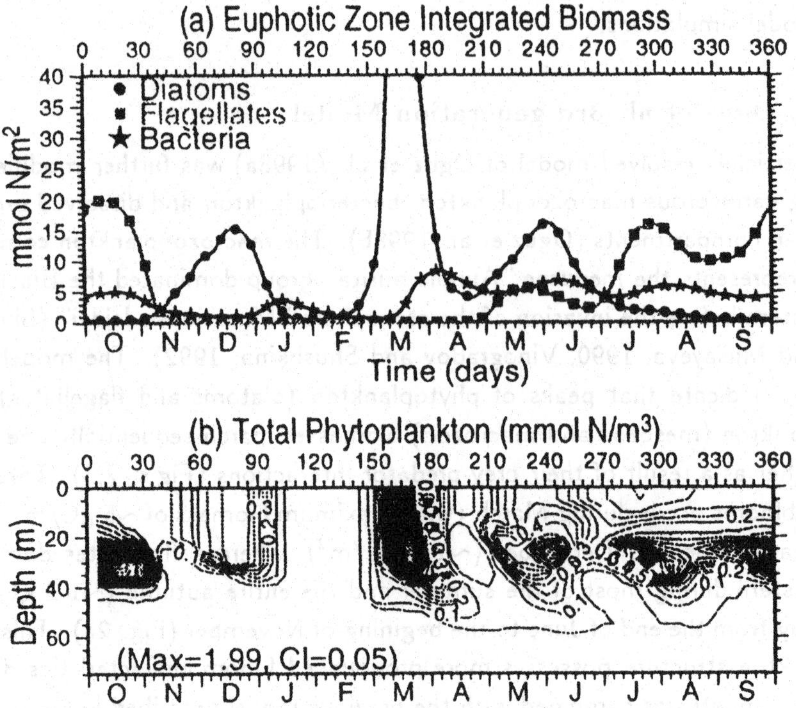

Figure 2. Annual distributions of the (a) euphotic zone integrated diatom, flagellate and bacterioplankton biomasses, (b) vertical structure of the total phytoplankton biomass within the upper layer water column computed by Oguz et al. (1998b)

As compared to diatoms and flagellates, bacterioplankton biomass exhibits a somewhat weaker distribution within the euphotic zone (Fig. 2a). The stock is typically less than 0.5 gC/m^2 in the late autumn and winter months. It almost doubles itself after the March diatom bloom till the end of summer. The summer bacterioplankton population is located mainly below the seasonal thermocline at the same levels with the flagellates. The simulated annual bacterioplankton distribution seems to be consistent with the data which indicate slightly higher biomasses after the Mnemiopsis invasion. This is, however, expected since increasing the biomass of gelatinous species should ultimately cause an increase on the particulate and dissolved organic matter contents.

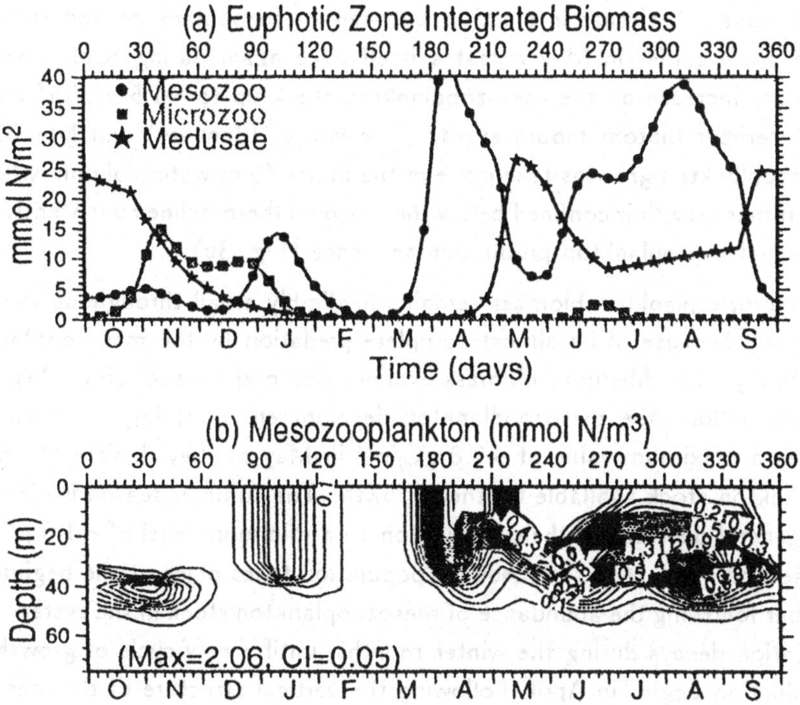

Figure 3. Annual distributions of the (a) euphotic zone integrated zooplankton biomasses, (b) vertical structure of the mesozooplankton biomass within the upper layer water column computed by Oguz et al. (1998b)

This in turn should lead to some increase in the bacterioplankton population in the system.

As soon as the March diatom bloom degrades, the mesozooplankton biomass starts increasing as they assimilate the diatoms (Fig. 3a). As the grazing pressure introduced by the mesozooplankton decreases the diatom population towards the end of March, mesozooplankton biomass keeps increasing in the euphotic zone. Their biomass tends to decline during May, which coincides with the period of Medusae growth. The summer mesozooplankton growth is principally caused by the reduction in medusae population, with additional contributions by the degradation of the phytoplankton blooms towards the end of June. A similar interaction between mesozooplankton and medusae taking place earlier in May repeates itself once again during September. A secondary increase on the mesozooplankton stock up to ~1.5 gC/m^2 follows the December diatom bloom event. The winter (January) and late March mesozooplankton growths take place in the upper 40 m water column, whereas the summer growth is confined below the seasonal thermocline consistently with the annual phytoplankton production sequence (Fig. 3b).

The microzooplankton biomass remains negligibly small throughout the year (Fig. 3a), because of its almost complete predation by the mesozooplankton community. The Medusae biomass exhibits two major peaks (Fig. 3a). The first one follows the mesozooplankton development in spring. The biomass reaches a maximum value of ~2.8 gC/m^2 in May as they deplete the mesozooplankton stock available for their growth. The summer season is identified by a general decrease in their population to a minimum level of ~1.0 gC/m^2. The second increase in the medusae population takes place at the begining of October following the abundance of mesozooplankton stock in the system. The population decays during the winter months, until a new cycle of growth and reproduction begins in April. Following the vertical structure of the mesozooplankton community, the first medusae growth event is distributed uniformly within the upper 40 m, whereas the second event is confined below the seasonal thermocline. The form of the annual medusae distribution predicted by the model thus agrees reasonably well with the data.

3. Modeling the Upper Layer Biochemical Structure

3.1. YAKUSHEV MODEL

Modeling the nitrogen and sulfur cycles across the oxic-anoxic interface region was described by Yakushev and Neretin (1997). The model considers the water column between 50 m and 200 m depths. It is forced by a constant dissolved organic matter concentration from the upper boundary, and by a fixed sulphide concentration from the lower boundary. Organic matter is decomposed to the form of ammonium under aerobic, subaerobic and unaerobic conditions at the expense of oxygen, nitrate and sulfate, respectively. Nitrification (ammonium to nitrate conversion) and denitrification (nitrate loss to the form of nitrogen gas) are two other processes of the nitrogen cycle. The main feature of the

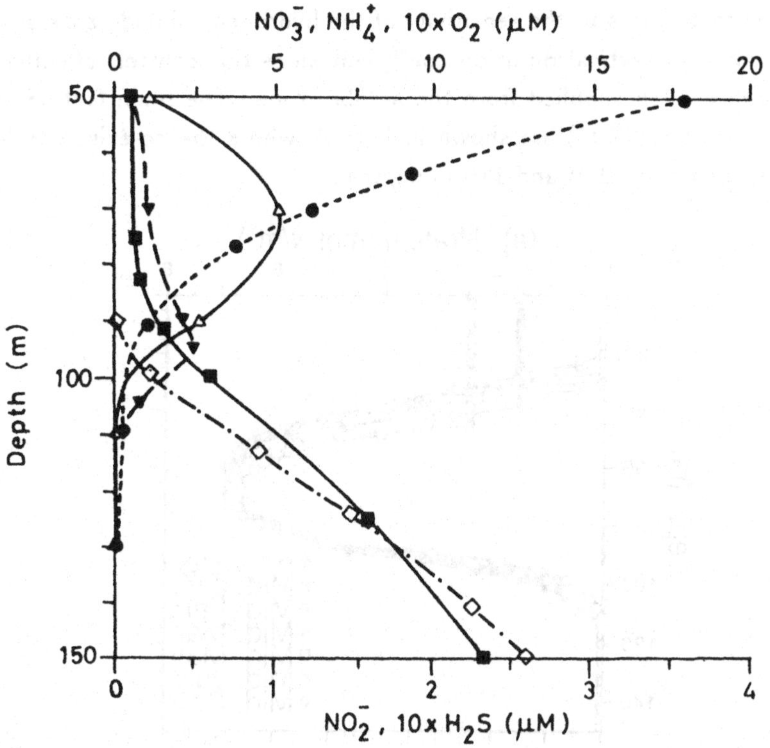

Figure 4. Vertical Profiles of hydrogen sulphide, oxygen, nitrate, ammonium and nitrite simulated by Yakushev and Neretin (1997)

sulfur cycle is the oxidation of H_2S primarily by the oxygen available at the anoxic interface zone. Thiodenitrification involving H_2S oxidation by nitrate is considered as a complementary process. The oxygen is the main oxidizer in the upper part of this layer, whereas the nitrate is in the lower part. The amount of oxygen necessary to maintain the H_2S oxidation and the rest of the sulfur cycle seems to be supplied there by specifying a rather high downward diffusive oxygen flux. The vertical diffusion coefficient taken as 0.1 cm^2/s is approximately an order of magnitude higher than its typical values estimated from the Gargett (1984) formula for the pycnocline region of the Black Sea. This is also apparent from the presence of 50 μM oxygen concentrations in the vicinity of nitrate maximum (Fig. 4), contrary to about 10 μM values suggested by observations. A similar model given by Lyubartseva and Lyubartsev

(1997) in fact points to the sensitivity of the hydrogen sulphide zone structure to the values of vertical diffusion coefficient since this controls effectively the amount of oxygen supplied from the aerobic zone. The main features of the Yakushev model (1997) are shown in Fig. 4, where the co-existence layer is indicated between 90 m and 110 m depths.

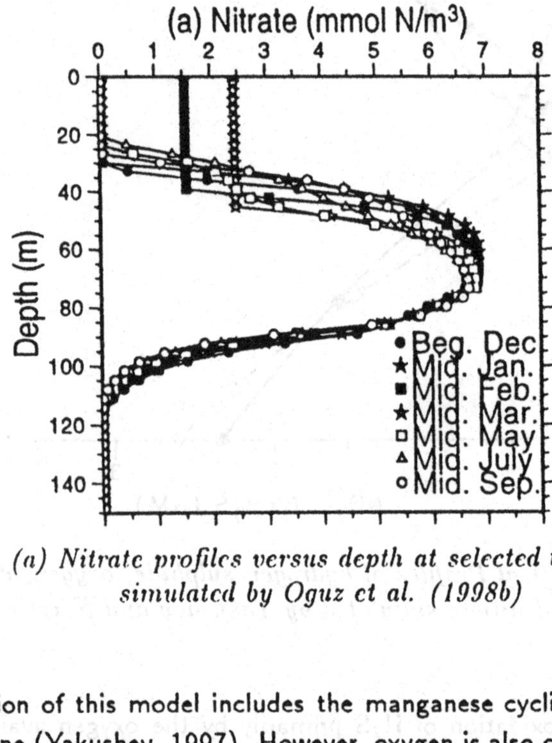

Figure 5. (a) Nitrate profiles versus depth at selected times of the year simulated by Oguz et al. (1998b)

One extension of this model includes the manganese cycling near the anoxic interface zone (Yakushev, 1997). However, oxygen is also considered to be the principal oxidizing agent in the manganese oxidation-reduction reactions. In this repect, the manganese cycling is nothing more than an intermediate step of the H_2S oxidation by the oxygen. On the other hand, as proposed by Murray et al. (1998) the manganese cycle plays a different role, and is used to explain the sulphide oxidation in oxygen depleted environment where the dissolved manganese is oxidized by the nitrate, the particulate manganese formed by this reaction is then utilized in the oxidation of hydrogen sulphide. For that reason, it is quite reasonable to obtain similar vertical structures of H_2S and oxygen in two different version of the models, as reported by Yakushev (1997). This was however interpreted as a negligible contribution of the manganese cycling to the sulphide oxidation process.

3.2. OGUZ ET AL. MODEL

The model describing the upper layer biogeochemical structure for the central Black Sea waters constitutes an extention of the pelagic ecosystem model given by Oguz et al. (1998b). The food web model was modified to include oxygen dynamics and its role in the processes of particulate matter decomposition and nitrogen transformations, as well as denitrification and a simplified representation of the hydrogen sulphide oxidation processes in the suboxic zone. This model is able to simulate many observed features of the upper layer biogeochemical structure, and to provide some understanding on the mechanisms controling the suboxic zone dynamics.

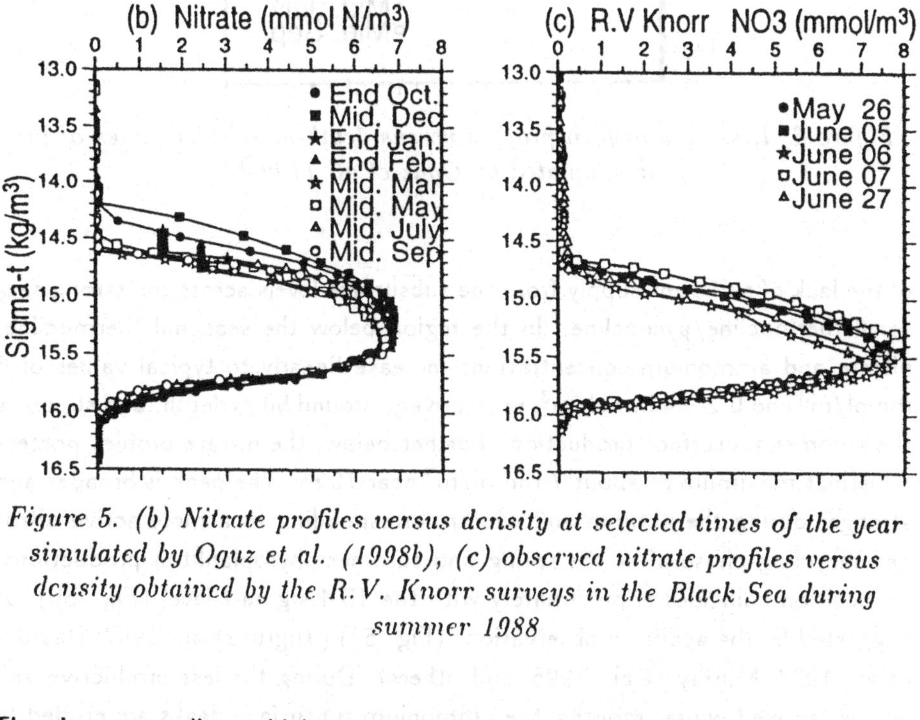

Figure 5. (b) Nitrate profiles versus density at selected times of the year simulated by Oguz et al. (1998b), (c) observed nitrate profiles versus density obtained by the R.V. Knorr surveys in the Black Sea during summer 1988

The nitrogen cycling, which supports the plankton productivity within the interior Black Sea, seems to occur over the uppermost 75 m of the water column. During the winter months, prior to the March diatom bloom event, intense vertical convective mixing gives rise to enhanced nitrate concentrations more than 2 mmol/m^3 within the upper 50 m (Fig. 5a). The summer mixed layer, on the other hand, is characterized by depleted nitrate and ammonium stocks, because

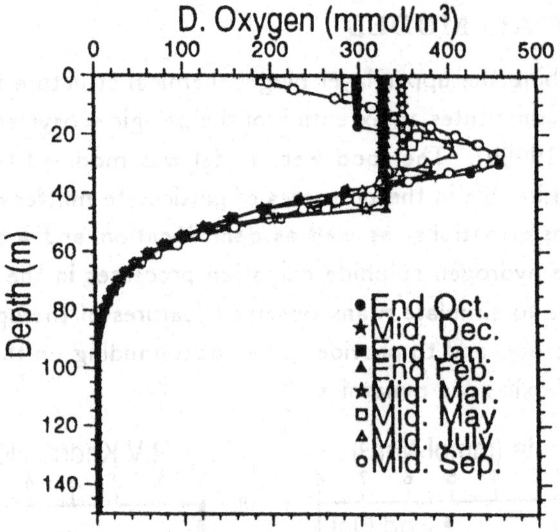

Figure 6. Dissolved oxygen profiles versus depth at selected times of the year simulated by Oguz et al. (1998b)

of the lack of sufficient supply from the subsurface levels across the strong seasonal thermocline/pycnocline. In the region below the seasonal thermocline, nitrate and ammonium concentrations increase linearly to typical values of 6 mmol/m^3 and 0.2-0.4 mmol/m^3, respectively, around 50 m depth, and supports the summer subsurface production. Further below, the nitrate profiles possess a distinct maximum of about 7 mmol/m^3 near 70 m. The peak is broader and stronger during the autumn and winter corresponding to a more active nitrogen recycling phase after the spring and summer phytoplankton productions. Its position coincides approximately with the 15.4 sigma-t level (Fig. 5b), as suggested by the available observations (Fig. 5c) (Tugrul et al., 1992; Basturk et al., 1994; Murray et al., 1995; and others). During the less productive and poorly-recycled winter months, the ammonium subsurface peaks are eroded to a large extent as it is oxidized to the nitrate form. The nitrate concentrations do not possess any seasonal variability below the peak. They tend to decrease uniformly to their trace level values around 16.0 sigma-t level, which is located roughly 40 m below the position of the nitrate maximum.

The oxygen, generated photosythetically within the euphotic zone and modulated by the ocean-atmosphere interactions at the near-surface levels, are

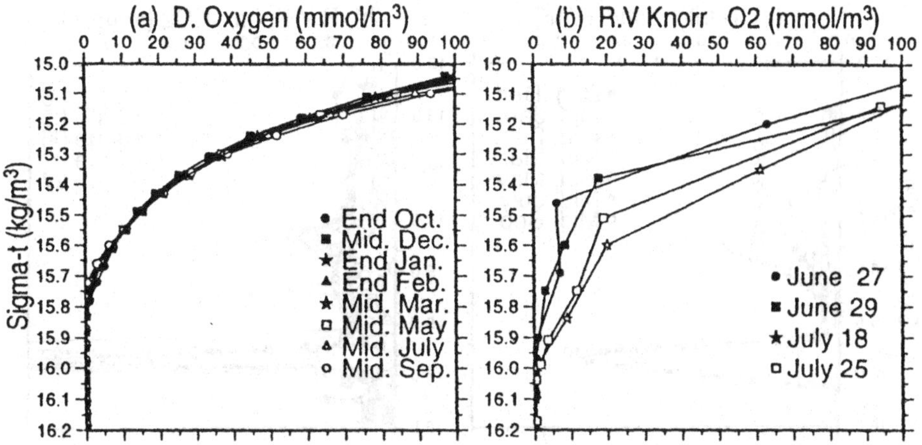

Figure 7. (a) Dissolved oxygen profiles versus density at selected times of the year simulated by Oguz et al. (1998b), (b) observed dissolved oxygen profiles versus density obtained by the R.V. Knorr surveys in the Black Sea during summer 1988

consumed during the organic matter decomposition and nitrification at subsurface levels which normally reveal very restricted ventilation below 50 m in the central Black Sea. The euphotic layer oxygen concentrations vary seasonally within a broad range of extremum values from 200 to 450 μM (Fig. 6a). The winter profiles exhibit vertically uniform mixed layer concentrations of about 350 μM, ventilating at most the upper 50 m part of the water column as a result of the winter convective overturning process. After March, once the cooling season comes to an end, oxygen is lost to the atmosphere, and the surface concentrations can reduce up to 200 μM in the summer. Depending on the strength of summer phytoplankton productivity, the sub-thermocline concentrations vary from 350 μM in mid-May to 450 μM during July and October at the times of flagellate blooms. Both the form of the euphotic layer oxygen structures and the range of variability of the oxygen concentrations agree well with the observations.

Below the euphotic zone, the vertical oxygen structure undergoes very steep variations with almost two order of magnitude changes in their concentrations

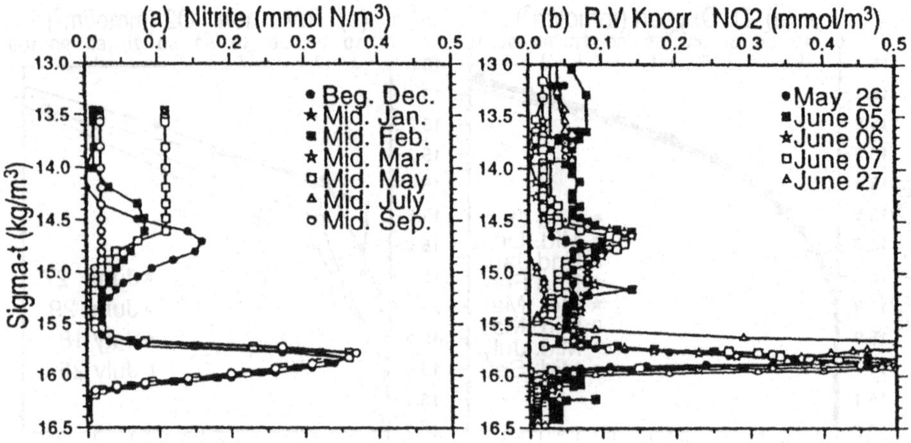

Figure 8. (a) Nitrite profiles versus density at selected times of the year simulated by Oguz et al. (1998b), (b) observed nitrite profiles versus density obtained by the R.V. Knorr surveys in the Black Sea during summer 1988

within about 25 m interval (oxycline) (Fig. 6). Typically, the ~10 μM oxygen level,corresponding roughly to the 15.6 sigma-t level, identifies the position of vanishing aerobic mineralization and nitrification processes in the water column (Fig. 7a) . The oxycline thus coincides with strong nitrate variations (upper nitracline) in which the nitrate concentrations increases from their trace level values in the mixed layer up to the maximum values of 6-8 μM across this zone. These model predictions on the position and slope of the oxycline zone, as well as its structure within the suboxic layer agree reasonably well with the available observations (Fig. 7b).

In the oxygen deficient part of the water column (i.e. $O_2 < 10\mu M$), the organic matter decomposition occurs via the denitrification process. This causes excessive nitrate consumption within a narrow layer adjacent to the oxycline and associated strong reduction in the nitrate concentrations. A distinguishing signature of the denitrification in the model is the formation of a narrow nitrite peak located at about 15.8-15.9 sigma-t levels (Fig. 8).

Oguz et al. (1998b) model also considers interaction of hydrogen sulphide with oxygen and nitrate near the anoxic interface in a more simplified form of the sulfur cycle than given by Yakushev and Neretin (1997). As oxygen maintains the vertical structure shown in Fig. 6 in the absence of its oxidation by oxygen and nitrate, hydrogen sulphide evolves the form shown in Fig. 9a as a result of the diffusive transport from the prescribed bottom source at the end of 4 years of time integration of the model. It is noted that the subsurface (below the 75 m depth) value of the vertical diffusion coefficient of 2.0x10^{-6} m^2/s is approximately an order of magnitude smaller than the one utilized by Yakushev and Neretin (1997) and is computed from the Gargett (1984) formula. It is shown that, under such conditions, H^2S can penetrate up to the 15.4 sigma-t level and co-exists with oxygen (Fig. 9a). Once the oxidation of the H_2S with the dissolved oxygen and NO_3 is allowed in the model, the oxygen and hydrogen sulphide overlapping layer is eroded and the oxygen and H^2S profiles are separated gradually within the subsequent year (Fig. 9b). During the following year of time integration, an anoxic-nonsulfidic layer is finally established between 15.65 and 16.10 sigma-t levels (Fig. 9c), which resembles the suboxic layer structure inferred from the observations. Clearly, choices of higher a thiodenitrification rate would lead to more rapid depletion of sulphide and generate the suboxic zone much earlier.

In the case of slightly higher choices of the vertical diffusion coeeficient (e.g. 5.0x10^{-6} m^2/s in the experiment described herein), the model was shown to exhibit a much deeper penetration of oxygen, in the absence of its utilization in the H2S oxidation (Fig. 10). The two July 1988 profiles shown in Fig. 7b provides an observational support for this type of oxygen structure. In this case, it is characterized by 30 μM concentrations at 15.6 sigma-t level and 10 μM around 16.0 sigma-t level. H^2S attains the values of about 5 μM near the anoxic interface and decays exponentially towards 15.6 sigma-t level (Fig. 10). Once the oxidation of H_2S by O_2 and NO_3 is allowed in the model, for the case of realistic oxidation rates about 0.05-to-0.1 day^{-1}, the entire H_2S present in the oxygenated part of the water column is oxidized rapidly within a few weeks (Fig. 11a, b). H_2S remains to exist only below $\sigma_t \sim 16.05$ level where the oxygen and nitrate are no longer available due to their consumption up to this level. These numerical experiments therefore suggest that H_2S and O_2 cannot exist together since the fast reaction rate of the oxidation process will quickly deplete the H_2S. At most, H_2S can penetrate up to a level of vanishing oxygen

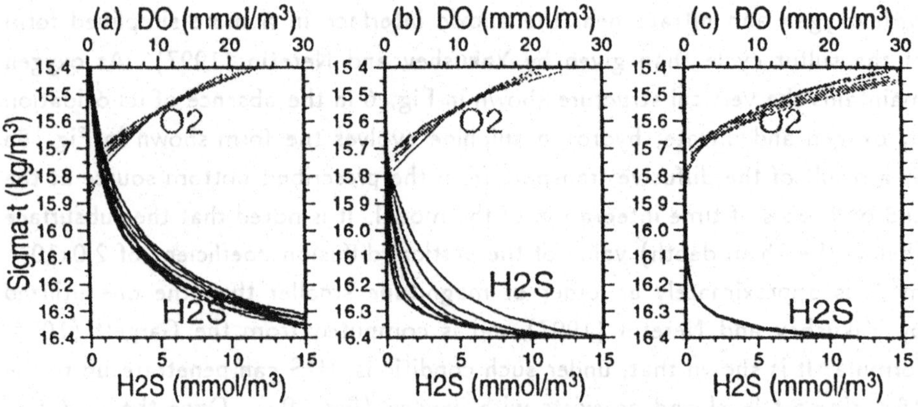

Figure 9. Dissolved oxygen and hydrogen sulphide profiles versus density near the anoxic interface zone simulated by Oguz et al. (1998b). These profiles show the evolution of the oxygen-H_2S structure shown in Figure 6 (a) within the fourth years of integration in the absence of oxygen-sulphide interactions, (b) within the subsequent year of integration after the sulphide oxidation by oxygen and nitrate is allowed, (c) during the following year of integration

concentrations. These results, contradicting with the findings of Yakushev and Neretin (1997), suggest that even this simple model of H^2S oxidation is sufficient to provide a quantitative confirmation for the existence of the oxygen-sulphide free suboxic zone and for the absence of the coexistence layer in the Black Sea. The manganese and iron cycles might further refine this structure by contributing to the oxidation of H^2S near the anoxic interface.

4. Summary and Discussion

The present paper describes an overview of the existing numerical modeling studies on the structure and functioning the ecosystem as well as the biogeochemical structure of the upper layer water column in the Black Sea. Some of the pelagic ecosystem models consider only the mixed layer integrated properties of the system (Lebedeva and Shushkina, 1994; Eeckhout and Lancelot, 1997; Cokasar and Ozsoy, 1998). Even if their simplicity of implementation and experimentation, they have deficiency of neglecting the plankton produc-

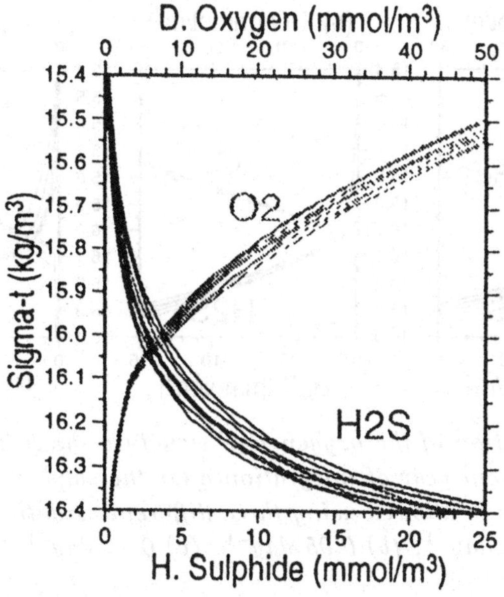

Figure 10. Evolution of the dissolved oxygen and hydrogen sulphide profiles versus density near the anoxic interface zone during the fourth years of integration (after Oguz et al., 1998b).

tion below the seasonal thermocline when the mixed layer is shallower than the euphotic zone during late spring, summer and early autumn periods. The success of these models crucially depends on the specification of the daily mixed layer depth and temperature as well as the subsurface nutrient structure which controls the nutrient flux across the thermocline. These models therefore rely on the quality of external input coming from the observations. Despite these drawbacks, this type of biological models were useful for understanding the ecosystem changes during the last several decades. Lebedeva and Shushkina (1994) model suggested that recent increase in the gelatinous carnivore population in the interior Black Sea results in approximately five-fold decrease in mesozooplankton biomass and subsequently considerable increase in the winter phytoplankton biomass. It was however not supported by the findings of Cokasar and Ozsoy (1998) and Eeckhout and Lancelot (1997) which seem to be less successful in simulating the observed features of the annual plankton cycle.

The mixed layer based biological models were complemented by the vertically-resolved models (Oguz et al., 1996, 1998a,b; Staneva et al., 1998). They

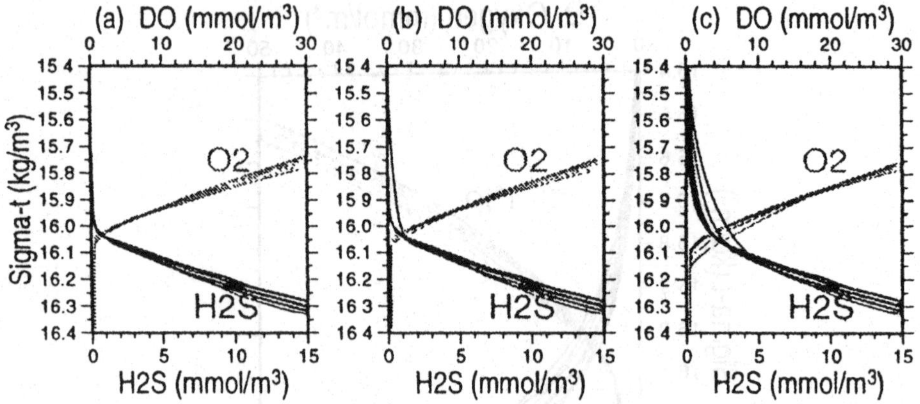

Figure 11. Evolution of the oxygen-H_2S structure shown in Figure 10 within the subsequent year of integration after the sulphide oxidation by oxygen and nitrate is allowed using three different oxidation rates of (a) 0.1 day^{-1}, (b) 0.05 day^{-1}, (c) 0.01 day^{-1}

consider biogeochemistry of the entire upper layer water column with varying degree of complexity, depending on the specific purposes of these models. The most simplified, five compartment, version (Oguz et al., 1996; Staneva et al., 1998) explored the most robust biological features of the ecosystem and the role of upper layer physics on the evolution of the euphotic zone biological processes. For example, Oguz et al (1996) described quite interestingly how the early spring bloom was triggered immediately after the weakening of the convective mixing in the water column well before the formation of the seasonal thermocline which is traditionally accepted as a prerequisite for the spring bloom formation. The other models (Oguz et al., 1998a and 1998b) introduced some biological complexities to the five compartment model and explored how these biologically structured models were more capable of simulating intensified subsurface summer production, and more dynamic plankton structure arising after the increasing role of the gelatinous carnivores in the ecosystem during 1980's. The sensitivity studies carried out with these models were instrumental in exploring the possible causes of regional as well as year-to-year variabilities of the system, and how the internal structure of the biological pump is in action under different scenarios.

The vertically-resolved model (1998b) reveals fairly sophisticated nitrogen cycling in the water column characterized by remineralization, ammonification, nitrification and denitrification processes with the water column. These processes are complemented by the oxygen dynamics in the oxygenated part of the water column and a simplified sulfur cycle near the anoxic interface zone. The model, which was principally similar to the one given by Yakushev and Neretin (1997) but employed different parameter setting, provided a quantitative evidence for the presence of the oxygen and sulphide depleted suboxic layer between the anoxic pool and the oxycline. The model suggested that, under given realistic oxidation and vertical diffusion rates, the oxygen and hydrogen sulphide can not exist together because of the fast reaction rate of the oxidation mechanism. While the lower boundary of the suboxic zone has a fairly stable character, the upper boundary (i.e. the slope of the oxycline) might change up and down depending on the local, internal mixing characteristics. The oxygen dynamics, on the other hand, suggest that the subsurface levels are very poorly ventilated by the surface layer processes. This is the key element for the permanency of the suboxic layer in the Black Sea.

Acknowledgements

This work was carried out within the scope of the TU-Black Sea Project sponsored by the NATO Science for Stability Program. It is supported in part by the National Science Foundation Grant OCE-9633145.

References

Armstrong R.A. (1994) Grazing limitation and nutrient limitation in marine ecosystems: Steady state solutions of an ecosystem model with multiple food chains. Limnol. Oceanogr., 39, 597-608.

Basturk, O., C. Saydam, I. Salihoglu, L. V. Eremeeva, S. K. Konovalov, A. Stoyanov, A. Dimitrov, A. Cociasu, L. Dorogan, M .Altabet (1994) "Vertical variations in the principle chemical properties of the Black Sea in the autumn of 1991". J. Marine Chemistry, 45, 149-165.

Cokasar, T. and E. Ozsoy (1998) "Comparative analyses and modeling for regional ecosystems of the Black Sea". to appear in: *NATO ASI Series on the Proceedings of the Symposium on the Scientific Results of the NATO*

TU-Black Sea Project, Crimea-Ukraine, June 15-19, 1997.

Eeckhout, D.V. and C. Lancelot (1997) "Modeling the functioning of the Northwestern Black Sea ecosystem from 1960 to present". to appear in: *NATO ASI Series on the Proceedings of the NATO Advanced Research Workshop "Sensitivity to Change: Black Sea, Baltic Sea and North Sea, E. Ozsoy and A. Mikaelyan (Editors), NATO ASI Series 2, Environment-Vol.27, 455-469.*

Gargett, A.E. (1984) "Vertical eddy diffusivity in the ocean interior". J. Marine Research, 42, 359-393.

Lebedeva, L.P., and E.A. Shushkina (1991) "Evaluation of population characteristics of the medusae *Aurelia aurita* in the Black Sea. Oceanology (English transl.), 31, 314-319.

Murray, J.W., L.A. Codispoti, G.E. Friederich (1995) "Oxidation-reduction environments: The suboxic zone in the Black Sea". In: *Aquatic chemistry:Interfacial and interspecies precosses. ACS Advances in Chemistry Series No.224. C.P. Huang, C.R.O'Melia, and J.J. Morgan (Editors), 157-176.*

Murray, J.W., B.S. Lee, J. Bullister, G.W. Luther III (1998) The Suboxic Zone of the Black Sea". to appear in: *NATO ASI Series on the Proceedings of the NATO Advanced Research Workshop "Environmental Degradation of the Black Sea: Challenges and Remedies, Constantza, Romania, 6-10 October 1997.*

Oguz, T., H. Ducklow, P. Malanotte-Rizzoli, S. Tugrul, N. Nezlin, U. Unluata (1996) "Simulation of annual plankton productivity cycle in the Black Sea by a one-dimensional physical-biological model". J. Geophysical Research, 101, 16585-16599.

Oguz, T., H. Ducklow, P. Malanotte-Rizzoli, J.W. Murray, E.A. Shuskina, V.I. Vedernikov, U. Unluata (1998a) "A physical-biochemical model of plankton productivity and nitrogen cycling in the Black Sea". to appear in Deep Sea Research.

Oguz, T., H. Ducklow, E. A. Shuskina, P. Malanotte-Rizzoli, S. Tugrul, L.P. Lebedeva (1998b) "Simulation of upper layer biogeochemical structure in the Black Sea". to appear in: *NATO ASI Series on the Proceedings of the*

Symposium on the Scientific Results of the NATO TU-Black Sea Project, Crimea-Ukraine, June 15-19, 1997.

Sarmiento, J.L., R.D. Slatter, M.J.R. Fasham, H.W. Ducklow, J.R. Toggweiler, G.T. Evans (1993) A seasonal three-dimensional ecosystem model of nitrogen cycling in the North Atlantic euphotic zone. Global Biogeochemical Cycles, 7, 417-450.

Shushkina,E.A., and E.I. Musayeva (1990) "Structure of planktic community of the Black Sea epipelagic zone and its variation caused by invasion of a new ctenophore species". Oceanology (English transl.), 30, 225-228.

Staneva, J., E. Stanev, T. Oguz (1998) "On the sensitivity of the planktonic cycle to physical forcing: Model study on the time variability of the Black Sea ecological system". to appear in: *NATO ASI Series on the Proceedings of the Symposium on the Scientific Results of the NATO TU-Black Sea Project, Crimea-Ukraine, June 15-19, 1997.*

Tugrul, S., O. Basturk, C. Saydam, A. Yilmaz (1992) "The use of water density values as a label of chemical depth in the Black Sea". Nature, 359, 137-139.

Vinogradov, M. E. (1992) Long-term variability of the pelagic community structure in the open Black Sea. Paper presented at the "Problems of Black Sea International Conference", Sevastopol, Ukraine, Nov. 10-15, 1992.

Vinogradov, M.E. and E.A. Shushkina (1992) "Temporal changes in community structure in the open Black Sea". Oceanology (English transl.), 32, 485-491.

Yakushev,E.V. and L.N. Neretin (1997) "One dimensional modeling of nitrogen and sulfur cycles in the aphotic zone of the Black and Arabian Seas". to appear in J. Global Biogeochemical Cycles.

SATELLITE ALTIMETRY OBSERVATIONS OF THE BLACK SEA

G.K.KOROTAEV*, O.A.SAENKO*, C.J.KOBLINSKY#, V.V.KNYSH*
*Marine Hydrophysical Institute, National Academy of Sciences of Ukraine (MHI) 335000 Ukraine, Crimea, Sevastopol, Kapitanskaya 2
#Goddard Space Flight Center, NASA, Code971, Greenbelt MD20771, USA

Abstract

Long data set of altimetric measurements is available for the Black Sea as the result of Pathfinder Project of Goddard Space Flight Center. Strong seasonal signal obviously presented in the satellite altimetry data Satellite-derived signal reproduces well amplitudes and phases of the sea level variability estimated through the leading term of the water budget for the Black Sea. The method of the reconstruction of dynamical level from satellite altimetry is proposed in the paper. The altimetric dynamical sea level correlates well with corresponding sea level resulted from the CoMSBlack data. Analysis of seasonal variability of the general circulation based on data of ERS-1 and TOPEX/POSEIDON, showed that its minimum corresponds to July- October. Maximum intensity of the Rim current falls on the period from December-January to May. Assimilation of the sea level in the model of the Black Sea circulation permits to reproduce the mesoscale dynamics.

1. Introduction

Industrial activity of countries of the Black Sea basin provides increasing anthropogenic pressure on the marine environment. A number of international programs were established during the last decades to evaluate the degradation of the Black Sea ecosystem (JEF Black Sea Environmental Program, The IOC Black Sea Regional Program, TU-Black Sea - NATO Science for Stability Program, EROS-2000 of EC, etc.). However all programs are short of basin-wide regular observations. Even physical component of ecosystem is ill-covered by observations during the last few years due to the bad shape of economy of the Black Sea countries. Obviously new sources of information should be used to fill up the lack of traditional observations. Operational and near-operational satellites offer a means of organizing of continuous, long-term observation of a sea surface to best advantage.

Application of IR and color scanners to observation of mesoscale variability of the Black Sea is well-know [6,16,17]. However another very efficient type of information

S. Beşiktepe et al. (eds.),
***Environmental Degradation of the Black Sea: Challenges and Remedies*, 225–244.**

about the sea state namely, satellite altimetry widely used for the investigation of the World Ocean is not so popular in the study of the Black Sea dynamics. The problem is probably in the specific of usage of satellite altimetry data. Many corrections of different type should be applied to the raw data before they will have appropriate accuracy. Now long data set of altimetric measurements from several missions (SeaSat, GeoSat, ERS-1, Topex/Poseidon) is available for the Black Sea as the result of Pathfinder Project of Goddard Space Flight Center. Our paper presents part of this data and shows how to apply them to the study of the Black Sea dynamics. We investigate the seasonal variability of the sea level induced by the volume transport through the basin boundaries and basin-wide circulation into the Black Sea, its mesoscale and seasonal variability.

2. Seasonal variability of the Black Sea level from satellite altimetry

We use in this study the satellite altimetry from the mission ERS-1 of European Space Agency and the joint US-France project Topex/Poseidon. The scheme of tracks of ERS-1 and Topex/Poseidon over the Black Sea area is shown in Figure 1. Satellite orbit conditions and the general characteristics of collected data are summarized in the Table1.

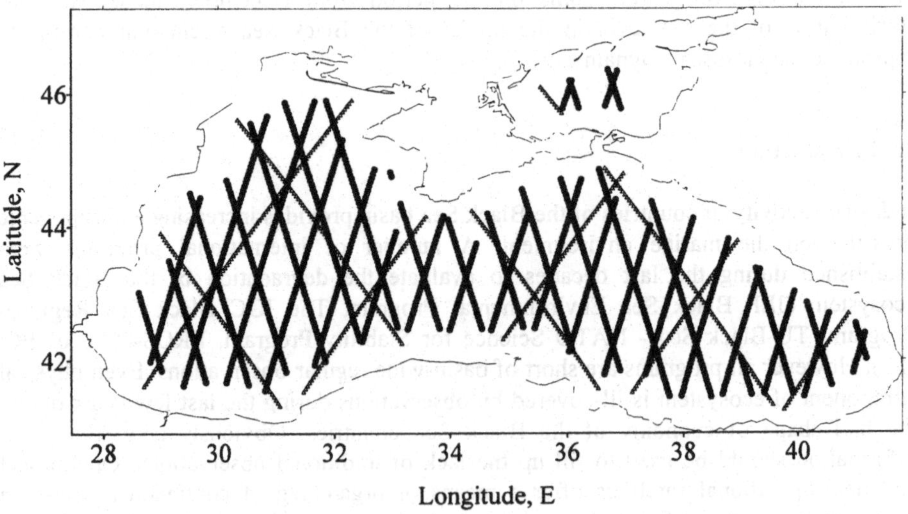

Figure 1. Tracks of TOPEX/POSEIDON (gray lines) and ERS-1 (dark lines) for the Black Sea area.

TABLE 1. Satellite orbit conditions and the general characteristics of collected data

Satellite	Number of days in the repeating cycle	Number of repeating cycles	Span of data
Topex/Poseidon	9.92	154	92/09/23-96/11/29
ERS-1	35.00	18	92/04/14-93/12/20

The Black Sea level manifests the strong seasonal variability induced by the river runoff, precipitation/evaporation and water exchange through strait cycle. The amplitude of the annual oscillation of the Black Sea level due to variability of fresh water budget is up to 25 cm [14]. There is the same order of magnitude as the overall sea level difference induced by the general circulation of the Black Sea.

Strong seasonal signal obviously presented in the ERS-1 as well in the Topex/Poseidon data (Figure 2a). Lines in Figure 2a show the evolution with time of the sea level calculated by averaging of all ERS-1 or Topex/Poseidon data obtained for the one-cycle. Spatial averaging is reasonable because the size of the Black Sea is significantly less than the barotropic Rossby radius. In this case the response of the sea to the variability of the outer volume transport should be spatially uniform. Satellite data confirms this assumption since the sea level averaged along each track differs slightly for the sea level averaged over the whole cycle. Figure 2a shows very well correspondence of the sea level variability for 1992-1993 derived from the data of ERS-1 and Topex/Poseidon.

Unfortunately, we cannot to carry out accurate calculation of the water budget for the Black Sea due to the lack of data about precipitation/evaporation and water exchange through straits. However we would like to show a reasonable agreement between the sea level variations during 1992-1996 calculated from the Topex/Poseidon altimetry and from the data about the volume transport by rivers (Figure 2b). The sea level variability induced by the river runoff is calculated from continuity equation using the volume transport of the Danube and the Dniper rivers estimated by Sevastopol Branch of Ukrainian Hydrometeorological Institute and kindly presented for this study by Dr. Vladimir Belokopytov. Satellite-derived signal reproduces well amplitudes and phases of the sea level variability estimated through the river runoff data. It encourages to use satellite altimetry in future for the high accuracy estimation of seasonal variability of the Black Sea fresh water budget.

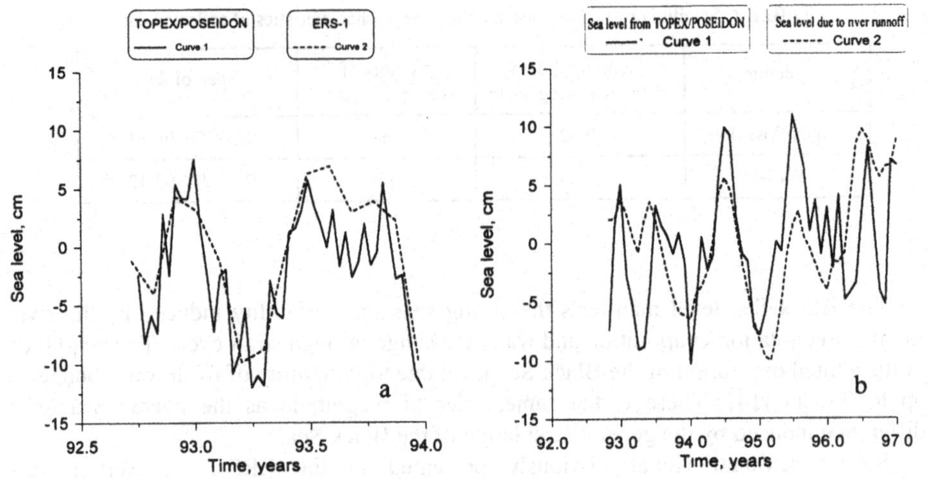

Figure 2. Evolution of the averaged Black Sea level: a - based on data of TOPEX/POSEIDON (solid line) and ERS-1 (dotted line); b - based on data of TOPEX/POSEIDON (solid line) and the component of the average sea level calculated based on river volume transports (dotted line).

3. Comparison of altimetric data with hydrology and analysis of seasonal variability

Application of altimetric data for the analysis of baroclinic circulation requires extraction of the water-exchange component of the sea level, i.e., obtaining of the dynamical sea level. The water exchange component of the sea level is uniform spatially as we discussed above. We use this circumstance for calculation of the dynamical part of sea level. Let us find for each track the sea level deviation from the along track averaged value using altimetry data. In that way we distract from the altimetry a sea level part given by the volume transport through the boundary of the basin. However we also distract different for each track time-dependant constant which is determined by seasonal variability of general circulation of the Black Sea. To evaluate the time-dependant constants we calculate the climatic dynamical sea level by the method of hydrodynamical adaptation [13] based on monthly climatic data on temperature and salinity [1]. This dynamical sea level is interpolated in time and its mean value is evaluated along each satellite altimeter track. The obtained earlier from altimetry the sea level anomaly is supplemented by the averaged along a track climatic level corresponding to the time of altimetric measurements and is used in the following as the altimetric dynamical sea level. The procedure of reconstruction of the dynamical sea level of the Black Sea based on the altimetric data is discussed in details in [8].

There is an opportunity to compare the dynamical sea level resulted from altimetric measurements with the corresponding sea level calculated from the CoMSBlack hydrology. Four large-scale hydrological surveys of the Black Sea were carried out within the framework of that program in 1992-1995. Figures 3-6 represent topography of corresponding sea level surfaces for the periods of hydrological surveys: July 1992, CoMSBlack-92; April 1993, CoMSBlack-93; April-May 1994, CoMSBlack-94 and March-April 1995 CoMSBlack-95. We used the procedure of objective mapping [5] to obtain the maps of the sea level. The sea level based on the CoMSBlack surveys, was reconstructed using the method of hydrodynamical adaptation [13].

The altimetric sea level correlates well with corresponding sea level resulted from the CoMSBlack data. Both fields describe main features of the Black Sea general circulation. Namely the cyclonic circulation of the Rim Current and anticyclonic mesoscale eddies on its periphery are clearly seen in Figures 3-6. At the same time the altimetric sea level contains much more small-scale features. It is connected with rough resolution of hydrological data and filtering effect of the model in the calculation of the hydrological sea level as well as with noise in the altimetric data or. Presence of the near-Bosporus anticyclone in the altimetric dynamical sea level and its practical absence in the hydrological data seems to be the major another difference.

Up to now there is no unified opinion on the seasonal variability of the Black Sea. Some authors suppose that the general circulation in the Black Sea achieves its highest intensity in winter and in summer. Other authors assume that the maximum intensity takes place in winter and the summer circulation is weak. The paper of [2] gives more detailed review of this problem. The reason of contradictions consists in irregularity of hydrological information. At the same time spatial and temporal distribution of satellite information makes it unique in investigations of basin-wide variability of various hydrophysical processes. Analysis of seasonal variability of the general circulation based on data of ERS-1 and Topex/Poseidon shows that its minimum corresponds to July- October. Attenuation of the Rim current jet is accompanied by growth of a number of different sign vortices and enforcement of mesoscale variability practically around the whole area of the sea (Figure 7e,f). Here we can see a greatest difference between ERS-1 and Topex/Poseidon maps due to difference in spatial and temporal resolutions of missions. Starting from approximately October-November the mesoscale variability weakens and the Rim current is recovered gradually (Figure 7g,h). Maximum intensity of the Rim current falls on the period from December-January to May (Figure 7a-d). The described seasonal cycle of the Black Sea general circulation resulted from the analysis of altimetric data is in agreement with recent numerical simulations of [15].

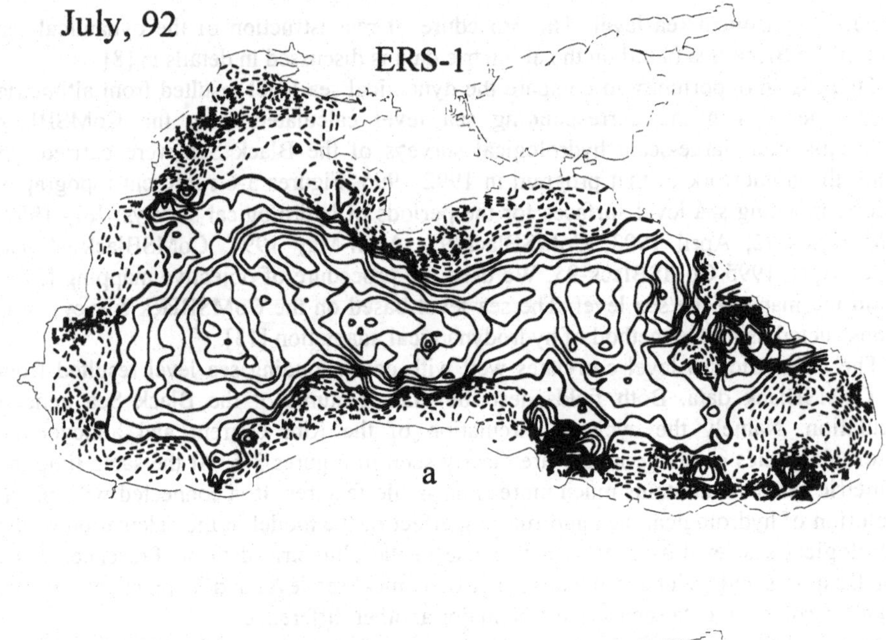

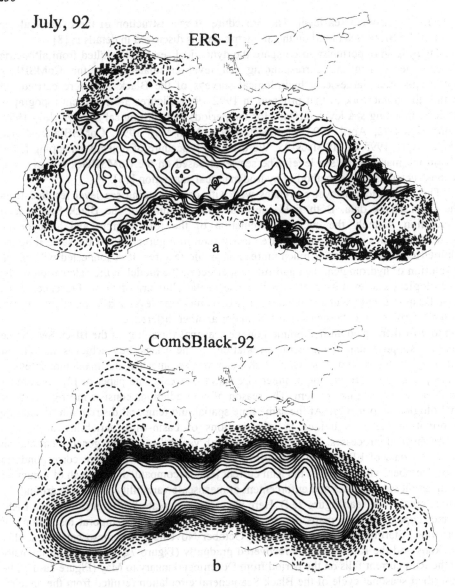

Figure 3. Dynamical level of the Black Sea in July, 1992 obtained (a) from the altimetric data of ERS-1 and (b) from the hydrographic data of CoMSBlack-92 as a result of the adaptation by the model Dotted line shows positive values of the sea level; contour interval is 1 cm

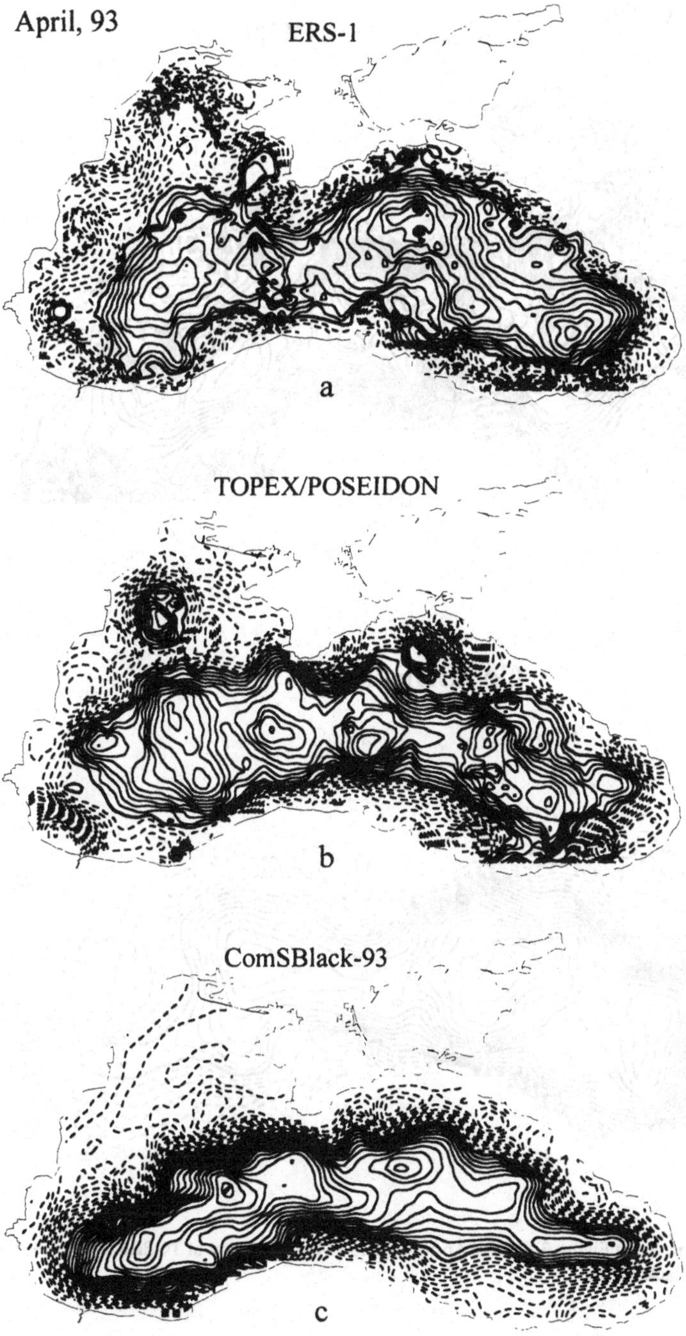

Figure 4 The same as on Figure 3, but for April,1993 based on data of ERS-1 (a), TOPEX/POSEIDON (b) and CoMSBlack-93 (c).

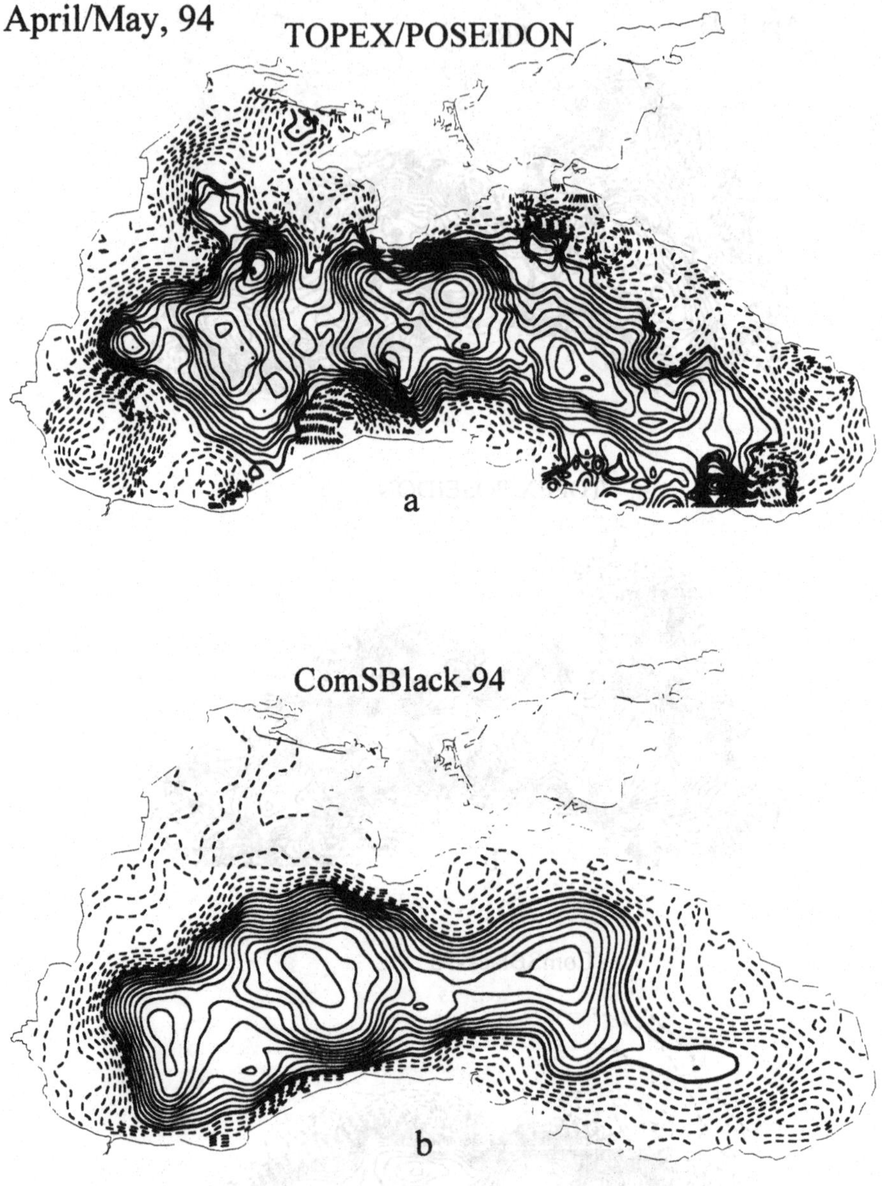

Figure 5. The same as on Figure 3, but for April-May, 1994 based on data of TOPEX/POSEIDON (a) and CoMSBlack-94 (b).

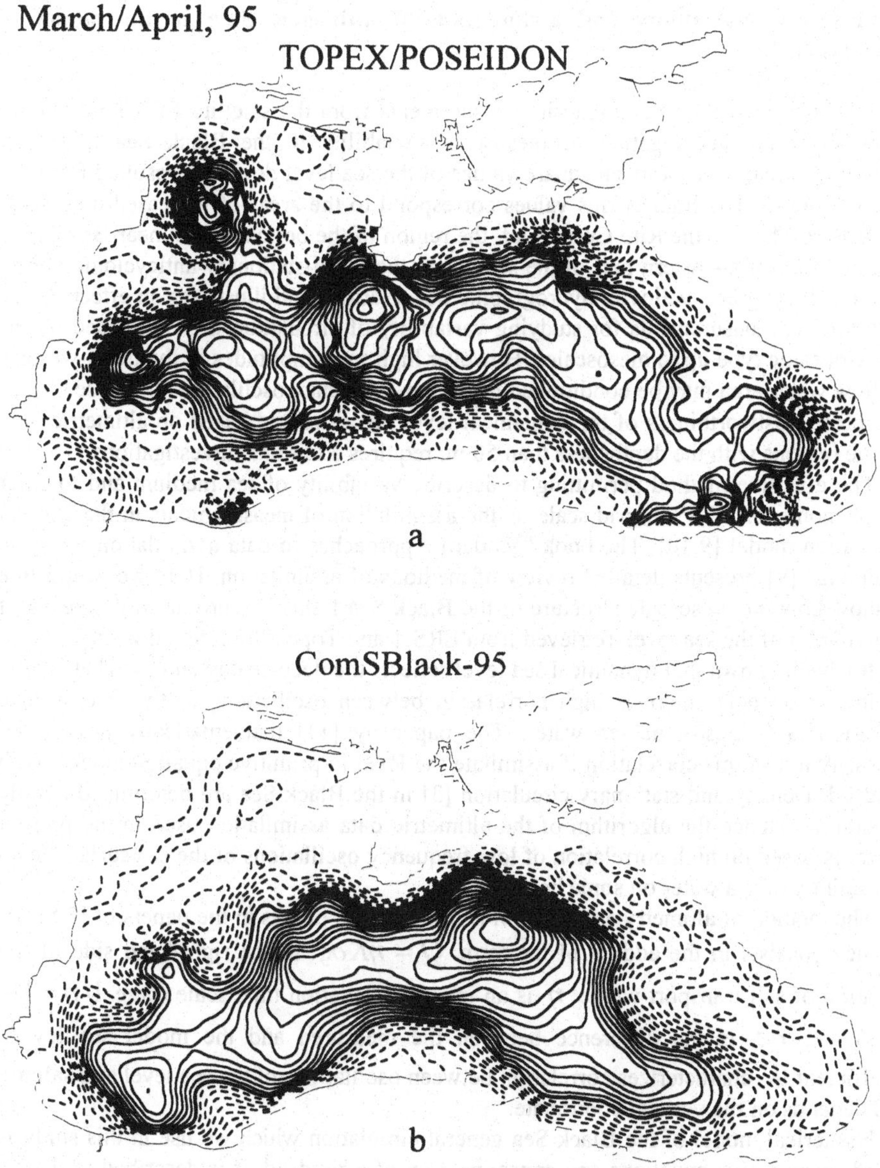

Figure 6 The same as on Figure 3, but for March/April, 1995 based on data TOPEX/POSEIDON (a) and CoMSBlack-95 (b).

4. Mesoscale variability and assimilation of altimetric data in the model of circulation

High spatial resolution of altimetric measurements from the satellite ERS-1 permits to apply them for investigation of mesoscale variability in the Black Sea. Figure 8 represents a map of root mean square values of the sea level deviation resulted from the data of ERS-1. The highest rms values correspond to the areas of increased mesoscale variability. There is the Rim Current jet, the region of the bottom slope, near-Bosphorus area and the south-eastern part of the sea where the quasi-stationary anticyclonic vortex is situated. It agrees well with hydrographic observations and illustrates an effective use of altimetric measurements for studying variability of circulation in semi-closed basins of the oceanic type up to mesoscales. However high rms amplitudes in the north-western shallow water area may be conditioned by higher-frequency oscillations of the sea level. Therefore trustworthiness of the dynamical level reconstructed based on altimetric data for the regions with the depths less than 50 m, requires additional investigations.

The effective method permitting to describe variability of sea medium and to filter out phenomenon of a sub-grid scale, is the assimilation of measurements in the general circulation model [9,10]. The book "Modern Approaches to data assimilation in ocean modeling" [9] presents detailed review of methods of assimilation. Here we would like to show how the mesoscale structure of the Black Sea fields is reproduced by means of assimilation of the sea level, retrieved from ERS-1 and Topex/Poseidon altimetry.

It is well known that dynamical sea level reflects the deep-sea dynamics. That is why on low frequencies there is high correlation between oscillations of the free surface elevation and density of sea water. The paper by [11] mathematically proves the necessity to use this correlation if assimilate the level in primitive equation models with rigid lid. Density and stationary circulation [3] in the Black Sea are determined mainly by salinity. Hence the algorithm of the altimetric data assimilation used in the present paper, is based on high correlation of low-frequency oscillations of the sea surface level and salinity of sea water on some horizons.

The method of altimetry assimilation is described in details in the papers of [7,8]. Its essence consists in the adding of the term $Q = \eta K \delta \varsigma$ to the right-hand side of the equation of salt transport where η is an inverse relaxation time scale (here $1/\eta = 10$ days) and $\delta\varsigma$ is the difference between the measured and the model sea levels. Coefficient K characterizes correlation between oscillations of the sea level and salinity and depends on the vertical coordinate.

Numerical model of the Black Sea general circulation which we use in this study is based on primitive equations and approximation of a rigid lid. It is described in details in the paper by [4]. The sea level is used in the model as the integral function. It is more sensitive to the upper layer density variations than the stream function of the volume transport [12]. The use of the sea level is comfortable for the Black Sea where the seasonal signal is very weak after 200 m due to the strong halocline.

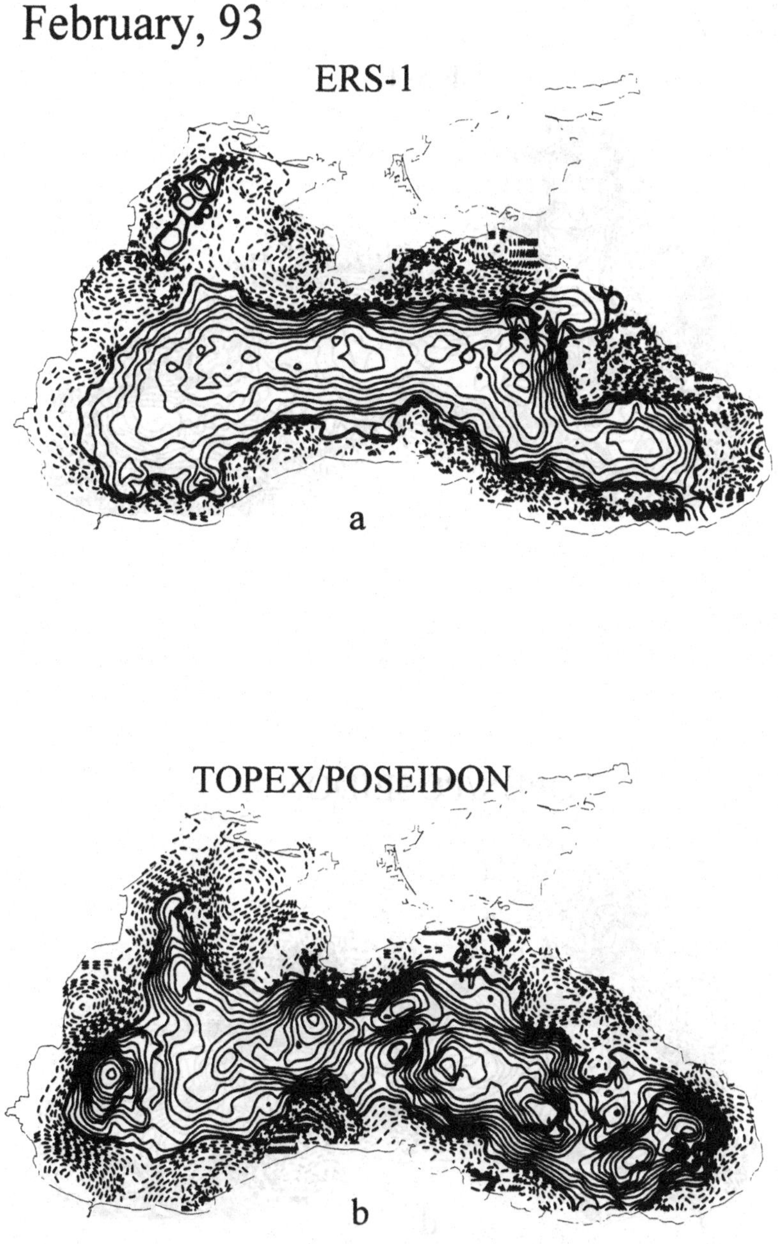

Figure 7(a,b). Dynamical level of the Black Sea resulted from the objective interpolation of altimetric data of ERS-1 and TOPEX/POSEIDON, 1993: February Dotted line shows positive values of the level; contour interval is 1 cm

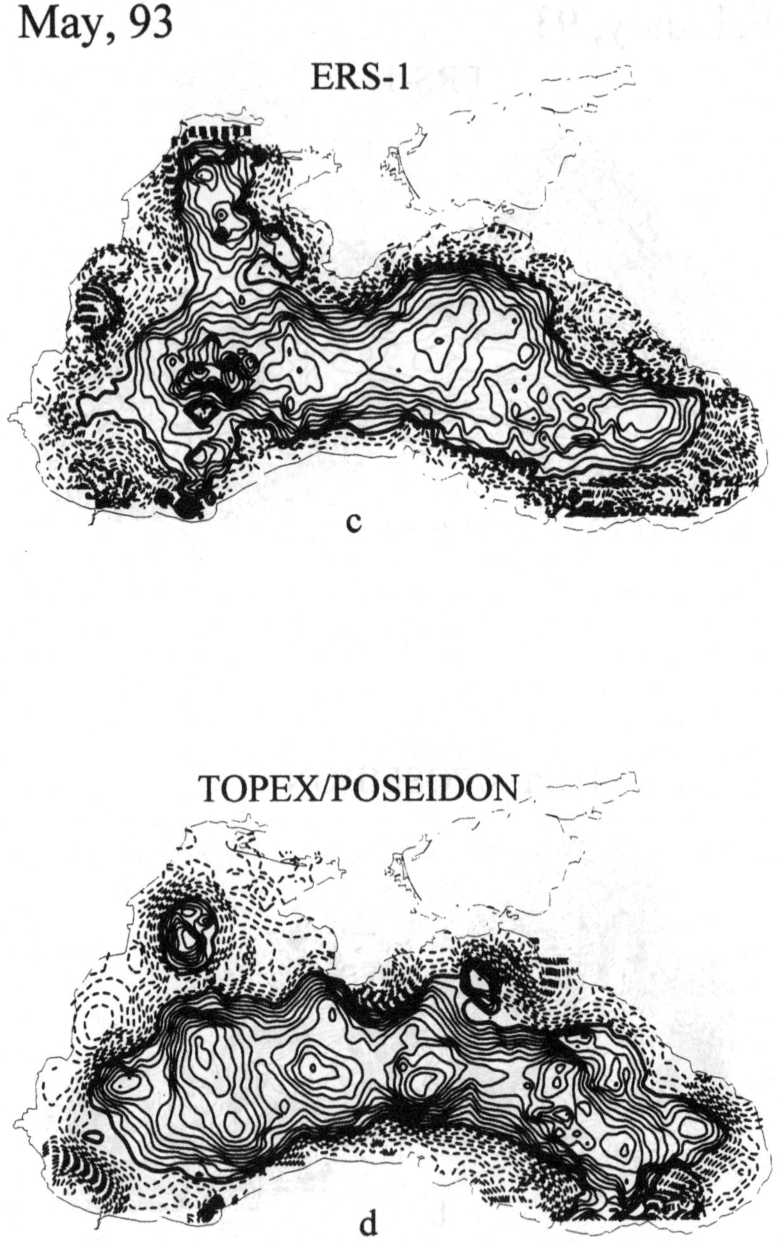

Figure 7(c,d) Dynamical level of the Black Sea resulted from the objective interpolation of altimetric data of ERS-1 and TOPEX/POSEIDON, 1993, May. Dotted line shows positive values of the level; contour interval is 1 cm

August, 93

ERS-1

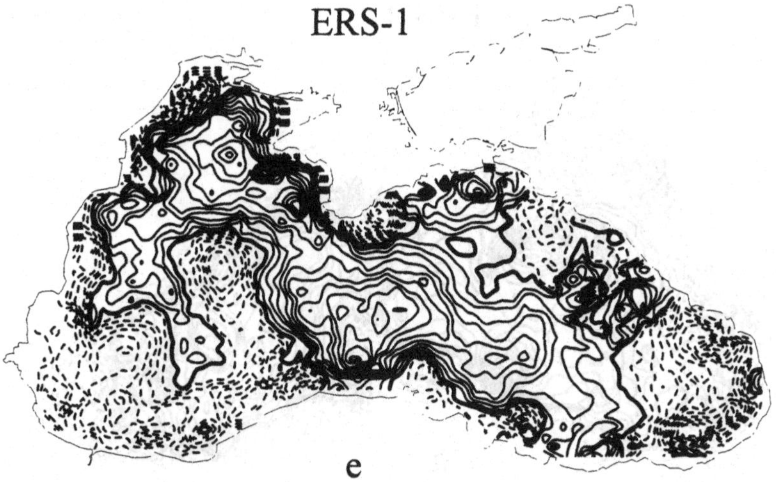

e

TOPEX/POSEIDON

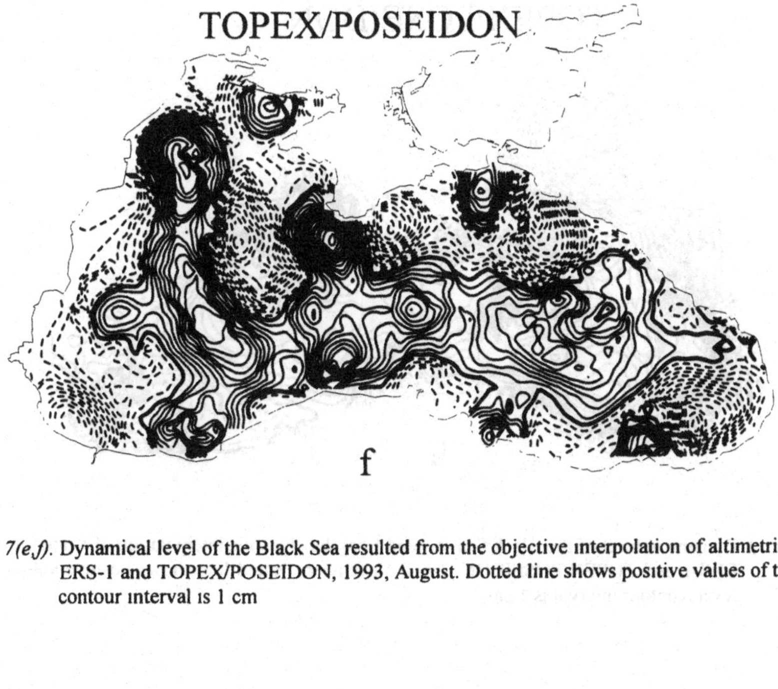

f

Figure 7(e,f). Dynamical level of the Black Sea resulted from the objective interpolation of altimetric data of ERS-1 and TOPEX/POSEIDON, 1993, August. Dotted line shows positive values of the level; contour interval is 1 cm

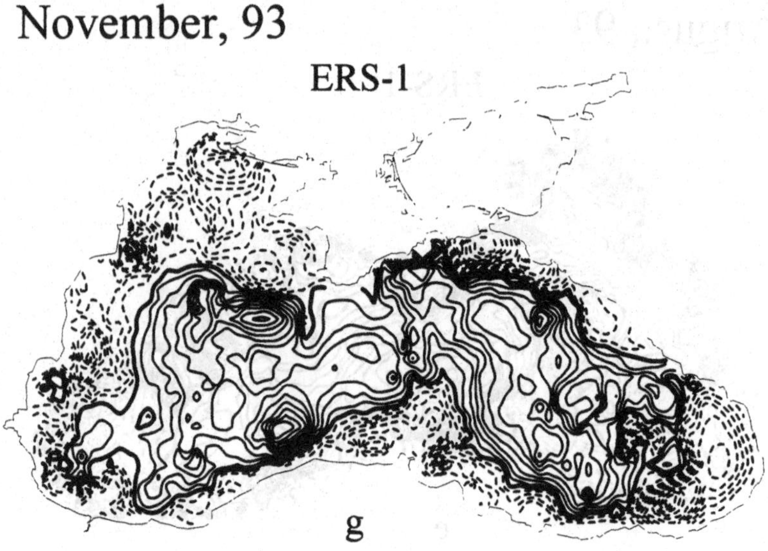

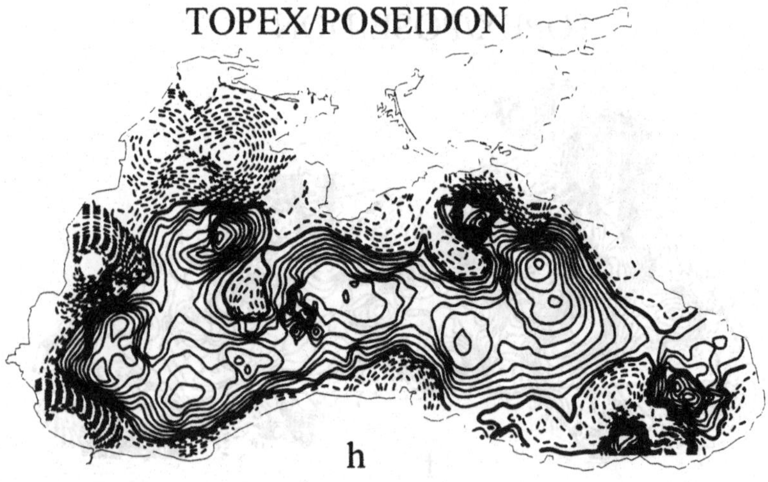

Figure 7(g,h) Dynamical level of the Black Sea resulted from the objective interpolation of altimetric data of ERS-1 and TOPEX/POSEIDON, 1993, November. Dotted line shows positive values of the level; contour interval is 1 cm.

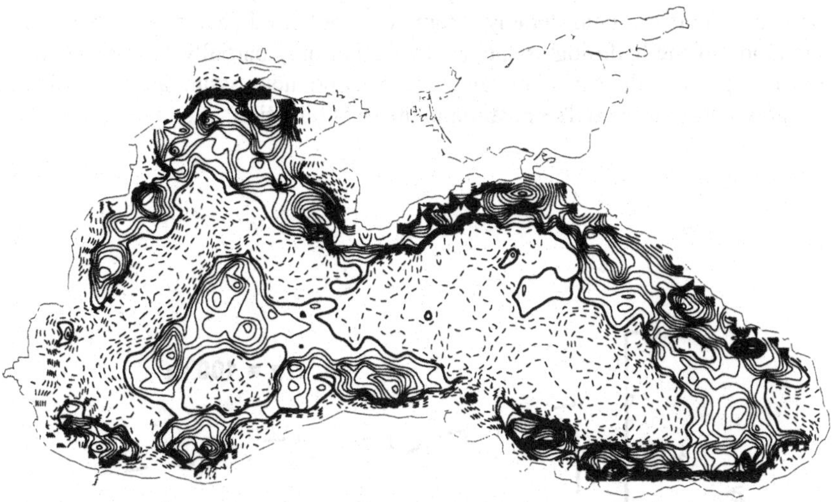

Figure 8. Root mean square sea level deviation based on data of ERS-1. Values smaller than 3 cm are denoted by a dotted line, contour interval is 0 25 cm.

The model is adapted to the physical-geographical conditions of the Black Sea. Spatial grid steps in zonal and meridional directions are11' and 8', respectively. Twenty horizons are used in the vertical direction with a grid step from 5m near the surface to 500m in the deep sea. Coefficients of vertical turbulent exchange of heat and salt are preset variable over depth. They have the same values and decrease from 1.5cm^2/s in the upper 5m layer to 0.1cm^2/s between 5-10m depths and to 0.03cm^2/s below 10m. Affects of evaporation/precipitation and river run-offs is taken into account indirectly through the salinity flux on the sea surface which is assumed to be proportional to the difference between the known climatic salinity value on the sea surface and calculated by the model. A water exchange through straits is neglected in present calculations. Climatic distribution of the wind stress and the heat flux are prescribed on the surface using data arrays prepared by Prof. Emil Stanev. The model runs initially with periodical boundary conditions for 12.5 years without data assimilation. Simulated fields of temperature, salinity and current after 12.5 year run are taken as the initial fields in the experiment with data assimilation.

It is known that after a long-term integration the model tends to achieve its climate. Features of the applied model are demonstrated by Figure 9 which shows evolution of area-mean salinity on the level of 100 m. Three experiments are carried out: without assimilation of any data (dotted line), with assimilation of CoMSBlack-92-95 hydrology (dashed line) and with assimilation of ERS-1 Topex/Poseidon altimetry. It is seen that during 12.5 years of integration salinity in the halocline, and on the level of 100 m in particular, deviates considerably from the observed average value. Successive assimilation of the hydrology removes this deficit only partially as less than one survey per year implies on the period of calculations. Continuous assimilation of altimetric data noticeably improves model's operation in the right direction.

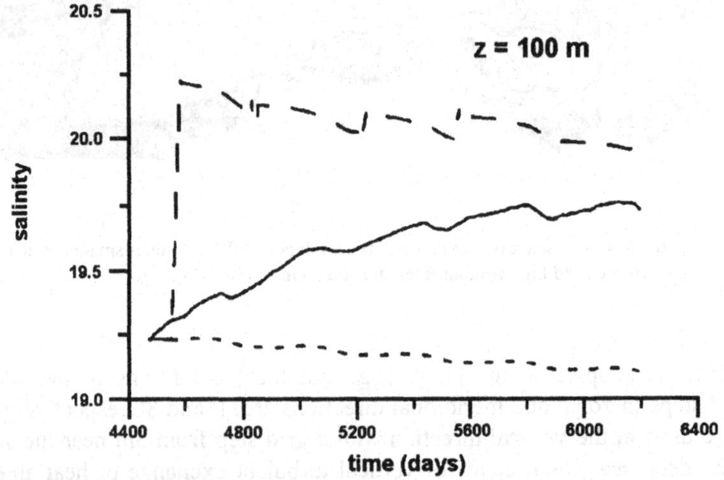

Figure 9. Temporal variations of salinity on the 100 m horizon for the calculation without data assimilation (dotted line), with assimilation of CoMSBlack-92-95 data on temperature and salinity (dashed line) and with continuous assimilation of altimetric data of ERS-1 and TOPEX/POSEIDON from April, 1992 to December, 1996.

Figure 10a-d shows two-month fragment of the sea level simulation with assimilation of altimetry. It demonstrates the possibility to follow the evolution of mesoscale vortex in the Black Sea, particularly the dynamics of two intensive vortices in the eastern and north-western parts of the sea. We can see that during summer-autumn attenuation of Rim Current such vortices can exist in the Black Sea for 3-4 months starting from the moment of their origin. The fact of reproducing a vortex evolution for such long periods using assimilation of ERS-1 and Topex/Poseidon altimetry increases degree of trustworthiness of results. Figure 10 also demonstrates a possibility of filtering out noise from the altimetric data after the assimilation in a numerical model.

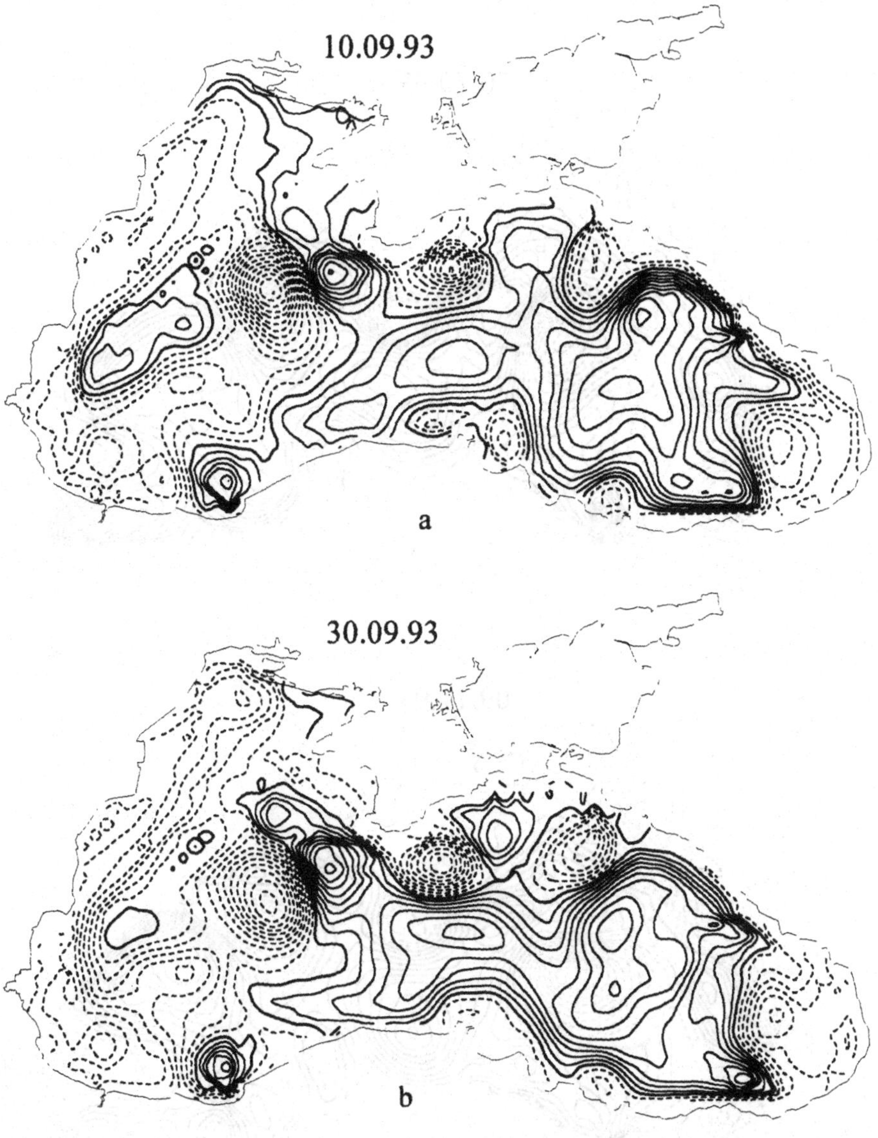

Figure 10 (a,b). Dynamical level resulted from assimilation of ERS-1 and TOPEX/POSEIDON data. Shown is the fragment from September 10, 1993 (a) and September 30, 1993 (b). Dotted line shows positive values of the level; contour interval is 1 cm

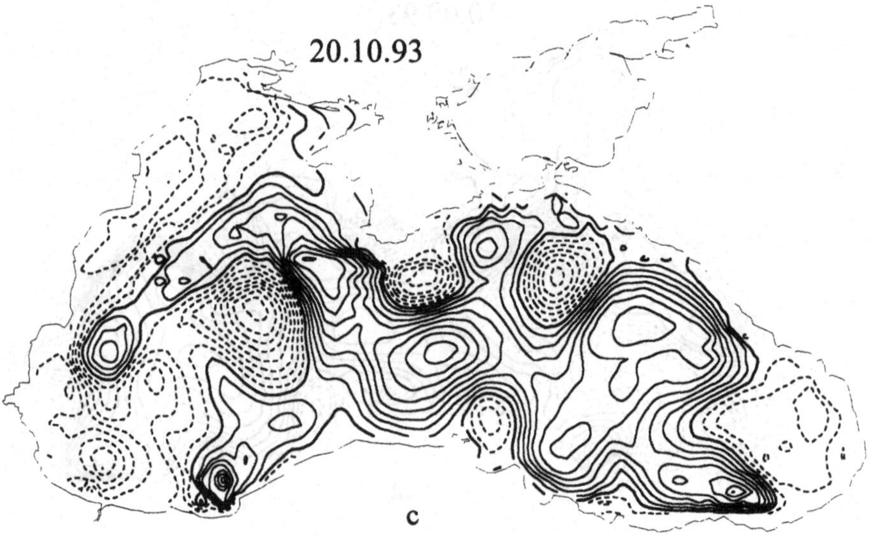

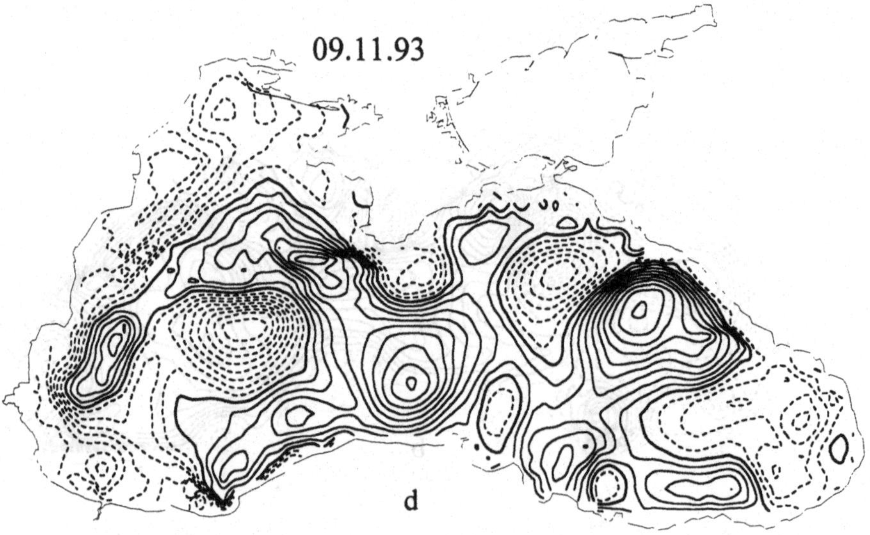

Figure 10 (c,d). Dynamical level resulted from assimilation of ERS-1 and TOPEX/POSEIDON data. Shown is the fragment from October 10, 1993 (c) and November 9, 1993 (d). Dotted line shows positive values of the level; contour interval is 1 cm.

5. Conclusions

Accurate processing of the ERS-1 and the Topex/Poseidon data carried out in frame of GSFC Pathfinder project, leads to very encouraging application of satellite altimetry for the study of general circulation and the sea level variability of the Black Sea. Semi-closed nature of the Black Sea strongly simplifies the usual problem of the open ocean study with boundary conditions on the open boundaries. Thus, specifying adequately atmospheric forcing on the free-surface and improving the model for the direct account of rivers runoff and water exchange through straits, we obtain ideal conditions for the verification of assimilation procedures and the physical content of existed oceanic circulation models. Applied aspect of the problem consists in the possibility to create near-real time procedure of the control of the three-dimensional state of the Black Sea.

Acknowledgements

Authors are grateful to Academician Artem Sarkisyan for the possibility to use the numerical model of the Black Sea circulation elaborated in the Institute of Numerical Mathematics of Academy of Sciences of Russia. We also grateful to Professor Emil Stanev who kindly presented for us climatic heat fluxes and wind stress and Ms. Larisa Sukhih who processed data and prepared Figure 2b.

References

1 Altman, E.N., Gertman, I.F. and Golubeva, Z.A. (1987) *The climatic fields of salinity and temperature of the Black Sea water*, Sevastopol's Department of SOI, Sevastopol (in Russian).
2. Blatov, A.S , Bulgakov, N.P , Ivanov, V.A , Kosarev, A.N. and Tujilkin, V.S. (1984) *Variability of the hydrophysical fields in the Black Sea*, Gidrometeoizdat, Leningrad (in Russian).
3. Bulgakov, S.N. and Korotaev, G.K (1984) *Possible mechanism of water stationary circulation in the Black Sea*, Complex research on the Black Sea. Sevastopol, 32-40 (in Russian).
4 Ibraev, R A. (1993) Reconstruction of Climatic Characteristics of the Gulf Stream Current, *Izv. Atm Ocean.Phys.*, **29, N6**, 803-814 (in Russian).
5 Gandin, L.S. (1963) *Objective analysis of meteorological fields*, Gidrometeoizdat, Leningrad (in Russian).
6. Ginzburg, A.I., Kostianoy, A.G., Soloviev, D.V. and Stanichny, S.V. (1996) Anticyclonic eddies evolution in noth-western Black Sea, *J Earth RES. Space* , **N 4**, 67-76 (in Russian).
7. Knysh, V V , Saenko, O A. and Sarkisyan, A S. (1996) A method of assimilation of altimeter data and its testing in the tropical North Atlantic, *Russ. J. Numer. Anal. Math. Modell* ,**11, N5**, 393-409
8. 8.Korotaev,G.K., Saenko,O.A , Koblinsky,C.J., Demyshev,S G. and Knysh, V V. (in press) An accyracy estimation, methodology and some results of assimilation of the TOPEX/POSEIDON altimeter data into the Black Sea general circulation model, J. Earth Res. Space, Moscow (in Russian)
9 Malanotte-Rizzoli, P. (ed) (1996) Modern approaches to data assimilation in ocean modeling, *Elsevier Oceanography Series*, **61**.
10 Mellor, G L. and Ezer, T (1991) A Gulf Stream model and an antimetry assimilation scheme, *J Geophys. Res* , **96**, 8779-8795
11 Pinardi,N ,Rosati, A and Pacanovski, R C (1995) The sea surface pressure formulation of the rigid lid models. Implications for altimeter data assimilation studies, *J. Mar. Syst.*, **6**,109-119
12 Sarkisyan, A.S (1996) On some results and problams of ocean modelling, *Oceanology* ,**36**, **N5**, 647-658 (in Russian)

13 13.Sarkısyan, A.S. and Demın, Yu. L (1983) A semidiagnostic method of sea currents calculation, *Large scale oceanographıc experıments in the WCRP*, **2, N1**, 201-214.
14 Simonov, A.I., Altman, E.N. (eds.) (1991) *Reference book "The USSR Seas" Project, Hydrometeorology and hydrochemıstry of USSR seas. Volume IV. The Black Sea,* Jssue 1. Hydrometeorological conditions, Gidrometeoizdat, St.Petersburg, (in Russian).
15. Stanev, E.V., Roussenov, V.M., Rachev, N.H. and Staneva, J.V. (1995) Sea response to atmospheric variability. Model study for the Black Sea, *J. Mar. Syst.*, 241-267.
16. Suetın, V.S. and Korotaev, G K. (1997) MOS: data processing algorithm development and satellite experiments in the Black Sea, *Proceedıngs of the 1-st Internatıonal Worshop on MOS-IRS and Ocean Colour, Berlin*, 166-170.
17 Sur, H.I., Ozsoy, E. and Unluata, U. (1994) Boundary current instabilities, upwelling, shelf mixing and eutrophication processes ın the Black Sea, *Progr in Oceanogr.*,**33**, N4 249-302.

HYDRO-OPTICAL STUDIES OF THE BLACK SEA: HISTORY AND STATUS

V.L. VLADIMIROV*, V.I. MANKOVSKY *, M.V. SOLOV'EV*,
A.V. MISHONOV*, and S.T. BESIKTEPE*
(*)Marine Hydrophysical Institute, NASU, 2 Kapitanskaya St., Sevastopol, 335000, Ukraine
(*)Middle East Technical University, Institute of Marine Sciences, P.O. Box 28, 33731 Erdemli, Icel, Turkey

Abstract

A review of the hydro-optical studies in the Black Sea from 1922 till now is presented. Seasonal and long-term variability of the Black Sea optical parameters are analysed using data sets from the data bases of Marine Hydrophysical Institute (Ukraine) and Institute of Marine Sciences (Turkey). A dramatic decrease in the water transparency was observed during 1986-1992. This coincided with significant changes of the spectral distribution of optical parameters. The main reasons of such changes are eutrophication, influence of the biological invader *Mnemiopsis leidyi* on the sea ecosystem, and the natural 11 year solar cycle.

1. Introduction

In situ measurements of optical characteristics have been conducted in the Black Sea since the 1960s and Secchi disk measurements since the 1920s. A valuable data set of the optical characteristics is available at present at the Marine Hydrophysical Institute (MHI) and the Institute of Marine Sciences (IMS). The data sets contain more than 13000 Secchi disk measurements, 2500 profiles of beam attenuation coefficient, 200 scattering functions, and other optical parameters. These data permit us to estimate vertical, spatial, seasonal and long-term variability of some optical characteristics of the Black Sea and to analyse their relations with biological variables and hydro-dynamics.

2. Review of the optical investigations in the Black Sea

The first measurements of optical properties of the Black Sea water were performed using Secchi disk and the Forele-Ule scale for the determination of the water colour. Systematic measurements were commenced in 1922 by the Azov - Black Sea Scientific - Fishery expedition organised by the All-Union Research Institute of Fishery of USSR

S. Beşiktepe et al. (eds.),
***Environmental Degradation of the Black Sea: Challenges and Remedies*, 245–256.**

(VNIRO). These expeditions continued until 1927, and the results were summarised by its chief - N.M. Knipovich [1], together with results of other expeditions (Main Hydrographic Board and Sevastopol Biological Station). These data, which covered the north part of the sea approximately to the limits of the economical zone of the former USSR, permitted estimation of the distribution of water transparency in the region, with some information on the seasonal variability using data from hydrometeorological stations in ports. The maximum values of transparency ever observed were: $Z_d = 30.5$m - at a point opposite the Kerch Strait (44°50.5'N, 36°49'E, March 30 1924) and $Z_d = 30$m - in the south part of the sea (42°39'N, 33°30'E, May 04 1925).

Secchi disk measurements were performed in different expeditions in the following years, giving extensive information about spatial distribution and time variability of the water transparency. However, the data do not provide any information on the vertical optical structure of the water column.

In 1929, academician V.V. Shuleykin, well-known by his fundamental work on physics of the sea, founded the Black Sea Hydrophysical Station in Crimea. Hydro-optical research was started there under his leadership. At first the research was directed towards theoretical analysis of light penetration in the sea, and laboratory verification of theoretical results in turbid media, which modelled sea water. In the 1940s, the Black Sea Hydrophysical Station began *in situ* hydro-optical investigations directed towards verification of theoretical results under natural conditions. In 1948, measurements of angular distribution of light radiance at different depths confirmed *in situ* the existence of the asymptotic radiance distribution, which had been predicted by Shuleykin. In 1959 the first measurements of underwater irradiance were conducted in the deep part of the sea down to 100 meters at three spectral bands, and water transparency measurements down to 200 meters; the scattering function *in vitro* was also measured for surface water. In 1960, investigations of sunlight polarisation in the sea were started, showing that asymptotic radiance distribution characterised not only the steady radiance, but also the degree of polarisation.

Field research conducted by the Black Sea Hydrophysical Station (reviewed in [2]) gave some idea about vertical distribution of optical properties of the sea. But the technical capabilities in those years were severely limited, so many questions of optical properties of the Black Sea water, including their vertical and horizontal distribution, remained open.

A new stage in optical investigations in the Black Sea came after the relocation of the Marine Hydrophysical Institute (MHI) from Moscow to Sevastopol in 1963. The Marine Optics Department was established under the leadership of professor G.G. Neuymin. The department began a broad based work program, including construction of modern instruments for field hydro-optical measurements. Up to the 1970's a number of sounding and towed instruments were constructed, which permitted measurements of both inherent and apparent (light field parameters) optical properties of the sea: the attenuation and scattering coefficients, the scattering function, underwater irradiance, angular distribution of the underwater radiance and polarisation, colour index, bioluminescence, and fluorescence. During the next decade, instruments were constructed for remote studies of the sea using the spectra of sea radiance. The Department initiated a long-term program of measurements, for the study of the optical structure of water and its relationship with hydrological, chemical and biological fields. The beam attenuation coefficient was chosen as the main parameter and its vertical profiles were measured in expeditions. The measurements of Secchi disk depth were continued.

The quantity of hydro-optical investigations conducted by MHI in these years surpasses many times those conducted by other institutions. These investigations focused on the optical structure of the sea and processes of its formation, such as the following example: During August - September 1964, the vertical structure of water transparency was studied down to 1000 meters (spectral band 640nm) and down to 1700 meters at some stations using a wireless transmissometer with acoustical communications. The research was performed along two transects crossing the entire sea - along 43°N and 34°E. Besides the turbid layers of biological origin, there is an additional interesting layer in the Black Sea below the euphotic zone at depths of 100-200 meters, near the upper boundary of the anoxic zone. The mechanisms of its formation are discussed in [3,4]. It corresponds with intense bacterial activity and with complex geo-chemical processes occurring in the oxic/anoxic interface.

In 1967 the sounding transmissometer with cable-rope connection was completed, which permitted measurements of the beam attenuation coefficient in the visible and near ultraviolet spectral bands. Measurements using this instrument showed the spectral difference in vertical profiles of transparency in the deep anoxic layer. The attenuation coefficient in this layer is constant in the red band but it increases with depth in violet, and especially in the near-ultraviolet spectral band. This regularity suggests that the concentration of so-called "yellow substance" (gelbstoff), which is included in dissolved organic matter, increases with depth. This substance absorbs light strongly in the short wave band of the spectrum and does not affect the red band.

In 1970, the first measurements of the scattering function were conducted *in situ* using an underwater nephelometer at a meridian section along 34°E. Measurements to 200 meter depth clarified the mistake of early views about scattering of light in deep water layers, which were based on laboratory measurements using water samples. In such samples, if one does not take special efforts, the hydrogen sulphide is oxidised and a turbid colloid is formed from sulphur particles, which scatter the light considerably.

In 1972-1973, investigations of the vertical bioluminescence light field structure in the regions with different hydrological conditions [5] showed that bioluminescence in the Black Sea exists only in the upper oxic zone and is absent in deeper anoxic layers. As distinct from other seas and oceans, the daily vertical migration of bioluminescent layers is absent. The *in situ* first measurements of bioluminescence in the Black Sea using an underwater photometer were performed by Institute of Biology of the Southern Seas in 1965-1966 [6]. However, these limited measurements at a fixed point near Crimea at small depth (near 60 m) did not provide information about the spatial structure of bioluminescence.

Since 1974, measurements of colour index have been conducted in Black Sea expeditions. The new research vessel "Akademik Vernadsky" (obtained by MHI in 1969) was a very effective vessel for this purpose. A vertical hull hole permitted measurement of colour index and other parameters while she was sailing, and permitted the study of the small-scale variability of optical parameters in the surface layer of the sea.

During hydro-optical research near the Bosphorus in 1976 some peculiarities in the vertical distribution of transparency in deep waters, connected with Marmara Sea water inflow, were observed. Strongly turbid layers were found here at depths of 200-400 meters, within the anoxic zone. The beam attenuation coefficient was sometimes 1.6 1/m in the violet band in these layers. Research conducted here in 1978 showed the intensity of the turbid layers decreased rapidly with distance from the Bosphorus, but it could be observed at the depth of Marmara Sea water penetration easterly along the Turkish coast up to Caucasus.

In the early 1980's MHI started the research on spectra of sea radiance in various optical bands in conjunction with the satellite oceanography program. The multiple measurements of the radiance spectra of the Black Sea were conducted in 1983-1985 as part of the experiment "Inter-Cosmos." The measurements were performed from ship, aircraft and the orbital space station "Salut-7." Algorithms were constructed to extract information about optical and biological properties of sea water from the sea radiance spectrum.

By the middle of the 1980s, the large data set accumulated in MHI permitted analysis of the climatological structure of the optical fields in the Black Sea. The first maps were constructed of the mean distribution of the beam attenuation coefficient for 1977- 1985 for different depths down to 400 meters, and the depth of characteristic optical layers for summer and winter seasons. These maps showed the main features of space - time variability of water transparency on climatological scales, and the close relationships of optical fields with other oceanographic fields.

At the end of the 1980's, the data base of the Black Sea optical parameters was created by MHI. The stimulus for this research was the extremely low values of water transparency which appeared in the Black Sea in the late 1980's.

During the last few years, the hydro-optical investigations conducted by MHI in the Black Sea have been directed on the study of the time and space variability of optical parameters in the framework of international monitoring programs. A specific focus is optical structure in the near-shore zone, where the anthropogenic loading is maximum.

The main optical parameters measured now by MHI expeditions include: vertical profiles of the beam attenuation coefficient down to 250 meters in different bands of visible and near-ultraviolet spectra, spectral distribution of the radiance index of the sea, colour index, Secchi disk depth, and colour of the sea. The ultimate goal of the optical investigations in the Black Sea is to construct a hydro-optical model of the basin, from which one can make operative diagnosis and forecast of the optical sea state.

Results of optical research in the Black Sea till the end of the 1970's were reviewed in [7,5], and to the end of the 1980's in [8]. The results of remote sensing of the Black Sea and ground truth measurements were described in [9].

Another major institution studying the hydro-optics of the Black Sea is the South Branch of the Oceanology Institute (Russian Academy of Science) located in Gelendzhik. This Branch performed research primarily in the eastern near-shore part of the Black Sea. In 1989-91 it carried out long-term (18 months) optical measurements in the Caucasus near-shore zone using a series of transects between Gelendzhik and

Novorossiysk. Results of these measurements (reviewed in [10]), showed typical profiles of the attenuation coefficient in the near-shore zone and its seasonal variability. They also showed the possibility of optical monitoring of cyclonic and anticyclonic eddies on a synoptic scale. In winter 1991 (cruise 21 of R/V *Vityaz*), the South Branch performed hydro-optical research in the central and eastern part of the sea. The optical structure of the water (using beam attenuation coefficient) and its connections with water dynamics were determined for the end of the winter season [11]. They also studied the dependence between quanta irradiance of photosynthetic-available radiation and spectral values of the downward irradiance [12], recommending that now, since transparency has decreased significantly for Black Sea, one must use the spectral band of 520-555 nm, rather than 465 nm as was recommended by N. Jerlov earlier, for the calculation of photosynthetic-available radiation.

Some optical research in the Black Sea also was conducted by the Institute of Biology of the Southern Seas (Ukrainian Academy of Sciences). Except for bioluminescence measurements [5,13], the research conducted in 1986-1989 for the determination of the euphotic layer depth (1% of surface irradiance) used Secchi disk depth in the Northwest part of the sea (in press).

In 1986 optical research in the Black Sea was started by the Institute of Marine Sciences (IMS METU, Turkey). The measurements of the water transparency, Secchi disk depth, fluorescence and PAR were performed in the south part of the Black Sea.

A notable event in Black Sea optical research was the measurements of vertical profiles of water transparency in the red spectral band up to full depths conducted in autumn 1988 by R/V *Knorr* (USA). This research covered the southern part of the sea in the Turkish economic zone. Previous measurements to such depths were performed only once in 1964 by R/V *Mikhail Lomonosov*. Interesting results were obtained by the R/V *Knorr* expedition about the fine structure vertical distribution of transparency in the intermediate layer and about its connections with thermohaline parameters. The transparency decreased in layers having thermohaline anomalies. Using these data they traced the evolution of anomalies on transects from the coast to the sea and discussed transport of particulate matter from the Turkish shelf to the deep zone of the sea in the intermediate water layer [14].

3. Relationships between the Secchi disk depth and bio-optical parameters of the Black Sea surface waters

The simplicity of Secchi disk measurement attracted researchers' attention long before knowledge of the connections between the Secchi disk depth and different hydro-optical parameters measured by precise optical instruments. A Complete examination of the different factors of underwater vision was provided by [15,16]. For specific external conditions (irradiance, sea surface state, etc.) the maximum Secchi disk depth (Z_d) depends on optical water properties (so-called inherent optical properties) and on parameters of the underwater light field (apparent optical properties). Although Z_d depends on the combination of inherent and apparent optical water parameters, there is a high correlation between Z_d and these parameters taken separately. This correlation results because hydro-optical parameters in the sea are strongly connected. The main factor that causes this correlation is phytoplankton and its life products, which influence both absorption and scattering characteristics of light. Thus, there is a high correlation between Secchi disk depth and biological parameters (phytoplankton). Different bio-geo-chemical conditions in different regions of the World Ocean, especially the changes in the species composition of phytoplankton populations, influence the relationships between optical and biological parameters. Therefore regional optical relations are more reliable than global relationships. Table 1 lists empirical regression equations which connect the different bio-optical parameters of the Black Sea:

c_m $(0\text{-}Z_d)$ - mean value of the beam attenuation coefficient in the layer from surface down to Z_d, m^{-1};

$k_m(0\text{-}Z_d)$ - mean value of the vertical attenuation coefficient for the daylight in the layer from surface down to Z_d, m^{-1};

I_c $(540,440)=B^{540}/B^{440}$ - colour index, where B^{540} and B^{440} are the spectral radiance in upward light under the sea surface at wavelengths of 540 and 440 nm;

$L(R_{max})$ - the wavelength corresponding to the maximum on the spectral curve of the radiance index of the sea, nm;

L_d - the dominant wavelength in spectral distribution of the sea radiance index calculated in the X, Y, Z system of colour co-ordinates, nm;

C_o - chlorophyll "a" concentration near the sea surface, $mg{\cdot}m^{-3}$;

P_o - primary production near the surface, $mgC{\cdot}{\cdot}m^{-3}$ day^{-1};

S - integrated production in the euphotic zone of the sea, mgC $\cdot m^{-2}$ day^{-1};

M_o - concentration of particulate matter in the near surface layer, $mg{\cdot}l^{-1}$.

TABLE 1. Relationships between Secchi disk depth and bio-optical parameters for the Black Sea

#	Equation y=f(x)	Wave length (nm)	Limits for x / y	# of point	σ^2 of regr.	Correlation coefficient r	Ref.
1	$c_m(0\text{-}Z_d)=(8.75/Z_d)-0.11$	422	4.5-27/0.1-0.93	302	0.17	0.80±0.05	20
2	$k_m(0\text{-}Z_d)=1.43/Z_d$	490	10-26/0.02-0.14	20	0.018	0.81±0.05	@
3	$I_c=123{\cdot}Z_d^{-2.02}$	540, 440	3-44/0.05-16	124	-	0.93±0.02	21
4	$L(R_{max})=593-37{\cdot}\ln(Z_d)$	-	1.5-28/475-580	89	-	0.95±0.02	9
5	$L_d=463+285/Z_d$	-	3-40/470-560	121	-	0.89±0.02	9
6	$C_o=52.5{\cdot}Z_d^{-2.11}$	-	1-27/0.02-79	307	0.33	-0.85	22
7	$P_o=2400{\cdot}Z_d^{-2.24}$	-	1-24/0.7-2700	156	0.37	-0.85	22
8	$S=5012{\cdot}Z_d^{-1.28}$	-	1-24/22-3585	156	0.28	-0.77	22
9	$M_o=14.8{\cdot}Z_d^{-1.29}$	-	0.05-29/0.12-1390	284	0.24	-0.97	23

@ - unpublished data of the MHI

The equations for I_c, $L(R_{max})$, L_d and M_o were calculated using data obtained not only in the Black Sea, but in other marine basins [9, 21, 23]. Data from the Black Sea alone are in good agreement with these equations. High correlation coefficients for the equations for Table 1 permits the use of Secchi disk depth for estimating the bio-optical state of the Black Sea upper layer. Strong changes which occurred in the Black Sea ecosystem during the last years suggest that equations 3-8 (Table 1) can be applied only to data obtained till 1985; these equations can give high deviations for the recent data.

For comparison, analogous equations obtained in basins other than the Black Sea have nearly the same range of variability of the Secchi disk depth. We shall not discuss here the difference between the equations for different basins.

For Norway and Barents seas [17]: k_m $(0\text{-}Z_d)$=1.5/Z_d; L=465nm, Z_d=5-18m.
For Lake Baikal [18]: c_m $(0\text{-}Z_d)$=7.3/Z_d; L=480nm, Z_d =5-28m.
For the region of Peru up-welling [19]: $C_o=670\cdot(Z_d)^{-3}$; Z_d =4-15m.

4. Seasonal and long-term variability of water transparency

Figure 1 shows the seasonal variability of the Secchi disk depth (Z_d) analysed using monthly mean values, calculated for 1922-1985 for the central deep part of the Black Sea, limited by the latitudes 42°20' and 44°15'N and longitudes 31° and 38° E. There are two minima in the inter-annual variability of Z_d, namely in spring and at the end of autumn and two maxima, namely in summer and at the end of winter. The difference between maximum and minimum mean values is about 6.2 meters or 37% of mean annual value (16.8 m).

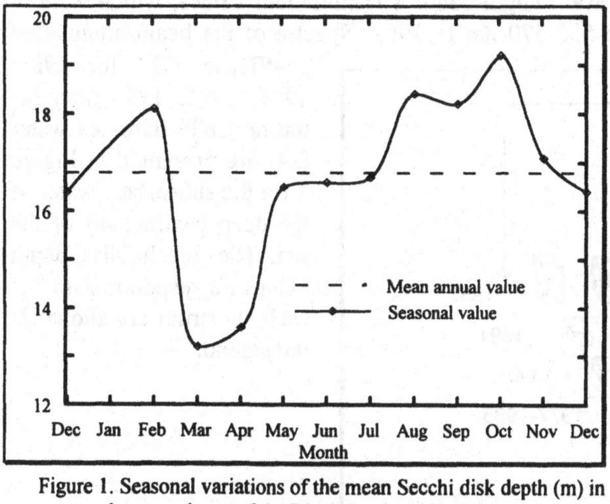

Figure 1. Seasonal variations of the mean Secchi disk depth (m) in the central part of the Black Sea in 1922-1985.

The strong changes in the optical properties of the Black Sea which have taken place in recent years are of great interest [8]. From the early 1920's to the mid-1980's, only a weakly pronounced transparency decrease was observed. Over 60 years, the Secchi disk depth had decreased from 20-21 to 15-16 meters (mean values in the deep central part of the sea), however, single values up to 25 meters were occasionally observed. The Black Sea water transparency has decreased dramatically since 1985. In 1990-1993, values in excess of 15 meters were no longer observed and mean values were only 6-10 metres. It is worth to mention that the water transparency in the Black Sea started to increase after 1993 (Figure 2).

Water transparency measurements using *in situ* instruments, which were performed by MHI during the Black Sea surveys, also indicate the same decrease of transparency. Mean, maximum, minimum values and standard deviation of the attenuation coefficients (wavelength 410-420 nm) in the surface layer

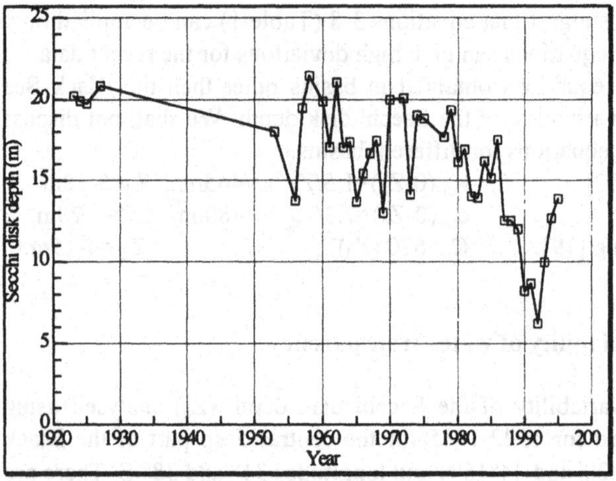

Figure 2. The long-term variability of annual mean Secchi disk depths in the central deep part of the Black Sea in 1922-1995.

of the central, deep part of the sea for the summer period are shown in Figure 3. Values obtained in the period 1977-1985, which are considered as the "background" data, show significant changes compared to 1990, 1991, and 1992 data. Specifically, a continuos increase in the attenuation coefficient is observed.

Spectral optical properties of the Black Sea water have also changed in the same periods. In the 1990s compared with 1984, not only the values of the beam attenuation coefficient strongly increased, the shape of spectral curves changed as well, due to the initial enhanced increase in the short wavelength band. The minimum values, which were at 480 nm in 1984, shifted to the 550-570 nm in 1992. Spectra of the beam attenuation coefficient c(λ) for 1984, 1991, 1992, 1995, and for the optically pure sea water [24] are presented in Figure 4 for the subsurface water of the deep central part of the sea. The Secchi disk depth values corresponding to the each spectrum are shown in the legend.

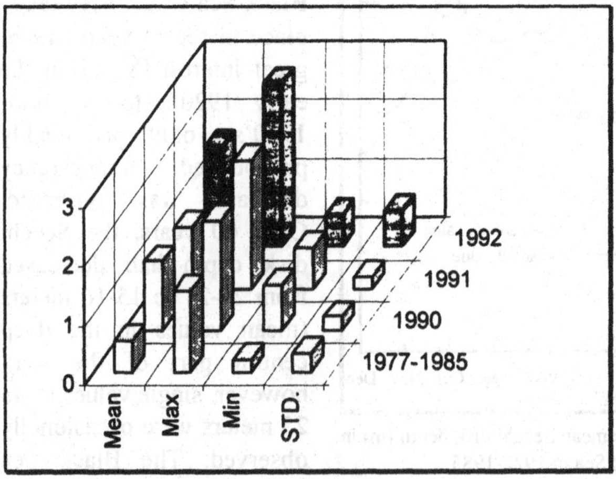

Figure 3. Beam attenuation coefficient, c (m^{-1}) in 1977-1992.

5. Factors influencing the drastic decrease of water transparency

The detail analysis of the collected data has helped to understand the main reasons for the drastic transparency decrease in the deep basin of the Black Sea in the 1986 - 1992 period and to reach the conclusions stated below [25].

Significant eutrophication occurs both in the near-shore areas of the Black Sea and in the deep basins, being connected with the long-term increase in the input of nutrients of anthropogenic origin.

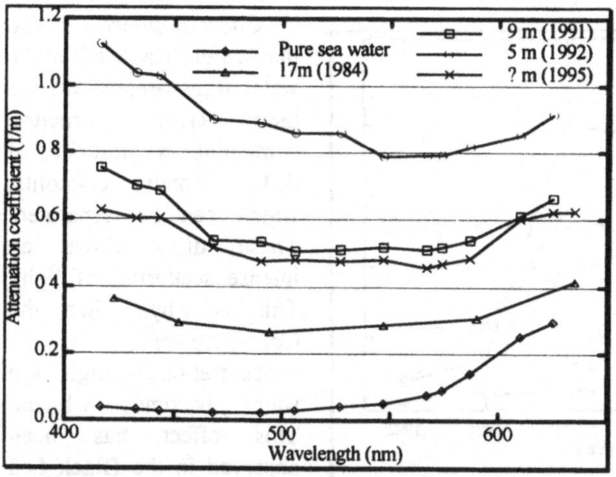

Figure 4. Spectra of the beam attenuation coefficient c(λ) for the subsurface water of the deep central part of the sea for the different years and for the pure sea water. The Secchi disk depth values measured at the same station are given in the legend (it was not measured in 1995).

For example, input of biogenic compounds of nitrogen and phosphorus by the Dnieper and Danube rivers, which are responsible for 3/4 of the overall riverine input to the Black Sea, increased 5-7 times from the 1960s to the 1980's [26]. The mean chlorophyll concentration has become one and a half time larger in the deep part of the sea during the same period [27].

Water transparency gradually declined due to the increase of the content of optically active matter in the sea. Over 20 years, from the mid-1960s to the mid-1980s, the mean annual decrease of the Secchi disk depth has attained about two meters.

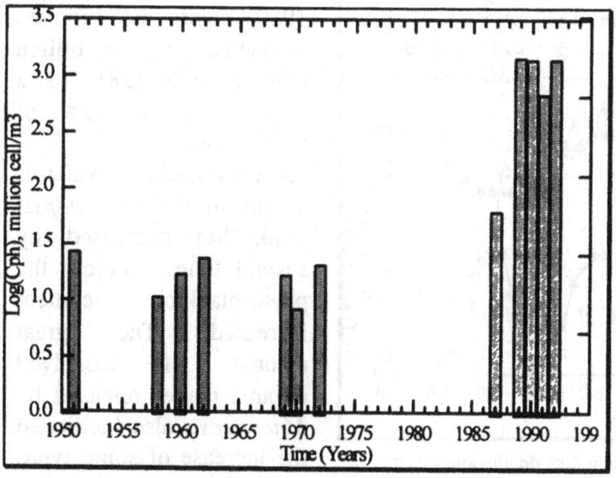

Figure 5. Dynamics of the phyto-plankton amount in the euphotic layer of the Black Sea western deep area in 1950 - 1992.

However, the main reason for the drastic transparency decrease from 1986 till 1992 was the enhanced bloom of *Peridinium* and *Coccolitophores*. Their number in the Black Sea was 1.5 - 2 orders of magnitude larger than the in previous years and has reached 2-3 billions per cubic meter (Figure 5). A significant increase in the biomass of these plankton organisms has changed the structure of the plankton community. Nanoplankton, which contributes immensely to the light scattering, accounted for 90% of the phytoplankton content in this period.

The intense blooms of *Peridinium* might have been also caused by the increased concentration of particulate organic matter, as *Peridinium* can switch to heterotrophyc nutrition.

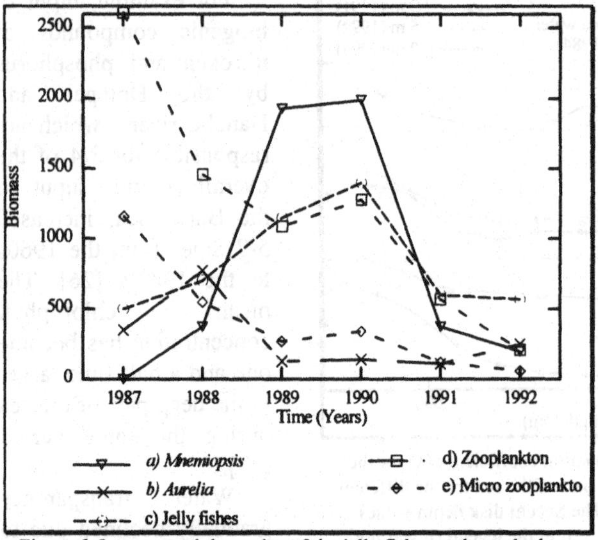

Figure 6. Inter-annual dynamics of the jelly fishes and zooplankton bio-mass in the Black Sea in 1987 - 1992 (Jelly fishes - 10^3 g·C·m^{-2}, Aurelia and Mnemiopsis - g·m^{-2}, others - 10^2 g·m^{-2}).

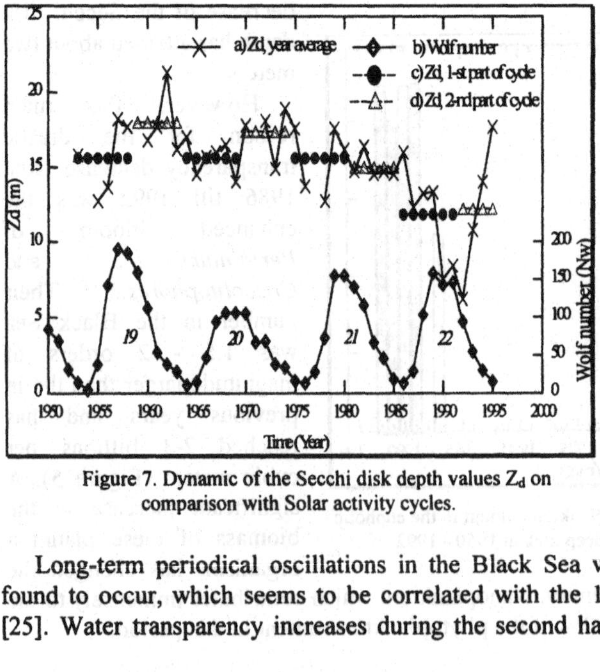

Figure 7. Dynamic of the Secchi disk depth values Z_d on comparison with Solar activity cycles.

Coccolitophores also play a significant role in the water transparency, due to their crusts structure containing a number of disks, namely, coccolits, which can be separated. These disks cause an intense scattering of light. This is why, when the *Coccolitophores* concentration is high, sea water becomes whitish. This effect has been observed in the Black Sea over the last few years.

Phytoplankton structure was also strongly impacted by the biological invaders, such as *Mnemiopsis Leidyi* which appeared in the Black Sea in the early 1980s and its biomass amounted to one billion tons in 1989 [28]. As a result, the amount of herbivorous microzooplankton, which is a part of the *Mnemiopsis* food, has decreased by several times, whereas the phyto-plankton content increased. The great amount of dissolved organic matter released by *Mnemiopsis* also facilitated the increase of some types of phytoplankton and bacteria (Figure 6).

Long-term periodical oscillations in the Black Sea water transparency have been found to occur, which seems to be correlated with the 11-year cycle of solar activity [25]. Water transparency increases during the second half of the cycle and decreases during the first one. Drastic decrease in the transparency in late 1980s coincided with the second half of the 1980-1991 solar activity cycle. However, the magnitude of the transparency decrease was more intense during this cycle than those during the previous cycles (Figure 7).

From the analyses of the collected data, it may be concluded that the catastrophic transparency decrease in 1990-1992 took place due to the combined effect of three factors coinciding in time and sign: (1) the natural 11-year cycle, (2) increased eutrophication, and (3) the influence of the biological invader *Mnemiopsis Leidyi* on the ecosystem structure.

It is important to emphasize that the intense decrease in water transparency ended in 1992, and in 1993 water transparency started to improve. As a result, already by the end of 1995, the mean Secchi disk depth in the deep central part of the basin had reached the levels observed for the early 1980s, i.e. about 17 meters. Future research will give the evidence on the changes in the ecosystem of the sea during these years, which caused this transparency increase, but it coincides with the beginning of a new cycle of solar activity.

Acknowledgements

We thank all the scientists and crew of the research vessels which participated in data collecting in the Black sea. We wish to thank also Drs. O.Yunev, L.Georgieva and B.Anninsky from IBSS NASU for the very useful discussion on the reasons of catastrophic transparency decrease in the Black Sea. The research described in this publication was supported from the NATO Linkage Grant under reference ENVIR.LG 951190 for project "ENVIRONMENTAL TRENDS FOR THE BLACK SEA".

References

1. Knipovich, N.M. (1932) Hydrological investigations in the Black Sea. *Sci. work of the Azov-Black-Sea scientific producers expedition,* Iss. **10**, Moscow, 272 (in Russian).
2. Timofeeva, V.A. (1979) On study of optical phenomena in the Black Sea. *Multidisciplinary investigations in the Black Sea,* 77-82, MHI of the Ukr.SSR Acad. of Sci., Sevastopol (in Russian).
3. Sovga, E.E., Mankovsky, V.I, Prokhorenko, Yu. A. *et. al* (1987) The nature of the deep turbid layer in the Black Sea. *Rep. of the Ukr. Acad. of Sci.,* ser. B, No. **6**, 32-35 (in Russian).
4. Bezborodov, A.A. (1989) Fine hydrochemical structure of interaction zone between aerobic and anaerobic water in the Black Sea. *Complex oceanographic investigations of the Black Sea,* 131-152, MHI of the Ukr.SSR Acad. of Sci., Sevastopol (in Russian).
5. Urdenko, V.A., Zemlyanaya, L.A., and Vladimirov, V.L. (1978) The regional peculiarities of bioluminescence fields in the Black Sea. *Hydrobiological Journal,* v. **14**, No. **2**, 17-23 (in Russian).
6. Bityukov, E.P., Rybakov, V.P., and Shaida, V.G. (1967) The annual variations in intensity of bioluminescence field in the Black Sea neritic zone. *Oceanology,* v.7, Iss.**6**, 1089-1099 (in Russian).
7. Neuymin, G.G. (1980) The optical characteristics of the Black Sea waters. *The multidisciplinary oceanographic investigations of the Black Sea,* 198-215, Naukova dumka, Kiev, (in Russian).
8. Mankovsky, V.I. (1992) Optical structure of the Black Sea waters and regularities of its formation. *Hydrophysical and hydrochemical investigations of the Black Sea,* 7-30, MHI of the Ukr. Acad. of Sci, Sevastopol (in Russian).

9. Urdenko, V.A., and Zimmerman, G. (eds.) (1985) *Remote sensing of the sea and the influence of atmosphere*. USSR Acad. of Sci. - GDR Acad. of Sci -Ukr.SSR Acad. of Sci. Moscow-Berlin-Sevastopol (in Russian).
10. Nikolaev, V.P., and Salin, V.D. (1992) On spatial- temporal variability of water transparency in the Black Sea coastal zone. *IO RAS South. Branch, Gelendzhik, Deposit VINITI, file* **1307-B92** (in Russian).
11. Nikolaev, V.P. (1992) Peculiarities of the water optical structure in February-March, 1991. *IO RAS South. Branch, Gelendzhik, Deposit VINITI, file* **1306-B92** (in Russian).
12. Levin, I.M., and Nikolaev, V.P. (1992) On estimation of the vertical attenuation of quantum irradiance in the region of photosynthetic active radiation in the Black Sea. *Oceanology*, v. **32**, Iss .**2**, 240-245 (in Russian).
13. Bityukov, E.P. (1971) Bioluminescence in the Atlantic Ocean and in the Seas of the Mediterranean basin. *Problems of the sea biology*, Naukova dumka, Kiev, (in Russian).
14. Ozsoy, E., Top, Z., White, G. and Murray, J.W. (1991) Double diffusive intrusions, mixing and deep sea convection processes in the Black Sea. *Black Sea Oceanography, E. Izdar and J.W. Murray (eds.)*, Kluwer Press, Dordrecht.
15. Levin, I.M. (1980) On the white disk theory. *Izv. USSR Acad. of Sci., FAO*, v. **16**, 926-932 (in Russian).
16. Preisendorfer, R.W. (1986) Secchi disk science: Visual optics of natural waters. *Limnol. & Oceanogr.*, v. **31**, 909-926.
17. Aas, Eyvind (1980) Relations between total quanta blue irradiance and Secchi disk observation in the Norwegion and Barents Seas. *Rep. Inst. fer Fysisk Oceanografi Kobenhavns Universitet*, Copenhagen, No. **42**, 11-27.
18. Sherstyankin, P.P., Bondarenko, N.A., Stepanova, V.V., Tarasova, E.N., and Shur, L.A. (1988) Interrelation between the beam attenuation coefficient and hydrophysical and biological indexes of the Baikal water. *Optics of sea and atmosphere, Theses of rep. of 10 plenary Session of WG "Ocean optics"*, 223-224, Leningrad (in Russian).
19. Kullenberg, G. (1980) Relationships between optical parameters in different oceanic areas. *Rep. Inst. fer Fysisk Oceanografi Kobenhavns Universitet*, Copenhagen, No. **42**, 57-79.
20. Mankovsky, V.I., and Zemlyanaya, L.A. (1989) The relation between the white disk visibility depth and the beam attenuation coefficient for the Black Sea waters. *Multidisciplinary Oceanographical Investigations of the Black Sea*, 82-85, MHI Ukr.SSR Acad. of Sci., Sevastopol (in Russian).
21. Shemshura, V.E., and Fedirko, V.I. (1981) On relationships between some hydrooptical parameters. *Oceanology*, v. **21**, iss. **1**, 51-54 (in Russian).
22. Shemshura, V.E., Finenko, Z.Z., Burlakova, Z.P. *et al.* (1990) The relationship between chlorophyll "a" concentration and primary production and relative water transparency in the Black Sea. *Optics of sea and atmosphere. Reports theses of the 11 plenary session on Ocean Optics*, part **1**, SO of the USSR Acad. of Sci., Krasnoyarsk, 176-177 (in Russian).
23. Shemshura, V.E., and Vladimirov, V.L. (1989) Estimation of suspended material concentration in the sea over the white disk visibility depth and upward radiation spectra. *Oceanology*, v. **29**, iss. **6**, 946-950 (in Russian).
24. Ivanov, A. (1975) *Introduction a l'oceanographie. Proprıetes physiques et Chimiques des eaux de mer*. Vol. **II**, Librairie Vuibert, Paris.
25. Mankovsky, V.I., Vladimirov, V.L. Afonin, E.I. *et al.* (1996) *Long-term variability of the Black Sea water transparency and factors determined its strong decrease in the late 1980s early 1990s.*, MHI of the Ukr. Acad. of Sci., Sevastopol, 32 pp., (in Russian).
26. Altman, E.N., Bezborodov, A.A., Bogatova, Yu.I., *et al.* (1990) Black Sea Hydrobiology. *Practical ecology of sea regions. Black Sea.* V.P. Keondgyan, A.M. Kudin, Yu.A. Terekhin, (eds.), Nauk. dumka, Kiev,.
27. V.I. Vedernikov and A.B. Demidov, (1993) Primary production and chlorophyll in the deep regions of the Black Sea *Oceanology*, v. **33**, iss. 2, 229-235
28. Vinogradov, M.E., Sapozhnikov, V.V., and Shushkina, E.A. (1992) *The Black Sea ecosystem*, Nauka, Moscow.

INFLUENCE OF ANTHROPOGENIC IMPACT ON THE PHYSIOLOGY OF SOME BLACK SEA FISH SPECIES

G. E. SHULMAN , A. YA. STOLBOV, E. V. IVLEVA ,
V. YA. SHCHEPKIN, G. S. MINYUK
Institute of Biology of the Southern Seas National Academy of Sciences of Ukraine 2, Nakhimov Ave., Sevastopol, Crimea, Ukraine.

Abstract

The effect of the two major factors of aquatic environment (oxygen deficiency and food supply) on the physiology of Black Sea fish was investigated.

A considerable decrease of the rate of the oxygen uptake and sharp increase of the rate of the nitrogen excretion was observed in three fish species (*Trachurus mediterraneus ponticus Aleev, Diplodus annularis Linne and Scorpaena porcus Linne)* during experiments of short-term and long-term hypoxia. The decline of the value of O:N quotient from 40-80 to 2-7 was also observed. The data obtained indicated the involvement of proteins and nitrogenous compounds in the energy metabolism of fish under low oxygen concentration in water. The results are discussed in relation to the adaptation of fish to oxygen deficiency mostly of anthropogenic origin.

Moreover, the disintegrating effect of the long-term hypoxia on the structure of muscle fiber was found in golden grey mullet (*Liza aurata Risso)* and round goby *(Neogobius melanostomus Pallas*) in histological assay. The presumable scheme of the muscle destruction mechanism is suggested.

Additionally, data on the fatness of the Black Sea sprat by the end of the feeding season for the period from 1960 to1996 are presented. These data are discussed in connection with monitoring of the nutritive base and food supply of planktivorous fish and changes in pelagic ecosystem caused by anthropogenic impact in the Black Sea.

1. Introduction

The Black Sea has been most dramatically afflicted by pollution due to the river inflow carrying ever increasing amounts of products of human activity [1] and the extreme "narrowness " of the biotic zone which lies between the surface and water layer containing H_2S and seems to be sensitive to harmful impacts. Populations of mass Black Sea fishes seem to be also susceptible to the noxious influences that may be explained by two main factors :

1) hypoxic conditions (oxygen deficiency) of the environment and
2) worsening food supply (food deficiency).

S. Beşiktepe et al. (eds.),
***Environmental Degradation of the Black Sea: Challenges and Remedies*, 257–272.**

The first factor was generated primarily by eutrophication and leads to persistent deficiency of oxygen in fish and the majority of water organisms inhabiting shelf areas, especially in the north -western part of the sea [2]. Oxygen deficiency frequently develops into complete anoxic conditions inducing mass mortality of benthic organisms and fish [2]. However, the hypoxia may be provoked not by low dissolved oxygen content (DO) but also by a variety of pollutants e.g. oil and oil products, heavy metals, pesticides which may hamper oxygen supply to tissues [3].

The second factor decreases the condition factor (well-being) and hence influences the reproduction capacity of population, resulting in a reduction of abundance [4].

It should be noted that the effect of the two factors on aquatic organisms of the Azov Sea is even more pronounced [4]. The reasons might be the small capacity, shallowness and relatively large river inflow, characteristic of the Azov Sea, that makes it even more vulnerable to anthropogenic impact than the Black Sea [4]. Therefore, it is of special interest to investigate the effect of the hypoxic conditions and the variability of food supply of fish to understand the related physiological mechanisms and possible adaptation to the adverse environmental conditions. Investigations in this direction have been conducted for several years at the Department of Physiology of Animals, Institute of Biology of the Southern Seas (IBSS) NAS Ukraine [3, 5-9].

This article is a review of our published and unpublished works on the problem and summarizes the results we have obtained and the research prospects which we believe may conspicuously contribute both to scientific and practical knowledge of the rational utilization of fish stock.

2. Materials and methods

2.1. STUDIES OF THE HYPOXIA.

Three Black Sea fish were chosen for the investigations:, the scorpion fish *Scorpaena porcus Linne*, the annular bream *Diplodus annularis Linne* and the horse mackerel *Trachurus mediterraneus ponticus Aleev*. The scorpionfish is an ambuscader displaying low motility, the horse mackerel an active pelagic fish adhering to nutrition of mixed type, and the annular bream an intermediate moderately motile fish inhabiting near bottom water. In nature these three species (as many others) are subject to the impact of hypoxic conditions. The study of oxygen deficiency involved performing experiments of two varieties - short-term (45 to 150 min) and long -term (14-h and more). In experiments of the first series all three species were examined and oxygen deficiency developed by means of an autogenous technique, through continuous oxygen consumption by fish whereby DO decreased from 8.8 to 1.5 $mg \cdot l^{-1}$. , that is from, 100 to 20% of the saturation. During the second series of experiments the research object was scorpionfish only, and the content of oxygen was kept at 1.5 $mg \cdot l^{-1}$ (20% of saturation). The fish for the experiments were captured by seine (horse-mackerel, annular bream) and drift-net (scorpion fish) near the southeastern coast of the Black Sea (the Karadag reserve, which is almost without pollution). Only healthy specimens, not infected by external parasites, were selected. The number of internal parasites was as usual for the studied species [10]. Collected fish was transported by boat in tanks (volume 0,1 m^3) for 10-15 min and placed in flow-through tanks (volume $2m^3$) for 5-7 days for acclimation to experimental conditions. The fish were then brought to a respirometer. The exposition time prior to the beginning of experiment was 1,5-2 hours. The fish exhibited no excitement at start time. Temperature during acclimation and experiments was similar to that in the sea. Data on the experimental material are given in Table 1.

Table 1. Experimental material (1991-1994)

Species	Data	n	l, cm	Stage of maturity	t^0 C	Duration of experiment
T. mediterraneus	09-10. 1991	11	13,5-15,5	VI-II*	18-20	45-55 min
ponticus	05-06. 1992	12	11,0-17,5	III-IV	15-16	45-55 min
D. annularis	09-10. 1991	5	12,5-18,5	VI-II	18-20	50-65 min
	05-06. 1992	6	18,0-20,0	III-IV	15-16	50-65 min
S. porcus	09-10.1991	10	14,0-19,0	VI-II	18-20	65-150 min
	05-06. 1992	10	14,0-23,0	III-IV	15-17	65-150 min
	05-06. 1994	10	17,5-24,5	II-III	15-17	14 h
	06. 1994	8	16,5-24,0	III-IV	19-22	14 h

* Stage of maturity was determined according to [4].
All experiments were carried out on one fish specimen only.

The studies were conducted at the aquarial laboratory of the Karadag Branch of IBSS during 1991-1994. After the above two series of experiments had been completed, experiments on the long-term effects of hypoxic conditions on two Black Sea fishes, the golden grey mullet *Liza aurata* Risso and the round goby Neogobius *melanostomus Pallas*, were conducted in 1995 at IBSS .

Table 2. Experiment material (1995).

Species	Data	n	l, cm	Stage of maturity	t^0 C	Duration of experiment
Liza aurata Risso	08-10. 1995	6	17,0-20,0	juvenile	13-16	50 days
Gobius melanostomus Pallas	08-10. 1995	8	12,0-17,0	II	18-20	22 days

Fishing, transportation and acclimation techniques during the experiment were the same as mentioned above. The DO was decreased from 9.2 to 2.6 mg·l^{-1}. (35-25 % of the oxygen saturation). Scorpionfish, annular bream and horse mackerel were kept in a 2.6 l container-respirometer of special design (Figure 1), and the golden grey mullet and goby in a common pressurized 45 l aquarium .

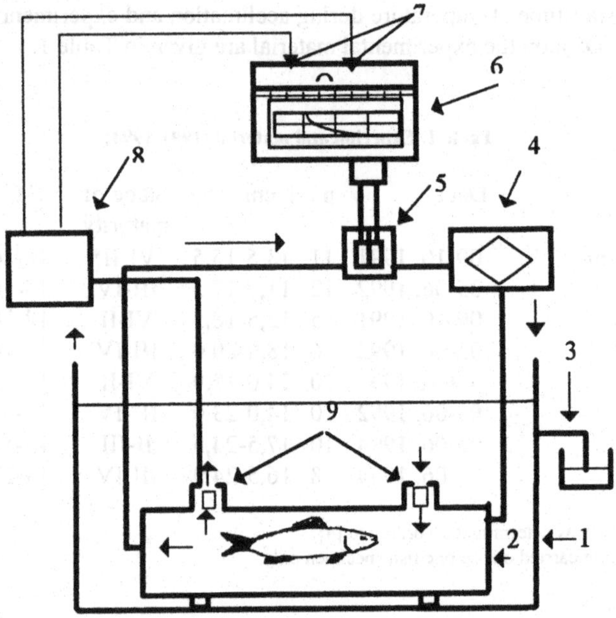

Figure 1. Respirometer for hypoxia experiments. 1. Thermostatic bath. 2. Camera for fishes. 3. Sample collector. 4. Peristaltic pump. 5. Oxygen electrodes (cells). 6. Oxymeter. 7. Recorder and magnetic contact. 8. Air compressor (airlift). 9. Automatic valve (inlet and outlet).

In physiological descriptions of the fish the following metabolic parameters were used: 1)consumption of oxygen by the whole organism; 2) excretion of ammonium nitrogen;3) atomic ratio between consumed oxygen and excretion of nitrogen, O:N or the ammonium quotient, AQ. For examination of the condition of fish placed in aquarium for long time, we made histological assay of their muscular tissue. Oxygen content was determined by polarographic method [7], ammonium nitrogen content according to Solorzano [11] and ammonium quotient by the equation AQ= (O:N) K, where K is the ratio between atomic weight of O and N mg·l^{-1} [12]. The technique has been described in detail elsewhere [7,8]. For histological assay we took parts of lateral white muscle; histological preparations were made in conformity with traditional procedures [13].

2.2.THE STUDY OF FOOD SUPPLY

In this context we understand food supply as the ratio between food consumed and that essential for maintaining, normal vital activity [4, 14].It is almost impossible to directly determine this ratio in nature, therefore we assume the consumption effect to be the adequate relevant characteristic. For several decades we have been using estimate of the fatness attained by fish population by the end of feeding season as a reliable index of the effect of food consumption [4, 6, 14]. For estimating the impact of anthropogenic factors on fat content of fish, we used data obtained for sprat predominantly from the north-western Black Sea where the main stock concentrates. Obtained since 1960, data about sprat completing the fat accumulation period (June- July) have been used for the analysis. Methods and techniques for collection of the material and data processing have been described in detail [6].

3. Results and discussion

3.1. OXYGEN DEFICIENCY (HYPOXIA)

Data from the study of short-term (fast) hypoxia in Black Sea fishes are given in Figures. 2-4 [7].

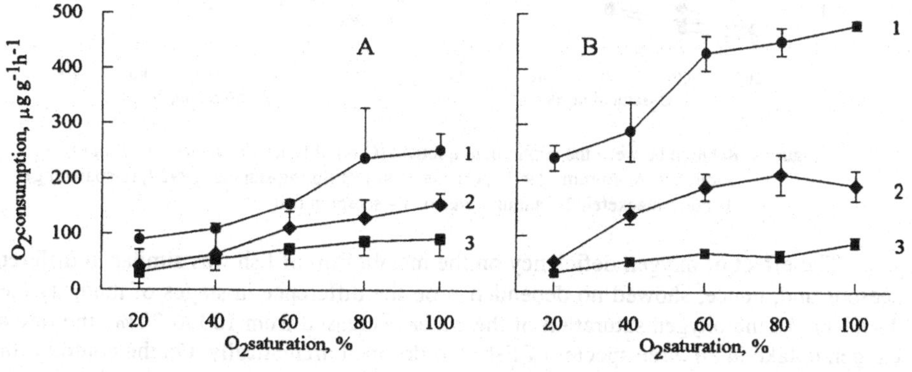

Figure 2. Relation between O_2 consumption of Black Sea fishes and % of O_2 saturation in the water.
A-Autumn 1991 (post spawning). B -Spring-summer (pre-spawning).
1 - horse mackerel. 2 - annular bream. 3 - scorpion fish.

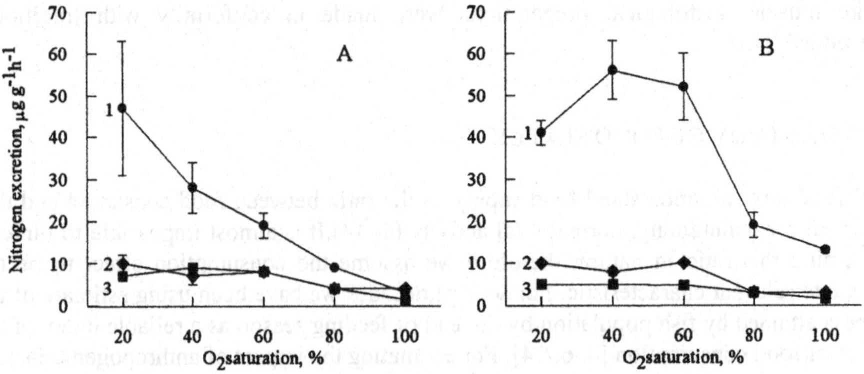

Figure 3. Relation between nitrogen excretion of Black Sea fishes and % of O_2 saturation in the water. A-Autumn 1991 (post spawning). B- Spring-summer 1992 (pre-spawning). 1- horse mackerel.. 2 -annular bream.. 3 - scorpion fish.

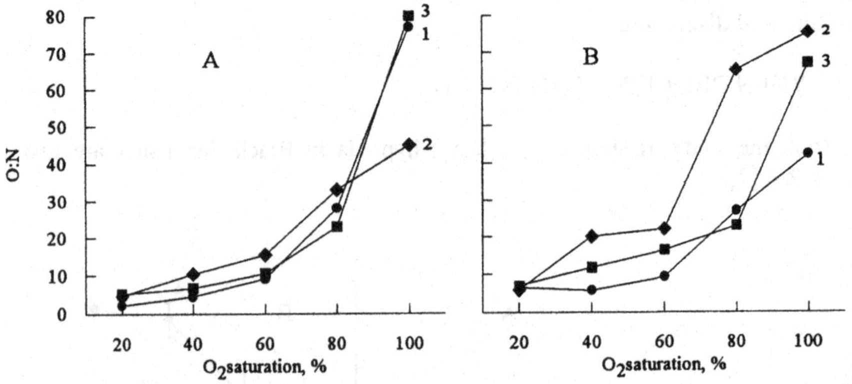

Figure 4. Relation between the ammonium quotient (O:N) of Black Sea fishes and % of O_2 saturation A- Autumn 1991 (post spawning). B-Spring-summer 1992 (pre-spawning). 1- horse mackerel. 2 – annular bream. 3 – scorpion fish.

The effect of oxygen deficiency on the metabolism of fish was similar in different seasons and, hence, showed no dependence on the difference in stages of maturity (see Table 1). As the oxygen saturation of the water decreased from 100 to 20% , the rate of oxygen uptake in all three species of fish also dropped dramatically. On the contrary, the rate of nitrogen excretion sharply rose. There was also a decline from 40-80 to 2-7 in the value of AQ.

It is known that AQ >30 indicates predominantly protein - free energy metabolism at which lipids (nonesterified fatty acids, triacylglycerols) and carbohydrates (glycogen, glucose) are the basic biochemical substrates providing catabolism [8]. At 20 < AQ <30 both protein- free substrates and proteins are equally used, at AQ < 20 proteins along with lipids and free amino acids become prevailing substrates. For ammonium quotient less than 8.76, proteins and nitrogenous compounds are involved in anaerobic metabolism because the oxygen consumed does not suffice to support catabolic processes. Thus, we may observe the switch from lipids and carbohydrates to proteins and nitrogen employment that takes place in the metabolism of fish under hypoxic environmental conditions (Figure 5).

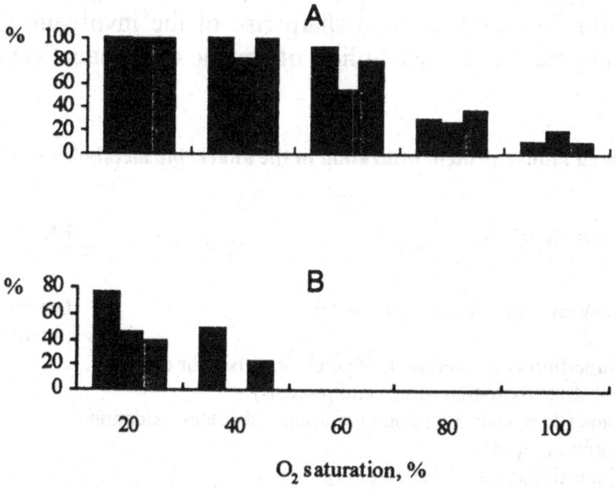

Figure 5. Relation between protein utilization by Black Sea fishes and % of O_2 saturation in the water of respirometer A. Part of O_2 used for protein oxidation. B. Part of protein used in energy metabolism anaerobically. Shaded bars - horse mackerel. Spot bars - annular bream. Empty bars - scorpion fish

The sharp increase in the involvement of nitrogenous substrates, in amounts both relative and absolute, is of special interest. Apparently, such intensive mobilization of protein and nitrogenous substrates is an adaptation developed in response to the stress provoked by the drastically altered environment.

Not only fish manifest this sort of response. In experiments performed in the Department of Physiology of Animals, IBSS, similar reaction was found in bottom (the shrimp *Palaemon elegans*) and planktonic (*Moina micrura and Calanus euxinus*)

crustaceans [15, 16, 17]. However, planktonic crustaceans did not display absolute increase of the excretion of nitrogen [16].

By now extensive knowledge has been gained and reported about the leading role of nitrogenous substrates in the metabolism of water organisms (fish and invertebrates) under oxygen deficiency [5, 18-22] and about the possible involvement of proteins of peptides and free amino acids in the anaerobic metabolism [23, 24].

Over the World Ocean peculiar zones with depths varying from 100 to 700 m were found in which oxygen deficiency was as grave as 0.3-1.0 ml O_2 l^{-1}[25-27]. The Arabian Sea and the north-east Pacific ocean near California are the illustrative examples. Despite very low DO estimates, these waters exhibit surprisingly rich diversity of life - planktonic crustaceans, squids and fish, primarily lightfishes (*Gonostomidae*) and lanternfishes (*Myctophidae).* In the Black Sea, in the interface between redox zone and H_2S-zone where oxygen content drops dramatically (0.3 - 0.5 mg·l^{-1}) the numbers of planktonic crustaceans is also unusually high [1].

Using original data and relevant literature (see review [3]), we have identified those hypoxic conditions which cause a sharp rise of the involvement of protein and nitrogen catabolism in the energy metabolism of aquatic organisms (Table 3.).

Table 3. Factors which induce protein utilization in the anaerobic metabolism of aquatic animals.

N	Type of hypoxia	Limiting factors	Effect	Referen ce
1.	External hypoxia	Low level of oxygen in the water	Deficiency of O_2 in tissues	[3]
2.	Functional hypoxia	Superfluous consumption of O_2 by organism for energy substrate oxidation (lipids and proteins)	- « -	[28]
3.	Overfeeding	Superfluous consumption O_2 nutrient substrates oxidation (protein, lipids)	- « -	[29, 30]
4.	Stress	Superfluous consumption of O_2	- « -	[31]
5.	Toxicoses	Difficulties for oxygen delivery to tissues, cells and mitochondria (eutrophication, heavy metals, pesticides, oil products)	- « -	[32]

Oxygen deficiency deserves special consideration in relation to the effect of toxicants. Many toxicants hamper the supply and access of oxygen to tissues [32] that evokes "inner hypoxia" of the organism at which the metabolism of cellular and subcellular structures (e.g. mitochondria, membranes) are disturbed. Then it is again proteins and nitrogen catabolism products, the essential substrates of energy metabolism that resist the ruinous impact. It has been shown [33] that under the impact of heavy metals, organic toxicants, pesticides and oil products the activity of malatdeghydrogenase, the enzyme responsible for the decay of nitrogenous compounds, sharply rises in mussels.

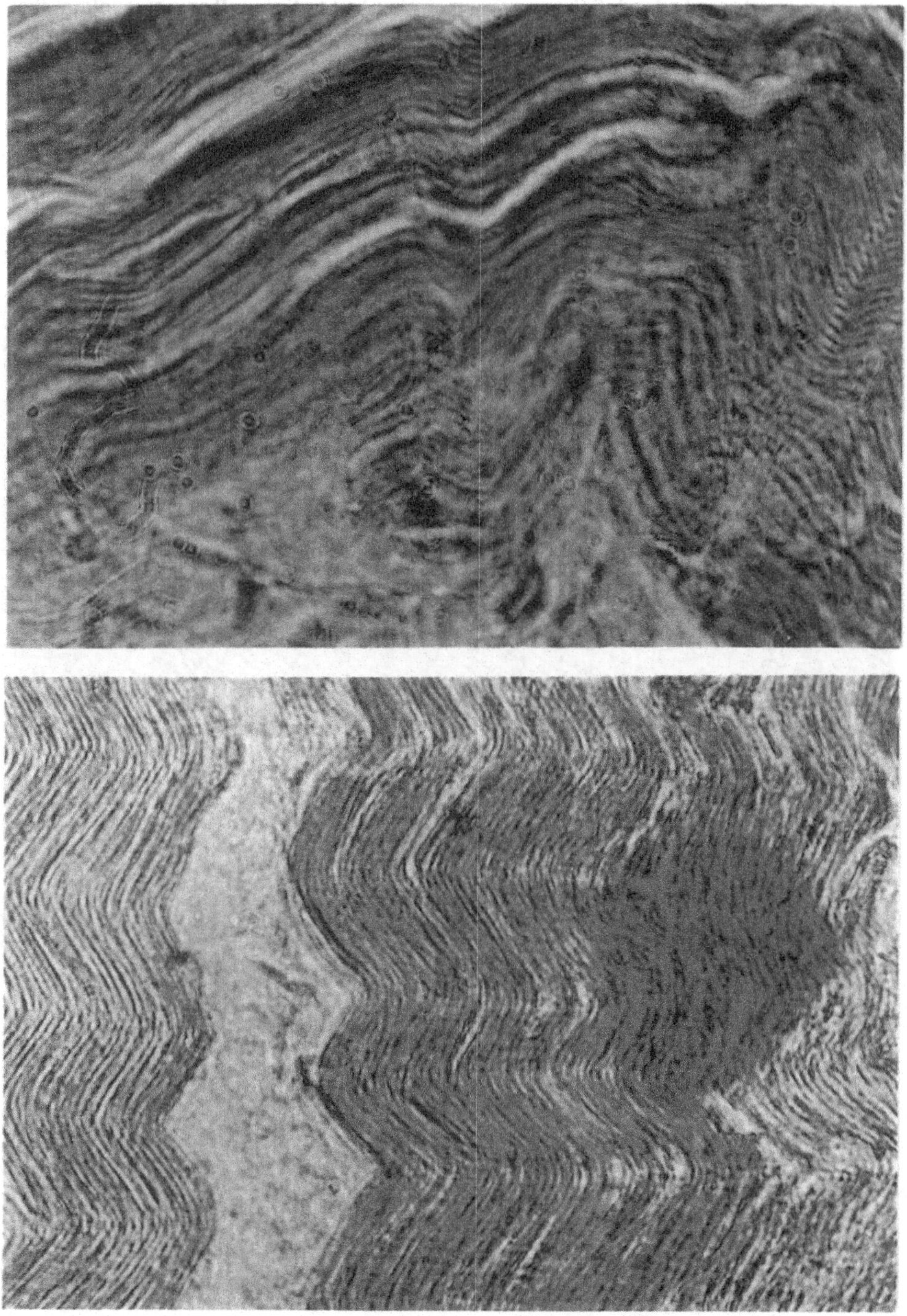

Figure 6. Influence of hypoxia on the muscular tissue structure in Black Sea fishes:
A Goby: severe deformatıon. B. Golden grey mullet: temperate deformation.

Apparently , it is the deficiency of oxygen supplied to tissues that may explain the muscle fiber disintegration frequently found in recent years in sturgeons of the Volga and Caspian region and in some other seas and freshwater bodies [34]. We reproduced this phenomenon in experiments with reduced DO (Figure 6) and made a comparison with normal muscle structure (Figure 7)

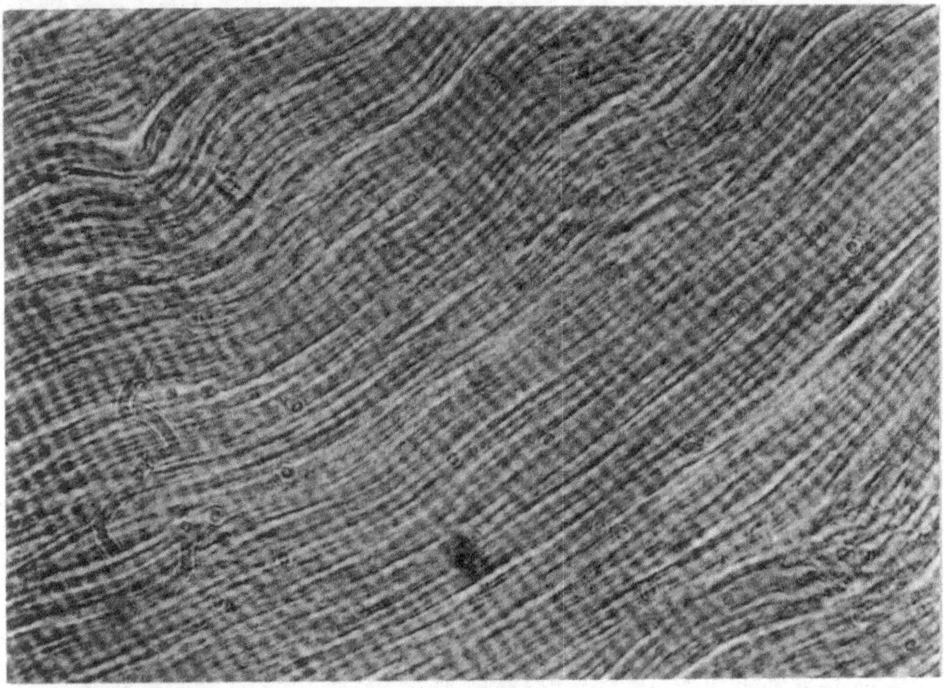

Figure 7. Control fishes: unchanged muscular tissue.

Figure 8 shows the hypothetical mechanism of the muscle disintegration in fish. The background history of this hypothesis is in [3], [5] and [32].

In general outline it is as follows: the shortage of oxygen supply to tissues increases the concentration of peroxides (free radicals) which results in disruption of the structure and functioning of the double phospholipid layer of cellular membrane [32]. In turn, this evokes a "collapse" of the protein base of cellular membrane [32]. Furthermore, lactate, accumulated in the tissues, shifts the pH in " acid direction" which

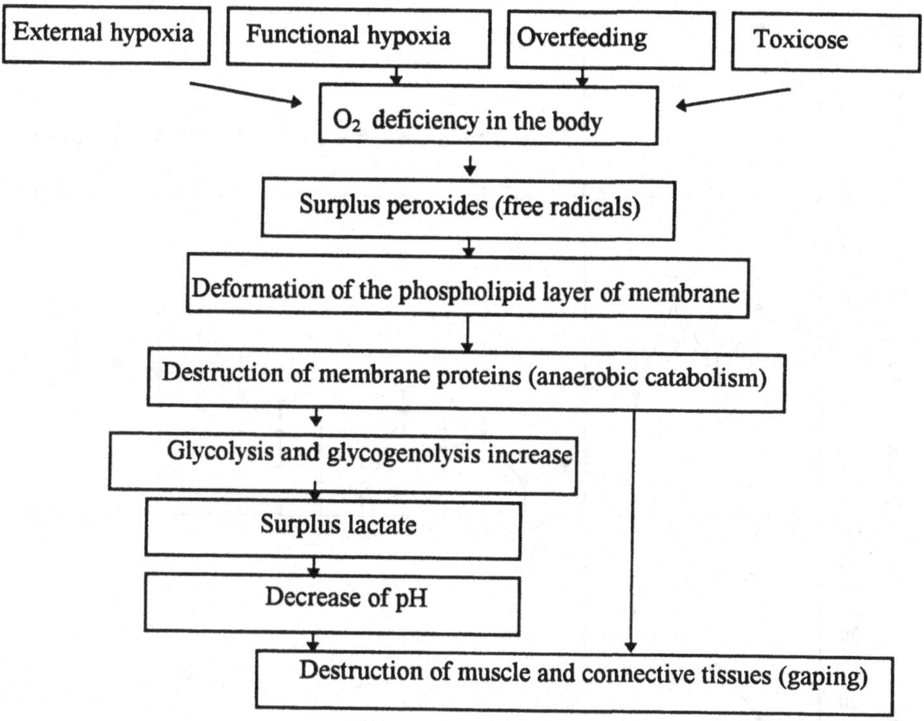

Figure 8. Hypothetical scheme of the destruction of muscles during hypoxia.

adds to the destruction of tissue proteins [35]. The menacing phenomenon of muscle disintegration is a distinct indicator of unhealthy state of the fish population in a water body [3]. Regular observations coupled with histological assay of muscular tissue may form the basis of a monitoring of the condition of fish stock in the polluted environment. Experiments on the long- term oxygen deficiency in fish were performed in order to understand the part played by nitrogen metabolism; their design simulated persistent hypoxic conditions in nature. Results obtained show (Figure 9) that long-term oxygen deficiency considerably reduces the rate of energy (oxygen uptake) and nitrogenous (nitrogen excretion) metabolism in fish.

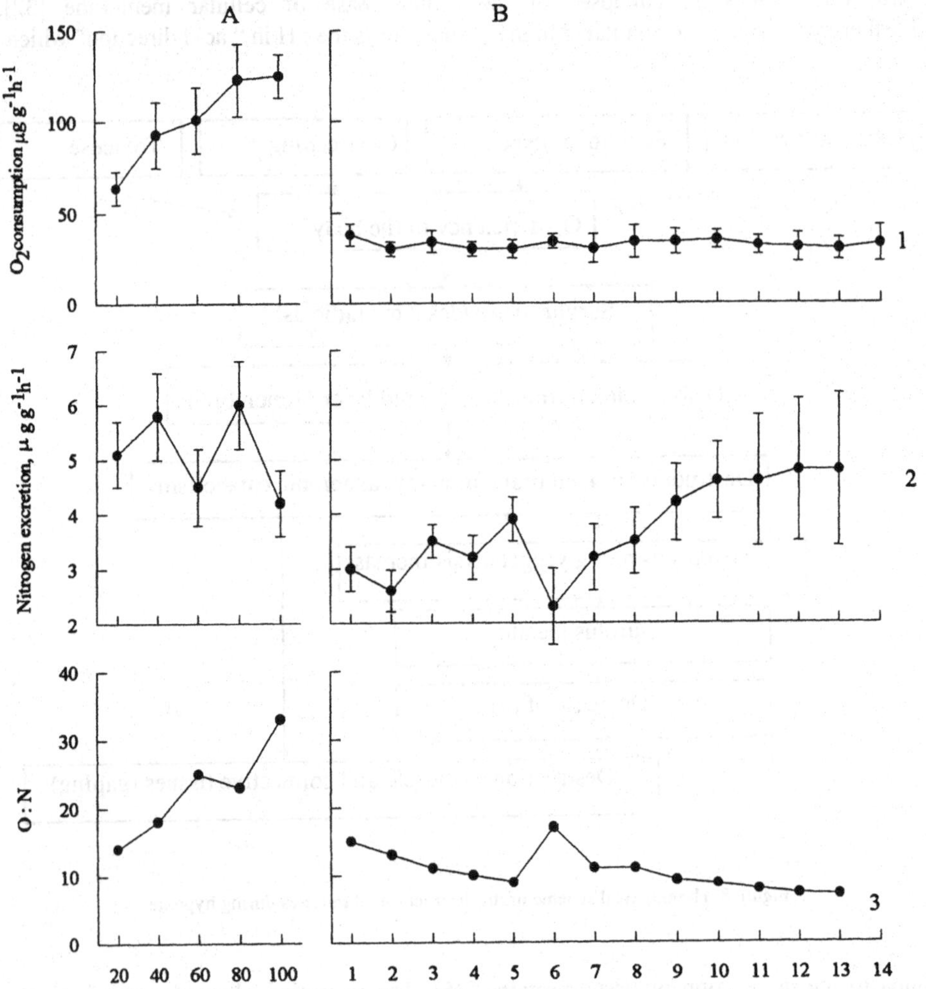

Figure 9. The energy and nitrogen metabolism of the Scorpaena porcus at long-term hypoxia. [8]
1. Oxygen consumption. 2. Nitrogen excretion. 3. Ammonium quotient (O : N).
A. Short-term hypoxia. B. Long-term hypoxia (20% oxygen saturation).

Simultaneously, the involvement of proteins and nitrogenous compounds in the energy metabolism sharply rises. Actually, it is protein and nitrogen metabolism that maintain energy processes in fish surviving through the hypoxia. The experiment has also proved that under persistent oxygen deficiency some portion of proteins was used anaerobically; estimates of the ammonium quotient were less than 8 for most of the time.

We find it very important that proteins and products of their catabolism allow fish to live under persistent hypoxic conditions. Similarly, the importance of nitrogenous products has been proven through a series of experiments carried out at our Department on the long-term oxygen deficiency in planktonic crustaceans *Moina micrura* and *Calanus euxinus* [17].

3.2. FOOD SUPPLY

Figure 10 presents data about the peak of fat accumulation in sprat [6, 9] (June-July) and the abundance of the Black Sea stock estimated through acoustic surveys[6].

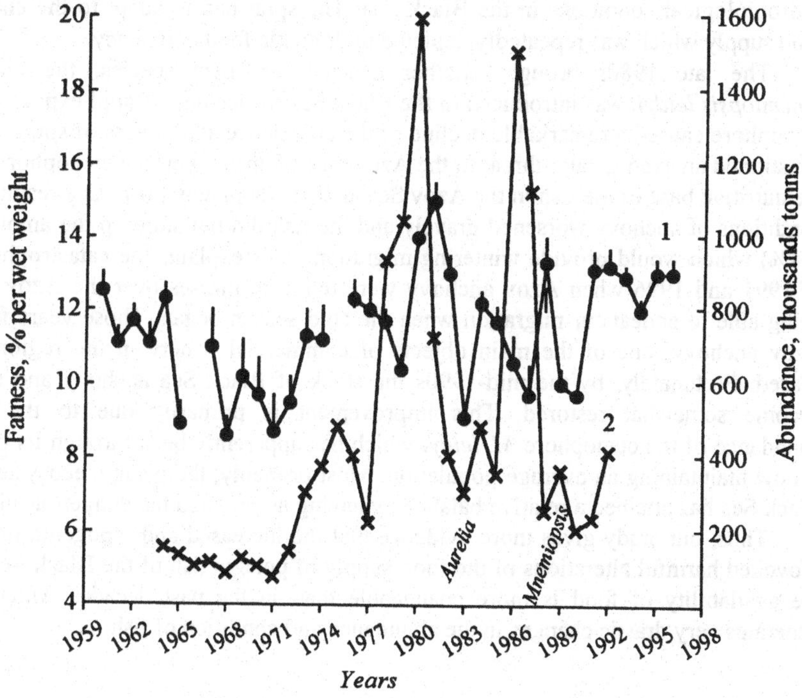

Figure 10. Average fatness(1) and abundance (2) of the whole stock of Black Sea sprat [6, 9].

It follows from these results that since the 1960s till the early1970s the content of fat in sprat was decreasing, probably because of climatic changes. Beginning from the early 1970s the fatness was increasing rapidly owing to growing nutrient-rich river inflow to the sea related to the development of industry and the scope of fertilizers applied in agriculture. The increased content of phosphates, nitrates and other nutrients had two interrelated effects. Firstly, it brought about eutrophication the shelf, especially in the north-western part of the Black Sea [1, 2]. Secondly, it stimulated productivity in the pelagic area and the nutritive base of plankton-eating fish such as sprat and anchovy. The stock of the two fish had rapidly grown in numbers (unfortunately, we do not have data about the fatness of Black Sea anchovy).

However, the generous enrichment of Black Sea pelagic ecosystem with nutrients was beneficial not only for planktivorous fish but for their food competitor, the predatory medusa *Aurelia aurita*, as well. Although an outbreak of *A. aurita* began after 1997, maximal abundance was reached in the first half of the 1980s [36]. The nutritive base of plankton-eating fish was destroyed [6, 9]. The fatness and numbers of sprat decreased dramatically as well as the abundance of anchovy [37]. During the 1980s, the content of fat and the numbers of the sprat fluctuated, which was due to unsteady hydrobiological conditions in the Black Sea. The sprat had to adapt to the changeable food supply which was repeatedly depleted to dramatic food deficiency.

The late 1980s brought another menace: an alien species, the ctenophore *Mnemiopsis leidyi*, was introduced to the Black Sea by accident. Rapid expansion of the ctenophore caused a remarkable decline of the abundance and fatness of sprat. *M. leidyi* constituted an even greater threat to the Azov race of anchovy as the ctenophores ate out the nutritive base of the fish in the Azov Sea almost completely [38]. As a result, feeding conditions of anchovy worsened gravely and the fish did not store up the amount of fat (14%) which would provide wintering migration. This explains the catastrophic events of 1991 and 1996 when Azov anchovy were dying in masses over the Azov Sea, not being able to endeavour migration when the cold season began. Those years fishing of Azov anchovy, one of the main objects of commercial fishery in the region, almost ceased. Fortunately, by the mid-1990s the stock of Black Sea anchovy and sprat has become somewhat restored. This improvement is primarily due to the reduced abundance of the ctenophore *M. leidyi* which has apparently been through its peak, and is now maintaining an average population. Most probably, the pelagic ecosystem of the Black Sea has attained a relative balance by having assimilated the dangerous alien.

Thus, our study gives more evidence that the increased anthropogenic impact has provoked harmful alterations of the food supply of pelagic fish of the Black Sea. Today the availability of food is more changeable than in the past decades which in turn generates very drastic changes in the abundance and condition of fish.

References

1. Vinogradov, M.E., Sapozhnikov, V.V. and Shushkina, E.A. (1992) *Black Sea Ecosystem*, Nauka Publishing House, Mockow. (In Russ).
2. Zaitsev, Yu. P. (1992) The ecology of the Black Sea shelf along the Ukrainian coastline (a review), *Gidrobiol. Zhurn.* **28**, 3-18.
3. Shulman, G. E., Abolmasova, G. I. and Stolbov, A. Ya. (1993) The use of protein in energy metabolism of water animals, *Uspekhi Sovremennoi Biologii*, **113**, 576-586. (In Russ.)
4. Shulman, G.E. (1972) *Physiological and Biochemical Features of Annual Cycles of Fish*, Pishcheprom Publishing House, Moscow. (In Russ.).
5. Shulman, G. E., Stolbov, A. Ya. and Abolmasova, G.I. (1993) The possible cause of muscle destruction in sturgeons, *Rybnoe Khozaistvo*. **4**, 6-29. (In Russ.).
6. Shulman, G.E., Chaschin, A.K., Minyuk, G.S., Shchepkin, V.Ya., Nikolsky, V.N., Dobrovolov, I.S., Dobrovolova, S.G. and Zhigunenko, A.S. (1994) Long-term monitoring of the condition of Black Sea sprat, *Doklady Akademii Nauk* **335**, 124-126. (In Russ.).
7. Stolbov, A. Ya., Stavitskaya, E.N. and Shulman, G.E. (1995) Oxygen consumption and nitrogen excretion in Black Sea fishes of different ecological specialization, *Gydrobiol .Zhurn.* **31**, 73-78. (In Russ.).
8. Stolbov A.Ya., Stavitskaya, E.N. and Shulman G.E. (1997) The dynamics of oxygen consumption and nitrogen excretion in Black Sea scorpion fish at short-term and long-term hypoxia, *Doklady Akademii Nauk*, **356**, 569-572. (In Russ.).
9. Minyuk, G.S., Shulman, G.E., Shchepkin, V.Ya. and Yuneva, T.V. (1997) *Black Sea Sprat*, EKOSI-Gydrophysika, Sevastopol. (In Russ).
10. *Determinative Plates and Description of Vertebrate Parasites of the Black and Azov Seas* (1975), Naukova Dumka Publishing House, Kiev (In Russ.)
11. Solorzano, L. (1969) Determination of ammonia in natural waters by phenolhypochlorite method, *Limnol. Oceanogr.* **14**, 799-801.
12. Prosser, L. and Brown, F. (1967) *Comparative Physiology of Animals*, Mir, Moscow. (In Russ).
13. Romeis, B. (1953) *Microscopic Techniques*, Inostrannaya literatura Publishing House, Moscow. (In Russ.).
14. Shulman, G.E. (1974) *Life Cycles of Fish. Physiology and Biochemistry*, John Willey and Sons, N.Y.-Toronto.
15. Stolbov A.Ya., Stavitskaya, E.N. and Shulman, G.E. (1997) The metabolism of Black Sea shrimp at short-term hypoxia, *Doklady Akademii Nauk*, **356**, 141-142. (In Russ.).
16. Svetlichny, L.S., Yuneva, T.V., Shulman, G.E. and Houseman, J.A. (1994) The use of protein in energy metabolism of *Cladocera crustacea Moina micrura* at different oxygen content in water, *Doklady Akademii Nauk* **337**, 428-430. (In Russ.).
17. Svetlichny, L.S., Gubareva, E.S. and Arashkevich, E.G. (in press) Physiological and behavioral response to hypoxia in active and diapausing copepodites stage V of *Calanus euxinus*, *Arch. Hydrobiol. Adv. Limn.*
18. Kutty, M.N. (1972) Respiratory quotient and ammonia excretion in *Tilapia mossambica*, *Mar. Biol.* **16**, 126-133.
19. Kutty, M.N. and Mohamed, P.M. (1975) Metabolic adaptations of mullet *Rhinomugil corsula* with special reference to energy utilization, *Aquaculture* No 5, 253-270.
20. Mathur, G.B. (1967) Anaerobic respiration in cyprinoid fish *Rasbora daniconius*, *Nature*, **214**, 318-319.
21. Thillart, G. and Kesbeke, F. (1978) Anaerobic production of carbon dioxide and ammonia by gold fish *Carassius auratus*, *Comp. Biochem. Physiol* **59 A**, 393-400.
22. Waarde, A. (1983) Aerobic and anaerobic ammonia production by fish, *Comp. Biochem. Physiol.* **74 B**, 675-684.
23. Hochachka, P.W. and Mustafa, T. (1972) Invertebrate facultative anaerobiosis, *Science* **178**, 1056-1060
24. Savina, M.V. (1992) *Mechanisms of Tissue Metabolism Adaptation in the Evolution of Vertebrates*, Nauka Publishing House, Sankt- Petersburg. (In Russ.).
25. Douglas, E.L., Friede, W.A. and Pickwell, G V. (1976) Fishes in oxygen-minimum zones: blood oxygenation characteristics, *Science* **191**, 957-959.
26. Kukharev, N.N., Rebik, S.T. and Trushin, Yu. K. (1988) On the feeding activity of schooling pelagic and demersal fishes from hypoxic zone of the western Arabian Sea, *Materials of All-Union Conference*

"Feeding of Marine Fishes and Utilization of Nutritive Base as Elements of Fisheries Prediction", Murmansk pp. 124-125. (In Russ).
27. Vinogradov, M.E. and Shushkina, E.A. (1989) The macroscale distribution of quantitative patterns of plankton in Pacific Ocean, *Okeanologia* **29**, 121-126. (In Russ.).
28. Lukyanenko,V..I. (1987) *Ecological Aspects of Ichthyotoxicology*, Agropromizdat, Moscow. (In Russ.).
29. Shulman, G..E., Abolmasova, G.I., Stolbov, A.Ya. (1992) On the utilization of protein in energy metabolism of epipelagic squids of the genus *Stenotheuthis*, *Doclady Akaademii Nauk*, **325**, 630-632. (In Russ.).
30. Sukumaran, N and Kutty, M.N. (1977) Oxygen consumption and ammonia excretion in the cat fish with special reference to swimming speed and ambient oxygen, *Proc. Indian Acad. Sci. B.* **86**, 195-206.
31. Klyashtorin, L.B. (1982) *Aquatic Respiration and Oxygen Requirements of Fish*, Legkaya i Pishchevaya Promyshlennost, Moscow. (In Russ.).
32. Connell, D.W. and Miller, G.J. (1984) *Chemistry and ecotoxicology of pollution*, John Willey and Sons, N.Y.
33. Goromosova, S.A. and Shapiro, A.Z. (1984) *The Biochemistry of Energy Metabolism of Mussels*, Legkaya i Pishchevaya Promyshlennost, Moscow. (In Russ).
34. Altufiev, Yu.V., Romanov, A.A. and Sheveleva, N.N. (1992) Hystopathology of cross-striated muscle tissue and liver in Caspian stugerons, *Vopr. Ichthyology* **32**, 157-171 (In Russ.).
35. Hochachka, P.V., Somero, G.N. (1984) *Biochemical Adaptation*, Princeton University Press, Princeton, New Jersey.
36. Anninsky, B.E. (1988) Physiological requirements and real possibilities of *Aurelia aurita (l)*. Nutrition under conditions of the Black Sea, *Ecologya Morya* **29**, 38-45 (In Russ.)
37. Domashenko, G.P., Mikhailyuk, A.N., Chashchin, A.K., Shlyakhov, V.A. and Yuryev, G.S. (1985) Present condition of industrial stocks of anchovy, sprat and whiting in Black Sea. *Oceanological and Fisheries Researches of the Black Sea*, Agropromizdat, Mockow, pp. 97-100 (In Russ.).
38. Volovik, S.P., Mirzoyan, Z.A. and Studenikina, E.I. (1996) Assessment of consequences of the ctenophore *Mnemiopsis leidyi* introduction in the Azov Sea, *Rybnoye Khozyistvo* **1**, 48-51 (In Russ.).

INTERNATIONAL MUSSEL WATCH (UNESCO/IOC) IN THE BLACK SEA: A PILOT STUDY FOR BIOLOGICAL EFFECTS AND CONTAMINANT RESIDUES

M. N. MOORE[1], D. M. LOWE[1], R. J. WEDDERBURN[1,2], T. WADE[3], G. BALASHOV[4], H. BÜYÜKGÜNGÖR[5], Y. DAUROVA,[6], Y. DENGA.[7], E. KOSTYLEV[7], P. MIHNEA[8], S. MONCHEVA[9], S. TABAGARI[10], C. CIOCAN[8], H. ÖZKOC[5], M. H. DEPLEDGE[2]

[1]*Plymouth Marine Laboratory (CCMS-NERC), Citadel Hill, The Hoe, Plymouth, PL1 2PB, U.K.*

[2]*Plymouth Environmental Research Centre, University of Plymouth, Drake Circus, Plymouth, PL4 8AA, U.K.*

[3]*Geochemical & Environmental Research Group, Texas A&M University 833, Graham Road, College Station, Texas 77845, USA.*

[4]*Ministry Of Environment, Research institute of Shipping, 1, Slaveikov str., 9001 Varna, Bulgaria.*

[5]*19 Mayis University, Engineering Faculty, 55139 Kurupelit, Samsun, Turkey.*

[6]*Centre for Hydrometeorology and Environmental Monitoring, 25, Sevastopolskaya, Sochi, 354000, Russian Federation.*

[7]*Ukrainian Scientific Centre of the Ecology of Sea, 89, Frantsuzki Blvd, Odessa, 270009, Ukraine.*

[8]*Romanian Marine Research Inst, B-dul Mamaia, 300, 8700, Constanta, Romania.*

[9]*Institute of Oceanology-BAS, P.O. Box 152, 9000 Varna, Bulgaria.*

[10]*Sanitary and Hygienic Scientific Research Institute, P.O. Box 172, 380079, Tbilisi, Georgia.*

Abstract

The Black Sea is under increasing stress as a result of inputs of contaminants and eutrophying discharges. This study was an attempt to implement a Pilot "Mussel Watch" programme (supported by UNESCO-IOC, International Mussel Watch) to assess the health of mussel populations in the Black Sea. By utilising mussels for biological monitoring those areas suffering poor environmental quality can be identified. A simple non-injurious test using blood cells as a biological marker (lysosomal integrity/neutral red retention) of pollutant effect was deployed. Contaminants (PAHs, PCBs, selected pesticides and trace metals) were also measured in the tissues of mussels from some sites (Bulgaria and Ukraine). This data was

S. Beşiktepe et al. (eds.),
Environmental Degradation of the Black Sea: Challenges and Remedies, 273–289.

supplemented with the results of a rapid source inventory of land based discharges (World Health Organisation) into the Black Sea.

The results of the "Mussel Watch Pilot Study" clearly showed that there were harmful effects at sites where there were known anthropogenic inputs, as identified in a WHO Inventory. Samples taken from recreational sites and sites well removed from significant anthropogenic influences showed no evidence of pathological perturbation. The effects measurement was based on the intracellular retention of the dye neutral red. Retention time of this dye is markedly reduced when the mussels are exposed to toxic chemicals. The results showed a strong correlation between the biological effect, Biochemical Oxygen Demand (BOD) and Total Suspended Sediment (TSS). Contaminant data for polycyclic aromatic hydrocarbons was available: this showed a direct correlational trend with BOD and TSS; but this was not significant due to the limited number of samples analysed. Chemical data was only available from the sampling sites in Russia, Ukraine and Bulgaria.

The conclusions are that the biological effects data clearly identify sites where there are significant harmful impacts on the mussels. These sites were all subject to significant anthropogenic inputs as evidenced by the measurements of BOD, TSS and chemical contaminants. Despite the problems encountered in obtaining comprehensive chemical data it is recommended that the Mussel Watch should be continued using the neutral red retention procedure as an indicator of harmful effect, coupled with chemical contaminant measurements. Efforts to develop the analytical capabilities in the region should continue to be supported. This strategy will help to develop the regional, national and international infrastructures for assessment of environmental impact to the Black Sea.

1. Introduction and Background

Much of the waste generated by human activities finds its way into the oceans where some of it may present a threat to marine life and possibly to man as a consumer of seafood [1, 2]. Considerable past effort in environmental monitoring has focused on the determination of residue levels. Unfortunately, a large gap exists in our ability either to quantify the exposure to toxic chemicals in the environment or to assess the biological significance of such exposure [3]. Exposure cannot always be quantified by measuring the concentration of contaminants in tissues; since many toxic chemicals are metabolised, especially in fish. However, this problem is considerably reduced in molluscs which have limited capacity for biotransforming organic chemicals and thus tend to reflect exposure more closely. Even then, measurement of levels at one point in time tells us little about the pattern of exposure that resulted in those levels. In addition, the relationship between tissue concentration and toxic response is complex, as is assessing the significance of exposure to complex mixtures where there may be possible interactions that can invalidate predictions that are based on the toxicity of individual chemicals.

An approach to the question of whether or not marine organisms and ecosystems are being endangered has involved the development and deployment of **indices of biological effect and exposure (known as "biomarkers")** as early warning systems of adverse environmental change [2, 4, 5]. Biomarkers can demonstrate that environmental chemicals have entered an organism, reached sites of toxic action, and are exerting harmful effects on the organism [3]. In fact, the organisms are functioning as integrators of exposure, accounting for abiotic and physiological factors that can modulate the dose of chemical taken up. Biomarkers can be used to quantify exposure to toxic chemicals and to detect **distress signals** from the organisms. Such methods are being used in combination with analytical chemistry on a rapidly increasing basis and on a world-wide scale [4].

The organisms of choice for this type of environmental monitoring have frequently been sedentary filter-feeding molluscs such as mussels and oysters. These animals accumulate chemical contaminants both from the seawater and particulate food material filtered from the water. The tissue concentrations of many environmental xenobiotics can reach very high levels, thus making them useful tools for chemical monitoring [6, 7]. Mussels in particular appear to be relatively tolerant to many metals and organic xenobiotics. This tolerance, however, does not mean that the animals are unresponsive; in fact there is considerable evidence for pathological reactions to even low concentrations of contaminants. Such pathological reactions have been described at all levels of biological organisation ranging from the molecular to the physiology of the whole animal [8-11].

For instance, at the molecular level 7-ethoxyresorufin-*o*-deethylase (EROD/CYP 1A1), metallothionein and inhibition of acetylcholine esterase can provide relatively specific information about exposure to several major classes of chemical contaminants, while at the cellular level the lysosomal system have been identified as a particular target for the toxic effects of many contaminants which they can readily accumulate. Changes in this latter system are useful since they can contribute directly to pathology [12], so alterations in lysosomes have been especially useful in the identification of adverse environmental impact, since many of the tissues in molluscs and other invertebrates, as well as fish liver are extremely rich in lysosomes [8, 9, 13-15]. In the Black Sea region, marine laboratories generally have the facility to carry out standard lethal toxicity tests, with a more restricted capability to perform ecophysiological and other sublethal tests [16]. While the deployment of biomarker tests (biochemical and cellular) is not widespread in the region, this capability is being actively developed with the assistance of the UNESCO-IOC.

In this paper the rationale is developed for an integrated programme which will allow scientists to measure adverse biological effects and contaminant levels in the Black Sea using biomarkers and sentinel organisms. In designing this programme, we have drawn on the recommendations for monitoring and associated quality assurance work developed by the International Mussel Watch Programme/Caribbean-American Phase (UNESCO-IOC, UNEP, US-NOAA), IMO/FAO/UNESCO-IOC/WMO/WHO/IAEA /UNEP [17, 18], and ICES [19].

This strategy can incorporate biological effects measurements at all organisational levels and is sufficiently flexible to accomodate new diagnostic tests for both exposure and pathology as future additions.

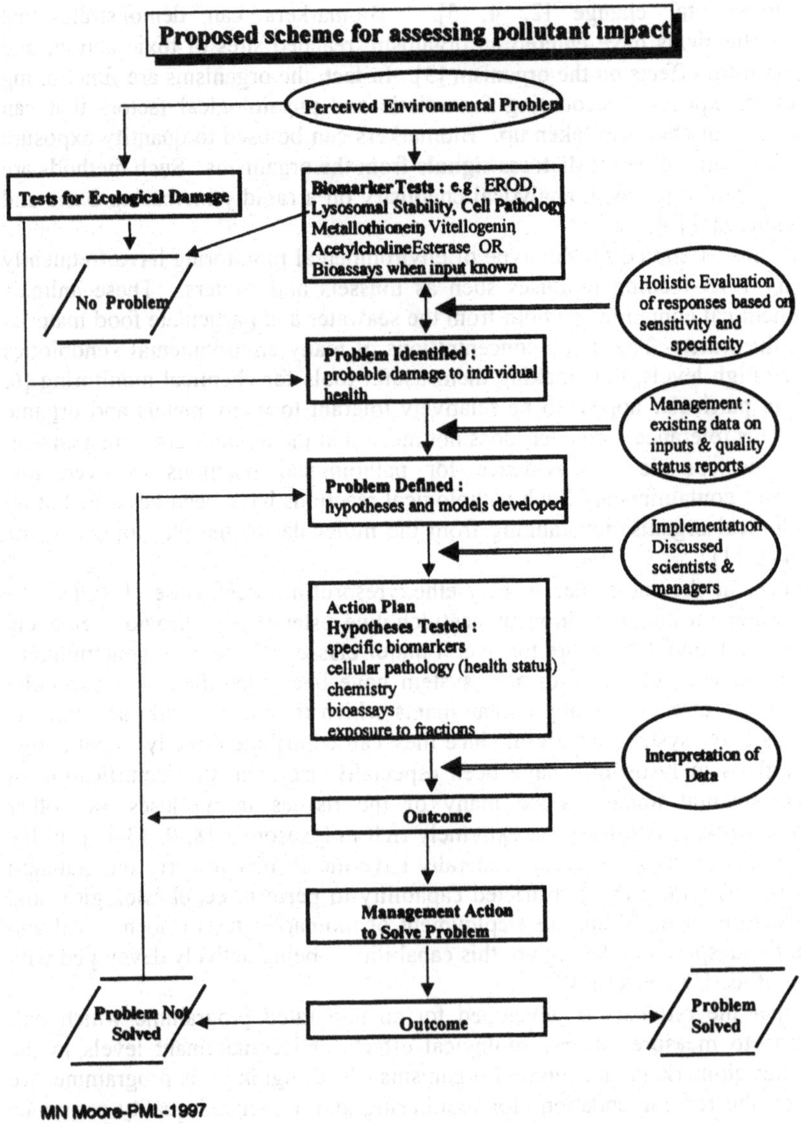

Fig. 1 Proposed environmental management tool for using biological effect monitoring to resolve impact assessment problems.

The key feature of this approach is that biological effects measurements are carried out prior to any programme of intensive chemical monitoring. If the initial biological effects measurements identify a real problem then, in addition to more focussed biological effects measurements being carried out to further investigate the situation, the source of this problem can be investigated through chemical monitoring work. By working this way, the investigating institute focuses its limited resources to provide a better system for indentifying 'real problems' from a biological viewpoint rather than 'perceived problems' which are often based on simply the presence of chemical substances in the marine environment (Fig. 1).

Lessons learned from other international monitoring programmes
There has been a tendency to adopt and implement international monitoring programmes with very broad and ill-defined objectives and, as a consequence, the results arising from these programmes were frequently difficult to evaluate. With the exception of the most recent work in the North Sea, most of the cooperative international programmes were conducted along similar general lines and were not designed for specific sub-regional issues and characteristics. In most cases, very little effort was devoted to the design of monitoring by comparison with the effort devoted to the monitoring work.

Other reasons for the lack of a successful outcome of monitoring work have included the lack of adherence to agreed protocols for sampling, quality assurance (QA) work and data assessment procedures. Such protocols are specifically designed to ensure that data collected by different laboratories and countries are comparable and that there is agreement between participants at the outset of the monitoring programme on potentially contentious issues, e.g. "Do the data from individual participating laboratories conform to the required QA criteria and if not should the data be rejected from the set of data to be assessed?". If protocols are ignored, the aims of the monitoring programme are undermined and valuable resources are wasted. It is also clear that only during the data assessment stages were such deficiencies revealed. Had they been detected, and corrected, at an earlier stage in the monitoring programme it would have allowed the organisers to meet their objectives and enhanced the cost-effectiveness of the programme.

Rationale for a "Black Sea Mussel Watch"
Sedentary filter feeders such as bivalve molluscs, accumulate chemical contaminants directly from the sea water or via particulate food filtered from the water column. The use of mussels, oysters and clams as monitors of anthropogenic pollution in the coastal environment is becoming increasingly established [20, 21].

The application of mussels as integrators of chemical contaminants in their tissue and shells [22] is the underlying concept of the UNESCO-IOC International Mussel Watch Project. This activity was formed in the early 1970's with the goals of assessing the extent and severity of contamination of the coastal waters of the world with respect to selected contaminants [23].

The residues of contaminants within the organisms are identified by means of analytical chemistry. However, although this indicates what types of contaminants are present in the environment, it does not provide information on their biological availability once within the organism, nor does it demonstrate any resultant toxicological effect.

Investigation of the endangered state of marine ecosystems has more recently involved the development of monitoring techniques utilising indices of biological response. These are known as biomarkers, which can identify the biological reaction occurring in an organism, as a result of contaminant exposure, and could act as "early warning distress signals" of adverse ecological change [5]. Biomarkers can demonstrate that chemicals have entered an organism, that they have reached sites of toxic action [3]; and, that they are instigating a pathological response [2, 5].

Pathological reactions to contaminants have been described at all levels of biological organisation ranging from the molecular and cellular level to the physiology of the whole animal [8, 9, 10, 11]. The sub-cellular lysosomal system in invertebrate and vertebrate cells has been confirmed as a target for the toxic action of xenobiotics in a number of different studies [2, 8, 24-30]. This is a result of the unique ability of lysosomes to accumulate a diverse range of trace metals and organic chemicals [2]. In phagocytic blood cells lysosomes have an important role as a crucial component in the cellular immune response. They can perform a controlled release of the acid hydrolyses they contain to break down, for example, unwanted cellular components and foreign material that has entered the animal. The accumulation of contaminants within lysosomes will induce damage in the lysosomal membrane [10], which will in turn affect the efficiency of intracellular digestion, protein turnover, and other aspects of vesicular traffic. Severe membrane damage may also result in the unscheduled release of lysosomal enzymes into the cytosol [10], where they will exert a degradative effect on other cellular systems and processes. Impairment of the lysosomal system is well documented as being directly responsible for many pathological effects [2] leading to impairment of the "health" of the whole organism. For instance, reduction in scope for growth has been shown to be functionally linked with impairment of the lysosomal system [31, 32]. Also, a number of laboratory studies have shown that lysosomal function and membrane stability are affected by exposure to organic [5, 28, 33] and inorganic substances [34, 35]. Molluscan digestive cells have been shown to undergo lysosomal destabilisation, after contaminant exposure, in a wide variety of studies [8, 10, 24, 27, 36]. The lysosomes of mussel blood cells (haemocytes) have been shown to be affected by contaminants in a smaller number of studies [28, 30, 37]. However, Lowe *et al.* [10] demonstrated a similar response in the lysosomes of mussel digestive cells and blood cells following whole animal exposure to the hydrocarbon fluoranthene.

Only a few studies have utilised the lysosomal biomarker in indigenous mussels to reflect the effects of chronic contaminant exposure [4, 8, 10, 29, 35, 38]. Only Lowe *et al.* [37] has previously demonstrated the field application of the lysosomal biomarker in mussel blood cells. The prime advantage of using blood cells is their relatively simple means of extraction from the mussels and that the organism need not be destroyed. This has ethical implications but also allows repeated, and long term, sampling of blood cells.

Contaminant induced lysosomal damage has been demonstrated in a number of studies by use of the neutral red assay [39, 40]. This technique has been used in a number of different studies on the blood cells of marine invertebrates such as the blue mussel [9, 28, 30] and the common shore crab, *Carcinus maenas* [41]. The neutral red assay, as described by Lowe *et al.* [9] utilises the visual assessment of cellular lysosomes, by light microscopy, to test their ability to retain the neutral red dye. By determining the lysosomal retention of the dye over a measured time period, an assessment can be made of the integrity of the lysosomal membrane and an index of lysosomal health obtained. The simplicity of this technique represents a significant departure from the complexity of more commonly used methods to assess the lysosomal response, such as the histological sectioning and staining of tissue. It has the added advantage that severe reduction in dye retention is a prognostic biomarker for the development of degenerative pathology [2, 10]; and, hence, it can be used as a direct measurement of animal health status.

This project was undertaken as a pilot study for a larger scale "Mussel Watch" monitoring programme in the Black Sea. The Black Sea, a partially enclosed sea in Eastern Europe, has suffered catastrophic ecological damage as a result of human activity, principally from land based sources [42]. This has resulted in eutrophication, loss of fisheries and sea mammals and contamination by pathogenic microbes and toxic chemicals [42].

This study attempts to not only integrate the biomarker approach and the chemical analysis of mussel tissue, but also to supplement this data with the results from a rapid source inventory of land based inputs to the Black Sea. The rapid source inventory was conducted under a protocol obtained from the World Health Organisation by the Black Sea Environment Programme [43]. This inventory provided information on amounts of oil, suspended sediment (TSS), nitrogen (TN), phosphorus (TP), and the biochemical oxygen demand (BOD) of the inputs into the Black Sea. A high BOD indicates the increased activity of heterotrophic organisms and thus heavy pollution [44].

The importance of including biological measurements in 'mussel watch' programmes has been previously emphasised [45]. Consequently, the dye retention information has been used to assess the status of the mussel populations of the Black Sea littoral states. This represents a novel adaptation of the 'mussel watch' concept and allowed the effective implementation of this study in regional laboratories located in the six littoral countries bordering the Black Sea. The lysosomal results obtained, by means of the neutral red assay, were used to examine evidence of linkages between the contaminant exposure of the mussel populations and adverse cellular reactions. The overall aim of the work was to rapidly generate a meaningful data set, which will be used to aid the management and remediation of the polluted areas of the Black Sea coastline. The specific objectives of the mussel watch were as follows:

- To determine health status of mussels
- To estimate exposure to selected pollutants based on body burdens
- To provide an integrated approach for assessing the quality of an important and sentinel component of the ecosystem.

2. Results of the Black Sea Mussel Watch Pilot Study

The Mussel watch was implemented from October 1996 to February 1997 and full details of sampling and methods have been reported elsewhere [46]. Lysosomal dye retention times, reflecting cellular health status, were measured in mussel blood cells from all six participating Black Sea countries (Fig. 2). The WHO Rapid Inventory Assessment data [43] were obtained for all the participating countries, with the exception of Georgia (Fig. 3).

Analytical chemical analysis of mussel tissue was conducted at the Bulgarian and Ukrainian sites for tissue burden of hydrocarbons, PCBs and pesticides [46].

Significant differences ($P \leq 0.05$) in lysosomal retention times were demonstrated in the blood cells of mussels (*Mytilus galloprovincialis)* between the various selected sites.

The rapid inventory assessment indicated considerable differences in levels of biochemical oxygen demand, suspended sediment, nitrogen and phosphorous at the surveyed sites, while residue data indicated sites where the indigenous mussels had an increased body burden of hydrocarbons, PCBs or pesticides [46].

Lysosomal Data (Fig. 2)

Mean lysosomal retention times are summarised in Figure 2, for each individual country [46]. The Romanian data set contained the lowest retention values, with results in the range of 0 to 60 minutes. Lysosomal membrane destabilisation was most pronounced in mussels collected from the Romanian sites of Navodari and IRCM, as shown by the low retention times of 5.4 and 12 minutes, respectively. These are exceptionally low values, when compared with the results presented here and data from other field studies [37]. Mussels sampled from the Ukrainian port of Odessa had a similarly depressed lysosomal retention time of 17.5 minutes. The Bulgarian site of Port Varna had a low retention time of 11 minutes, while Shkorpilovtsy had a much greater retention time of 84 minutes. The highest retention times were recorded in mussels from the Russian site of Lazarevck, 108 minutes, the Georgian sites of Kobultre, 117 minutes, and Poti Recreational site, 120 minutes, and the Ukrainian site of Koblevo, 150 minutes.

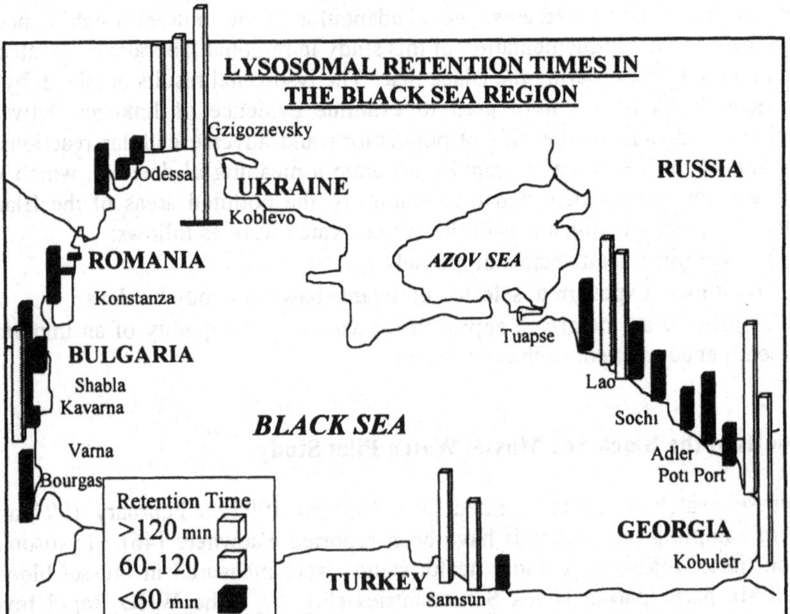

NB Koblevo is located in the Ukraine, it has been moved to be kept within the map confines

Figure 2. Lysosomal dye retention times for the Black Sea Mussel Watch.

WHO Rapid inventory Data/Tissue Residue Data (Figs. 3 & 4)
The Romanian study sites showed the greatest loading of BOD with an estimated input of 38,840 tonnes per year at the Navodari site alone [43, 46]. The Romanian land based sources also produce the largest inputs of phenol by discharging 1,254 tonnes per year into the Black Sea [43, 46]. The Ukraine inputs the largest amount of oil with 62,700 tonnes per year entering the Black Sea from the Odessa Port [43, 46]. The Odessa site also inputs the greatest amounts of total suspended sediment (TSS), with 50,730 tonnes per year entering the Black Sea [43, 46]. The Romanian site of Navodari receives the greatest amount of total nitrogen (TN) input by discharging 50,730 tonnes per year [43, 46]. The Ukrainian and Romanian study sites are heavily influenced by the presence of the Danube, Dniester, and Dnieper rivers which are responsible for the greatest input of water, into the Black Sea, with elevated levels of oil, BOD, and phenol [43]. In total these rivers are responsible for 54,600 t/y of oil, 242 t/y of phenol, 1,493 t/y of copper, and 844,464 t/y of BOD into the Black Sea [43]. The Odessa site receives a further 626.6 t/y of phenol from industrial sources [43, 46].

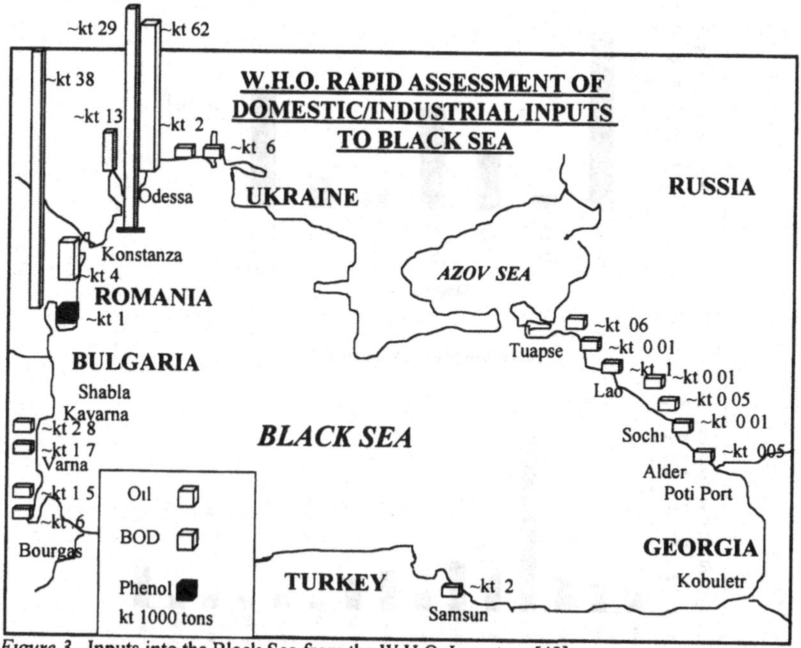

Figure 3. Inputs into the Black Sea from the W.H.O. Inventory [43].

Of the Bulgarian sample sites, the Port Varna site receives the highest loading of BOD, 1,740 t/y, oil, 2,823 t/y, and TSS 3,170 t/y; while the sample sites of Kavarna and Shkorpilovtzy receive the lowest amounts of land based discharges of the Bulgarian sites [43, 46]. The Russian site of Lazarevck has the lowest inputs of BOD, TSS, oil, TN, and TP, with values all less than 1.5 tonnes per year [43, 46].

The tissue burden analysis of the mussels indicated those animals at the Ukrainian Odessa site had the greatest burden of polycyclic aromatic hydrocarbons, 6,882 ng/g and pesticides, 243.3 ng/g DDT (Fig. 4; [46]). These animals also had high levels of

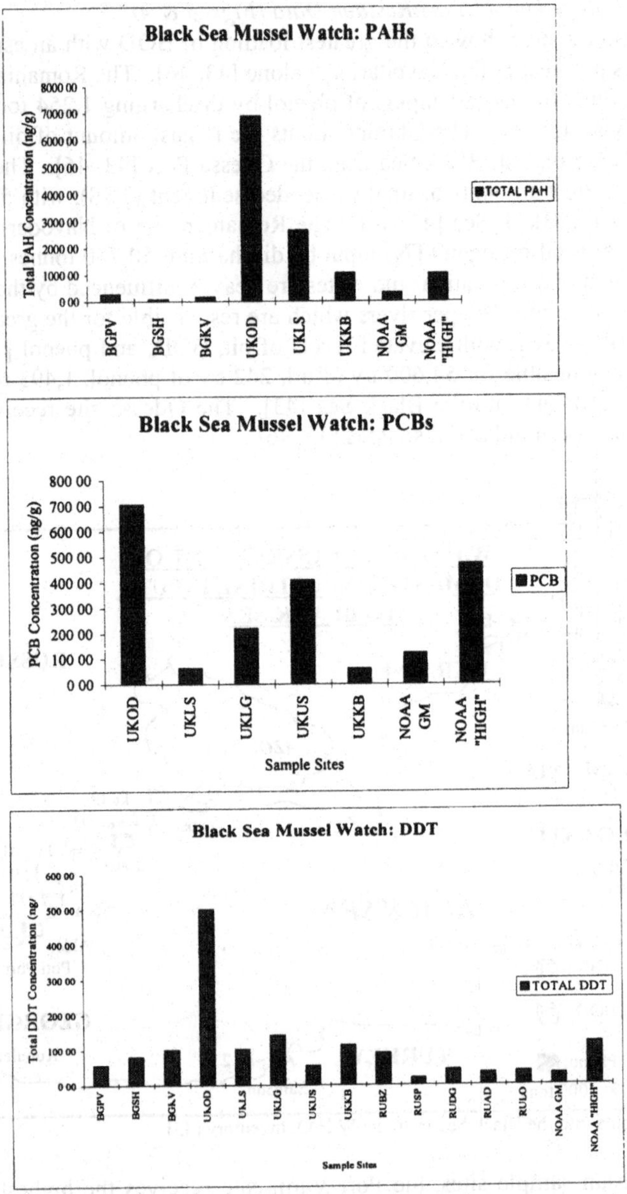

Figure 4. Concentrations of PAHs, PCBs & total DDT in mussel tissues (BGPV-Port Varna; BGSH-Shabla; BGKV-Kavarna; UKOD-Odessa Port; UKLG-Liman Grygoresvsky; UKLS-Liman Sukchoy; UKKB-Kobleva; UKUS-UkrSCES; NOAA GM-NOAA geometric mean; NOAA "HIGH"-NOAA highest value; RUBZ-Bzugo; RUSP-Sochi-Port; RUDG-Dagomis; RUAD-Adler, RULO-Lao).

PCBs, 700 ng/g in their tissue. Animals sacrificed from the Port Varna population, Bulgaria, had a tissue residue of 1565 ng/g (Fig. 4; [46]). The lowest concentration of tissue residue for polyccyclic aromatic hydrocarbons was at the Bulgarian site of Kavarna, with a value of 63 ng/g (Fig. 4; [46]).

Statistical Analysis
One-way analyses of variance tests were conducted on the individual lysosomal integrity data sets from the participating countries to identify significant differences in lysosomal retention times (calculated to the 95% confidence interval).

Regression analysis of the individual components of the rapid assessment data was carried out to determine to what extent the BOD, oil, TSS, TN, and TP data could account for the variability of the lysosomal data [46]. The results show that BOD and TSS accounts for 70% ($p < 0.0002$) and 76% ($p < 0.0002$) respectively of the biological variability. TN and TP account for 65% ($p < 0.0009$) of the biological variability [46]. The rapid assessment data for discharged oil only accounted for 46% of the variability in the lysosomal data and was not significant ($p < 0.08$).

The limited data sets for the detected tissue burden of either polycyclic aromatic hydrocarbons or pesticides did not produce a significant correlation with the lysosomal integrity ($p = 0.09$ and $p = 0.82$ respectively) [46].

A regression analysis of the rapid source inventory components with PAH tissue burden in the mussels was carried out [46]. The results indicate that TSS and BOD accounted for 70% ($p > 0.10$, 90% confidence interval, correlation coefficient of $r = 0.84$), and 82% ($p > 0.10$, 90% confidence interval, correlation coefficient of $r = 0.90$) respectively of the variability in the PAH tissue burden.

3. Discussion

The low lysosomal neutral red retention times exhibited by mussel blood cells sampled from the highly industrialised sites on the Black Sea coastal zone is consistent with the findings of other studies, which have examined lysosomal integrity in indigenous, chronically exposed, mussel populations [8, 37, 29]. A strong relationship has been recorded in this study between reduced lysosomal dye retention time and elevated inputs of contaminants, as indicated by the rapid source inventory data [46]. For example, the mussels from the Ukrainian sites of Odessa Port and Liman Sukchoy demonstrate greatly reduced lysosomal retention times in conjunction with massive inputs of organic micropollutants [46]. This trend is repeated in mussels from the Romanian sites of Navodari and IRCM, as well as the Bulgarian site of Port Varna. These study areas are impacted by high quantities of industrial waste, in particular BOD and phenol [43] and the lowest retention times for the entire study were recorded at these sites. Conversely, sites which have indigenous mussels exhibiting higher retention times are also shown to have reduced inputs of organic and inorganic contaminants [46]. For example, the Bulgarian sites of Kavarna and Skorpilotsyvky show increased retention times in conjunction with relatively reduced inputs of BOD and oil; and the Russian site of Lazarevck shows a similar trend [46].

An association of low retention times with high tissue residues of hydrocarbons at the Odessa Port site and Port Varna site is indicated [46]. Mussels at the Odessa site were also shown to have extremely high levels of PCBs with a tissue burden of 700 ng/g recorded [46]. In contrast, mussels collected from the Ukrainian site of Koblevo, where higher retention times were observed, had a PCB tissue burden of 29 ng/g. However, no significant relationship with retention could be obtained with the tissue residue values obtained at other sites.

No studies have attempted to investigate the lysosomal biomarker on the ambitious scale shown here, nor with the use of such a rapid, simple means of assessment. This study has shown that the lysosomal biomarker, as investigated with the neutral red technique, is a rapid, simple means of determining the health of indigenous mussel populations. This represents a novel approach in environmental impact assessment as it is both simple and cheap, yet rapidly generates relevant and useful biological information by the indication of those mussel populations which are demonstrating an adverse, contaminant linked, biological response.

In this study, consistent pathological responses were observed in blood cell lysosomes of mussels collected from urban and industrial associated sites [46]. A clear relationship is indicated, at certain sites, between the presence of hydrocarbons in the tissue and water, and low lysosomal retention times [46]. Lysosomal retention times were shown to be particularly depressed in areas with a high input of hydrocarbons, for example the Navadori and Varna sites in, respectively, Romania and Bulgaria [46]. Laboratory studies have shown that PAHs, such as phenanthrene and anthracene, can induce dose-dependent lysosomal destabilisation in the digestive cells of bivalves [24, 27, 33, 36, 48]. However, in this study there is only an association of lysosomal response and hydrocarbon tissue burden in sites which have an extremely high level of background hydrocarbon contamination [46].

The strong association of lysosomal perturbation with the rapid assessment data indicates the impact that anoxia, as a result of elevated BOD, may be having on the mussel populations. Konsulova [48] indicated the susceptibility of the Black Sea zoobenthos to the lack of oxygen caused by anthropogenic eutrophication. Tracy [49] demonstrated a reduction in the growth of *M. edulis* in the presence of increasing eutrophication, while Losovskaya *et al.* [50] reported a decline in the population of *M. galloprovincialis* after significant discharges of eutrophic waters in the region between the Dnieper and Dniester rivers. Indigenous populations of mussels located near the Danube have also been reported to decrease in size during periods of heavy river flow [50]. There may also be additive effects occurring between elevated levels of organic material and organic and inorganic pollutants. Gilek *et al.* [51] reported an increase of PCB accumulation in the blue mussel, *M. edulis*, after exposure in conjunction with high levels of algal enrichment. Tracy [49] reported a modification of copper uptake in *M. edulis* after exposure with varying levels of nutrient enrichment.

The lysosomal integrity of the mussels can be strongly correlated with the presence of large inputs of suspended sediment, which may have associated particulate bound contaminants. These contaminants may only become available under certain coastal conditions or after uptake by the biota. The input of suspended sediment, being largely composed of sediment derived from the water shed will no doubt include organic and inorganic contaminants. Suspension filter feeders, such as mussels, will respond to contaminants in both the dissolved and suspended phase [52] and higher levels of suspended sediment will presumably increase the potential for the uptake of greater amounts of contaminants. Nelson *et al.* [53] found that scope for growth decreased in mussels exposed to increasing concentrations of suspended sediment from a contaminated harbour while Prins and Smaal [54] reported a net carbon loss in mollusca exposed to high particle concentrations, possibly due to an increase in mucus production.

The evidence for the eutrophic nature of the north western shelf of the Black Sea is overwhelming [42]. This part of the Black Sea is noted for the impact of the Danube, Dniester, and Dnieper rivers and their considerable contribution of contaminants [43]. The impact of the Danube river on BOD and the large inputs of oil from the Romanian coastline are almost certainly contributing to the severely reduced retention times observed in this area (see fig 3). For example, the Romanian oil refineries at Novadari are solely responsible for 25% of the total BOD to the Black Sea [43].

The demonstration here, in a major component of the coastal ecosystem, of a biological response that has previously been shown to be associated with toxicological stress demonstrates the sensitivity of the Black Sea's ecosystems to continual pollutant exposure. Indeed, in the north western shelf, the population of *M. galloprovincialis* has been shown to be reduced in numbers, and this has been attributed to the effects of eutrophication [55].

Given the restricted nature of the available information, and the complex nature of many discharges, it is difficult to assign the pathological response observed in the mussel blood cells to the toxic effect of a single polluting discharge, chemical, or group of chemicals. Depledge [56] suggested that whilst biomarkers which indicated exposure to pollutants are useful, those which signify that organisms are experiencing adverse effects are more ecological relevant. To function, in this manner, it is necessary for a biomarker to demonstrate a long term response. Regoli [35] demonstrated no recovery in lysosomal stability, in *M. galloprovincialis*, after a four month depuration period, following metal exposure. This would indicate that the lysosomal response does not appear to be transient. Lysosomal damage has direct implications for the future growth and reproductive potential of mussels [37]. Moore and Viarengo [57] indicated that lysosomal membrane fragility in mussels is directly related to enhanced protein catabolism.. Krishnakumar *et al.* [29] showed that mussels, sampled from urban associated sites, exhibited decreased lysosomal stability and a corresponding reduction in size and somatic tissue weight. A number of studies have shown a functional relationship between adverse alterations in lysosomal stability and scope for growth [31, 32], following contaminant exposure.

4. Conclusions & Recommendations

In conclusion, this study examined the contaminant impact on mussels sampled from indigenous populations around the Black Sea coastline, using a simple lysosomal integrity test as a biomarker of harmful effect on mussel blood cells. In addition, using regression analysis, the relationship between the probe (neutral red) retention time and polluting discharges into the coastal zone was examined. In mussels from certain sites the hydrocarbon tissue burden was also examined [46].

Mussels from some areas of the Black Sea coastline are clearly impacted, as indicated by the extreme perturbation of the lysosomal component within their blood cells. The presence of oil, inputs of suspended sediment and elevated biochemical oxygen demand appear to be the major contributing factors to the pathological response observed in the mussel blood cells. This study indicates an adverse biological response from a major benthic component of the Black Sea ecosystem, as well as the use of a simple biomarker test to demonstrate contaminant induced effect. It has also clearly demonstrated that tests for pathological effects can be used to rapidly pinpoint problem sites with complex contamination situations, which can then be subject to more intensive chemical and biological assessment. This will permit the more focused and effective use of limited analytical resources in future surveys by eliminating the need for expensive blanket chemical analysis. The results obtained in this study will add to the knowledge of those responsible for the management of the coastal and marine resources of the Black Sea. This work also demonstrates the ease with which sensitive biological effects monitoring can in the future be integrated with other "Mussel Watch" programmes.

Based on the findings of the feasibility study, the recommendations for a monitoring programme for the Black Sea are as follows:

- Establish a Black Sea Mussel Watch as a full component of International Mussel Watch
- Harmonise appropriate elements of the national monitoring programmes for the region
- Continued training in analytical chemistry and biological effects procedures through the UNESCO/IOC GIPME expert groups, with additional support from TACIS and PHARE
- Establish collaborative research projects between the laboratories involved in monitoring to facilitate the creation of a regional network
- Form links with research groups involved in the FAO-MEDPOL programme and in ICES related activities, in order to encourage capacity building in the Black Sea region

Acknowledgements

The authors would like to acknowledge the assistance of Dr.Laurence Mee and Dr. Vladimer Mamaiev at the (G.E.F. Black Sea Environmental Programme P.C.U., Istanbul); Dr. Neil Andersen (CEES Horn Point Laboratory, University of Maryland); and Dr. Robin Harger (UNESCO-IOC, Paris). This work was supported by UNESCO-IOC as part of the International Mussel Watch.

References

1. Goldberg, E.D., Bowen, V.T., Farrington, J.W., Harvey, G., Martin, J.H., Parker, P.L., Risebrough, R.W , Robertson, W., Schneider, E. & Gamble, E. (1978) The mussel watch. *Environ. Conserv.* **5**, 101-26.
2. Moore, M. N. (1990). Lysosomal cytochemistry in marine environmental monitoring. *Histochem. J.* **22**, 187-191.
3. Depledge, M. H., Amaral-Mendes, J.J., Daniel, B., Halbrook, R S., Kloepper-Sams, P., Moore, M. N , and Peakall, D. P. (1993). The conceptual basis of the biomarker approach. In: Biomarkers-Research and application in the assessment of environmental health, (Eds. D. G. Peakall, and L. R. Shugart), pp. 15-29. Springer, Berlin, Heidelberg.
4. Bayne, B.K., Addison, R.F., Capuzzo,J.M., Clarke, K.R., Gray, J.S., Moore, M.N., & Warwick, R. M , (1988) An overview of the GEEP Workshop. *Mar. Ecol. Prog. Ser.* **46**, 235-43.
5. Moore, M. N. (1985). Cellular responses to pollutants. *Mar. Poll. Bull.* **16**, 134-9.
6. Topping, G. (1983). Guidelines For The Use Of Biological Material. In: *The First Order Pollution Assessment And Trend Monitoring* Dept of Agriculture and Fisheries for Scotland, Marine Laboratory, Scottish Fisheries Research Report No 28. ISSN 0308 8022, 28 p.
7. UNEP (1995). Guidelines For Monitoring Chemical Contaminants In The Sea Using Marine Organisms. UNEP Reference Methods For Marine Pollution Studies No. 6 .
8. Moore, M N. (1988). Cytochemical responses of the lysosomal system and NADPH-ferrihemoprotein reductase in molluscan digestive cells to environmental and experimental exposure to xenobiotics. *Mar. Ecol. Prog.* **46**, 81-89.
9 Lowe, D.M., Moore, M. N, Evans, B. M (1992) Contaminant impact on interactions of molecular probes with lysosomes in living hepatocytes from dab *limanda limanda*. *Mar Ecol. Prog. Ser.* **91**, 135-140
10. Lowe, D. M., Soverchia, C., and Moore, M. N. (1995). Lysosomal membrane responses in mussels to experimental contaminant exposure. *Aquatic Toxicol.* **33**, 105-112.
11. Widdows, J , and Johnson, D. (1988). Physiological energetics of *Mytilus edulis*: scope for growth. *Mar. Ecol. Prog. Ser* **46**, 113-21.
12. Hawkins, H.K. (1980) Reactions of lysosomes to cell injury. In *Pathology of Cell Membranes,* Vol. 2 (edited by TRUMP, B.F. & ARSTILA, A.V.), pp. 252-85. Academic Press .New York, San Francisco, London.
13. Lowe, D.M. (1988) Alterations in cellular structure of *Mytilus edulis* resulting from exposure to environmental contaminants under field and experimental conditions. *Mar. Ecol. Prog. Ser.* **46**, 91-100.
14. Moore, M.N., Pipe, R.K., Farrar, S.V., Thomson, S. & Donkin, P. (1986) Lysosomal and misrosomal responses in *Littorina littorea:* further investigations of environmental effects in the vicinity of the Sullom Voe Oil Terminal and the effects of experimental exposure to phenanthrene. In: *Oceanic Processes in Marine Pollution - Biological Processes and Waste in the Ocean,* Vol.1 (edited by Capuzzo, J M & Kester, D.R.), pp. 89-96. Krieger Publishing, Melbourne, Florida.
15. Moore, M.N., Livingstone, D.R & Widdows, J (1988) Hydrocarbons in marine molluscs: biological effects and ecological consequences. In: *Metabolism of Polycyclic Aromatic Hydrocarbons in the Aquatic Environment* (edited by VARANASI, U.), pp 291-328. CRC Press, Boca Raton, Florida.
16. Mihnea P.E., 1995 Black sea Action Plan and the future legislation on sea water quality. Regional Conference, Varna, Bulgaria 13-15 June 1995, Conf. Preprints, 411-419.
17. GESAMP (1994). *Guidelines for Marine Environmental Assessment*, GESAMP Reports and studies No 54. IMO, London, 1994
18 UNEP (1995). *Guidelines For Monitoring Chemical Contaminants In The Seausing Marine Organisms.* UNEP Reference Methods For Marine Pollution Studies No 6
19 ICES (1992). *Report of the ICES Advisory Committee on Marine Pollution 1992.* ICES Cooperative Research Report No. 190, 203pp. International Council for the Exploration of the Sea, Copenhagen.
20. Phillips, D. J. H. (1981). A comparative evaluation of oysters, mussels and sediments as indicators of trace metals in Hong Kong waters *Mar Ecol. Prog. Ser* **3**, 285-293.
21 Phillips, D. J. H. (1988). Barnacles and mussels as indicators of trace elements: A comparative study. *Mar Ecol. Prog. Ser.* **49**, 83-93.
22. Goldberg, E. D (1986) The Mussel Watch concept. *Environ Monit. Assess.* **7**, 1.
23. Tripp, B. W., Farrington, J. W., Goldberg, E. D., Sericano, J. (1992) International Mussel Watch· The initial implementation phase. *Mar Pollut. Bull.* **7**, 371-373.
24. Cajaraville, M. P., Marigomez, J. A., Angulo, E. (1989). A sterological study of lysosomal structure alterations in *Littorina littorea* exposed to 1-naphthol. *Comp. Biochem. Physiol.* **93**, 231-237.

25. Hinton, D. E. (1989). Environmental contamination and cancer in fish. *Mar. Env. Res.* **28**, 411-416.
26. Köhler, A. (1991). Lysosomal perturbations in fish liver as indicators for toxic effects of environmnental pollution. *Comp. Biochem. Physiol.* **100**, 123-127.
27. Viarengo, A., Moore, M. N., Pertica, M., Mancinelli, G., Accomando, R. (1992). A simple procedure for evaluating the protein degradation rate in the mussel (*M. galloprovincialis* Lam.) tissues and its application in the study of phenanthrene effects on protein catabolism. *Comp. Biochem. Physiol.* **103**, 27-32.
28. Lowe, D. M., Pipe R. K. (1994). Contaminant induced lysosomal membrane damage in marine mussel digestive cells: an *in vitro* study. *Aquat. Toxicol.* **30**, 357-365.
29. Krishnakumar, P. K., Casillas, E., Varanasi, U. (1994). Effect of environmental contaminants on the health of *Mytilus edulis* from Puget Sound, Washington, USA. I. Cytochemical measures of lysosomal responses in the digestive cells using automatic image analysis. *Mar. Ecol. Prog. Ser. Vol.* **106**, 249-261.
30 Moore, M. N., Wedderburn, R. J., Lowe, D. M., Depledge, M. H. (1996). Lysosomal reaction to xenobiotics in mussel hemocytes using BODIPY-FL-Verapamil. *Mar. Environ. Res.* **42**, 99-105.
31. Widdows, J., Bakke, T., Bayne, B. L., Donkin, P., Livingstone, D. R., Lowe, D. M., Moore, M. N., Evans, S. V., Moore, S. L. (1982). Responses of *Mytilus edulis* on exposure to the water-accomodated fraction on North Sea oil. *Mar. Biol.* **67**, 15-31.
32. Moore, M. N., Livingstone, D. R., Widdows, J., Lowe, D. M., Pipe, R. K. (1987). Molecular, cellular, and physiological effects of oil-derived hydrocarbons on molluscs and their use in environmental impact assessment. *Phi. Trans. R. Soc. Lond.* **316**, 603-623.
33. Nott, J. A., Moore, M. N. (1987). Effects of polycyclic aromatic hydrocarbons on molluscan lysosomes and endoplasmic reticulum. *Histochem. J.* **19**, 357-368.
34. Viarengo, A., Moore, M. N., Mancinelli, G., Mazzucotelli, A., Pipe, R K., Farrar, S. V. (1987). Metallothioneins and lysosomes in metal toxicity and homeostasis in marine mussels: the effects of cadium in the presence and absence of phenanthrene. *Mar. Biol.* **94**, 251-257.
35. Regoli, F. (1992). Lysosomal responses as a sensitive stress index in biomonitoring heavy metal pollution. *Mar. Ecol. Prog. Ser.* **84**, 63-69.
36. Winston, G. W, Moore, M N., Straatsburg, I., Kirchin, M. A. (1991). Decreased stability of digestive gland lysosomes from the common mussel *M. edulis* L. by *in vitro* generation of oxygen-free radicals. Arch. environ. *Contam. Toxicol.* **21**, 401-408.
37. Lowe, D. M., Fossato, V U., Depledge, M. H. (1995b). Contaminant induced lysosomal membrane damage in blood cells of mussels *M. galloprovincialis* from the Venice Lagoon: an *in vitro* study. *Mar. Ecol. Prog Ser.* **129**, 189-196.
38. Moore, M. N., Livingstone, D. R., Widdows, J., Lowe, D. M., Pipe, R. K. (1982). Molecular, cellular, and physiological effects of oil derived hydrocarbons on molluscs and their use in impact assessment. *Phil. Trans. R. Soc. Lond.* **B316**, 603-623.
39. Finter, N. B. (1969). Dye uptake methods for assessing viral cytopathogenicity and their application to interferon assays. *J. Gen. Virol.* **5**, 419-427.
40. Borenfreund, E., Puerner, J. A. (1985). Toxicity determined *in vitro* by morphological alterations and neutral red absorption. *Toxicol. Lett.* **24**,199-124.
41. Wedderburn, R. J., Cheung, V., Bamber, S., Bloxham. M.J., Depledge, M. H. (1997). Biomarkers of biochemical and cellular stress in *Carcinus maenas*: An *in situ* field study. *Mar. Env. Res.* (Submitted).
42. Mee, L. D. (1992).The Black Sea in crisis: A need for concerted international action. *Ambio.* **21**, 278-286.
43. Sarikaya, H. Z., Oitil, E., Sevimli, M. F., Germirli, F., Oktem, Y. (1997). Region wide assessment of the land based sources of pollution to the Black Sea. Black Sea Environmental Programme Technical Reports 7. UNDP-UNEP-World Bank, Istanbul, 140pp.
44. Lawrence, E. (1995). Dictionary of biological terms.Longman Scientific and Technical, Longman House, London.
45. Gray, J. S. (1992). Biological and ecological effects of marine pollutants and their detection. *Mar Pollut. Bull.* **25**, 48-50.
46. Wedderburn, R. J., Moore, M. N., Wade, T., Lowe, D. M., Wedderburn, R. J., Wade, T., Balashov. G., Büyükgüngör, H., Daurova, Y, Denga, Y., Kostylev, E., Mihnea, P., Moncheva, S., Özkoc, H., Tabagari, S , Depledge, M. H. (1998). The Black Sea Mussel Watch: biological effects and contaminant residues. Mar Poll. Bull., In Press.
47. Axiak, V., George, J. J , Moore, M. N. (1988). Petroleum hydrocarbons in the marine bivalve *Venus verrucosa*: accumulation and cellular responses. *Mar. Biol.* **97**, 225-230.
48. Konsoulova, T. (1993). Marine macrozoobenthic community structures in relation to some environmental factors. *Comp. Rend. Acad. Sci. Bulg.* **5 (46)**, 115-118.

49. Tracy, G. A. (1990). Test for effects of eutrophication on the toxicity of copper to the blue mussel *Mytilus edulis*. *J. Shellfish. Res.* **8**, 439.
50. Losovskaya, G. V., Garkavaya, G. P., Salskij, V. A. (1990). Changes in the benthic communities and fluctuations in the number of dominant species under conditions of eutrophication in the north-western part of the Black sea. *Ehkol. Morya*. **35**, 22-28.
51. Gilek, M., Bjoerk, M., Broman, D., Kautsky, N., Naef, C. (1996). Enhanced accumulation of PCB congeners by Baltic Sea blue mussels, *Mytilus edulis*, with increased algal enrichment. *Environ. Toxicol. Chem.* **15**,1597-1605.
52 Rainbow, P. S. (1995). Biomonitoring of heavy metal availability in the marine environment. In: International conference on marine pollution and ecotoxicology held in Hong Kong, 22, 26, January 1995. Wu, R. S. S., Atlas, R. M., Goldberg, E. D , Sheppard, C., Chapman., P. M., Conell, D. W., McIntyre, A. D., Rainbow, P. S. -eds. **31**,183-192.
53. Nelson, W. G., Phelps, D. K., Galloway, W. B., Rogerson, P F., Pruell, R. J. (1987). Effects of Black Rock harbour dredged material on the scope for growth of the blue mussel, *Mytilus edulis* after laboratory and field exposures. Tech. Rep. U. S. Army. Eng. Waterways. Exp. Stn. 122 pp.
54 Prins, T. C., Smaal, A. C. (1987). Carbon and nitrogen budgets of the mussel *Mylus edulis* and the cockle *Cerastoderma edule* in relation to food quality. Proceedings of the 22nd European Marine Biology Symposium. Ros, J. ed. Barcelona, Spain, Inst. De Ciencias Del Mar. *Top. Mar. Biol.* **53**, 477-482
55. Zaitsev, Yu, and Mamaiev, V. (1997). Marine Biological Diversity of the Black Sea. United Nations Publications. New York.
56. Depledge, M. H., Aagaard, A., Gyoerkoes, P. (1994). Assessment of trace metal toxicity using molecular, physiological and behavioural biomarkers. Trace Metals in the Aquatic Environment. Proceedings of the Third International Conference held in Aarhus, Denmark. Phillips, D J. H., Rainbow, P. S.-eds. **31**, 19-27
57. Moore, M. N., Viarengo, A. Lysosomal membrane fragility and catabolism of cytosolic proteins: Evidence for a direct relationship. *Experientia*. **43**, 320-322.

DATABASE AND DATABASE MANAGEMENT SYSTEM OF THE TU-BLACK SEA PROJECT

V. L. VLADIMIROV
Marine Hydrophysical Institute,
NASU, 2 Kapitanskaya St.,
Sevastopol, 33500 Ukraine

S. T. BESIKTEPE
Middle East Technical University,
Institute of Marine Sciences, P.O. Box 28,
Erdemli, Icel, Turkey

D. G. AUBREY
Woods Hole Oceanographic Institution,
Woods Hole, MA 02543, USA

Abstract

Development of the unique regional, multi-disciplinary database created within the framework of the NATO TU-Black Sea project is given. Database contains principal physical, chemical and biological variables for the entire Black Sea basin and covers the most crucial period in the history of the Black Sea ecosystem starting from the "background" situation in 1960-th till the drastic changes occurred recent years. Database is supplied with the special powerful database management system.

1. Introduction

One of the main objective of the NATO TU-Black Sea Project was to establish a database management system (DBMS) for environmental and oceanographic data, pertinent to the goals of the Programme. A Working Group, with participants from all collaborators, was formed to coordinate the data base development. The requirements of the data base and it's management system were defined by the group. This group coordinated its work with national and regional efforts which shared part of the work. Another task of the group was to collect contaminant input data, especially focusing on nutrients and eutrophication.

S. Beşiktepe et al. (eds.),
Environmental Degradation of the Black Sea: Challenges and Remedies, 291–302.

The Working Group met in Moscow in November, 1993 to layout the framework for the DBMS and to determine the strategy for establishment of the data base. It was decided that the data included in the data base be restricted to the period from 1963 to the present, as this period encompasses the major changes in the ecosystem of the Black Sea. It was recommended that data entry and preliminary quality control procedures should preferably be made at the country of origin.

Tasks in the creation of the data base and the data base management system were defined in the chronological order as follows;

- preparation of the data base inventory ,
- compilation of the data reported in the data base inventory,
- quality control of the data,
- compilation the first ASCII version of the database,
- development of the DBMS for the final version, and:
- loading data into data base and checking the final version.

Two main groups were established in order to accomplish the above tasks; a data base task team and expert groups. A Data Base Task Team was formed from software engineers, oceanographic data base specialists and oceanographers to deal with the technical issues in the development of the data base. Expert Groups were formed from originators of data as a responsible on the flow of the data to the data base. Expert Groups was working closely with Data Base Task Team during the compilation and quality control of the data. DBMS were also developed considering requests of the expert groups.

This paper discusses realization of the above tasks and gives overview of the final products; NATO TU-Black Sea data base and its data base management system. The paper is organised as follows. Initiation of the data base development and the chronological progression are given in Section 1. Production of the data base inventory is presented in Section 2. Data set formats and list of data base variables for data supply to the data base are given in Sections 3-4. Data quality control procedures applied by the expert groups are presented in Section 5. Facilities of the DBMS are given in Section 6. Structure of the data base is given in Section 7. Comparisons of the NATO TU-Black Sea data base with other main oceanographic data bases are done in Section 8.

2. Data Base Inventory

In order to determine an inventory of available data, forms prepared according to the requirements of the data base were sent to the cooperation partners in April 1994. It was requested that the inventory forms be returned by February , 1995. The data inventories prepared by the participants, were edited and printed in the "Inventory of Time Series and Station Data". in February, 1995. It was also decided to prepare a computerized version for distribution to the participants, to provide easier access to the information .

The development of the version of the Computerized Data Base Inventory (CDBI) was based on a similar system in use at the Marine Hydrophysical Institute (MHI). It has a Windows-like interface and provides the following features: ability to run under both DOS and Windows on a personal computer; adjustment to hardware configurations; customizable screen based on user preferences; context-sensitive, and on-line help based on hyper-texts.

The CDBI contained information on datasets delivered to the database and a report on the status of data which were either not delivered or not digitized. It was distributed to all participants in September 1996.

The CDBI provides the following capabilities:

- display of maps:
    - selection of geographic regions;
    - selections of data sets or stations on the map;
    - display statistical information on data coverage;
    - printing and saving in bitmap format;
    - customization of map view;
- data selection from the internal data base:
    - country;
    - institution;
    - name of research vessel;
    - year;
    - month;
    - data storage media;
    - type of measurement;
- access to the information from the "Inventory of Time Series and Station Data" and creation of brief reports on data selection.

The CDBI includes information on 377 data sets. The latest version of the Inventory has been placed on the anonymous FTP server of IMS-METU, available to project participants and the public.

3. Data set formats

Simple formats were used for the data sets to make digitization easier for the originators. The formats were developed taking into account the CoMSBlack data formats. The data sets are organized in ASCII files as follows:

Information File: contains detailed description of the data set (in English)

Brief Description File: contains concise information on origin of the data, instrumentation, methods and units, aspects of data processing and other relevant items. The file either has the same name as the data file(s) or the data directory if it relates to a directory.

Data File: These contain one of the following files :(1) station data, containing measurements at a station, usually obtained from a vertical cast; (2) surface data, containing observations at the sea surface; or (3) time series data, containing time series data at fixed points or stations.

4. List of principal ecosystem variables

The following 116 variables and group of variables were loaded into the final data base:

- *Physics*: Temperature, Salinity (CTD-profiles, Nansen bottles, and bathythermograph), Fluorescence, Light attenuation (407nm, 422nm, 427nm, 457nm, 465nm, 540nm, 660nm), Photosynthetic active radiance (PAR), Secchi disk depth.
- *Chemistry*: Dissolved oxygen, Hydrogen sulfide, Total sulfides, Thiosulfate, Sulfite, Sulfur-o, Silicic acid, Phosphate, Nitrite, Nitrate, Total amount of Nitrite+Nitrate, Urea, Ammonia, Total Alkalinity, Carbonate Alkalinity, pH, RedOX potential, Particulate organic carbon, Total organic carbon, Particulate organic nitrogen, Total nitrogen, Particulate phosphorus, Total phosphorus, Humic matter, Total suspended matter, Manganese.
- *Biology*: Chlorophyll-a, Phaeopigments-a, Primary production, Bacterioplankton, Benthos, Microphytoplankton, Nanophytoplankton, Picophytoplankton, Total phytoplankton, Macrozooplankton, Mesozooplankton, Noctiluca, Total zooplankton.

The following variables were kept in the form of ASCII data sets

- Climatic and averaged variables.
- Meteorological variables,
- Sea level,
- Spectra of transparency,
- Currents.

5. Data Quality Control, Expert Groups

In order to ensure exclusion of erroneous data, strict quality control measures were taken. Two levels of data quality control were applied; the originating institution data quality control and the task team control. Overall data quality control was done with the Expert Groups studies on the pooled data.

Originating Institution Data Quality Control:
The following two methods were used for data quality control at the origin of the data;

1. Each data file is prepared separately by two persons. Differences between the files are checked by computer and corrected.

2. The investigator who is the source of data checks each data entry in a file manually against the original records to eliminate errors.

Task Team Quality Control:
All data sets were also evaluated by the Task Team, by comparing them against each other and identifying outliners and trends.

Among the various schemes used to flag data quality [2] which may be equally acceptable, the scheme employed for data reported in real-time, as given in the GETADE Formatting Guidelines for Oceanographic Data Exchange [3], devised by the IOC's Group of Experts on the Technical Aspects of Data Exchange (GETADE) was selected.. This scheme is universal and can be applied to all types of data. It uses a one character field for data quality with the following interpretation:

0 = data are not checked
1 = data are checked and appear correct
2 = data are checked and appear inconsistent but correct
3 = data are checked and appear doubtful
4 = data are checked and appear to be wrong
5 = data are checked and the value has been altered
6 = data are checked and appear to be assigned to the wrong depth

Flag 6 was added according to request of the Expert Groups.

Expert Groups
When a valuable data set on any type of characteristics are delivered, the Task Team met in Erdemli, Turkey together with a group of experts to merge and make quality control of the combined data set, and produce a technical report. A series of Expert Groups had therefore been identified. It was agreed that the Expert Groups include the scientists which participated in the collecting and primary processing of the data subsets provided to the database. Five Expert Groups were formed according to the characteristics of the data sets:

1. Physics
2. Chemistry
3. Biology 1 (zooplankton)
4. Biology 2 (phytoplankton)
5. Biooptics (chlorophyll, PP, fluorescence, light attenuation, PAR)

Each Expert group met two times for two weeks. The first meeting was held in October-December 1996 and the second one was in January-March 1997. They evaluated all delivered data and metadata, put the quality flags for physical, chemical and biooptical data and prepared technical reports.

The CruBase DBMS [1] designed by the MHI for the oceanographic data and intended for interdisciplinary multi-parameter data of oceanographic cruise or multi-ship survey was used by the Task Team and Expert Groups for the quality control and preparation of the final data set. Special subsystem of the CruBase DBMS was developed (CruBase Data Control System - DCS) and improved during the Task Team meetings to provide the possibility of the effective quality control of the data loaded into the CruBase system. The format of the CruBase files was updated. Now this format permits to store quality flag with each data value. During the loading the quality flag 0 is assigned to each data value (data was not checked).

6. OceanBase-II DBMS

A special DBMS (OceanBase-II) was developed for the project data base, on the basis of the OceanBase DBMS system developed by the MHI Data Base Laboratory [4]. It was designed using Borland Delphi Developer to run under Windows-95. OceanBase-II is a specialized data base application to work with large sets of interdisciplinary oceanographic data.

The OceanBase supports three types of data according to their relation with depth:

- Standard - profiles with uniform or random depth grid, with values at points,
- Surface - measurements at only one depth (surface or bottom), for example, Secchi disk visibility depth or zoobenthos biomass,
- Layer - measurements representing an average value for a layer, for example, zooplankton net tows.

The DBMS allows the user quick and easy access to the entire data base. One can view, sort, select and export all necessary data and metadata using a user friendly interface.

The main features of OceanBase are:

- Selection of stations based on institutions, data sets, time (year, month, day, hour), region, sea depth, measured variables, etc. ,
- Display of station positions on maps, with a possibility to choose required stations through the graphical interface,
- Options for smoothing, interpolation, decimation, etc.,
- Possibility to merge data from different stations,
- Calculation of mean profiles,
- Displaying data profiles in tables and plots,
- Filters to select data based on their quality flags,
- Calculation of averages in a layer and its the spatial distribution. The depth of this layer can be fixed or variable linked with special requirements at a station (extreme or fixed values of a variable),
- Calculation of time variability (yearly, seasonal, or daily),
- Convenient navigation through data sets and stations,
- Data export to files in ASCII formats,

- Calculation of various statistics: distribution of stations and data values, contribution of institutions, data sets, regions, time, quality, depth, etc.,
- Several kinds of histograms: values, depths, maximums, minimums, data amounts, etc.,
- Many tools providing reference information,
- Generation of various reports,
- Possibility to save results in both text and graphic format.

The user can open as many windows as desired. Data and information in these windows are cross-linked. For example, clicking at a station position on the map, the user obtains the following links: corresponding dataset name in the Data Sets window; station information line in the Stations window; values of data at the station in the Values window, and plots of selected variables in the Plot window.

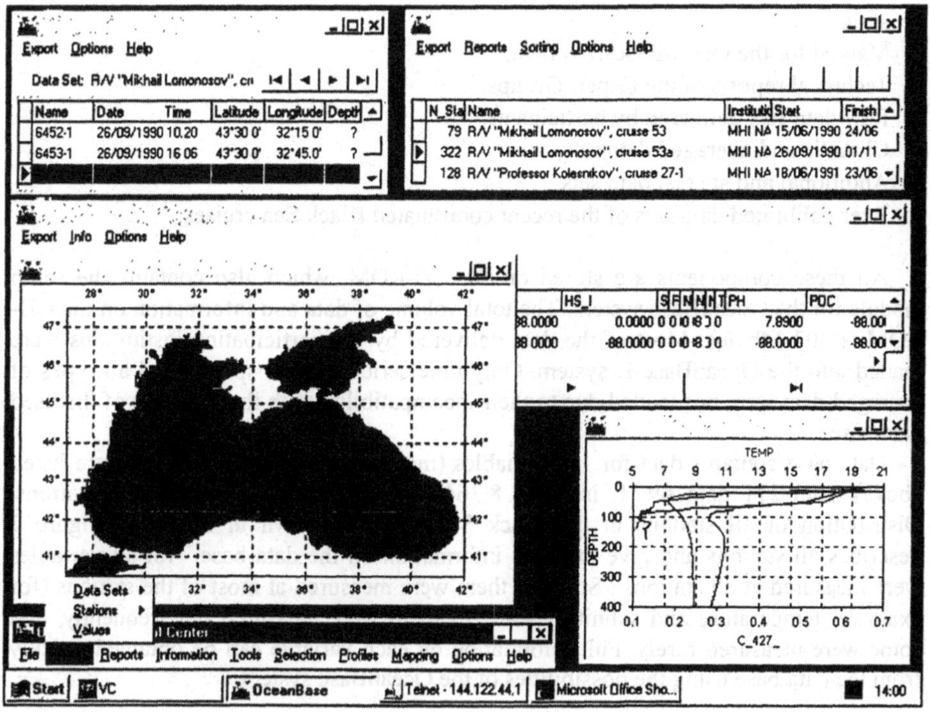

Figure 1. Sample OceanBase-II working desktops

Figure 1 shows an example of the OceanBase working desktops which can be created by the user. It includes cross-linked windows with information on the selected stations, values of selected variables for current station, map with the stations position, graph of the temperature, salinity and light attenuation coefficient for current station, and histogram with distribution of the salinity values for selected stations. The lower window (stripe) is the OceanBase control center which gives the possibility to manage all data and information.

7. Structure of the data base

The compilation of the data base was completed in April, 1997. It contains data for 26,035 stations from 271 data sets. The data delivered by the contributors were checked, transformed to the NATO TU-Black Sea formats and included in the joint data set. The data are arranged according to country and institution of origin. Brief description files accompany each data file (or directory) based on the information provided by the originator.

The data base contains:

- TU-Black Sea Data Base (quality checked and loaded into the OceanBase-II system),
- Manual for the OceanBase-II system,
- Technical reports of the Expert Groups,
- Raw data sets delivered by participants,
- Climatic and averaged data sets,
- Additional and special data sets,
- Inter-calibrated data sets of the recent coordinated Black Sea cruises.

All these components are stored on the CD-ROM, which also contains the setup module for the OceanBase system. The total volume of data and information on the CD-ROM is 302 Mbytes. Most of the data delivered by the participating institutions were loaded into the OceanBase-II system. Only time series, optical spectra and all types of averaged data were not loaded due to their incompatibility with the structure of the used data base.

Data base contains data for 116 variables (main groups of these variables are listed above) from 271 data set. It includes 8,364,731 data value for the 26,035 stations. Distribution of all stations in the Black Sea basin is shown in Figure 2. Figure 2 describes all stations and give the total information on the data base. Not all variables were measured at all stations. Some of them were measured at most of the stations (for example, temperature and salinity). Some of them were measured not frequently, and some were measured rarely. Full information on each variable can be obtained quickly from the data base using the possibilities of the OceanBase system.

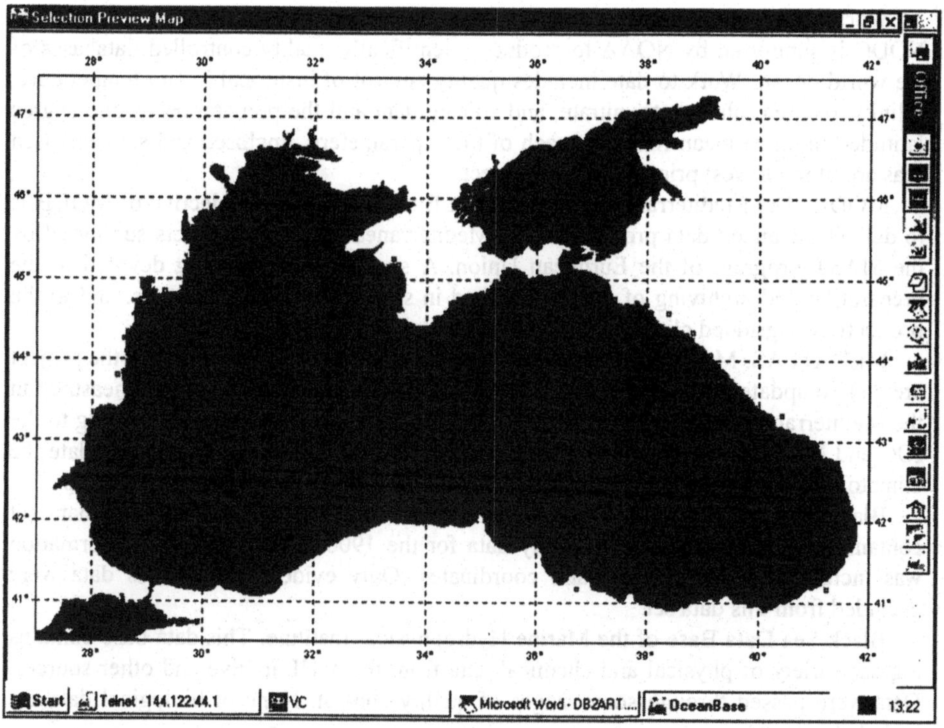

Figure 2. Positions of oceanographic stations exists in the database (the picture was obtained using OceanBase).

8. Comparison with other data bases

Brief information is given below on some well-known recent regional and global data sets to compare them with the TU-Black Sea Data Base.

The North Sea Tidal Data CD-ROM is a pilot project within the MAST Program of the Commission of European Communities. The CD-ROM contains tidal data (water levels, currents and sets of tidal constants) of the North Sea, with emphasis on data from the 10-years period 1984-1993. The data have been collected from relevant data sources in Netherlands and by a search of international gray literature. All data have been quality controlled and formatted.

NERC North Sea project Data Set CD-ROM was created by British Oceanographic Data Center in 1992. It contains the multi-disciplinary data set for the region for the 1987-1992 period.

The BOFS CD-ROM (the Biogeochemical Ocean Flux Study) contains 98% of all data series collected during the BOFS field program (1989-1993) in the North Atlantic and covers a great diversity of biological, chemical and physical measurements both in the water column and in the air-sea and benthic boundary layers.

World Ocean Atlas CD-ROM Series 1994. The Ocean Climate Laboratory (OCL) at NODC is supported by NOAA to produce scientifically quality controlled databases of the world ocean. Work to date includes quality control of historical in situ temperature, salinity, oxygen, phosphate, nitrate, and silicate data and the preparation of one-degree latitude-longitude mean fields for each of these parameters. Enclosed and semienclosed seas are of the lowest priority for this project.

MODB. The Mediterranean Oceanic Data Base. The general objective of MODB is to deliver advanced data products to the Mediterranean research projects supported by the MAST program of the European Union. A special effort is being devoted to the preparation and archiving of quality checked in situ hydrographic data sets, and to the production of gridded climatological fields for temperature and salinity.

MEDATLAS. Mediterranean Hydrographic Atlas. The main objectives of the project are: (1) To update the available data sets of temperature and salinity profiles measured in the Mediterranean Sea. (2) To check the data quality for scientific use according to the IOC and MAST recommendations. (3) To merge the compiled data sets and update the climatological statistics with a resolution adapted to the Mediterranean space scales.

Black Sea Hydrographic Data Set of the Moscow State University. This data set contains only temperature and salinity data for the 1900-1990. No header information was included except the date and coordinates. Only evidently erroneous data were excluded from this data set.

Black Sea Data Base of the Marine Hydrophysical Institute. This data base contains a great variety of physical and chemical data from the MHI archive and other sources. Data were passed through some stages of quality control. None of biological data are included.

Comparing the TU-Black Sea Data Base and these data bases one can conclude that it is the first successful attempt to create a regional historical interdisciplinary multipurpose database supplied with the special powerful DBMS. The development of the TU-Black Sea Data Base is a unique event in modern regional oceanography.

9. Conclusions

In relation to interdisciplinary data, the data base created through the NATO TU-Black Sea Project is a first in ocean science. The main features of the TU-Black Sea Data Base are as follows:

- The database includes all main physical, chemical and biological (including plankton) variables for the entire Black Sea basin; beginning from 1963, with extensive data sets for the period 1973-1994. It contains data from 26,000 stations.;
- It encompasses a crucial period in the history of the Black Sea ecosystem, from the background situation in the 1960s to the present, and thus includes the period of drastic changes in recent years;
- It includes data from all main regional and international sources; All data included into the database have been quality checked by groups of regional experts.; Each

value of physical, chemical and biooptical data is accompanied with a quality flag, as in the MEDATLAS project, and provides regional compatibility).

It can be roughly estimated that the TU-Black Sea database includes now from about 10% (e.g., temperature and salinity) to 30% (e.g., transparency) of all existing data of measurements performed in the Black Sea since the beginning of Ocean Science in the basin.

A special and powerful DBMS has been designed for to provide easy and quick access. While the main objective of the database is to serve the development and implementation of ecosystem models, it can be used for other oceanographic and environmental purposes. It includes the initial set of ASCII files and hence provides the possibility of creating an alternative database, using different quality control standards. The database has been delivered to participants on a CD-ROM.

Acknowledgments

We would like to thank Vladimir Miroshnichenko and Vyacheslav Lyubartsev for the titanic work on the new software creation. We wish to thank also all members of the Task Team: Dr. A. Mishonov, N. Kapustina, V. Tsiganok, L. Galkovskaya; heads of the Expert groups: L. Ivanov, S Konovalov, V. Melnikov, A. Mikaelyan, and O. Yunev; all members of Expert groups and all scientists, who performed the huge work on the data base compilation and data quality control. Our special thanks to Professors Umit Unluata and Valery Eremeev for their permanent support of this work.

References

1 Vladimirov, V.L. (1992) Integrated data bank for the cruise data of the research vessel, in *Automated systems for the monitoring of the marine environment*, Marine Hydrophysical Institute, Sevastopol, pp 126-131 (in Russian)

2. UNESCO (1993) *Manual of Quality Control Procedures for Validation of Oceanographic Data* Prepared by: CEC: DG. XII, MAST and IOC/IODE. Manual and Guides No.24.

3 ICES/IOC (1995) GETADE Formatting Guidelines for Oceanographic Data Exchange. Preliminary report

4. Vladimirov V.L., Miroshnichenko V.V. (1997) Multipurpose database management systems for the marine environmental research, in Nilgun B. Harmancioglu, M. Necdet Alpaslan, Sevinc D. Ozgul and Vijay P. Singh (Eds.), *Integrated Approach to Environmental Data Management Systems Proceedings of the NATO ARW*. Kluwer Academic Publishers, Dordrecht, pp 355-364.

GIS FOR REGIONAL SEAS PROGRAMMES: A CASE STUDY: THE BLACK SEA

Vladimir O. Mamaev
Guest Investigator
Woods Hole Oceanographic Institution
Woods Hole, MA 02543

David G. Aubrey
Senior Scientist
Woods Hole Oceanographic Institution
Woods Hole, MA 02543

Oleg Musin
Senior Scientist
Moscow State University
Moscow, Leninskie gory 119 888

Abstract

The Geographic Information System (GIS) has become a valuable tool for management of environmental resources. Recent applications to the marine realm have shown how a variety of diverse types of information can be presented simply and clearly, bringing together political, social, economic, scientific and management data in a visual format. The Black Sea GIS was developed as part of the Black Sea Environmental Programme of the Global Environmental Facility, taking some two years to develop with assistance of all Black Sea country specialists. The goal of the GIS was to present data in a graphic fashion, using a sophisticated Navigator to allow computer non-specialists to access a myriad of graphical data types. The IBM-compatible platform was chosen as the most frequent in the region, and a "home-grown" GIS was developed to be distributed free to users, to preclude the need for users to acquire expensive GIS software. The expense of GIS software is a specific impediment to its widespread use, particularly in the cash-poor Black Sea region where economic shifts inhibit free access to sophisticated software. The resulting GIS, consisting of some 600 individual maps and layers, provides information on a variety of data of importance to the Black Sea region, for managers, policy-makers, students, and scientists alike. It is a model for application to other regions where expensive software, poor access to sophisticated computer facilities, and slow transition to tele-communications inhibit free data availability.

S. Beşikteppe et al. (eds.),
Environmental Degradation of the Black Sea: Challenges and Remedies, 303–316.

1. Introduction

1.1 THE BLACK SEA: A UNIQUE ENVIRONMENT IN CRISIS

Almost one third of the land area of continental Europe drains into the Black Sea. It is an area that includes major parts of seventeen countries, thirteen capital cities and some 160 million persons. The second, third and fourth most important European rivers discharge into this sea, but its only connection to the world's oceans is the narrow Bosphorus Channel. The Bosphorus is as shallow as 70 meters and in places is only 700 meters wide, whereas the depth of the Black Sea itself exceeds two kilometers in more than half of its basin. The Black Sea is the largest and most voluminous natural anoxic basin in the world. Despite this situation, for millennia its surface waters supported a rich and diverse marine life. Its coastal inhabitants have also prospered from the abundant fisheries and, more recently, from the millions of tourists who flocked from all over eastern and central Europe to bathe in its warm waters and enjoy the beauty of its shorelines, plains and mountains.

In a period of only three decades, the Black Sea has suffered the catastrophic degradation of a major part of its natural resources. Increased loads of nutrients from rivers caused an overproduction of tiny phytoplankton, which in turn blocked the light reaching the sea grasses and algae, essential components of the sensitive ecosystem of the northwestern shelf. Much of the coastal ecosystem began to collapse. This problem, coupled with pollution and irrational exploitation of fish stocks, started a sharp decline in fisheries resources. Poor planning has destroyed much of the aesthetic resources of the coastlines. Uncontrolled sewage pollution has led to frequent beach closures and considerable financial losses in the tourist industry. In some places, solid waste is being dumped directly in the sea or on valuable wetlands. Tanker accidents and operational discharges have often caused oil pollution. These problems have reached crisis proportion at a time when five of the Black Sea countries are facing an economic and social transition and therefore have difficulty in taking the necessary urgent remedial actions.

In order to make an early start to environmental action and to develop a longer-term Action Plan, the Black Sea countries requested support from the Global Environment Facility (GEF), a fund established in 1991 under the management of the World Bank, the United Nations Development Programme (UNDP) and the UN Environment Programme (UNEP). In June 1993, an initial Phase I three-year Black Sea Environmental Programme was established; later phases have assured its existence up to present.

1.2. BLACK SEA ENVIRONMENTAL PROGRAME

The GEF Black Sea Environmental Programme has three primary objectives:

- To strengthen and create regional capacities for managing the Black Sea ecosystem;
- To develop and implement an appropriate policy and legal framework for the

assessment, control and prevention of pollution and the maintenance and enhancement of biodiversity;

- To facilitate the preparation of sound environmental investments

The BSEP is being implemented through an interactive matrix of national coordinators, thematic regional activity centers and focal point institutions targeting various themes such as emergency response, routine pollution monitoring, special pollution monitoring, biodiversity protection, coastal zone management, environmental legislation and economics, data management and GIS, and fisheries. A Project Coordination Unit (PCU), based in Istanbul, Turkey conducted the overall programme coordination.

The Black Sea GIS was one of many products of the BSEP [1]. Earlier products included a thorough bibliography of the Black Sea for the period from 1974-1994 [2], bringing to light the extensive research published on the Black Sea during this particularly active score or time.

2. Methods

The Working Party on Data Management and GIS (GISWP) was established in 1993 by the PCU in order to ensure a region-wide compatibility in the generation and management of data bases and to promote data exchange. The working party included at least one expert (Contact Person) from each of the Black Sea countries, together with additional external expertise where appropriate. The working party paid particular attention to the Geographical Information System (GIS) as a means to communicate data to environmental managers, decision-makers, students, scientists, and the general public.

It was agreed that all existing maps, which will be collected by the GISWP, should be collated and organized at the Moscow State University (MSU), Department of Cartography and Geoinformation. The MSU was also responsible for the distribution of an in-house GIS software to the participants of the project as well as for the organization of the relevant training workshops.

The data were digitized in their institute of origin and transferred to MSU. In close association with the PCU and selected experts, the MSU staff was responsible for the accumulation and transformation of all data into a GIS and development of a user interface for the system. MSU staff were also responsible for development of the modeling system for the creation of the digital models of the GIS layers, development of the DBMS "Black Sea" thematic query, as well as for development of the Data Base Management System (DBMS) "Black Sea" geographic query.

Data quality assurance was an essential element in the success of the GIS strategy. Data gathered for incorporation in the GIS database were accompanied by full information to enable evaluation of its quality. This included information on sampling methodologies employed, equipment and algorithms used, data inter-comparison exercises (where relevant), and estimated errors and uncertainties.

The Black Sea Geographic Information System was developed for use by governments, students, scientists, the general public, NGOs and the media for the following purposes:

- Planning for marine environmental activities and impacts on a regional scale;
- Public awareness through training, education, workshops, lectures, and media;
- Scientific analysis, modeling, ecological impact assessment, science planning

The main GIS components are designed to perform the following functions:

- Data input
- Data storage and database management
- Data analysis and processing
- Interaction with the user (graphics/map editing); and
- Data output and presentation (plotting)

3. The Black Sea GIS

The development of the GIS involved two major activities:

1. Development of special software package for digitizing, editing and storing of cartographic information, for modeling of continuous geo-fields, for processing of digital models and creation of thematic maps, for transformation of varying geographic coordinates into Mercator projection, and finally for demonstrating data. This task was the responsibility of the Department of Cartography and Geoinformation of the Moscow State University.

2. Collection of basic historical data on the Black Sea, processing and integration of those data in the system as well as collection and processing new data produced by the different thematic Working Parties of the BSEP

The GIS consists of seven thematic blocks representing different aspects of the Black Sea ecosystem. In each thematic block there is a set of map layers describing different aspects of the functioning of the Black Sea ecosystem; for some maps a relational data base is available.

In order to understand and manage ecological and anthropogenic processes better, it is necessary to understand the physical processes which form the basis for the Black Sea ecosystem, composition of the landscape, and human distribution, as well as many other important processes which form the unique environment of the Black Sea.

3.1 THE NAVIGATOR

The GIS was designed to flow effortlessly to the user non-schooled in computers. Intended for the general populous and non-scientific users, the GIS had to be self-explanatory, contain a simple flow from program opening to data presentation, and provide self-help capabilities. All this functionality is contained in the GIS, which has been distributed widely on CD-ROM.

"Navigator" is a module with the help of which a user of the Black Sea GIS can perform the following operations:

1. To select one of the maps of the system and open it for examination and reading of information represented on it;
2. To choose and enter one of the modules of the system:
   a) To accomplish a quantitative comparative analysis and superposition of maps (overlay, correlation),
   b) To make visual analysis of changes of various dynamic indices (e.g., precipitation, water temperature, wind),
   c) To work with the information subsystem on the Black Sea basin rivers;
3. To call "Nastavnik" ("Tutorial") - a programme tutoring the user in the Black Sea GIS;
4. To get brief reference information on the system.

The Navigator entry screen (Figure 1) describes the GIS, and allows the user to select various options:

Tutorial
About (help, background information and descriptions on use)
Presentation (the data themselves)
Exit

The options are selected using standard button format for ease of selection; alternatively, the user can use the "tab" function to parse through the various options.

If the Presentation option is selected, a window (Figure 2) is opened allowing the user to select between thematic map presentations or analysis of maps. Selection of "Thematic Maps" sends the user directly into the screen allowing selection of any of the seven thematic areas (Figure 3). The seven thematic map areas are clearly labeled. At any point in this process, the user may request help from the system. In addition, most maps allow "balloons" which identify the functions of various icons representing specific map or data functions. The seven thematic areas are described in more detail below.

While in the Presentation window, selection of "Map Analysis" immediately sends the user to the screen "Map Analysis" (Figure 4). The Map Analysis allows selection of several options: correlation, overlay, or animation. Several types of data lend themselves to correlation, overlay or animation, but not all data types will permit

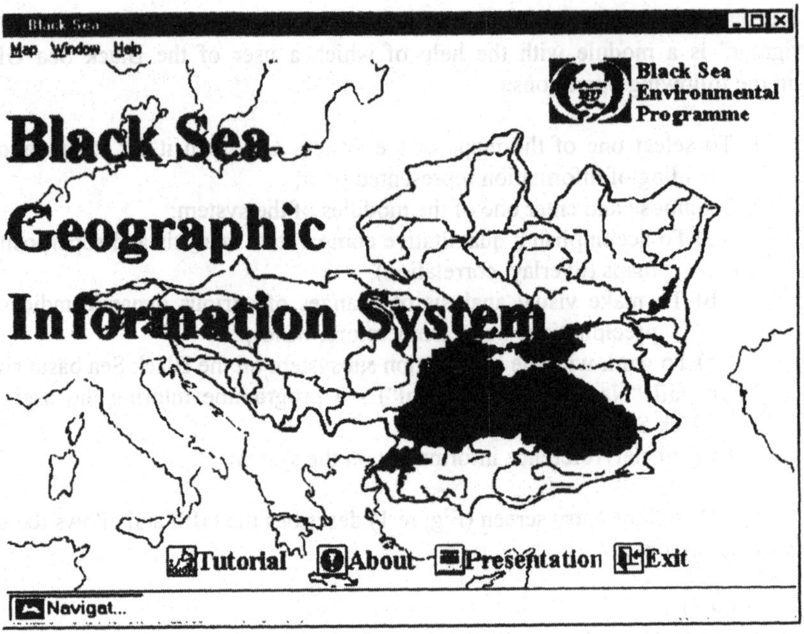

Fig.1 Navigator opening screen

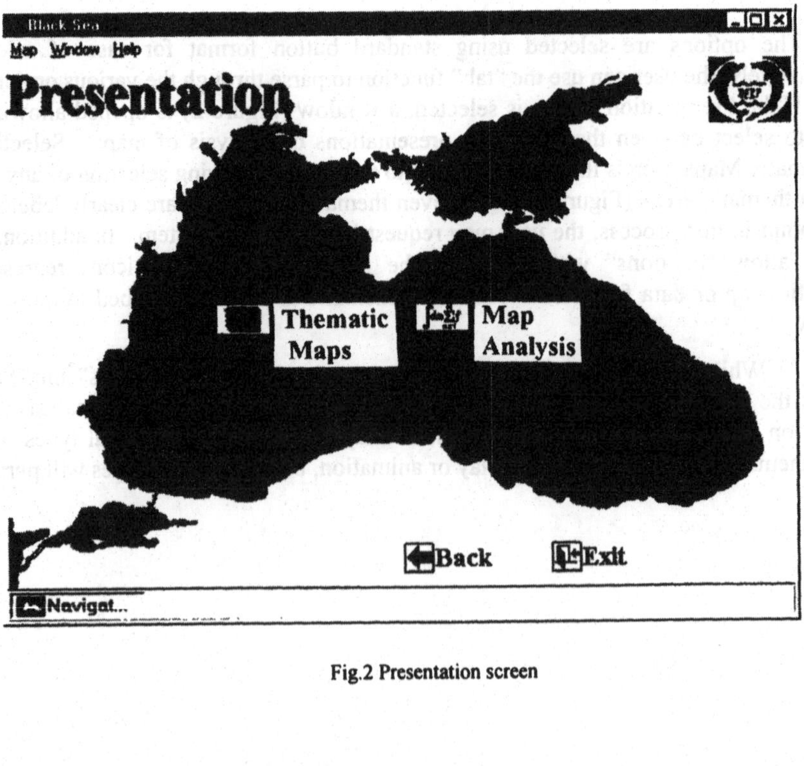

Fig.2 Presentation screen

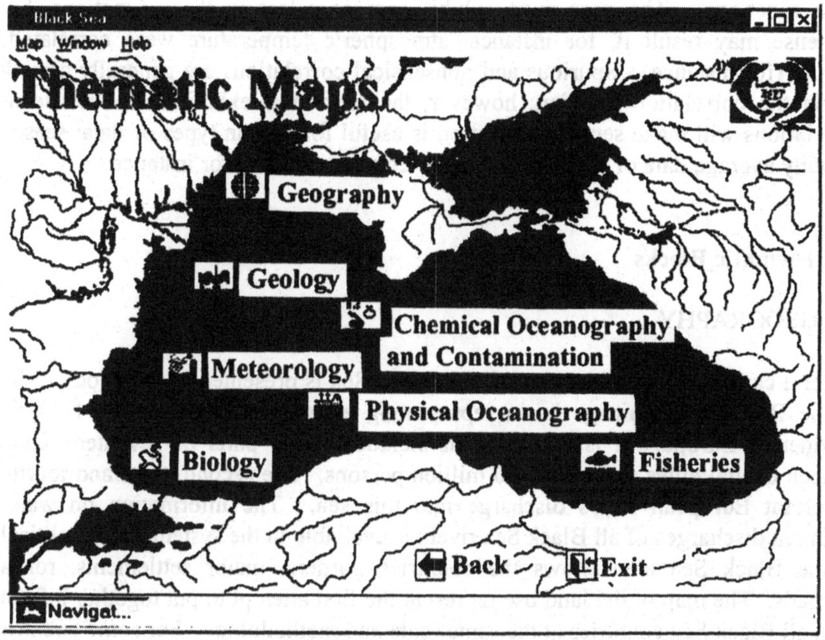

Figure 3. Thematic map screen

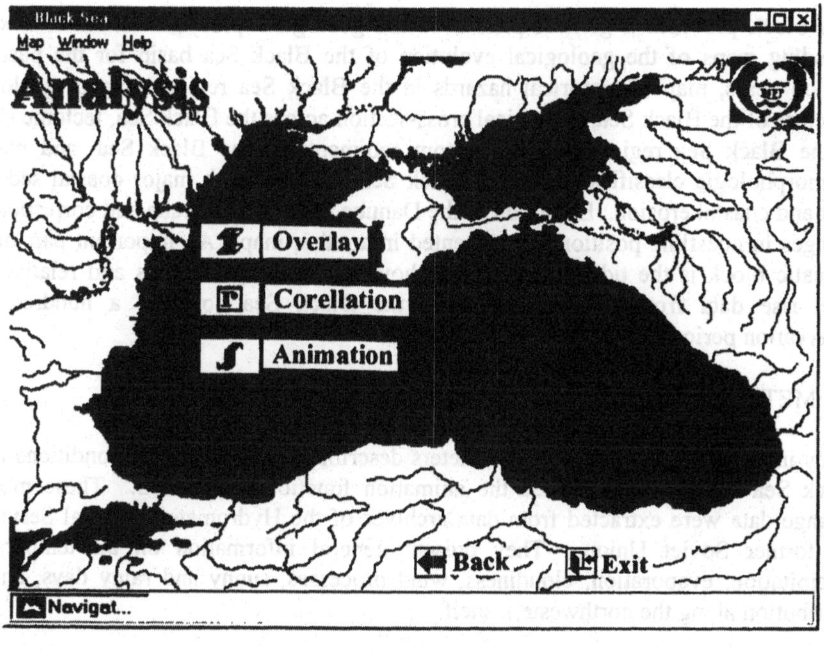

Figure 4. Map analysis screen

these functions. The user must exhibit care in selecting these functions, because nonsense may result if, for instance, atmospheric temperature were correlated with salinity (for instance). Spurious and nonsensical correlations are generally not allowed within the correlation function; however, the user must exhibit care because not all correlations will make sense. Animation is useful for certain types of data: seasonal or monthly average data of wind speed, water temperature, etc., for instance.

4. Thematic_Blocks

4.1 GEOGRAPHY

General cartographic information on the Black Sea is presented in this block. The map of the Black Sea drainage basin representing almost one third of the land area of continental Europe. It is an area that includes major parts of seventeen countries, thirteen capital cities and some 160 million persons. The second, third and fourth most important European rivers discharge into this sea. The information on water and sediment discharges of all Black Sea rivers is available in the system. The Political map of the Black Sea area shows the countries borders, major settlements, roads and railroads. The map of the land use represent the first attempt to put together information from all Black Sea countries at the same scale and methodology. For some countries the municipalities and country level administrative regions maps are accessible.

4.2 GEOLOGY

This block provides a general picture of the geological processes in the Black Sea including maps of the geological evolution of the Black Sea basin for the past 100 million years, map of historical hazards in the Black Sea region, map of geological structure of the Black Sea, geological cross-section across the Black Sea, tectonic sketch of the Black Sea region, map of bottom sediments of the Black Sea, and map of geomorphologic classification of the Black Sea coastline with major coastal sediment drift and coastal erosion. Evolution of the Danube Delta in Holocene and corresponding changes in coastline position are presented in another map. An important part of this thematic block is the tide-gauge section showing tide-gauge records and relative sea-level rise data from 48 station around the Black Sea covering a hundred year observation period.

4.3 METEOROLOGY

Many important meteorological parameters describing typical weather conditions in the Black Sea region are presented the animation function of the GIS. These monthly average data were extracted from data archives of the Hydrometeorological Service of the former Soviet Union. They present general information on air temperature, precipitation, evaporation, cloudiness, wind processes, sunny and rainy days, and ice distribution along the northwestern shelf.

4.4 PHYSICAL OCEANOGRAPHY

Physical processes occurring in the sea exert an important role in the formation of water masses of the sea as well as in the formation of specific hydrological features of the marine ecosystem. Black Sea oceanography was well studied during the last century. Many scientific cruises (map of oceanographic station network in the Black Sea) collected thousands of data records on many of the important parameters. Based on these data, a set of maps showing climatic oceanography of the Black Sea (temperature and salinity for each months and each season for 20 standard depths) as well as major water masses of the Black Sea were incorporated in the system.

Recent international programmes in the Black Sea were also conducting scientific interdisciplinary studies with main objective to collect physical and chemical data using common methodology, instruments, intercalibration and strict quality control. The most reliable data were collected during the Cooperative Marine Science Program for the Black Sea (CoMSBlack) surveys in 1992 and 1993. These data were used in the preparation of a set of maps presenting the distribution of temperature, salinity, dynamic topography, density anomaly, and cold intermediate layer in the Black Sea for 11 standard depths. Based on CoMSBlack data, Secchi Disk depth climatology maps were prepared. The map of seasonal mean circulation describes the general tendency of the Black Sea horizontal currents. The system is also supplied with satellite sea surface temperature data obtained by NOAA.

4.5 CHEMICAL OCEANOGRAPHY AND POLLUTION

Data set generated from cruises of CoMSBlack from 1992 and 1993, were used for the preparation of the maps showing the spatial distribution of the following parameters in the Black Sea for eleven standard depths: dissolved oxygen, hydrogen sulfide, inorganic nitrogen, inorganic phosphates and silicic acid.

In 1995 the Black Sea Environmental Programme organized a survey of the land-based sources of pollution in all six Black Sea countries using standard World Health Organization (WHO) methodology. Data collected during this survey are included in the system. Results of the 1995 sediment pollution survey in the open sea (Polygons study) are reported on the map of the state of the Black Sea pollution.

The location of the upper boundary of the hydrogen sulfide zone for different years and areas of hypoxia are demonstrated on the relevant maps. Specific information on the oil products and heavy metal pollution along the north-western Black Sea shelf are also available in the system.

4.6 BIOLOGY

The main objective of this thematic block is to demonstrate the richness of the Black Sea ecosystem in terms of biological diversity, present key species habitats, important protected areas as well as to illustrate the problems faced by the Black Sea ecosystem.

The map of Black Sea wetlands shows the distribution of wetlands around the Black Sea and gives detailed information on each of them. The system includes maps of existing natural reserves, sensitive species habitats, distribution of some exotic species (*Mnemiopsis* and *Aurelia*) in the Black Sea. Maps of primary production, distribution of phytoplankton, zooplankton and macrozoobenthos demonstrate the biological productivity of the Black Sea waters. Vegetation index and Chlorophyll A distribution maps are generated from CZCS data. Decrease of sea grass meadows on the Black Sea shelf and accidental and intentional introduction of species in the Black Sea concludes the biology block.

Two illustrative examples from biology include the disruption of the Black Sea ecosystem over time. Figure 5 shows the evolution of the *Phyllophora* sea grass meadows from 1950 through 1980. Phyllophora was the nucleus of a bio-community that included 118 species of invertebrates [3]. This field has declined dramatically as the Secchi disk depth has plummeted. In the 1950s, Zernov's *Phyllophora* field as the largest aggregation of red agar-bearing algae of this genus in the world, covering some 11,000 km2. By the early 1990s, the field was a mere 500 km2, and its biomass was reduced by more than 50-fold from early days.

A second example (Figure 6) describes one of the many invaders of the Black Sea, which has a long history of biological invasion, including both accidental and purposeful introductions. Introduced organisms include a long list, such as the sea barnacles *Balanus improvisus* and *Balanus eburneu*, the hydromedusa *Blackfordia virginica*, the polychaete *Mercierella enigmatica*, the hydromedusa *Bougainvillia megas*, the crab *Rhithropanopeus tridentata*, the gastropod *Rapana thomasiana*, the soft-shelled clam *Mya arenaria*, and the ctenophore *Mnemiopsis leidyi* (Mccradyi). Figure 6 shows the distribution of *Mnemiopsis* during a survey in August 1993, by Turkish, Ukrainian and Russian scientists.

4.7 FISHERIES

Based on archive data of Soviet Union fishery research activities in the Black Sea for 1980-1991, a set of digital maps was prepared showing the spatial distribution of two commercial species (anchovy and sprat) in the Black Sea. The following layers have been prepared: spawning stock distribution, eggs and larvae distribution.

Maps of distribution and migration of turbot, whiting, sturgeon, sprat, shad, red mullet, thornback ray, mullet, Mediterranean horse mackerel, mackerel, picked dogfish, bluefish, Atlantic bonito, anchovy in the Black Sea were prepared using FAO publications.

Two maps show examples of fish distribution. Figure 7 shows the distribution of sturgeon spawning and feeding areas throughout the Black and Azov seas. These extensive spawning areas have been adversely affected by development and pollution throughout the region. In Georgia, for instance, sturgeon numbers were estimated as

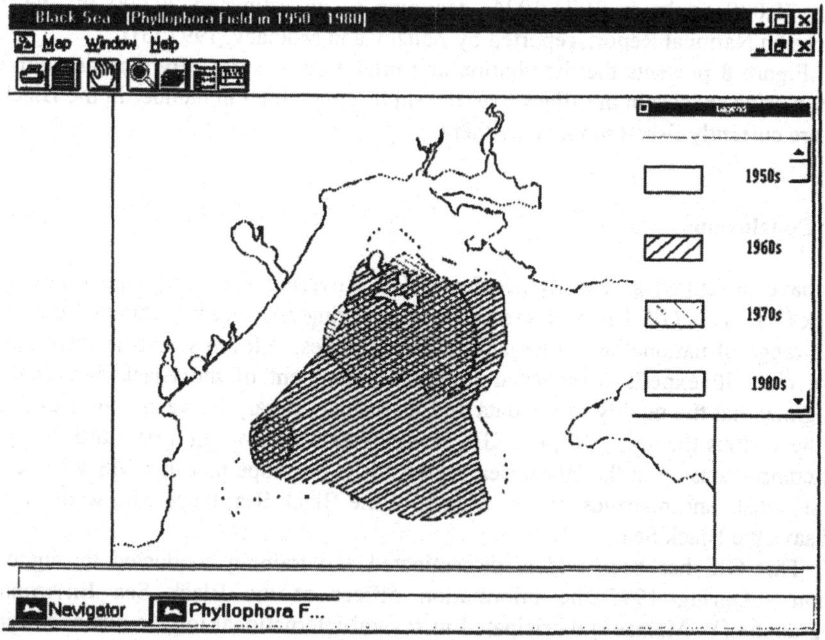

Fig. 5. Phyllophora fields, 1950-1980

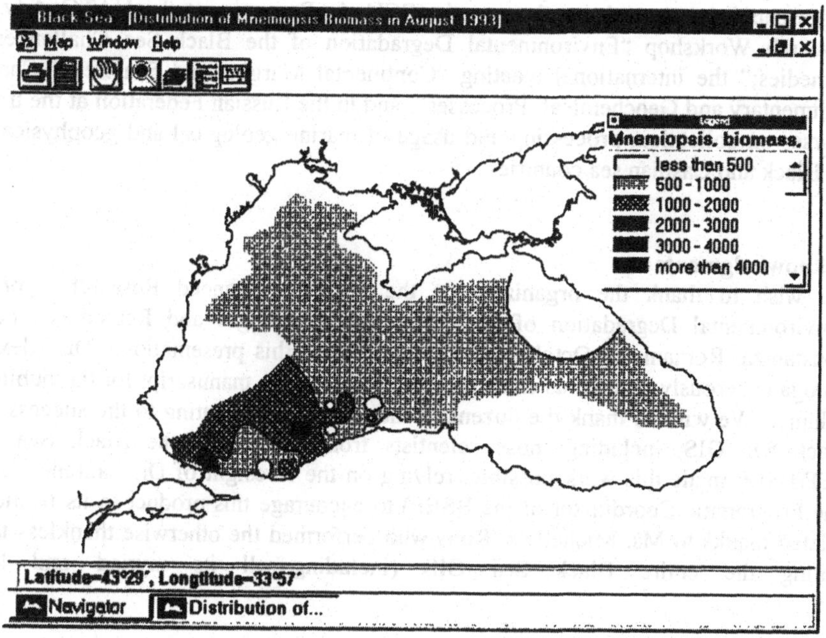

Fig. 6. Mnemiopsis biomass, Aug. 1993

some 75,000 adults in 1973-1974, declining to an estimated 20,000 by the 1990s (Georgian National Report, reported by Zaitzev and Mamaev, 1997 [3]).

Figure 8 presents the distribution and migration of sprat. By contrast with many other marine species in the Black Sea, the sprat remain high in number in the Black Sea, and are currently fished moderately there.

5. Conclusions

We have presented a recently-developed GIS covering the large-scale region of the Black Sea. This GIS has been created using existing and recently obtained data from a wide range of national and international data sources. Eleven scientific institution and more than 50 experts contributed to the development of the Black Sea GIS. We recognize that the quality of the data and its resolution vary between countries and also for the various themes. Still, this GIS is, as far as we know, the first multi-disciplinary and comprehensive in the Black Sea region. It is our hope that this GIS will be useful for scientists and manages, for all friends of the Black Sea, those who want to protect and save the Black Sea.

This GIS has been widely disseminated and training conducted throughout the region. During 1997 the information officer of the Black Sea Environmental Programme (V. Mamaev) participated in several international forums demonstrating the results of BSEP, sharing experience and knowledge with interested parties. Several demonstrations and training workshop on the Black Sea Geographic Information System were conducted in 1997: in Turkey - UN Office in Ankara, meeting of UNDP Resident Representatives, several meetings at the PCU; in Romania, at the NATO Advanced Research Workshop "Environmental Degradation of the Black Sea Challenges and Remedies;" the international meeting "Continental Margins and Sea Level Changes-Sedimentary and Geochemical Processes"; and in the Russian Federation at the training courses on collection, processing and usage of marine geological and geophysical data for Black and Caspian sea countries.

Acknowledgments
We wish to thank the organizers of the NATO Advanced Research Workshop "Environmental Degradation of the Black Sea Challenges and Remedies", held in Constantza, Romania in October, 1997, for inviting this presentation. Dr. Alexandru Bologa generously granted us extra time to complete this manuscript for the publication volume. We wish to thank the dozens of individuals contributing to the success of the Black Sea GIS, including those scientists from all around the Black Sea. The GEF/BSEP made this work possible, relying on the foresight of Dr. Laurence D. Mee (the Programme Coordinator of the BSEP) to encourage this product to its fruition. A belated thanks to Ms. Michelle J. Roos who performed the otherwise thankless task of editing the entire Black Sea GIS (including all its myriad and hidden

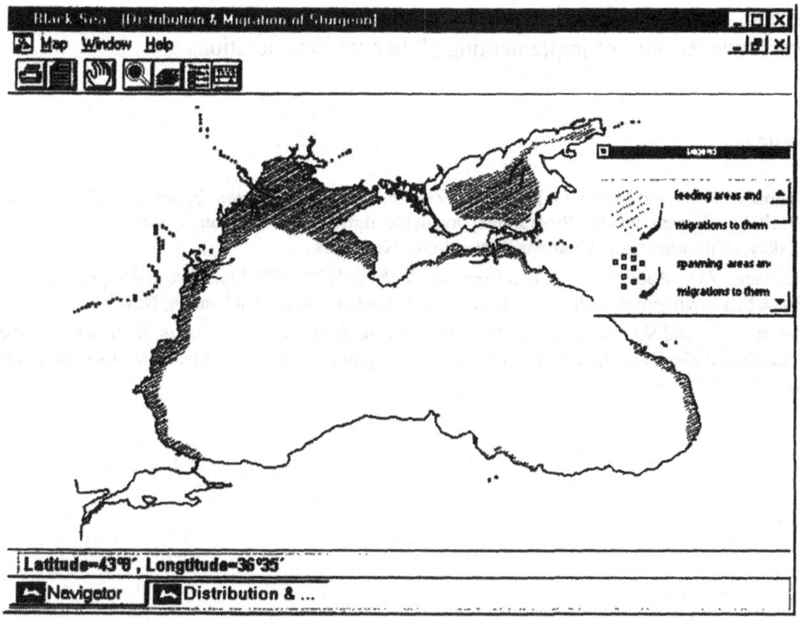

Fig. 7. Sturgeon spawning and feeding areas

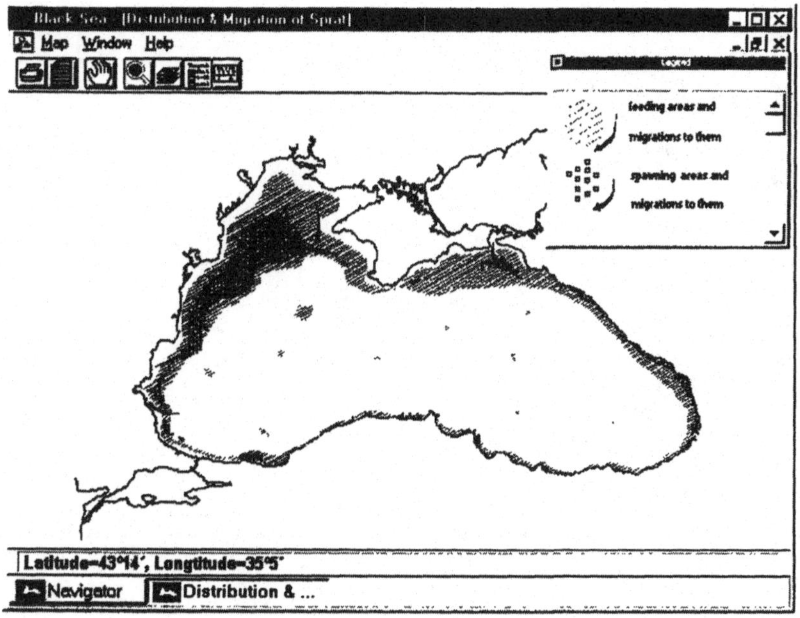

Fig. 8. Distribution and migration of sprat

layers and forks) for English usage; remaining errors are the responsibility of the authors, however, for not implementing all her recommendations.

References

1. Mamaev, V. O., and Musin, O. R. (eds.) (1997) *Black Sea Geographic Information System*, CD-ROM, Black Sea Environmental Programme, available through the Program Coordination Unit, Istanbul, Turkey, or through United Nations Publications, New York.
2. Mamaev, V.O., Aubrey, D. G., and Eremeev, V. N. (eds.) (1995) *Black Sea Bibliography, 1974-1994,* Black Sea Environmental Program Series, vol. 1, United Nations Publication, New York.
3. Zaitzev, Yu., and V.O. Mamaev, V. O. (1997) *Marine Biological Diversity in the Black Sea: A Study of Change and Decline,* Black Sea Environmental Program Series, v. 3, United Nations Publications, New York.

TOWARDS DEVELOPMENT OF AN OPERATIONAL MARINE SERVICES SYSTEM IN THE BLACK SEA

V. RYABININ[1], M. POPOVA[2], J. POITEVIN[3], P. DANIEL[3],
A. FROLOV[1] and G. KORTCHEV[2]
[1] *Hydrometeorological Research Centre of the Russian Federation*
9-13, Bol. Predtechensky Per, Moscow, 123242, Russia
[2] *National Institute of Meteorology and Hydrology*
66, Tzarigradsko chaussee, 1784 Sofia, Bulgaria
[3] *Meteo-France*
42, av. Gustave Coriolis, 31057 Toulouse Cedex, France

1. Introduction

In this paper a research and technological development program aimed at setting up a modern system of operational **BLACK** Sea **MAR**ine **S**ervices (BLACKMARS) is described. Its idea was expressed at the first session of the IOC Regional Committee for the Black Sea (Varna, September 1996). The program is seen as a regional project of the Global Ocean Observing System (GOOS) Program initiated and supported by the IOC of UNESCO and other international organisations. BLACKMARS is to be based on co-operative efforts of all riparian countries. The main activity within BLACKMARS will be the organisation of operational generation and distribution of real – time marine meteorological and oceanographic products based on observations and up-to-date numerical models in accordance with regional user requirements.

2. Goals and objectives of the program

The goals of BLACKMARS are:

<u>Creation of a modern regional operational marine services system in the Black Sea.</u> The system will help to ensure better safety and efficiency of marine activities in the region, will directly and indirectly benefit many users and will act as an important factor in mitigating present critical ecological situation in the region and maintaining ecological security. Operational marine meteorology and oceanography can create new jobs in the Black Sea countries.

<u>Development of a regional operational marine information system.</u> Strengthening of working contacts of relevant authorities in all countries around the Black Sea and initiation of operational exchange of data and products between them in real-time mode

S. Beşiktepe et al. (eds.),
***Environmental Degradation of the Black Sea: Challenges and Remedies*, 317–336.**

via reliable and fast telecommunication media will support all marine Programs and activities in the region.

Related development of marine sciences and technologies. National meteorological services and operational oceanographic agencies will require up-to-date products and technologies, and this need will stimulate corresponding scientific research. Regular flow of observational data and availability of analysed fields of oceanographic parameters will facilitate diagnostic studies and understanding of oceanographic processes.

In accordance with BLACKMARS goals the following objectives are specified:

1. Development and support of regional Black Sea forecasting system infrastructure.
2. Organisation of up-to-date marine meteorological information broadcasting.
3. Restoration and improvement of the Black Sea marine regional observing network with reliable data collection.
4. Creation of elements of the operational oceanography information system in the Black Sea with up-to-date telecommunications and distributed databases.
5. Development of regional capacity building and training in operational oceanography.
6. Development and validation of assimilation/processing schemes, forecasting models along with their eventual implementation in forecasting practice.
7. Establishing of a system for regional archiving of operational marine information.
8. Evaluation of observing network sufficiency and optimal development of the network.
9. Up-grading of national and regional capabilities to protect marine environment in case of accidental release of hydrocarbons and other pollutants.
10. Development of a subsystem for sea surface fluxes and sea surface temperature forecasts for forcing of oceanographic models.
11. Support of dedicated specialised oceanographic services for marine activities.

3. Relation to existing international conventions and Programs

The following conventions and international agreements form the basis for BLACKMARS activities:

- AGENDA for the 21st Age Report of UNCED
- Convention on Safety Of Life At Sea (SOLAS)
- UN Framework Convention on Climate Change
- Convention on Biological Diversity
- UN Convention of Law of the Sea (UNCLOS)

There are also regional conventions, which have direct relation to BLACKMARS. They are:

- Convention on the Protection of the Black Sea from pollution ("Bucharest Convention") and related protocols on land-based sources of pollution, oil pollution, buried pollutants.
- Ministerial declaration on the Black Sea protection (Odessa, 1993).

Development of BLACKMARS documentation and plans as well the Program implementation can benefit from interaction with many on-going international Programs. Most important of them are listed below:

- GOOS
- World Weather Watch (WWW) of the World Meteorological Organisation (WMO)
- The Integrated Global Ocean Services System (IGOSS) of the IOC and WMO
- Global Sea Level Observing System (GLOSS)
- International Oceanographic Data Exchange (IODE) System of the IOC

Major regional programs and projects pertaining to the BLACKMARS are:

- The Black Sea Environmental Program (BSEP) of the GEF, and Strategic Action Plan for the Rehabilitation and Protection of the Black Sea,
- The Environmental Program for the Danube River Basin of EU and GEF,
- The Co-operative Marine Science Program for the Black Sea (CoMSBlack),
- The TU-BLACK SEA Project of NATO,
- The TU-WAVES Project of NATO,
- EROS 2000 Project of the EU,
- Related activities under umbrella of IAEA, PHARE, and TACIS.

Some of these Programs have already ended, however, the data, experience, and the scientific and technological results obtained in the course of their execution will be of value for BLACKMARS. Development of BLACKMARS should benefit all riparian countries of the Black Sea. Therefore strong links need to be established to the related Programs on national, bilateral and multinational level.

4. Tentative time-table

BLACKMARS is to be developed during a five-year period, in three phases. The first two phases will span 4 years and will be devoted to the system design, development of the most appropriate forecasting models and observation methods, upgrading observational and telecommunication networks, creation of distributed database and testing the models. The third phase will be pre-operational. Each phase will end with some demonstration activities so that potential users will have an opportunity to acquire knowledge and understanding of the importance and usefulness of GOOS services for successful maritime activities in the Black Sea.

5. Need for the services and their users

BLACKMARS is urgently needed in the region for several reasons. First of all, the new political map of the area requires reconsideration of international responsibilities for the provision of maritime services for open sea and even coastal waters. Mandatory GMDSS standards for provision of safety information for shipping through INMARSAT SAFETYNet and NAVTEX systems, which should come into force in 1999, are not met in the region. Thus in its present state the safety of maritime activities in the sea is below common standards, and efficiency of the activities is suffering due to lack of adequate marine meteorological and oceanographic support. The main reason for this is, of course, inadequate attention to and financing of relevant services. At the same time, even if the funding *were* adequate, the degradation of marine services would still be inevitable because of considerable reduction of ship weather reporting during past 5-10 years.

In the existing critical situation of the economy of the region, every possibility should be used to minimise damage due to dangerous atmospheric or marine phenomena. Accordingly, of the *highest priority* for the safe maritime activities and sustainable use of the environment will be the development of a regional system for preparation and distribution *of real – time warnings about natural marine and atmospheric disasters*, such as storms, especially in the shelf zone. To be able to issue such products the National Meteorological Service (NMS) of the six riparian countries require operational *analyses and forecasts of wind, sea state (wind wave), current, storm surge, oil spill evolution, sea ice transport, and ship's icing.* Such products are of a general nature and should be issued on a regular basis by responsible national organisation(s).

At the same time, in the region, as everywhere in the world, there are some users requiring specific information. Their list includes:

- port/harbour operations;
- shelf research, engineering, exploration, and resources exploitation;
- recreation;
- weather sensitive operations at sea;
- towing of constructions with limited seaworthiness;
- fishing;
- mariculture;
- navies;
- special craft (hydrofoil, etc.) operations;
- coastal protection;
- pollution combat operations;
- search and rescue at sea;
- sporting events (such as yacht races);
- cable laying.

Very important users of the system will be scientific agencies involved in the research of the Black Sea ecosystem. In turn they are potential providers of operational oceanographic products, such as analyses and forecasts of ecosystem behaviour and, eventually, real – time consultation services for restoration of the sea health.

6. Basic directions of BLACKMARS research and technology development

BLACKMARS will co-ordinate related activities of institutes for meteorological, oceanographic, environmental research, emergency response agencies, etc. Feasibility of BLACKMARS is based upon advances in observations, real time data transmission and processing, and modelling of relevant processes. The concept of "real time" refers to intervals from hours to decades depending upon the characteristic time scale of the phenomenon variability, and on specific user needs.

6.1. BLACK SEA OBSERVATION SYSTEM

In accordance with GOOS principles, all BLACKMARS observations will be *long-term, systematic, relevant to the overall objectives, cost efficient and routine.* In addition, in provision of such observations, BLACKMARS should base its development on better co-ordination and integration of existing capabilities of participating countries rather than on development of new systems.

Most important elements of the Black Sea marine observing system are:

- coastal stations;
- voluntary observing ships;
- off-shore platforms;
- buoys;
- tide gauges and waveriders;
- satellite subsystems.

6.1.1. Coastal stations

The National Meteorological Services of the coastal countries maintain 46 coastal marine meteorological stations (Bulgaria – 6, Georgia – 8, Romania – 4, Russian Federation – 5, Turkey – 7, Ukraine – 16). These stations monitor with three hours interval the following parameters: air and sea water temperature, direction and speed of wind, height, period and direction of waves, atmospheric pressure, solar radiation, visibility, precipitation, humidity. A limited number of stations observe sea level, some chemical parameters, air pollution, etc. Some coastal marine meteorological stations have more than 50-year long series of observations. At all the stations observations are made manually. In addition, there are three automatic stations in Varna (Bulgaria), Constanta and Gloria (Romania) with limited number of sensors. Coastal stations report their observations with coded data, which are collected at National Meteorological Centres (NMC), as a rule no later than 20 minutes after the time of observation.

For BLACKMARS it will be very essential to install a small number of up-to-date marine automated stations at specially selected coastal locations, which will lead to a rapid improvement of the observation network performance.

6.1.2. Ship observations

Since 1987 a marked decline has been observed in the total number of the VOS reports from the Black Sea. It decreased from several hundreds to about 2-3 reports per day in 1997, which are, as a rule, available only at one NMC. The main reason for this is total decrease of the merchant shipping activity due to the adverse economic situation, and the refusal of ship's crew to make and report observations, mainly because transmission of reports to shore radio station is not free of charge for the ship.

Completely lacking in the Black Sea are operational observations of the sub-surface temperature versus depth obtained with XBT (Expendable Bathythermograph) and of sub-surface temperature and salinity versus depth obtained with CTD (conductivity / temperature / depth) or XCTD (expendable CTD) probes.

Urgent efforts are required for the restoration of at least minimally acceptable number of observations at open sea. *A possible solution to the problem of the observation lack in the open part of the Black Sea is related to use of regular schedule vessels such as ferries and tankers.* Ferries probably represent the optimal platforms for *in situ* observations because they operate like shuttles, adhering to a strict schedule.

An option is to equip a limited number of VOS with instrumentation enabling them to make discrete measurements of concentrations of various biogeochemical parameters, which are necessary for the assessment of water pollution and the Black Sea ecosystem status. The approach related to the use of remotely operated instruments on ferries and tankers seems to be the most doable in the current situation. It is likely to ensure data flow with at least minimally required spatial and temporal resolution.

6.1.3. Off-shore platforms

Some activities are underway in the region to install a number of off-shore oil and gas production platforms. As their work requires operational marine services, it is desirable that oil and gas platform operators carry out at least basic observations.

6.1.4. Buoys

During stormy periods buoy observations represent the only possible and reliable source of data from the open sea. Unfortunately, at present there are no operational buoys in the Black Sea although some buoys had been provided to research organisations of Bulgaria, Romania, the Russian Federation, and Turkey under the TU-Black Sea project of NATO. In addition, it would be relevant to deploy some low cost drifters in order to have some additional measurements of atmospheric pressure. It would also be desirable to deploy some moored buoys at several specially assigned locations to generate long series of observations and to use the corresponding data as widely as possible.

6.1.5. Tide gauges and waveriders

There are more than 24 sea level recorders in the coastal marine meteorological stations measuring sea level every 6 hours. It is also desirable to install a waverider close to entrance of every big harbour in the Black Sea.

6.1.6. Other types of in situ observations

It is expected that other operational oceanographic observations will be done by oceanographic agencies in the course of various international programs and projects. It will be very important to make buoy and other types of data, which are made in real time as parts of scientific Programs, available to regional operational services.

6.1.7. Satellite observations

All NMCs in the region have, as a rule, access to NOAA and METEOSAT satellite observations. In addition WMC Moscow has access to Russian polar orbiting satellites METEOR and OCEAN.

During recent years some new polar orbiting satellites such as ERS-1, ERS-2, TOPEX/POSEIDON, ADEOS were launched. They are equipped with altimeter, scatterometer, synthetic aperture radar and radiometer. These instruments allow measurements of wind strength and direction (with scatterometer), wave height and dynamic topography (with altimeter), wind wave spectrum (with synthetic aperture radar) and the sea surface temperature (with radiometer). *It will be relevant to use these data by meteorologists and oceanographers of all the countries around the Black Sea.*

6.1.8. Observational network development

The BLACKMARS observational network is seen as a combination of coastal marine meteorological stations, regular schedule vessels, such as ferry-boats and tankers, and also some buoys and waveriders. The land based network has to interact properly with satellite observations in the region. Network optimisation activities should ensure appropriate

- structure of the network;
- measurement locations;
- cost-effective telecommunications;
- feed-back connections;
- human participation;
- utilisation of existing observational networks;
- needed power supply;
- network management.

It is necessary to assess the impact of different observation types on the quality of analyses and forecasts. Observing system sensitivity experiments are a very useful tool for these studies.

6.2. TELECOMMUNICATIONS AND DATA MANAGEMENT

The basic organisation of data flow in BLACKMARS is shown in Figure 1. The system should give access to observational data for all agencies providing oceanographic services and should also ensure that all "users of oceanography" have access to desired services and data. Therefore the system needs good telecommunications and up-to-date data management.

6.2.1. Marine telecommunication network

The WMO Global Telecommunication System (GTS) is an appropriate telecommunication media for the BLACKMARS purposes. It is a three level structure connecting three World Meteorological Centres and fifteen Regional Telecommunication Hubs (RTH) at the first level, fourteen RTHs at the second level and more than 120 National Meteorological Centres (NMC) at the third level. All the GTS connections are implemented via leased two-ways data transmission circuits, 75 % of the circuits being operational at speeds from 9,6 kbps to 64 kbps. GTS was created for global exchange of all observational data (produced by surface, upper-air, ship, aircraft platforms, satellites, radars, buoys, etc.) as well as of all products (analyses and forecasts in alphanumeric or binary formats) created by the WMO WWW World and Regional Processing Centres. It ensures that every unit on the GTS has reliable access to necessary data and products in real-time and almost real-time. The total amount of GTS traffic is very large (for example at the first level it is more than 700 Mbytes per day). The system is designed to have at least 20% spare capacity and is generally open for operational exchange of data within other environmental programs (such as oceanographic data, seismic data, warnings and data of IAEA, etc.). National collection of data and distribution of products by NMCs is organised by all WMO members using cost-efficient telecommunication media able to accommodate the required traffic – dedicated circuits, such as meteorological radio networks, INTERNET, Public Package Switch Network (PPSN), TELEX.

In some regions of the world such as North America, Western Europe, etc. it may be recommendable to use the INTERNET for operational exchange of oceanographic data and products, but for the Black Sea region this is not acceptable. In 1995 and 1996 the Regional Telecommunication Hub (RTH) Sofia (one of the RTHs at the first level of the GTS) organised a two-month long experiment on using the INTERNET for *international* meteorological telecommunications between some countries in the south-eastern part of Europe and Near East. The results showed that 63% of the data volume was lost and more than 75% of received data had a delay unacceptable for the data operational use (more than 180 minutes). Poor performance was due to the lack of direct INTERNET links between adjacent countries so that INTERNET sessions between the countries (including use of such useful procedures as Telnet, FTP, etc.) needed re-routing via more than 20 servers in Europe and USA. Because of that processor – to - processor sessions were very slow and considerable loss of data occurred.

At the same time NMC Sofia has very positive experience of using INTERNET system for *national* meteorological telecommunications. Some other reviews also show that national INTERNET networks in all six Black Sea region countries are good enough at present and become better every year.

Taking into account availability of the WMO GTS and known shortages of direct INTERNET links between the BLACK Sea countries *the best composition of the marine telecommunication network in the Black Sea region is seen as a two-level structure incorporating*

- *the WMO GTS connections in the region for the international exchange of oceanographic data and products*

  and

- *national systems based on an appropriate combination of telecommunication media such as INTERNET, PPSN, VHF radio networks, etc.*

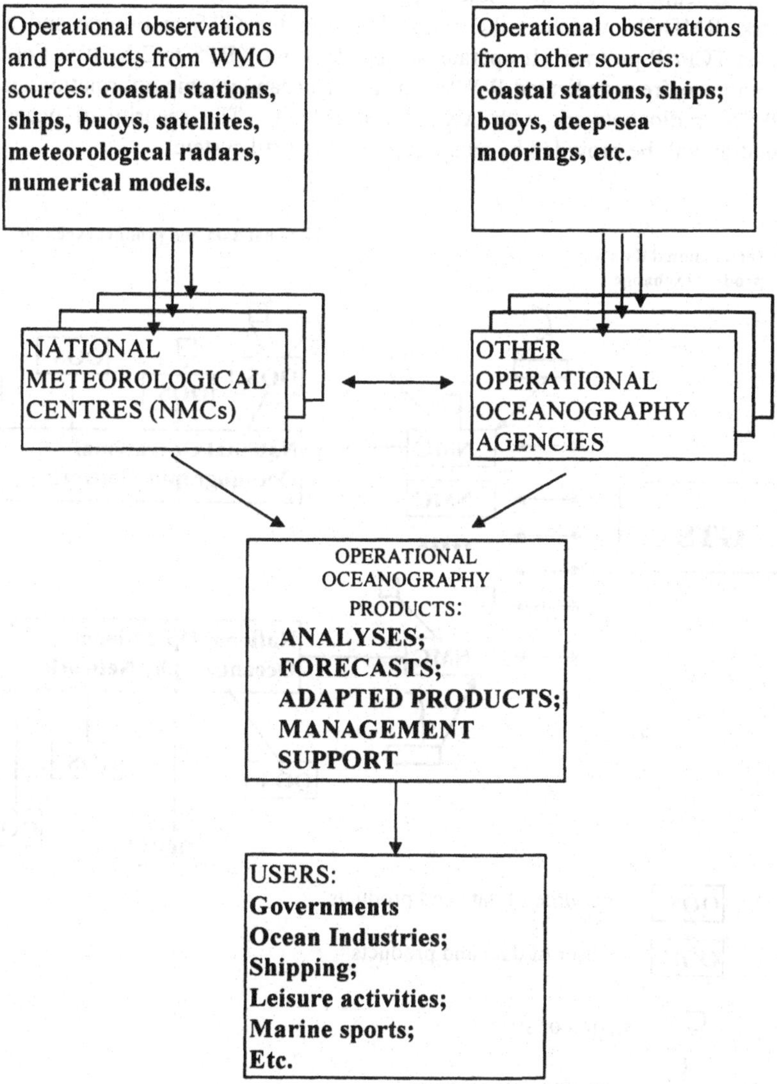

Figure 1. BLACKMARS data flow

Performance of the marine telecommunication network can be further analyzed and its structure optimized by use of Petri nets theory [17]. It will be possible to ensure that every unit on the network (data source or product user) has real – time access to all desirable units on the network such as data bases, processing centres, etc. Figure 2 shows the proposed structure of the Black Sea marine telecommunication network. In the near future it will be possible to accommodate regional real-time marine telecommunications on the new Regional Meteorological Data Communication Network, RMDCN, which is the regional part of the GTS in Europe and Near East based on TCP/IP protocol. It operates at speeds of 64 KBPS to 2 MBPS. For that each NMC will require a dedicated WWW server for oceanographic information and servers for PPSN, radio network with AX.25 protocol, etc. The security of oceanographic information will be provided by an appropriate fire wall system.

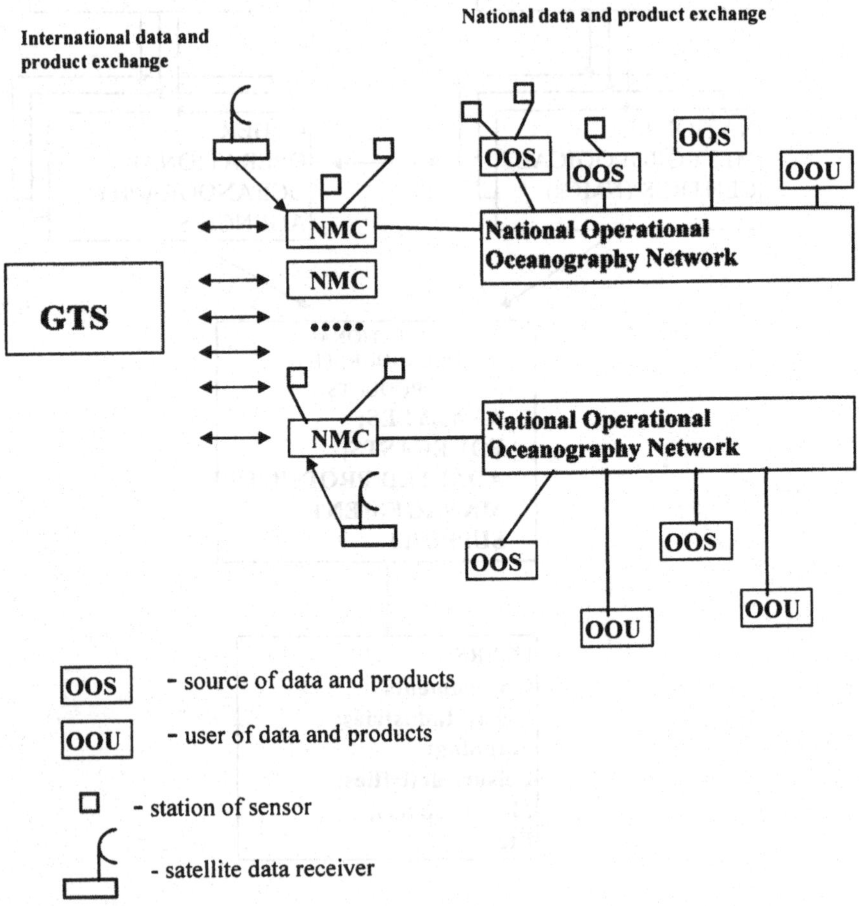

Figure 2. The Black Sea marine telecommunications arrangement

6.2.2. Data management

Data management functions of BLACKMARS include:

- Procedures for access to data and products;
- Codes and formats for the exchange of operational oceanographic data and products;
- Procedures for quality check and quality analysis of the collected data;
- Set-up of Distributed Data Base for the Black Sea operational marine services;
- Exchange of and access to metadata;
- Registering and exchange of developed software;
- Monitoring of the Black Sea marine services system operation.

Some codes and formats for operational oceanographic information exchange already exist (e.g., BATHY, TESAC, ...) but there is no doubt that many additional codes and formats will be developed in future, especially for new types of ecological oceanographic information. Binary codes such as BUFR, GRIB will be extensively used. Advantages of using formats of the IOC, ICES, JGOFS, GTSPP, such as GF3 should be studied in detail [11]. Obviously, such activities must be co-ordinated through all GOOS-related Programs and projects.

Procedures for real-time and non-real-time monitoring of the oceanographic information as well as procedures for quality control and quality analysis of data should be developed to improve the system performance.

Data management principles of BLACKMARS will be consistent with the guidelines of the International Oceanographic Data Exchange (IODE) Program of the IOC. Potential of IODE for support of GOOS activities was widely recognised and therefore should be fully used. In condition of observational data scarcity in the Black Sea it is crucial to ensure that information contained in each report is not lost. In doing so the concept of ETEDM, i.e. End-To-End-Data-Management, has to be applied.

Database set-up will be the core activity of BLACKMARS. A database is a set of data controlled by some Data Base Management System (DBMS), which prescribes a defined standard mechanism of data storage and retrieval. This mechanism should allow definition of logical associations and common retrieval of all associated data. The details of any physical structure (contrary to logical structure) should be hidden from the user. The DBMS also has to have (hidden to the user) procedures for backing up, archiving and restoration in a case of failure.

At present in the region there is a variety of oceanographic databases. Some of them store data in records and somewhere there are organised hierarchical and relational databases based on ORACLE, INGRESS, etc. Operational services will need access to data in different databases regularly and frequently. The old concept of access to and management of data was based on the exchange of files and storage of the same information in several databases. The introduction of open systems architecture and standards of open system interconnections (so called OSI model) as well as the progress in telecommunication network and data management technology made it possible to develop distributed data bases (DDBs). The DDBs concept better meets technical requirements for handling of data, which originates outside the WWW system [29].

As the international part of the Black Sea marine telecommunications network is to be incorporated on the RMDCN, the implementation of the higher (*Session, Presentation,* and *Application*) levels of the OSI model can be used for the development of DDBs. Such an approach is particularly attractive because it eliminates the need to set-up a large regional oceanographic database. This excludes a lot of complications resulting from accommodation of the regional database in a single country. Besides that it would be very expensive to set-up and maintain an enormous regional database like the one BLACMARS needs. With the DDB approach the whole system becomes more reliable, as each failure in it affects less number of users, etc.

The proposed Black Sea regional telecommunication network facilitates the set-up of the DDBs [19,20]. However, additional research will be needed for

- Creation of the upper "DDB plane";
- Selection of efficient regional operational DDBs structure;
- Elaboration of upper plane procedures and formats for a request-reply operation;
- Preparing appropriate interfaces and security features in the participant databases;
- Organisation of the regular metadata exchange.

Design of national operational oceanographic networks depends on many factors, and, therefore, it is desirable to have a list of equipment required in the activity. At each NMC the following facilities should be available:

- servers for access to telecommunication media required for operational oceanographic purposes;
- special oceanographic web servers;
- data management servers of oceanographic data and products;
- a security system (with routers, fire wall tools, etc.).

Every oceanographic agency acting as a source of operational oceanographic data and products should have:

- equipment (servers, modems, etc.) for accessing the national operational oceanography network;
- equipment for running DBMS.

6.3. DATA ASSIMILATION/ANALYSIS SCHEMES AND PREDICTIVE MODELS

There is considerable expertise in the region in all aspects of data assimilation and numerical modelling. A short description follows.

6.3.1. Objective analysis and data assimilation

A set of data analysis and assimilation schemes has been developed at the Hydrometcentre of Russia. The operational meteorological objective analysis system is described in [1]. At present a 4-D data assimilation system is about to be implemented [28], which includes a new meteorological data objective analysis system, normal mode initialisation, and an atmospheric GCM. A system for assimilation of temperature data in the ocean upper layer is also being developed [22]. It uses an ocean general circulation model and successive correction scheme. Two schemes for the objective analysis of sea surface temperature are in operation at the centre [2, 30].

Besides that there are some regional schemes of meteorological data analysis, and some methods of SST analysis using satellite infrared imagery. Wave spectrum parameters and integral wave characteristics are at present not assimilated but reconstructed using results of meteorological objective analysis.

Research is underway on the development of a joint marine and meteorological data assimilation system [4, 25].

6.3.2. Atmospheric general circulation models

The Black Sea marine forecasting system focuses mainly on time scales from several hours to several days (up to a week). It will be based on assimilation of observations and on prediction of basic marine parameter changes using hydrodynamic models forced by meteorological forecast data. At present quality of medium range meteorological forecast data remains sufficiently accurate for ranges of up to 5-7 days.

All countries in the Black Sea region have access to the output of numerical atmospheric models of main WWW meteorological centres such as WMC Moscow, WMC Washington, RMC Bracknell, RMC Toulouse, RMC Offenbach, and the European Centre for Medium Range Weather Forecasting (ECMWF). For example, WMC Moscow produces hemispheric medium range forecasts using a spectral atmospheric general circulation model SM15 [9] with comprehensive set of physical processes parameterisations [23]. Regional model of WMC Moscow produces forecasts up to 48 hours with 50 km resolution [5]. The model grid encompasses the whole Black Sea region.

At Meteo-France the ARPEGE model and its data assimilation counterpart are the global variable resolution applications of the IFS/ARPEGE joint numerical weather prediction (NWP) development of the ECMWF and Meteo-France. This common endeavour started in 1988. First partial operational applications were launched at Meteo-France in Toulouse and at ECMWF in Reading in 1992. Fully integrated character of the system allows it to cover now all NWP operational activities in Toulouse since 1995 and in Reading since 1996. Use of a common library for the codes allows both partners to benefit from the each other's progress with nearly no delay. This is reflected in very advanced status of the used methods, which include efficient spectral computations, two-time-level semi-Lagrangian time stepping, variational data assimilation, digital filter initialisation.

The ALADIN system is the limited area counterpart of ARPEGE. Both models have maximum compatibility thus ensuring the smoothest behaviour at the lateral boundary of the regional model. ALADIN was developed and is maintained as a partner activity of fourteen National Meteorological Services, four of which have a Mediterranean facade. Meteo-France provides the logistics of the project but the scientific and technical contributions of the other partners represent nearly three quarters of the common effort. In that sense the collaboration around ALADIN is anything but a "black-box" approach. This advantage is reinforced by the above-mentioned total compatibility with ARPEGE. ALADIN is the test-bed (currently at β-meso-scale) of the most advanced ideas for high resolution NWP contained in the IFS and ARPEGE developments. For example it has a fully transparent option for using either the primitive equations or the Euler equations (the latter at very high resolution, in research mode). There are currently six operational or pre-operational applications concerning 12 of the partners.

Both ARPEGE and ALADIN projects benefit from the stabilising influence of ECMWF excellency in NWP and from the long experience of Meteo-France in running sophisticated NWP systems on powerful computers under the severe clock constraints of short range weather forecasting applications. Such experience is very attractive for BLACKMARS. Some of participating countries, e.g. Bulgaria and Romania, are developing their own operational version of ALADIN for the Black Sea region named SELAM.

6.3.3. *Wind wave models*

In Bulgaria a second generation wave model VAGBULH was developed by the National Institute of Meteorology and Hydrology (NIMH). It has been in operational use at the Forecasting Department of NIMH since January 1996 [6-8]. The wave model VAGBULH is derived from the deep water VAG wave model that is operational at METEO-FRANCE for the North Atlantic and the Western Mediterranean. VAGBULH is a shallow water discrete spectral wave prediction model, which takes into account effects of shoaling, refraction, and bottom friction.

In Russia there is a large selection of wave models. At the Hydrometcentre of Russia a third generation Model for Operational And Scientific Computation of Ocean Waves (MOSCOW) has been in use since 1993 [24]. In 1997 a version of this model for the Mediterranean was prepared. A well-tuned integral parametric model is also operational at the Hydrometcentre of Russia [10]. The Arctic and Antarctic Research Institute (AARI) develops a new third generation wave model, and uses a second generation integral parametric model for operational prediction of waves in the Russian Arctic seas. Several versions of the first third generation WAM model are in use in the country.

In Turkey a large-scale activity was underway under umbrella of the NATO TU-WAVES project. A version of the first third generation model WAM was used in this study. The activity was mainly oriented towards development of comprehensive wave climatology for the Black Sea.

Several advanced wave models were developed at the Marine Hydrophysical Institute at Sevastopol, Ukraine.

6.3.4 *Storm surge models*

At the State Oceanographic Institute (SOI), Russia, two models are in operation for prediction of sea level [26]. One of them is a 2-D model with a specially designed numerical scheme, which ensures high stability of numerical solution. The model has been tested in case studies of the Russian Arctic seas. Besides that the model has demonstrated a good skill of storm surge prediction in the Sea of Azov [27].

A new multi-level model with free surface has also been tested with promising results at the SOI [18]. In addition to its demonstrated capabilities to predict sea level, the model also represents a powerful tool for calculation of currents.

A shallow water model developed by METEO-FRANCE has been configured and adapted for the coastal areas of Bulgaria. The model was verified using tide-gauge observations for the strongest storm observed during the last 10 years along the Bulgarian coast. It is used by the NIHM operationally.

6.3.5. Sea ice prediction tools

During winter season it becomes necessary to issue sea ice bulletins for the main ports and harbours in the northern part of the Black Sea. At present WMC Moscow provides sea ice data services for the Black Sea region. Besides that WMC Moscow issues long-term statistical forecasts of ice conditions (ice phases, ice edge, ice thickness) for six months ahead, which are updated each month. This information is valuable for long-range planning of operations in the northern part of the Black Sea. In the framework of BLACKMARS it is necessary to estimate the need and requirement for development of an operational numerical ice transport models for the northern part of the Black Sea. Such models are available at AARI and SOI.

6.3.6. Circulation modelling

There are a number of sophisticated circulation models that can be adjusted for BLACKMARS. For example, a global ocean circulation model is available at the WMC Moscow [21]. A multi-level circulation model with free surface is available at the SOI, Russia [18]. A set of models with different spatial resolution was developed in the course of several Black Sea environmental projects [12-14]. New perspectives for determination and further prediction of ocean currents, at least at the surface, are related to the use of satellite altimetry data. Software for corresponding data processing was developed at the Marine Hydrophysical Institute, Ukraine.

6.3.7. Oil spill modelling

An operational oil spill model of Meteo-France [3] has been configured and adapted for conditions of the Black Sea. It includes two models, namely a model for the current dynamics in the upper mixed layer of the Black Sea and a model for the computation of the oil slick. The oil slick evolution is represented as spreading of continuously released independent droplets moved by current shear, turbulence, and buoyancy. The model is also operational in NIMH and can be used in case of an accident, for contingency planning and risk assessment.

A sophisticated model for oil spill prediction was developed at SOI [15,16]. It utilises a special "particles in the box" numerical scheme, which allows it to simulate slicks with strongly variable depth, to reproduce interaction of slick with natural and artificial barriers including cases of oil stranding ashore, separation and joining of slicks, etc. The model is based on a full system of equations of fluid motion with account of all relevant physical processes. At present the model is extended to be able to simulate slick interacting with sea ice. The model software is presently upgraded to include a user-friendly interface, pleasant visualisation of results, and some elements of a GIS. The model has been tested using data of some oil accidents and was among the ones, which precisely predicted the oil slick position in 1991 during the Persian Gulf event.

6.3.8. Principles for forecasting system organisation

Efficient organisation of data flow between various elements in a real time meteorological and marine data processing system can be designed taking into account the following criteria [25]:

- characteristic spatial and temporal scales of processes variability;

- sensitivity of each system component output to different types of input data, their inherent errors and variations;
- organisation of several levels of data flow from central unit to regional and local forecasting centres (to ensure initial and boundary conditions for finer scale oceanographic problems);
- operational requirements of users, mostly in terms of timeliness of various services;
- peculiarities of procedure in use to generate the output;
- resources needed such as data acquisition means and computing facilities.

A sequence of model/data processing scheme runs is proposed in [25] ensuring that before any step of analysis (assimilation) or forecast is started, all information, to which the results of this step are sensitive, is supplied in its most accurate state. Regional distribution of analyses and forecasts from the central unit provides data for finer resolution analysis and prediction of wind wave, storm surge, and ice drift. Ecological models are largely to be used on a regional or local scale. Organisation of data assimilation and forecasting suite according to the above considerations leads to better quality of a broad range of real time products. In addition to forecast quality improvement, optimised forecasting suite excludes duplication in data processing and ensures regular data flow for archiving and diagnostic studies.

6.4. DATA AND PRODUCT ARCHIVING

The Black Sea Data and Product Archival System will consist of operational archives of observations, analyses, and forecasts stored in binary codes. The archives will be used for operational purposes and research in the field of the Black Sea monitoring and prediction. Continuous data flow to archives facilitate diagnostic studies, which are the basis for physical processes understanding, validation of numerical models, preparation of climatological data, and other research activities.

Each national archive will likely consist of two parts:

- a real-time database with 30-days storage capacity containing operational data, analyses, and forecasts
  and
- a delayed-mode database for selected interesting situations in the Black Sea.

Routine data such as meteorological observations by coastal stations and VOS will be stored in meteorological databases at NMCs.

A similar data archival structure is chosen by some other GOOS projects including the North East Asia Regional GOOS project, in which China, Japan, Republic of Korea, and the Russian Federation participate.

7. Expected products

In the course of BLACKMARS implementation, each participating country is expected to launch several kinds of forecasts and products. They will include:

<u>- safety-related products namely</u>

- co-ordinated real time marine safety information to be broadcast by NAVTEX stations,
- real time emergency pollution product for national agencies,
- real time storm surge advisories,
- real time ice transport advisories in particular for the northern part of the Black Sea;

<u>- specialised products namely</u>

- real time products for oil and gas companies,
- real time products for ferries and tankers,
- real time products for fisheries,
- etc.;

<u>- other products such as</u>

- real time fluxes across the air-sea interface,
- sea surface temperature analyses and forecasts,
- etc.

8. Capacity building

A series of workshops on capacity building and training has to be organised with the purpose to increase operational oceanography capabilities of participating countries. The following four workshops themes seem to be most useful:

- operational oceanography observational methods;
- data collection, management (including distributed database organisation), and product distribution;
- numerical forecast models in operational oceanography;
- marketing in operational oceanography (oriented towards users of BLACKMARS).

The following manuals will be needed for BLACKMARS participants:

- manual for marine forecaster;
- manual for marine data manager;
- user guide to operational oceanographic services;
- catalogue of marine products.

9. Conclusions and recommendations

1. Regional components of the WWW GOS including coastal marine meteorological

stations, satellite data reception stations represent an appropriate basis for the restoration and further development of the Black Sea observing system.

2. At present there are almost no operational open sea observations in the Black Sea.
3. Use of ferries and tankers as VOS platforms with installed meteorological and oceanographic automated stations is the only feasible approach to ensure at least a minimal number of regular open sea observations.
4. It is desirable to use coastal marine meteorological stations for making additional observations in coastal areas including measurement of ecological parameters.
5. Acquisition of tools for receiving ERS-1, -2, TOPEX/POSEIDON satellite data at some National Meteorological Centres is of great significance for development of operational oceanographic services in the Black Sea.
6. In case of a severe storm buoys remain the only source of reliable open sea observations.
7. The Black Sea regional fragment of the WWW GTS provides all necessary means for efficient international oceanographic data and products exchange. Its use will require almost no additional investments of the participating countries.
8. The INTERNET links in the Black Sea region are not suitable for the international operational exchange because direct connections between adjacent countries are underdeveloped. Conduct of a typical processor-to-processor session requires passing through more than 20 servers in Europe and USA and is usually too slow.
9. The proposed Black Sea telecommunication network is to be a combination of national operational oceanographic networks linked via the WMO GTS.
10. Set-up of regional operational oceanographic Distributed Data Base will offer a possibility for all participants to have the same level of data management technology. It excludes the need to purchase a powerful computer for a leading centre, maintains the same priority of real-time access to data for all participating countries, and increases the overall reliability of the system.
11. Modest investments will be needed for development of oceanographic telecommunications, installation of specialised data management servers at all participating NMCs, and for purchase of data management servers for all oceanographic agencies conducting operational work.
12. There are some global and regional atmospheric models capable to provide necessary forcing for the Black Sea marine modelling. However, further studies of usefulness and quality of atmospheric model forecast output for provision of real-time marine services are needed. It is important to encourage scientists to evaluate quality of atmospheric model output by using it as forcing of physical oceanographic models.
13. Some NMCs in the region have developed operational marine models for prediction of storm surge, wind wave, oil spill evolution and ice transport in limited areas of the Black Sea. Further activities should be aimed at co-ordinated implementation of the models for the whole area of the Black Sea and at their validation and tuning.
14. It is desirable to equip each national meteorological service in the region at least with two workstations for
    - running storm surge, wave, oil spill and ice transport models,
    - visualisation of observations and numerical model output.

References

1 Bagrov A.N., V.B. Shiliaev, E.A Loktionova (1986) Operational scheme for objective analysis of meteorological fields to be used in numerical hydrodynamic weather prediction. *Proceedings of the Hydrometcentre of the USSR.* **280,** 25-55 (in Russian).

2. Bagrov A.N., N.N. Kozhevnikova (1981) Objective analysis of ocean surface temperature in the Northern Hemisphere. *Meteorology and Hydrology*, No 12, 69-76 (in Russian).

3 Daniel P., J. Poitevin (1992) A numerical study of movements of an oil slick on the sea surface in the Persian Gulf from 25 January 1991 to 1 February 1991. *Proceedings of the first thematic conference on remote sensing for marine coastal environments* **1,** 249-260

4 Frolov A V, V E. Ryabinin A I. Vazhnik (1991) Atmospheric and ocean upper layer data assimilation system for medium range weather and marine forecasting. *Proceedings of Int Symposium on Assimilation of Observations in Meteorology and Oceanography Clermont-Ferrand*,. 297-302

5 Kadyshnikov V M., S.O. Krichak, V M. Losev (1991) Finite-difference scheme for regional short-range weather prediction with high resolution in s – co-ordinate system *Proceedings of the Hydrometcentre of the USSR* **310,** 27-44 (in Russian)

6. Kortcheva A., J -M Lefevre, G Kortchev, G Mungov (1994) Application of CH wave model for the Black Sea wind-wave operational forecasts in the NIMH *Proceedings of the International Conference Black Sea 94*, Varna, 12 - 17 September, 69 - 74

7 Kortcheva A (1996) Five hindcast studies of the Black Sea storms with shallow water version of the VAGBULH wave model Internal Report, Department of Marine Forecast - Meteo-France, Toulouse.

8 Kortcheva A, J.-M Lefevre (1996) Numerical real-time forecasting of the wave conditions in the Black Sea. *Proceedings of the International Workshop and Scientific Seminar on the IOC Regional Program for Complex Black Sea Investigations* Varna, 20-25 September

9 Kurbatkin G P, E D. Astakhova, V.N Krupchatnikov, V E. Ryabinin, V P. Salnik, V D. Smirnov (1987) A medium range forecast model. *Reports of the USSR Ac Sci.* **294**, No 2, 321-324 (in Russian)

10 Matushevsky G.V., I.M Kabatchenko, 1991. Unified integral parametric wind wave model and its use *Meteorology and Hydrology*, No. 5, 45-50 (in Russian).

11 Mikhailov N N (1997) Marine environmental data formatting systems and formats for data collection, accumulation and dissemination including international exchange In N B. Harmancioglu et al (eds), *Integrated Approach to Environmental Data Management Systems*, Kluwer Academic Publishers, 349-354

12 Oguz T et all. (1993) Circulation in the surface and intermediate layers of the Black Sea *Deep-sea Research I*, **40**, No 8, 1597-1612

13 Oguz T et al (1994) Meso-scale circulation and thermohaline structure of the Black Sea observed during HydroBlack'91 *Deep-sea Research I*, **41,** No.4, 603-628.

14 Oguz T., P Malanotte-Rizzoli, D Aubrey (1995) Wind and thermohaline circulation of the Black Sea driven by yearly mean climatological forcing *J Geophys Res* ,**100,** C4, 6845-6863

15 S Ovsienko, S. Zatsepa, A. Ivchenko (1993) Mathematical modelling of oil behaviour in ice covered sea. *Proceedings of Conference on Combating Marine Oil Spills in Ice and Cold Conditions*, Helsinki, Finland, 1-3 December 1992, 185-188

16. S Ovsienko, S. Zatsepa, A. Ivchenko (1993) A local operative model for oil drift and dispersion *Proceedings of Conference on Combating Marine Oil Spills in Ice and Cold Conditions*, Helsinki, Finland, 1-3 December 1992, 189-192

17 Peterson J L (1981) *Petri net theory and the modelling of systems.* Englewood - Cliffs

18 Popov S.K (1995) A three dimensional model for storm surge computation in the Sea of Azov *Proceedings of the State Oceanographic Institute, Jubilee Issue*, 205-214 (in Russian)

19 Popova M, S Haramiev (1996) Distributed Data Base in Black Sea region. *Proceedings of the International Workshop and scientific seminar on the IOC Regional program for complex Black Sea investigations* Varna, 20-25 September 1996

20 Popova M. (1997) OSI model and GTS. *Proceedings of the Workshop on Managed Data Communication Network Services in Region VI*, Vienna.

21 Resniansky Yu D, Zelenko A A. (1992) Numerical implementation of an ocean general circulation

model with a parameterisation of upper mixed layer. *Proceedings of the Hydrometcentre of the USSR*, **323,** 3-31 (in Russian).

22. Resniansky Yu. D., Zelenko A.A. (1995) Ocean upper layer thermal data assimilation system. In V.E. Ryabinin (ed.), *Development of elements of the Marine Analysis and Prediction System*. Technical Report of Research project I.7.6.1, Hydrometcentre of Russia (in Russian).
23. Ryabinin V.E. (1989) Physical parameterisations of a spectral atmospheric model at the Hydrometcentre of Russia. In P.P. Vasiliev (ed.), *Methods of Medium Range Weather Forecasting*, Hydrometeoizdat, Leningrad, 127-134 (in Russian)
24. Ryabinin V.E. (1992) "MOSCOW" - a third generation wind wave model. *Research Activities in Atmospheric and Oceanic Modelling*, **17**, 6.41-6.44
25 Ryabinin V.E. (1997) Organisation of marine data processing in real time mode. In N.B. Harmancioglu et al. (eds.), *Integrated Approach to Environmental Data Management Systems*, Kluwer Academic Publishers, 231-236.
26. V.E. Ryabinin and O.I. Zilberstein (1996) Numerical prediction of storm surges. WMO TD No. 779 *"Marine Meteorology and Related Oceanographic Activities"*, **33**, 1-60.
27. Tikhonova O.V., O.I. Zilberstein, G.F Safronov, O.A. Baranova (1995) Hydrodynamic simulation of storm surge in the Azov Sea and pollution transport between Azov and Black Seas. *Proceedings of the second International Conference on the Mediterranean Coastal Environment "MEDCOAST 95"*, Ankara, 3, 1633-1643.
28. Vazhnik A.I., A V. Frolov (1993) A system for four-dimensional discrete assimilation of data on the state of the atmosphere over the Northern Hemisphere *Meteorology and Hydrology*, No 11, 37-48 (in Russian).
29. WMO Guide on the World Weather Watch Data Management. WMO, TD - No.788. Geneva.
30. Zelenko A A., E.S. Nesterov, 1986: Objective analysis of the subsurface layer temperature in the North-eastern Atlantic. *Proceedings of the Hydrometcentre of the USSR*, **281,** 76-83 (in Russian).

OIL CONTINGENCY PLANS FOR NAVAL ACTIVITIES

PAUL HANKINS
Pollution Abatement Program Manager
United States Navy Office of the Supervisor of Salvage
Naval Sea Systems Command (SEA OOC)
2531 Jefferson Davis Hwy.
Arlington, VA 22242-5160, USA

FRED TOUCHSTONE and BILL HANION
PCCI Marine and Environmental Engineering
1201 E. Abingdon Drive, Suite 201
Alexandria, VA 22314, USA

Abstract

The key to successful mitigation of any oil spill is the quality of the response planning done prior to any oil escaping to the environment. A successful plan can mean a rapid response and quick mitigation. Bad planning, or no planning, is a sure means to failure.

The U.S. Navy began its part in oil spill contingency planning in the mid- I 970's in recognition of its worldwide oil spill risks. Since the inception of this program, Navy oil spill contingency planning has undergone significant growth and revision. The program was developed by assigning to major Naval commands having an assigned geographic area of responsibility the requirement to prepare for and respond to oil spills. Next, the program was modified to incorporate lessons learned during the implementation of these plans. Contingency plans were developed throughout the chain of command resulting in an umbrella of oil spill contingency plans, covering all the waters on which the US Navy operated.

Large changes occurred to the Navy's oil spill planning program after new U.S. laws came into effect as a result of the Exxon Valdez oil spill. Simultaneously, parallel efforts were made by the international community as a result of general awareness of the need to develop oil spill contingency plans for vessels. U.S. regulations resulting from these laws were very rigorous. Since there was already a global network of contingency plans in place, a shift in emphasis was made to develop a more thorough training program in which to exercise contingency plans. U.S. Navy contingency planning efforts are now focused on maintaining a high level of preparedness to execute the response plans. Furthermore, Navy contingency planning responds to changes in Navy mission. This includes the increased activity that the U.S. Navy anticipates with other Navies as exercises increase with Partnership for Peace (PFP) Navies in the Black Sea.

S. Beşiktepe et al. (eds.),
Environmental Degradation of the Black Sea: Challenges and Remedies, 337–349.

The Navy recognises this as an important mission and an expansion of its previous role in the area, and further realises that environmental concerns are very high on the Black Sea.

The U.S. Navy will examine the new spill risks it will face in the Black Sea and will incorporate into its existing contingency plan for its Sixth Fleet in the Mediterranean Sea the necessary response planning data. Of course, when operating with other Navies the problem of coordination becomes an important factor. It is appropriate that the needs of all Navies with an interest in the Black Sea be involved in this effort and that we collectively develop an oil spill contingency plan applicable to all Navies. Such a plan would be multinational and would be in a format acceptable to all participating Navies. The first step in developing this plan is to conduct an analysis of current Black Sea contingency planning related activities followed by a review of other environmental activities of NATO, the Black Sea Environmental Program, individual national contingency planning efforts, and other programs focusing on the Black Sea environment. Following this preliminary step, a risk analysis will take place, and a contingency plan will be developed to ensure compatibility with other planning areas. In this effort, it will be necessary to work closely with each participating Navy and other national contingency plan developers to ensure the plan is acceptable to everyone and that it accurately reflects all legal and operational requirements.

There are four key steps to be pursued in composing this first multinational Navy Contingency Plan;

- Develop the plan using new GIS technology to help create a coastal sensitivity map for the Black Sea.
- Work closely with those responsible for developing national contingency plans.
- The multinational spill contingency plan must be developed in concert with established international guidelines for contingency plans.
- The plan must provide for multi-national exercise of the plan, including training and drills, with Black Sea Navies.

This must be a multi-national effort. The U.S. Navy doesn't pretend to offer the only way to plan for spill response or to know all there is to know about how to develop a workable spill contingency plan. The U.S. Navy has much experience in oil spill contingency planning and pollution response. It can offer advice in the development of such plans based on experience gained in over 20 years of its contingency planning efforts. However, the Black Sea is a unique region where a number of Navies can cooperatively work on new ideas regarding oil spill contingency planning and response. Such an effort is an important forum to forge strong bonds with new partners at sea.

1. Introduction

In this NATO Advance Research Workshop, representatives from many of the Black Sea riparian countries, NATO, and other interested organisations have joined to discuss the environmental degradation of the Black Sea and programs to restore this body of water's environmental health. We all realise that the Black Sea is suffering from serious environmental problems and that only a collective effort can bring about a change.

One issue of important interest is that of oil spills. Oil spills are very visible events. They are, of course, unsightly and harm the environment. Additionally, however, an oil spill can ruin tourism, threaten drinking and industrial water supplies, destroy flora and fauna, and attract negative media attention. In recent years, oil spill incidents such as the "NASSIA" have highlighted the need for oil spill contingency planning on the Black Sea. Additionally, oil spill contingency planning development is required in accordance with the Bucharest Convention at both the national and international level.

Navies have an important role to play in this effort. First, naval operations inherently contain an element of risk. There is always the potential for an oil spill, either through collision, grounding, mishaps encountered during fuel transfer, or some other accident or error. Conversely, however, Navies can also prove to be a valuable asset, through use of specialised Navy equipment or personnel that can be quickly mobilised in an emergency situation. For both reasons, it is beneficial for Navies to create an oil spill planning and response program that acknowledges its risks and response capabilities.

The U.S. Navy has developed, over a period of many years, a spill response and preparedness program, which has allowed it to adequately prepare for its spill risks. Although ship and shore activity oil spills still occur, the Navy is better prepared to respond to spills by virtue of having an established oil spill contingency planning regime in place.

2. Navy Oil Spill Contingency Planning History

The U.S. Navy began its oil spill contingency planning efforts in the late 1970's before any national or international legal requirement for such plans existed. This Navy recognised that it had world wide risks of spilling oil so a comprehensive planning effort was undertaken to prepare the Navy for the response to such a large oil spill. This undertaking demonstrated the Navy's forward thinking regarding environmental matters and the responsibility it accepted concerning environmental issues. A significant consideration in entering into this early effort was the desire to help ensure entry into foreign ports during peacetime; as well as maintain good relations with coastal states. What was true 20 years ago is still true today. An oil spill from any Navy ship in today's environmentally friendly world would make that Navy or its nation, very unpopular.

2.1 THE BEGINNING

The first phase in the oil spill planning process began as an effort to geographically assess the U.S. Navy's oil spill risk. This included an analysis of the risk of a spill as well as the capability of a command to respond to such a risk. Once this analysis was accomplished, the Navy then created a network of Navy On-Scene Coordinators (NOSCS) to provide a worldwide, geographically based assignment of responsibility for oil spill response and to spill risks discovered in the assessment stage in each of the

Areas of Responsibility. Studies were also undertaken to build a database of local oil spill response capability existing in each of these areas.

The Navy also delegated the responsibility for management of the Navy's oil spill program to the Naval Sea Systems Command's Supervisor of Salvage (SUPSALV). This was a natural choice. SUPSALV regularly responded to maritime disasters such as ship strandings and sinkings, both of which usually result in release of oil into the sea. Thus, this office maintained both traditional salvage expertise and pollution response awareness. Additionally, SUPSALV maintained Emergency Ship Salvage Material bases, warehouses propositioned at several strategic locations around the world for the siting of ship salvage equipment. These proved to be ideal locations to also preposition spill response equipment for any large oil spill that might occur on the oceans of the world.

A Navy On-Scene Coordinator Oil and Hazardous Substance Pollution Contingency Plan was then developed for each NOSC Area of Responsibility. First a team of oil spills response and contingency planning experts travelled to the site and collected data. For the most part this assessment was conducted by the technical advisers from the International Tanker Owners Pollution Federation (ITOPF), under contract to the U.S. Navy. This information was then inserted into an NOSC contingency plan developed through the U.S. Navy Supervisor of Salvage (SUPSALV). The first contingency plan was then used as a model for the next series of plans so as to ensure consistency in all the Navy NOSC plans worldwide. Once each contingency plan was completed in draft form, the plan would be sent back to the NOSC for final review and acceptance.

To provide further contingency planning coverage, both ships and naval shore facilities were required to have spill contingency plans. Both of these plans were more specific in nature. Ship response plans considered the type of ship, and thus its volume and type of oil risk, and dealt with the limited response, which a ship's crew could bring to bear in responding to a spill. Shore facility plans varied greatly due to the varying degree of risks found at such facilities - a Naval Air Station with the requirement to store millions of gallons of aviation fuel poses a different risk than a Naval Base which must refuel ships at anchor or pier. Both ship and facility plans are supported by the NOSC for the geographic area in which they are located. This NOSC provides the higher capacity response effort and backup support. This linkage of different levels of command forms the interlocking worldwide network for our Navy's spill response program.

It should be noted that oil spill response equipment (boom, skimmer, etc.) development within the Navy followed a two tier approach. The first tier (local) provided equipment to respond to spills in harbours and in protected waters, resulting in smaller skimmer pumps and booms. The second tier consisted of equipment designed for Open Ocean and large spill response. Second tier (regional) response equipment was expensive and thus was normally stored at SUPSALV facilities, necessitating the NOSC to request assistance for the larger spills from a central location, SUPSALV. This was also due to the infrequency of large Navy oil spills and to the special expertise required to operate the equipment. Facilities and vessels (local) were provided with materials they would need to respond to an average small operational spill.

2.2 PROGPAM MODIFICATION

Certain problems surfaced in the Navy's oil spill planning organisation during these early years, and these had to be corrected. NOSC designation had been decided by the risk assessments, and in some situations, NOSCs encountered problems during a spill with higher echelon command elements. Therefore, area coordination and NOSC areas of responsibility needed to be realigned, resulting in a change to the Navy oil spill response and planning structure. By the early 80's there was a new designation of area coordinators and NOSC Areas of Responsibility. These Area Coordinators are depicted in Figure 1.

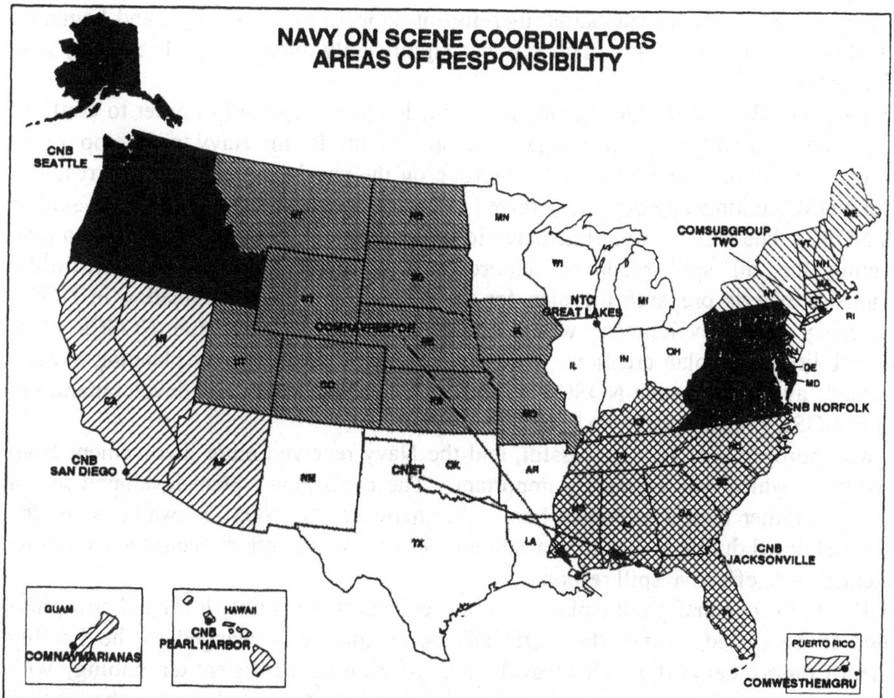

Figure 1

Figure 1 Navy On-Scene Coordinators areas of responsibility

Facility plan development was also now recognised as an essential planning element under this reformatted system. Navy On-Scene Commanders (NOSCDR) were designated, and they had responsibility of major risk areas. With this change, the gaps in the system began to close, and now all levels of responsibility were included in the planning. This system worked better once it followed the chain of command, and this action added to the strength of the system by enrolling top level support for the program.

Further to these activities, the U.S. Navy began a program of training, exercising and drilling the plans. This was seen as an integral part of the planning process. For example, a NOSC Oil Spill and Hazardous Substance Response was created and funded by the U.S. Naval Facilities Engineering Command (NAVFACENGCOM). Oil spill and contingency planning experts provided a two day course to the NOSC on the Oil Spill Contingency Plan. It became a vital part of the program, and it trained those who needed to understand the concept behind the plan itself and how they should follow it as a spill occurred.

2.3 POST OPA 90

The Exxon Valdez oil spill in Alaska resulted in the Oil Pollution Act of 1990. This law and its subsequent regulations resulted in major changes to the oil spill contingency planning regulation in the United States. U.S. Navy policy is to meet or exceed federal requirements and industry standards; therefore it adopted the regulations and began an OPA 90 compliance program. This law has significantly impacted the U.S. Navy's oil spill program.

First, all of the NOSC contingency plans needed to be replaced in order to conform with the new regulations. This also gave the opportunity for the Navy to develop a new contingency planning methodology to improve on the previous method. As a result, a model NOSC contingency plan was created in compliance with OPA 90 and was sent to each NOSC. The NOSCs were instructed to take the model plan and mold it into a plan reflecting the oil spill planning requirements of their Areas of Responsibility. Previously, experts prepared the plan for the NOSC, and this modification in NOSC contingency plan development was purposeful because it gave the NOSCs more responsibility in the plan creation. To assist the NOSC, contingency planning experts were made available to each NOSC by SUPSALV to assist in the plan development, but it is the NOSC's responsibility to complete the plan.

This approach was very successful, and the Navy received more involvement from the NOSCs, which was extremely important. The contingency plan developed into a useful tool rather than a reference book. Furthermore, the NOSC's awareness of the plan's usefulness during a spill increased and they realised its effectiveness in executing a quicker, more effective spill response.

OPA 90 further changed the planning requirements at the facility level and new plans needed to be issued. First, the NOSCDR designation was changed to the Facility Incident Commander (FIC). This was done so to maintain consistent terminology with the regulations. This was mostly a change in wording, but some changes in responsibility were also realised. Furthermore, the same planning method was employed as was utilised in the NOSC planning method. A standardised plan was created by NAVFACENGCOM and it was distributed to facilities, and more responsibility was placed on the FIC to prepare and complete the plan.

The OPA 90 regulations also focused on contingency plan exercising as a key element in preparing for an oil spill. This led to the creation of the U.S. Preparedness for Response Exercise Program (PREP) in the United States as part of the United States Coast Guard's (USCG) post Exxon Valdez oil spill prevention measures. Specifically, the program executes exercises throughout the United States and combines the efforts of the Federal government, industry, state governments, local authorities, and all other players who participate in large oil spill response to test local, regional, and national response capabilities. The scenarios are very realistic, and much valuable experience is gained from these exercises. For consistency, the U.S. Navy adopted the PPEP program and participates in it regularly.

A final change that occurred as a result of OPA 90 is the change in the Navy's response management structure. OPA 90 dictated a formal response management structure approach to an oil spill. This structure, called the Incident Command System (ICS), is based on a military command structure, but was originally adapted by fire departments all around the country to respond to fires. Therefore, it has been modified from its original military origins, and although it is similar to the command structure of the Navy, it does have differences. The Navy has adopted this structure for spills, and it has had to adjust accordingly to adapt to the differences between the two systems.

3. Current navy oil contingency planning activities

Oil spill response contingency planning is a continually evolving process, and the U.S. Navy routinely modifies its contingency plans and oil spill prevention and response measures to reflect the changes in the real world. Changing missions, national and international regulations, environmental perceptions, oil spill incidents, new technologies, and current contingency planning developments are just a few of the issues that planners must stay apprised of to incorporate into the system. This makes it a matter of routine to modify and adjust plans, ensuring that as lessons are learned during exercises, or modifications are required to integrate strategy with adjacent plan holders, the changes can efficiently be incorporated.

The Navy also sets a priority to maintain its vigorous oil spill exercise and training schedule to test and clarify these plans. Through this exercise program, the strengths and weaknesses of the plans, the planning methods and response techniques can be identified and reinforced or corrected as necessary.

3.1 CONTINGENCY PLAN DEVELOPMENTS

The U.S. Navy is still working to complete and issue all of the plans required from legislation written in the wake of EXXON VALDEZ, the Oil Pollution Act of 1990 (OPA 90). The regulations written to support this legislation required a wholesale rewrite of the plans the Navy already had in place. Under OPA 90, the Navy modified its two-tier plan approach to support local and regional planning requirements. These plans are included under the Navy On-Scene Coordinator (NOSC) plans and the Facility Incident Commander (FIC) Spill Contingency Plans. There are nine NOSCs within the United States, and many more Navy facilities subject to OPA90 regulation, and producing just one of these plans is very time consuming. Add on top of that the requirement to update the plans and it is easy to see how the system is continually under review.

To further add to these routine contingency planning revisions, the Navy must also adjust plans to adapt to changes in its operational organisation. Contingency plan writing increased substantially over the last few years as the Navy realigned its mission and organisational structure. The Black Sea region is an excellent example of the changes that have occurred. The increased frequency of U.S. Navy ships operating in the Black Sea will require a new emphasis on planning in this region to appropriately address the risks associated with the increased tempo of operations.

3.2 CONTINGENCY PLANNING UPDATES

Even as these plans are being rewritten, the planners are constantly thinking about the future of Navy oil spill contingency planning. As an example, the Navy is working on what has come to be called the "One Plan." This plan resulted from U.S. industries asking the federal government regulators to find a simple way for industry to comply with all federal requirements of the various federal agencies. With each regulatory agency having their own unique planning requirements, the burden of planning writing was, and is, onerous. A facility that had regulatory requirements from several different federal agencies could be required to write plans that take up 10- 1 2 volumes, an onerous prospect indeed. The 11 one plan" grew out of agreements from the federal agencies to allow the regulated community to submit one plan to fulfil all federal requirements. The U.S. Navy volunteered to lead the development of the first "one plan". The goal is to eliminate redundancies in contingency planning to make it a user friendly, cost effective document.

The Navy is also concurrently working on an automated contingency plan. Given the acronym "SNAP," this plan is a computer-friendly contingency plan whose goal is to be resident on an internet server and accessible by all those who need access to it, including other adjacent plan holders. The plan will link to appropriate databases so it will be much more up to date then a paper copy that may only get annual corrections. The "SNAP" plan can be produced almost instantaneously on the computer with the use of the INTERNET. This virtual plan would be of far greater use to both planners and operators than the current paper plans, whose static nature tends to make the plans stagnant The Navy believes this is the direction all contingency planning will take and is leading the way into the future with this vision.

As discussed earlier, among the requirements in the U.S. Navy's contingency plan program is guidance on plan review. In accordance with these oil and hazardous substances contingency planning guidelines, the Navy On Scene Coordinator (NOSC) contingency plan, Facility Response Plan (FRP), and shipboard plans are to be reviewed and updated annually. This encourages more frequent review and a better understanding of the plan contents.

3.3 TRAINING, DRILLS, AND EXERCISES

Perhaps the biggest change in the Navy's spill response program that developed from requirements of OPA 90 has been the way in which the program is tested. Training of personnel, drilling of procedures, and exercising equipment have received an entirely new emphasis to maintain and ensure the readiness to respond.

3.3.1 Training

Training, always an important component of any military system, has become even more important in the wake of OPA 90. The NOSCs must be trained on the responsibility and demands that are placed upon them. Training courses are sponsored every year to ensure the NOSC is provided the tools to properly prepare for his responsibilities. The courses, too, are continually updated to reflect current legal and planning requirements. In addition to training personnel, a classroom setting offers the opportunity for dissemination of new information and ideas concerning contingency planning and oil spill response issues.

SUPSALV also publishes a quarterly newsletter discussing its activities to all NOSC commands and spill response activities on its Navy spill response activities. It covers the latest in the Navy's training and drilling programs and acts as a discussion platform for the Navy to keep the fleet updated on current spill response issues.

3.3.2 Drills And Exercises

Drills and exercises are designed to test the procedures and equipment put in place to respond to a spill. Generally, drills and exercises are divided into two categories: internal and external.

Internal drills are conducted within the organisational structure of the Navy, and they occur at two levels of response (local ship or facility and regional). These drills are held in response to Navy guidelines incorporating U.S. Coast Guard (who regulates training requirements for spill response) exercise and drill guidelines. Essentially, if a facility/ship meets certain predetermined criteria involving the amount of oil handled or stored, than it must abide by these guidelines. The guidelines include two notification drills per year, one tabletop exercise per year (no equipment deployment), several annual equipment deployments and one full-scale equipment deployment exercise every three years. The exercise guidance is a three year continuous cycle; Navy policy is to abide by the guidelines in addition to requirements that may arise from internal exercise requirements. Often, to prepare for a major fleet exercise, spill drills will be conducted outside the published guidelines.

External drills occur in conjunction with other response organisations, local, state, or national governments and other organisations desiring to test their unified response capability. These drills occur at all levels of the response hierarchy. SUPSALV participates in approximately 12 major oil spill drills around the world every year. Obviously with this number of drills it is a major part of the overall response program. The more experience one has in drilling in a region, the better prepared that organisation is to respond to a spill there. Furthermore, strengths and weaknesses are located in response strategies, which can be corrected in subsequent plan changes.

The contingency plan holder discovers changes they can make in their operational structure to provide preventative measure that lower their risk of an oil spill as a result of these exercises. This may be as simple as increasing the monitoring frequency during fuel transfer operations or as complex as modifying an existing pipeline route. The Navy is a safer organisation and better prepared as a result of this emphasis on training, drills, and exercises.

4. Why Navy Involvement In The Black Sea Contingency Plan?

Given the background of the U.S. Navy's spill response planning history, it is a simple step to understand why there is multi-lateral interest in Black Sea contingency planning. There are three primary reasons driving this interest.

First, the U.S. Navy is finding that its operational presence on the Black Sea is increasing, and that there is an increased possibility of multi-national Navy operations in this area. The U.S. Navy, in coordination with other Navy's operating in the Black Sea must examine the oil spill risks and create contingency plans to cover these risks. The NATO Special Working Group Twelve (SWG 12) recognised this risk and formed a team of experts from participating Navies to develop this multinational Navy plan.

Secondly the Navies understand that the Black Sea is a sensitive environment. A large spill occurring in the Black Sea will impact the shoreline and adversely contribute to the history of environmental degradation already plaguing the Black Sea region. The Navy must be prepared for all possibilities, and be ready to learn from our new partners in the region in order to effectively plan.

Finally, the Black Sea represents an excellent opportunity to demonstrate multilateral Naval cooperation. By working together, the Navies of the Black Sea countries, Italy, the US, and others who have an operational presence there should be able to work together for a common goal. Such an effort would strengthen the resources and capabilities of oil spill response on the Black Sea, in general. This multinational Navy cooperative effort would be applicable to many other areas of the world.

5. Contingency Planning Activities on the Black Sea.

The participating Navies must in one form or another prepare a contingency plan for the Black Sea. Contingency planning is a complex initiative, and it relies on detailed environmental information from a specific region. Therefore, the success of tile Navies contingency planning efforts on the Black Sea are linked to the current environmental activities being conducted. As discussed earlier, the U.S. Navy, through the International Owners Pollution Federation (ITOPF), compiled country-specific oil spill response resource data for every region of the world in which the U.S. Navy operates. The Black Sea region was studied and the data is available for plan inclusion. This is the type of data all the Navies should draw upon on a single multi-lateral plan, eliminating the requirement for independent compilation.

The U.S. Navy is also currently involved in, or monitoring, a number of different activities that will relate to future plan creation.

First there is the effort of the riparian countries of the Black Sea to develop national oil spill contingency plans. These plans are being based on the International Maritime Organisation and other international guidelines. They are taking into consideration the requirements of the International Convention on Oil Pollution Preparedness, Response and Cooperation (OPRC) and the Bucharest Convention. The Navy must keep abreast of these developments so that this Black Sea Navy plan will reflect these local and regional contingency plans.

Secondly, the many NATO initiatives currently being taken in the region have a direct relevance to Navy contingency planning. For example, the Advance Research Workshops allow meeting many of the organisations whose knowledge is crucial to response plan development success. It is through the NATO Special Working Group 12 (SWG 12) that this multinational Navy contingency plan is being developed.

Also of importance is the Black Sea Environmental Programme's efforts in the Black Sea. It has done an excellent job in collecting scientific data and guiding development of national contingency plans that will be very helpful for Navy contingency planning efforts.

The multinational Navy Black Sea contingency planners need to stay current on all the other activities of national and international environmental and contingency planning efforts in general. An effective contingency plan must represent the most up to date information that exists in the region and effectively integrate with them.

6. Proposed Planning Activities

In support of the NATO Special Working Group 12 initiative, and in its role as team leader for this multinational Navy Black Sea Spill contingency Plan development, the U.S. Navy intends to use the experience it has gained from over 20 years of oil spill contingency plan writing to help write a comprehensive multinational contingency plan for the Navies operating in the Black Sea. Guidelines established by the IMO Marine Environment Protection Committee (MEPC 35/14/6 "Report of the correspondence group on proposed revisions to IMO Manual on Oil Pollution - Section 11 Contingency Planning" - Appendix 4) will be used to develop the outline and structure of this plan. Since the U.S. Navy spill contingency plans were developed long before the IMO guidelines were established, this Black Sea plan will not look like other spill contingency plans used by the U.S. Navy. However, since the IMO standard will be almost universally used, this will enable the plan to be truly international in nature and more acceptable to the countries and Navies that must accept this plan.

As a first step in developing this plan, meetings will be conducted with each Navy and appropriate government authority associated with every country wishing to participate. During these meetings the team will establish the criteria fundamental to each country's national contingency plan and will prepare the draft Navy spill contingency plan, integrating the mutual items from each national plan into a basic, core plan. Where there are significant differences between countries or Navies, those items will be separately identified, so that each country's interests are preserved in this multinational plan.

In the case of tile U.S. Navy, this internationally structured plan will be prepared so that there is an easily-identified linkage between the "standard" U.S. Navy NOSC plan and the multinational Navy plan. When the U.S. Navy operates in the Black Sea it will be a simple matter to shift plans. The spill response criteria will remain the same.

Several advancements in the development of oil spill contingency plans have recently been achieved, The use of geographic information systems (GIS) to record, display, and retrieve geo-referenced information related to the environment should add significantly to the usefulness of this plan. New technology and the application of remote satellite sensing coupled with GIS shows promise as an inexpensive method to map coastal areas and to visually display, in near-real time, the environmentally sensitive areas, which may be impacted by a spill. Commercially available satellite images, such as LANDSATT Thematic Mapping images, may be rectified and then, through analysis, coupled with ground data, an environmental sensitivity map can be created for the coastal area.

This effort will incorporate into the contingency plan framework the development of mapping information systems combining the best available information into a presentable format for use by the responder. Ecosystems, nationally protected areas, wildlife habitat, and areas especially protected by national and international laws and treaties can be reflected visually in this plan. The ability to display such information from digital images has been proven by the use of available multi-spectral digital images. Such data in a printed format or in an automated digitised database can provide the environmental planner or emergency responder with reliable information on protection strategies, environmental and public risks, and offer a basis by which to measure spill impacts and recovery techniques.

By linking the contingency plan to one of the existing data exchange networks, GIS data can be completely integrated into the plan. The NATO ECHS or the Partners for Peace Information Management System (PIMS) as well as other internationally available data exchange systems can be used. While the spill contingency plan structure will continue to follow the IMO guidelines, the supporting Appendices, reflecting the GIS/space imagery data, will provide the spill responder with up to date risk and area sensitivity information.

7. Conclusion

The past twenty years have demonstrated to the U.S. Navy that oil spill contingency planning works. Its success is due to the diligence of those that execute the plans as well as the way in which the plans are structured, by planning responses at the local, regional, and national level. The high state of preparedness and ability to respond to oil spills is an important mission to the Navy, and it is an essential element to our peacetime operations.

The Navies of Romania, Bulgaria, Turkey, Italy and the U.S. have agreed in principle to cooperate in development of a multilateral Naval Black Sea spill contingency plan. Other Black Sea Navies need to contribute to this planning effort or it will not be possible to protect the interests of the entire Sea.

Similarly, Navies elsewhere in the world that share the same body of water may do well to cooperate in spill management planning.

Furthermore, it is important to recognise Navies as an asset in a spill, and that they should be included in contingency planning efforts at all levels. Navy, national and international planning efforts must be coordinated to avoid wasteful redundancy and duplication.

Obviously, there are going to be many times during Black Sea contingency plan development that naval, national and international contingency planning will need to develop data or planning information that is crucial to all three. We need to keep informed of each others efforts so that we can all be successful in creating effective oil spill contingency plans that will best protect the Black Sea.

Finally, there are many innovative tools available to contingency planners resulting from recent advancements in technology. These tools will help make the area spill contingency plans, such as the Black Sea Navy plan, an excellent document. However, no planning document is of any use unless it is tested. After the multi-navy spill contingency plan is issued, success will only be achieved by conducting drills between Navies to train ourselves in international cooperation in preserving this environment, which belongs to everyone.

THE BLACK SEA CONTINGENCY PLANNING FOR MARINE OIL SPILLS

L. STOYANOV
Research Institute of Shipping,
1, Slavejkov Sq., Varna 9000,
Bulgaria

D. DOROGAN AND S. JELESCU
Romanian Marine Research Institute,
Bd.Mamaia 300,Constanta 8700
Romania

Abstract

It is recommend that the Istanbul Commission, upon the recommendation of its Advisory Group on the Environmental and Safety Aspects of Shipping based on MARPOL 1973/78, shall adopt and develop a Black Sea Strategy for contingency planning and emergency response. This strategy should provide a basis for assuring that the contingency plans developed within Black Sea states are sufficiently coordinated and will also serve as a basis for the development of the regional contingency plan. The paper presents the frames and the steps on the development of the Black Sea strategy for contingency planning and emergency response, pointing the following items : Black Sea Strategic Action Plan -- policies and actions, Advisory Group and Activity Center on the Environmental and Safety Aspects of Shipping -- presentation of its main objectives and tasks in the next years, National Contingency Planning (NCP), the draft of the Romanian National Contingency Planning for marine oil spills -- research and development activities, organizational structure.

1. Introduction

Accidental spills pose a very high level of risk for the marine ecosystems and the coast line. The Black Sea, as a closed basin, is very sensitive to such kind of rapid impacts.

Accidental spills have an acute effect as well on the all typical elements of the transboundary pollution. On rare occasions, they may create an ecological collapse and a long term crisis in large areas of the coast line and marine environment.

The data of the Transboundary Diagnostic Analysis (TDA) of the Black Sea has shown that oil discharge into the Black Sea amounts to more than 110,000 tones per annum (Table 1). In the case of a marine accident involving a large tanker, such a

S. Beşiktepe et al. (eds.),
***Environmental Degradation of the Black Sea: Challenges and Remedies*, 351–365.**

volume may be discharged into the sea in a single day only. This is one of the reasons the accidental oil spills to have a wide public response.

Accidental spills at sea differ from other sources of pollution in that decision making procedures and the subsequent response measures have to be made within a very short time.

Table 1. Oil inputs to the Black Sea (unit : t/year)

Source of Pollution	Bulgaria	Georgia	Romania	Russian Federation	Turkey	Ukraine	Total
Domestic	5649.00		3144.1		7.30	21215.9	30016.30
Industrial	2.72	78.0	4052.5	52.78	752.86	10441.0	15379.86
Land-based				4200.00		5169.2	9369.20
Rivers	1000.00			165.70		1473.0	2638.70
Total	6651.72	78.0	7196.6	4418.48	760.16	38299.1	57404.06

Total from the Black Sea Coastal Countries :	57404 t/y
Accidental oil spills *	136 t/y
From the Danube River	53300 t/y
TOTAL IN THE BLACK SEA	110840 t/y

Note : * The value for accidental oil spills is average for the last 10 years.
There is not included any information for illegal discharges from shipping.

Contingency plans integrate the coordination of the operations and all preliminary actions undertaken for pollution preparedness and response. The plans have to cover a variety of different factors, including: the responsible authorities and institutions, risk assessment, the channels for the information exchange, decision making systems, different databases such as the equipment available, models for forecasting oil movements, sensitive areas mapping, clean up technologies, restoration and rehabilitation, etc.

2.Emergency response and the Black Sea Strategic Action Plan

According to the Black Sea Strategic Action Plan (BS-SAP) the Advisory Group on the Environmental and Safety Aspects of Shipping coordinated by the Activity Center "will coordinate the regional approach to emergency response, particularly the international response to accidents involving the extraction, marine transport, handling and storage of oil and, where relevant, hazardous chemicals". It will also coordinate, on behalf of the Black Sea Commission, regional aspects of implementation of the MARPOL Convention defined in the BS- SAP. Furthermore, it will assist with the elaboration of port-state-control procedures, defined in the BS-SAP. Particular attention will be paid to developing a strong working relationship between Ministries of Environment and Transportation both internationally and within corresponding national focal points. It will collaborate closely with all relevant institutions and governmental bodies, international organizations (such as IMO, WMO, IOC) and the private sector (shipping, oil and gas industries).

2.1.MAJOR PROBLEMS IN THE PREVENTION OF EMERGENCIES AND CONTINGENCY PLANNING IN THE REGION

- Lack of contingency plans corresponding to the IMO guidelines at national and local level.
- Lack of a regional Black Sea contingency plan.
- Lack of regional and coordinated national classification and risk assessment approaches as well as vessel traffic control systems.
- Lack of appropriate national capabilities for emergency response and regional coordination including multi-lateral cooperation.
- Inadequate control and communication facilities.
- Lack of data and assessment of transport streams of crude oil, dangerous and harmful substances in the Black Sea.
- Inadequate national contingency planning legislation.

The main objectives of the Advisory Group are formulated on the base of TDA and the current status. Detailed description of principles, policies and actions are included in the SAP.

3. Advisory Group and Activity Center on Environmental and Safety Aspects of Shipping

3.1. MAIN OBJECTIVES

Development (and adoption by the Commission) of a Black Sea Strategy for contingency planning and emergency response.

Assessment of the existing capacity and needs of a development of harbor reception facilities to be enlarged to comply with MARPOL Special Area requirements.

Development and establishment of a Harmonized system of port state control.

Restriction of the vessel source pollution, including illegal discharges by vessels into the Black Sea. Development of a Harmonized system of enforcement, including fines.

Black Sea Contingency Plan.

3.2.THE STEPS THAT SHOULD BE UNDERTAKEN IN THE NEXT SEVERAL YEARS

Updating existing national plans and/or developing new ones according to international guidelines.

The legislation regarding the problems in the Black Sea Countries needs to be improved and brought into compliance with the international requirements (Preparation of national legislation necessary to ratify the international conventions related to emergency preparedness, response and CO-operation).

A campaign should be launched in the Black Sea states to ratify the CLC (1969), FUND (1972) and OPRC (1990) Conventions.

A local education and training program for all personnel involved in the area of Contingency Planning should be ensured.

Allocation of necessary national financial funds by the national authorities to operate the national focal points.

Harmonize not only contingency plans but also the separate components, as for example:

- the content of the reporting form (in accordance with the Annex to Article 6 of the Protocol on Cooperation in Combating Pollution of the Black Sea Marine Environment by Oil and Other Harmful Substances in Emergency Situations);
- the classification of the scales of spillage;
- methods for evaluation of the coastal line sensitivity;
- common methodologies on vulnerability of the coastal zone and the traffic in the region, etc.

4. National Contingency Planning

4.1. RESEARCH & DEVELOPMENT ACTIVITIES

4.1.1. Contingency planning.
Development of a Common Framework for National Contingency Planning.

A regional approach to the Black Sea Contingency Plan including as follows:

- development of the contingency plans in accordance with the IMO's and other international guidelines and requirements,
- organizing of training courses for the National Focal Points and representatives from the national authorities and institutions, responsible for the emergency response.

4.1.2. Risk assessment.
Develop a computerized expert system, including tools for scenario development and risk assessment, databases for contingency planning, decision matrices for predicting required response actions and analytical models for examining the adequacy of spill response resources.

4.1.3. Resource classification and inventory system for spill planning.
Develop universal classification schemes and inventory techniques that can be applied and used on different habitats and species. The final output will be an integrated

database that provides spill response managers with the information necessary to effectively deploy personal and equipment.

4.1.4. Spill decision support system.

Expert systems are to be developed for spill response based upon literature available worldwide on oil properties, spill effects, clean up techniques, as well as the experience of various oil spill experts. This includes development of databases, analytical models, decision matrices, and expert system for spill response planning and operation management.

Most of the projects last during two or three phases depend on their complexity.

4.2.STRUCTURE OF A NATIONAL CONTINGENCY PLAN

4.2.1. IMO guidelines and OPRC requirements

The increases of the sea traffic, especially oil tankers using sea ports, or passing in transit by shorelines, the prospecting for oil and gas in the continental shelf, represent a relatively high risk of pollution from blow out, collision, bunkering of fuel, or other marine activities. The pollution can affect beaches, tourist industry, fishery and maritime commerce, sea birds, the life in the sea, coastal installations and finally to cause considerable financial losses.

Having in mind the above mentioned and on the basis of the international law, the IMO guidelines, moreover the International Convention on Oil Pollution Preparedness, 1990 (OPRC), a framework of a National Contingency Plan was suggested under GEF BSEP.

The Contingency Plan structure aims to cover the actions on organizing, reports, alerts, communications, assessment, countermeasures, administration, finances, public relations and international CO-operation.

NCP is the result of a careful planning, considering the varied nature of the Black Sea and identifies all government authorities and structures, private agencies and organizations that are responsible to undertake response actions in case of pollution of the marine environment resulted of an emergency. The plan aims to cover not only the extreme but the lesser oil spillage.

NCP formulates the national policy, the strategy for offshore and onshore actions, the priorities for protection, the aims and the main tasks. It also identifies the national organizations responsible for the control and the response in case of oil pollution and other harmful substances, communication and operational systems, documenting and the step-by-step actions for immediate countermeasures to the expected situation.

The plan is intended to outline the whole national preparedness and relief actions incorporating the state and private response agencies in case of marine emergencies that may affect the marine environment.

The plan ensures on time and effective actions in the case of an oil spill or an alert of an oil spill or other harmful substances. The major actions include maintenance of the operational organization, establishing of the priority coastal zones for protection and clean-up and envisaging of minimal level of equipment in an acceptable type for response. To maintain a high preparedness level in the response organization, the plan outlines the framework of the basic principles for operational training on all levels in the agency.

4.2.2. Framework for a National Contingency Plan

Preamble:

- authorization
- list of content
- updating of the plan -- record of changes.

Introduction:

- legislation framework
- purpose and objectives on national marine pollution contingency plan.

Responsibility and organization:

- national authorities and points of contacts
- national response system
- national response priorities
- interagency participation and support.

Response planning and operations:

- alerting, reporting and communications
- pollution sources
- spill assessment, surveillance and forecasting
- sensitive areas and areas to be protected
- national combating strategy at sea and on land
- organization of national reporting center in spill situations
- documentation (log keeping);
- liaison with other authorities and to the private sector
- governmental and public equipment on hand
- assistance from other public and private sectors
- temporary and final disposal of recovered oil and oily materials
- international CO-operation and assistance
- termination of operation
- health and safety
- education and training.

Administration and logistic:

- funding, reimbursement, claims
- documentation of clean-up costs.

Action plan:

The role of the contingency planning process is emphasized:

- establish roles and responsibilities
- assess the risk
- develop appropriate response strategies

- identify and actualize resources
- prepare contingency/action plan
- train personnel
- exercise plan and equipment.

5. Romanian National Contingency Planning for marine oil spills.

5.1. GENERALS

The Romanian sea coast is 245 km long from Vama Veche (border with Bulgaria) up to Chilia (the Danube Delta) with 3 main harbors : Constanta-Midia and Agigea free zone system, Mangalia and Sulina-Danube Delta free port.

Oil figures especially prominent among our imports and it is an economical activity highly dependent on shipping. Particularly in Constanta and Midia harbors where a large petrochemical complex is located, it is most likely that vessels could cause large-scale oil spill incidents.

Fortunately, in recent years, no large-scale oil spill incidents involving tankers have occurred. Approximately 100 oil spills minor incidents have been reported and penalized in the last 7 years, mostly resulting from inept handling of ship's machinery and equipment. No major accident caused by ships running aground, or in collision with other vessels, has been reported.

The vulnerability index of Romanian shoreline types is given in Table 2.

5.2. NATIONAL OBLIGATIONS AND THE LEGAL SYSTEM CONTROLLING OIL POLLUTION

According to Bucharest Convention on the Protection of the Black Sea against Pollution ratified by Romania in April 21, 1992, particularly as the article 2 from the Protocol [7] stipulate, Romania has the major obligation : "to maintain and promote, either individually, on through bilateral on multilateral Cooperation, contingency plans for combating pollution of the sea by oil and other harmful substances. This shall include in particular, equipment, vessels, aircraft and work force prepared for operations in emergency situations."

The Ministerial Declaration on the Protection of the Black Sea signed by Romania in Odessa on 7th of April 1993, comes also to underlay the obligations stipulated by the Bucharest Convention with its 11th specification : "to develop national and regional contingency plans identified by the Convention on the Protection of the Black Sea against Pollution, for combating pollution in emergency situations".

In Romania the law concerning sea water protection ensures some of the content of MARPOL 73/78 regulations [9], the major point being" the Polluter Pays" principle.

TABLE 2 Vulnerability index of Romanian shoreline types

INDEX	SHORELINE TYPE	ZONAL IMPORTANCE: TOURISTIC	FISHING	INDUSTRIAL	NAVIGATIONAL	BIOLOGICAL	COMMENTS - concerning the consequences of oil impact
1	WAVE- CUT PLATFORMS (protection break waters constructions) ZONE						Wave swept, usually erosional Most of the oil is washed by natural process within weeks.
	1 1 Midia terminal harbor and petrochemical plant	-	X	X	X	X	
	1.2. Mamaia beach - "Tomis" recreational port, South-Constantza-Agigea harbor and littoral zone up to Mangalia areas (Harbor and Shipyard)	X	X	X	X	X	
2	EXPOSED FINE TO MEDIUM GRAINED SAND BEACH ZONES						Oil does not penetrate deep into beach sand, facilitating mechanical removal if necessary
	2 1 The Danube Delta area up to Midia Cap.	X	X	-	X	X	
	2 2 North Constantza Beach	X	X	-	-	X	Uncleaned oil may persist several months
3	COARSE GRAINED SAND BEACHES						Oil may penetrate, be buried rapidly, making clean-up difficult
	3.1.South Constantza Beach.	X	X	X	X	X	Uncleaned, most oil will be naturally removed within several months

The Romanian laws and governmental orders regarding the environmental protection are listed in Table 3. They work quite efficiently providing sanctions and penalties in case of sea water pollution. Unfortunately, the amount of penalties being levied is still too lenient.

In the event of an accidental oil spill, Romania still does not have an official system with a central organization and the necessary equipment, particularly for open sea.

5.3.RESEARCH & DEVELOPMENT ACTIVITY FOR THE NATIONAL CONTINGENCY PLAN, OF ROMANIA.

The Romanian Marine Research Institute based on the General National Contingency Planning [3] frame work, provided through Black Sea Global Environmental Facility Program [2, 4] by the International Maritime Organization experts, has developed the first draft the Romanian practical cooperation in emergency prevention and response.

The draft is applied in case of both major and minor oil spill pollution and it is based mainly on the principles stipulated by the International Convention on Oil Preparedness, Response and CO-operation 1990 [8] which has yet to be ratified by Romania as well as Civil Liability and Fund Convention [13]. The principles of the OPRC Convention were adapted to the Romanian specific particularities.

The draft received all the observations from the Ministries and bodies that possibly may be involved in an accidental oil spill pollution and it will enter in force in 1998 with a
governmental order of the Ministry of Waters Forests and Environmental Protection, becoming the legal base for this activity.

The draft contains:

5.3.1. Organization:

- general organization (Fig.1);
- responsibilities: - for Ministry of the Environment and Transportation;
  - for the Marine Pollution Executive Body;
  - for the Marine Pollution Advisory Board;
  - for the "Key Personnel."

5.3.2. Operation:

- reporting procedures (Fig.2);
- alerting procedures (Fig.3):
- sensitive areas and priorities for response:
- national combating strategy;
- available equipment and other sectors assistance;
- temporary and final disposal of the recovered oil;
- international CO-operation and assistance;
- termination of operation.

5.3.3. Administration:

- daily routine and spill situations;
- funding claims and compensations;
- public relations;
- collecting of evidences and oil samples;
- maintenance, and cleaning of the equipment;
- personal care, medical assistance, transport.

TABLE 3 Romanian laws, governmental orders, guidelines, regarding the environmental protection

No.	Name of the Document	Description	Into Force
1.	Law 27/1993	Marpol 73/78 Ratification	Mar. 1993
2.	Law 137/1995	Environment protection law	Jun. 1973
3.	Ministerial declaration (Odessa 1993)	Coordinated action to preserve the marine environment of the Black Sea	Apr. 1993
4.	Law 17/1990	Juridical aspects on national waters	1990
5	Law 13/1993	Ratification of Bern Convention '79,on wildlife conservation	Mar. 1994
6	Law 5/1991	Ratification of Paris protocol '82 Wetlands areas of international importance (birds habitats, etc.)	Jan 1991
7.	Gov. Order no 457/1994	Organization and functions of the Ministry of Waters, Forests and Environmental Protection	July 1994
8.	Law 5/1989	Water management & water quality assurance	Jun. 1989
9	Law 11/1994	Overcoming water standards	1994
10.	Law 8/1974	Water quality reglementations	Mar 1974
11	Law 98/1992	Ratification of the Convention on the protection of the Black Sea against pollution	Ian. 1994
12.	Gov. Order 138/1994	Penalties for environment pollution	Apr. 1994
13	Gov. Order 47/1994	Activities in case of disaster or calamities	Aug. 1994
14.	State council decree no. 37/ 1980	Penalties for the national waters pollution by ships	1980
15	Gov. Order 127/1994	Penalties for infringements in the environment field	Apr. 1994
16	Law 82/ 1993	Establishing Danube Delta Biosphere Reserve	1993
17.	Gov.order no.1001/1990	Penalties in the field of waters management	Sept.1990
18.	Oil Terminal order No.1/1994	Responsibilities & activities in case of an emergency response at Oil Terminal Constanta	Jan. 1994
19.	Harbor master order 54/1994	Communication and reporting system in case of emergency response for Romanian harbors	Sept. 1994
20.	Internal Navy notice	Environment protection responsibilities in the areas of the Romanian Navy activities	1991
21	Internal notice for Navigable Canals	Organization & instructions for emergency response actions on the Danube -- Black Sea Canal	1990
22	Notice for navigators No 1907/1994	Bunkerage, de-balasting, discharge bilge waters and sludge at port facilities	Jun. 1994
23	Gov. Order no. 531/ 1992	Civil defense in Romania -organization & functions	Sept. 1992
24	Instructions no. M4/ 1993	Application of gov.order no 531/1992 by the Ministry of National Defense	1993
25.	Law 69/ 1991	Local public administration -- organizational aspects	1991

5.3.4. Annexes:

- maps;
- addresses and phonelists;
- log keeping;
- inventory list of the equipment and forces;
- methodological guidelines;
- education, training and exercise programs;
- local marine oil pollution contingency plans.

Ministry of Transportation has a tradition and is permanently involved in this activity, being constantly informed with the latest IMO's Committee's resolutions. It's responsibilities are preventing, controlling and combating, as a body with response duties and full responsibility in case of an intervention in a marine accidental oil pollution.

The Ministry of Waters Forests and Environmental Protection, quite new born in Romania, is a support body having the responsibility of preventing the danger of an ecological disturbance for the marine environment, taking preventing and controlling measures.

Both ministries responsible (this specific Romanian bi-headed coordination), must be strongly supported by all the authorities involved.

To prepare the future functionality of this plan the Romanian Marine Research Institutes is also involved in the development of some aspects of national logistic support in case of accidental oil spill pollution.

- The last achievements are the development of some oil spill combating devices:
- offshore and coastal inflatable booms;
- portable tanks for temporary storage and gravitational separation of oily waste materials;
- emulsifiers.

The promotion of international cooperation is vital.

Encouraging the development of compatible standards for oil pollution combating techniques, equipment and training activities could form the basis for future common programs.

The main aspects presented, are related to oil as a major pollutant.

Romania has no experience, an organized system, or technical facilities to take proper countermeasures in case of marine accidents involving hazardous chemicals. This has to be the next priority.

Figure 1 Romanian responsible bodies structure in an emergency oil spill

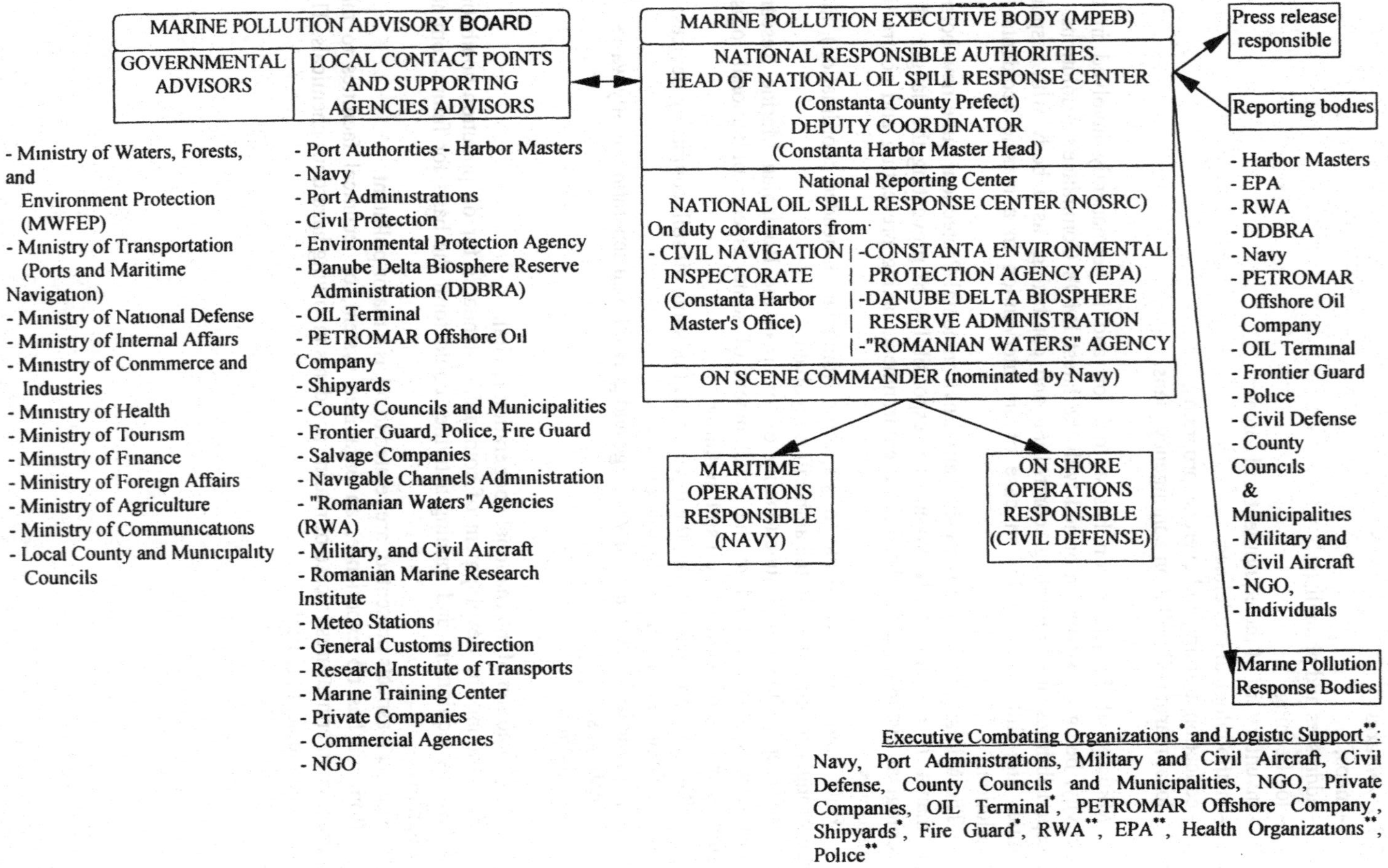
MARINE POLLUTION ADVISORY BOARD
GOVERNMENTAL ADVISORS
LOCAL CONTACT POINTS AND SUPPORTING AGENCIES ADVISORS
- Ministry of Waters, Forests, and Environment Protection (MWFEP)
- Ministry of Transportation (Ports and Maritime Navigation)
- Ministry of National Defense
- Ministry of Internal Affairs
- Ministry of Conmmerce and Industries
- Ministry of Health
- Ministry of Tourism
- Ministry of Finance
- Ministry of Foreign Affairs
- Ministry of Agriculture
- Ministry of Communications
- Local County and Municipality Councils
- Port Authorities - Harbor Masters
- Navy
- Port Administrations
- Civil Protection
- Environmental Protection Agency
- Danube Delta Biosphere Reserve Administration (DDBRA)
- OIL Terminal
- PETROMAR Offshore Oil Company
- Shipyards
- County Councils and Municipalities
- Frontier Guard, Police, Fire Guard
- Salvage Companies
- Navigable Channels Administration
- "Romanian Waters" Agencies (RWA)
- Military, and Civil Aircraft
- Romanian Marine Research Institute
- Meteo Stations
- General Customs Direction
- Research Institute of Transports
- Marine Training Center
- Private Companies
- Commercial Agencies
- NGO
MARINE POLLUTION EXECUTIVE BODY (MPEB)
NATIONAL RESPONSIBLE AUTHORITIES.
HEAD OF NATIONAL OIL SPILL RESPONSE CENTER
(Constanta County Prefect)
DEPUTY COORDINATOR
(Constanta Harbor Master Head)
National Reporting Center
NATIONAL OIL SPILL RESPONSE CENTER (NOSRC)
On duty coordinators from:
- CIVIL NAVIGATION INSPECTORATE (Constanta Harbor Master's Office)
-CONSTANTA ENVIRONMENTAL PROTECTION AGENCY (EPA)
-DANUBE DELTA BIOSPHERE RESERVE ADMINISTRATION
-"ROMANIAN WATERS" AGENCY
ON SCENE COMMANDER (nominated by Navy)
MARITIME OPERATIONS RESPONSIBLE (NAVY)
ON SHORE OPERATIONS RESPONSIBLE (CIVIL DEFENSE)
Press release responsible
Reporting bodies
- Harbor Masters
- EPA
- RWA
- DDBRA
- Navy
- PETROMAR Offshore Oil Company
- OIL Terminal
- Frontier Guard
- Police
- Civil Defense
- County Councils & Municipalities
- Military and Civil Aircraft
- NGO,
- Individuals
Marine Pollution Response Bodies
Executive Combating Organizations* and Logistic Support**:
Navy, Port Administrations, Military and Civil Aircraft, Civil Defense, County Councils and Municipalities, NGO, Private Companies, OIL Terminal*, PETROMAR Offshore Company*, Shipyards*, Fire Guard*, RWA**, EPA**, Health Organizations**, Police**

Figure 2. Marine pollution reporting system

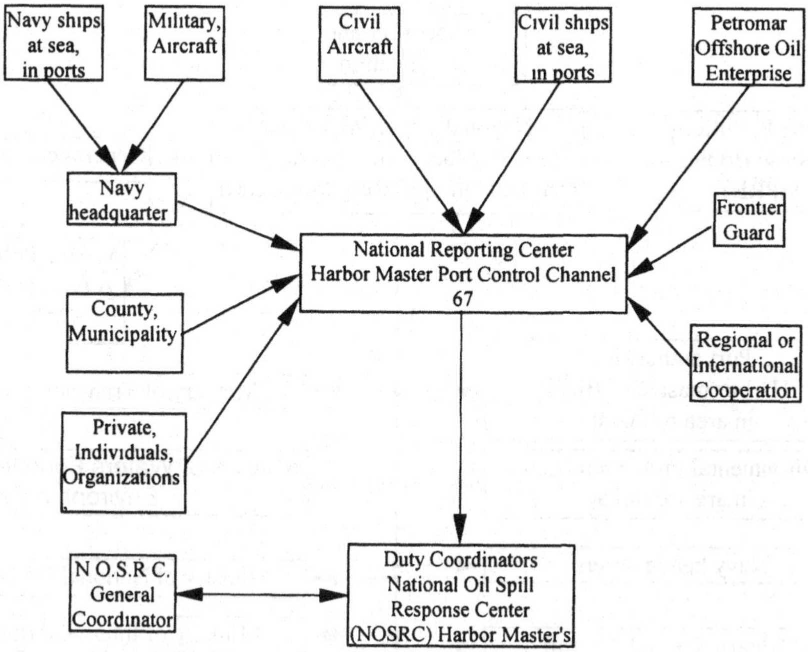

Figure 3. Alerting procedure in case of accidental marine pollution at the Romanian Littoral

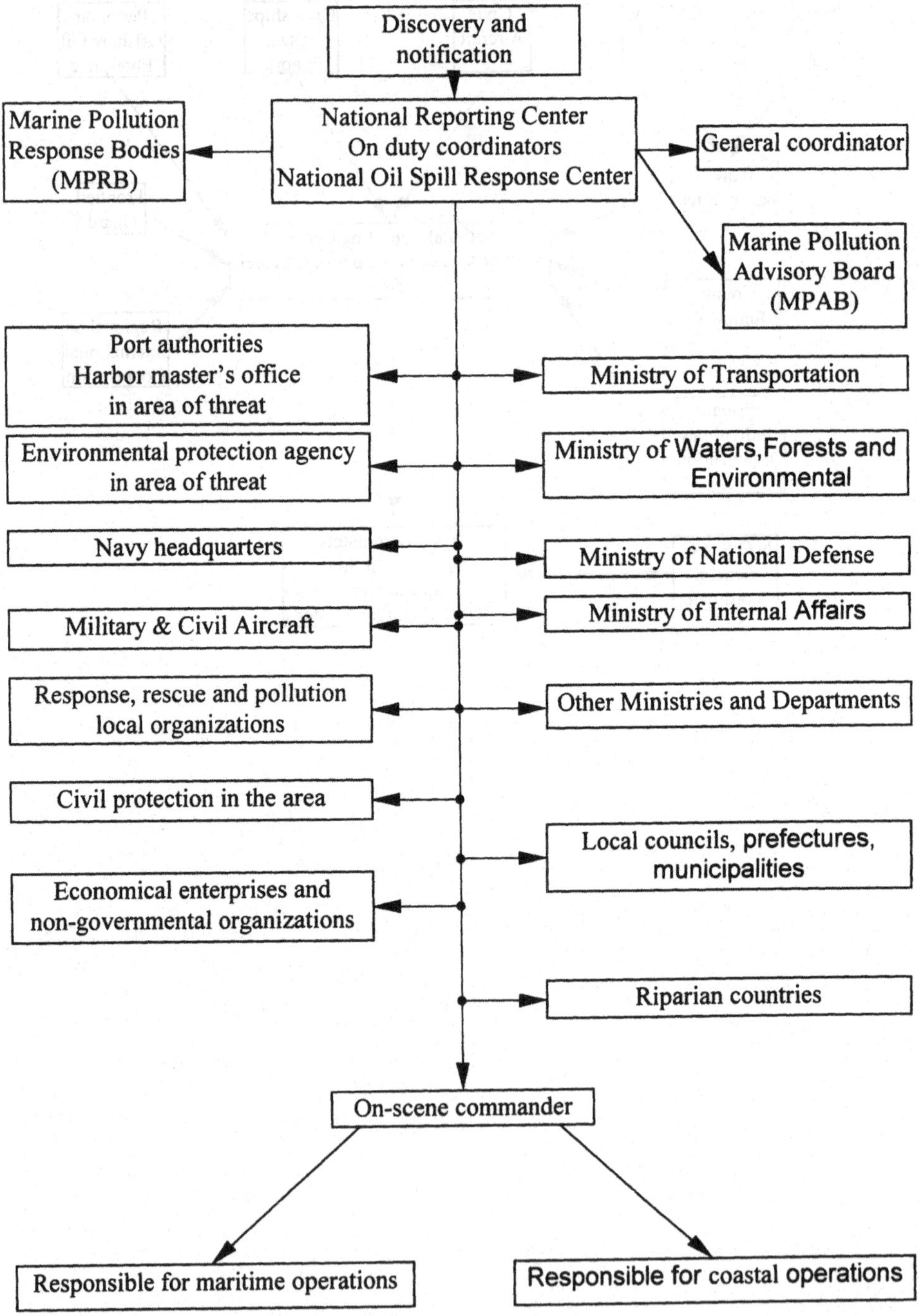

Acknowledgements

This work was carried out within the Global Environment Facility Black Sea Environmental Program through Emergency Response Center of Activity and Advisory Group on the Environmental and Safety Aspects of Shipping.

References

1 Strategic Action Plan for the Rehabilitation and Protection of the Black Sea,(1996) GEF-BSEP Istanbul, Turkey

2 Annual Report, GEF Black Sea Environmental Program,(1996) Istanbul, Turkey

3 Manual on Oil Pollution, Section II -- Contingency Planning,(1994)IMO London.

4 Emergency Response and Contingency Planning in the Black Sea, Current Status and Strategies for Improvement, (1995) BSEP,Istanbul, Turkey.

5 Black Sea Transboundary Diagnostic Analysis (1996), GEF, BSEP,PCU Istanbul, Turkey.

6.The Shell Company of Turkey Ltd. Presentation of Tier 3 Oil Spill Management Workshop, December 1996, Istanbul

7. Convention on the Protection of the Black Sea against pollution (1992) Bucharest

8 International Convention on Oil Pollution Preparedness Response and CO-operation -- OPRC - (1990) IMO, London

9.MARPOL 73/78 Convention (1978) IMO, London.

10.National Response Team of the National Oil and Hazardous Substances Contingency Plan (1989), Washington D.C. USA.

11.Response to Marine Oil Spill (1987), ITOPF, London.

12.Oil Spill Response -- Exxon -- Field Manual (1984), USA

13 International Oil Pollution Compensation Fund (1995) Claims Manual, London.

ROMANIAN CONTRIBUTIONS TO ONGOING BLACK SEA RESEARCH AND MANAGEMENT PROGRAMMES

A. S. BOLOGA
Romanian Marine Research Institute,
RO-8700 Constanta, Romania

Abstract

The extremely fragile ecological state of the NW Black Sea, under the direct influence of polluted inflows from the Danube, Dniestr, Dniepr and Bug, is well known. Significant changes have occurred in the benthic and pelagic subecosystems, as a consequence of increasing pollution and eutrophication of this marine environment which has originated from various human activities (agriculture, industry, transport, trade, tourism). These have led to a dramatic decrease of living resources and catches of economically important fishery resources. Accordingly, new research and management programmes for the Black Sea are obviously necessary at the multinational level. This is the best option for a more realistic and efficient way to ameliorate the existing situation.

The Government of Romania is concerned about developing adequate national policies in the field of ecology and environmental protection, which are in agreement with the modern concept of sustainable development. It has therefore encouraged - through the Ministry of Waters, Forests and Environment Protection, the Ministry of Research and Technology, and the Romanian Marine Research Institute (RMRI) participation of all concerned institutions in the ongoing Black Sea programmes which started since 1991 [6, 7, 8, 9, 10]. The main programmes in which the RMRI has participated are the following:

- Co-operative Marine Science Programme for the Black Sea (CoMSBlack);
- Ecosystem Modelling as a Management Tool for the Black Sea: A Regional Programme of Multi-Institutional Cooperation (NATO-TU Black Sea);
- NATO TU-Waves/Black Sea
- The Interaction between the River Danube and the North-western Black Sea (EROS 2000/21);
- Regional Center for the Black Sea (IOC) - Pilot project 1 and 2;
- Environmental Management and Protection of the Black Sea (GEF);
- National Integrated Monitoring System/Coastal Monitoring (EEC/PHARE).

1. Co-operative Marine Science Programme for the Black Sea (CoMSBlack), was carried out by the main oceanological institutes in the region, under the initial coordi-

S. Beşiktepe et al. (eds.),
Environmental Degradation of the Black Sea: Challenges and Remedies, 367–376.

nation of Woods Hole Oceanographic Institution / USA [1, 34, 35]. It was aimed at multidisciplinary research for the whole basin, in order to understand and measure the roles of fundamental physical processes and their impact on Black Sea chemistry and biology.

The results contributed to a better knowledge of pollutant transport and dispersion (pollution control included), assessment of biological productivity of the basin and recommendations regarding rational management of the environment and utilization of living resources. This programme was accepted by UNESCO (Intergovernmental Oceanographic Commission) as a part of the future "Action Plan for the Black Sea", and since 1993 it has been included in the NATO TU - "Black Sea" programme (Science for Stability).

Specialists from RMRI have participated in all activities of the programme since its beginning (Fig.1). Romanian researchers participated in the HYDROBlack '91 (September) and CoMSBlack '92 (July) oceanographic cruises on the R/V "AKADEMIK" under Bulgarian flag, and contributed to the CTD measurement programme (the first measurements of that type for the Romanian shelf) and the collection of new hydrochemical, hydrobiological and ichthyological data. A Romanian scientist participated in the CoMSBlack '93A (April) and CoMSBlack '94 (April) cruises on the R/V "BILIM" under Turkish flag and cooperated in determining the marine circulation in the northwestern Black Sea using (for the first time in this area) an apparatus for measuring the current vertical profile by an ADCP. 28 stations were occupied in Romanian waters in 1993. Current measurements (97 profiles) were carried out simultaneously with those of temperature, salinity, and depth (CTD). For the first time in this zone, 574 determinations of currents, with a vertical resolution of 2-4 m, were obtained between stations. In 1994, 16 CTD stations, 216 current profiles and 362 current measurements between stations were carried out on the Romanian shelf and offshore. Also in 1994 (April) the Romanian R/V "STEAUA DE MARE 1" conducted 25 oceanographic stations (on the network of 1991-1992) at which vertical CTD profiles were collected. During the CoMSBlack '95 cruise (March), Romanian specialists repeated the station network of the previous year on the R/V "STEAUA DE MARE 1" and carried out seven complete stations with the initially established parameters. During the same cruise (March-April), another Romanian specialist worked on the R/V "BILIM" and collected samples of phyto- and zooplankton, made some qualitative and quantitative determinations which were possible on board the vessel and completed in the laboratory.

The RMRI involvement has enabled participation of Romanian specialists not only in cruises on own and foreign research vessels, but also in working meetings, intercalibration exercises and scientific symposia. Through this programme, RMRI has benefited from the new GPS satellite navigation system and portable CTD sounder with external memory for the R/V "STEAUA DE MARE 1".

The processing of the results has resulted in scientific publications on the specific Black Sea processes and phenomena, in international journals with Romanian coauthors [3, 13, 18, 25]. The Reports of the Intercalibration Meetings CTD HYDROBlack '91 and CoMSBlack '92A have been re-published by UNESCO/IOC [19, 20].

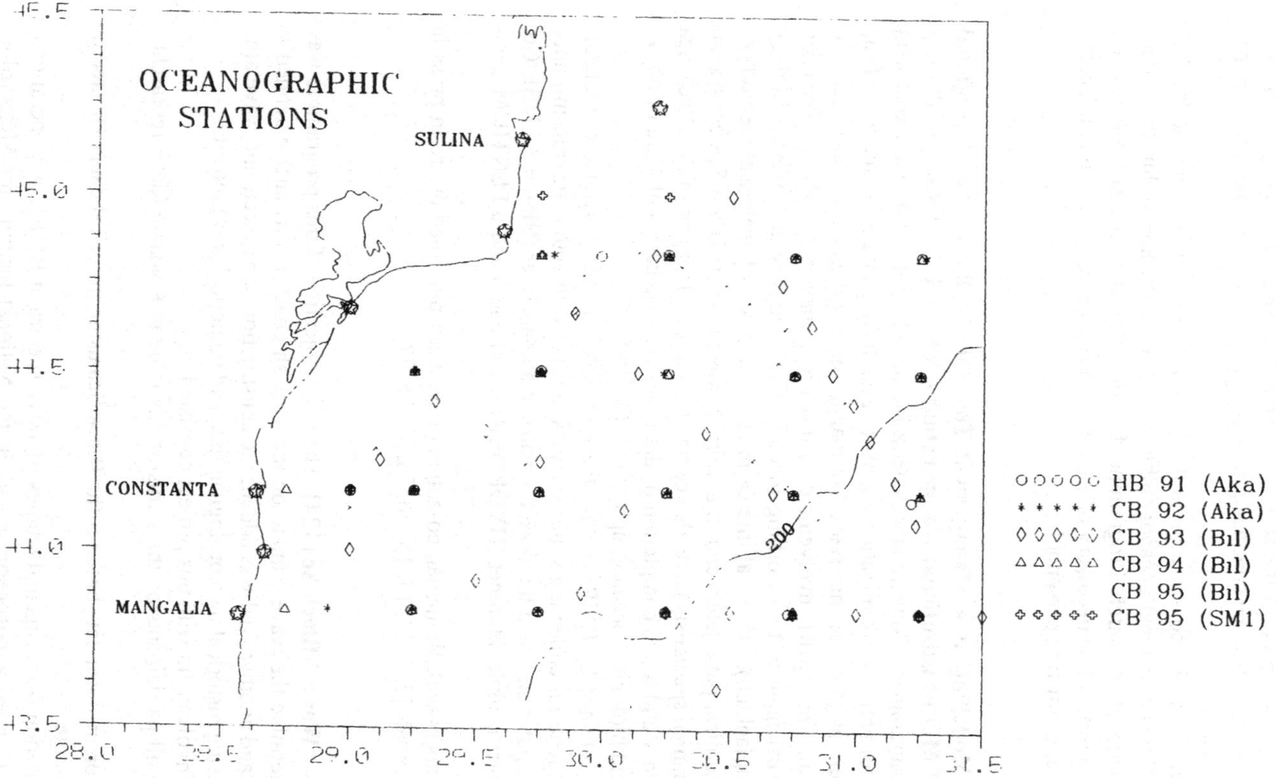

Fig. 1 Romanian participation in HydroBlack / CoMSBlack cruises (1991-1995)

According to the decision of the steering committee (Istanbul, January 1996), the CoMSBlack programme is to be continued in close cooperation with other programmes for another five years.

In the future, the RMRI contribution to this programme will depend on the level of funds allocated for the national oceanographic research programme within the framework of the National Research Programme for Environment or funds from other sources. This support will be necessary in order to ensure research vessel use and salary for specialized and assisting personnel.

2. Ecosystem Modelling as a Management Tool for the Black Sea: A Regional Programme of Multi-Institutional Co-operation / NATO TU - "Black Sea" [22], with the sub-programmes "Contaminant Penetration into the Sea", "Marine Ecosystem Response", "Information Dissemination", aims at establishing and using models of the marine ecosystem as tools for the Black Sea management and amelioration of its ecological condition. The RMRI involvement in this programme has enabled Romanian specialists to participate in two working meetings for chemistry and one for biology (phyto- and zooplankton), in several intercalibration exercises of analytical methods of physical and chemical parameters and in a training course for the Black Sea ecosystem modeling. Romanian specialists have also contributed to development of the Black Sea database inventory and to the completion of the Black Sea database with data of physical, chemical and biological oceanography for 1963-1995.

Equipment obtained by RMRI, within the framework of this programme, includes an automatic rosette for collecting water samples, a CTD with automatic transmittance of data to a computer on board the vessel, a nutrient autoanalyser (BRAN + LUBE Co. Germany), a macrozooplankton net (HYDROBIOS, Germany) and a PENTIUM computer.

Scientific papers with Romanian co-authors have been published or are in press in international journals [2, 5, 11, 14, 15, 26, 28, 33, 36, 37].

3. NATO TU - "Waves"/Black Sea [24]. The main objective of this programme was, initially, to determine the wave climate on the Turkish coast; it was later extended to the Black Sea, and another goal was added: the construction of an electronic wave atlas for the basin, using historical meteorological data and numerical wave models.

Romanian involvement with this project consisted in:

- informing all participants on the status of this issue in Romania (data, researches, etc.);
- contribution to storm data base for the Black Sea using historical data collected on the Romanian shore;
- implementation (experimental studies of wave spectra at RMRI, and operational, studies with a coupled atmospheric model, at the National Institute for Meteorology and Hydrology - NIMH) of the WAve Model - WAM, and sending comments on the performance and shortcomings of the model to the originators (DKRZ - German Climate Computer Center);

- implementation at RMRI, of wave studies, with the shallow water wave model Simulation WAves Nearshore - SWAN; comments on the performance and shortcomings of the model are sent to the originators (Delft Laboratories);
- use of the electronic Black Sea Wave Climate Atlas (in preparation) for RMRI and NIMH studies, and for the project objectives;
- use of data from a wave gauge (provided by the Marine Hydrophysics Institute of Ukraine) installed at the Romanian offshore drilling platform "Gloria" as input for the project data base and for indigenous studies (wave spectra theory, "tuning" of numerical wave models, historical nearshore data calibration and extrapolation for the open sea, etc.).

4. The Interaction between the River Danube and the north-western Black Sea / EROS 2000/21 [32]. This programme has the subprogrammes "Eutrophication", "Pollution", "River Outflows, Sedimentation Processes and Sediment Transfer" and "Biogases". It is a continuation of other programmes of the European Union for the Mediterranean Sea and for the Danube.

RMRI has been involved in the first two subprogrammes and its researchers participated in the complex survey of the north-western Black Sea sector on the R/V "PROFESOR VODIANITSKYI" under Ukrainian flag, between the 17 July and 1 August 1995, with specialists from Bulgaria, Belgium, France, Germany and United Kingdom. A total of 26 oceanographic stations were occupied and 240 water samples collected. Romanian specialists on board the vessel determined dissolved oxygen, silicate and marine phytoplankton (qualitatively and quantitatively). In the laboratory they analyzed another 40 samples for zooplankton (quantitatively and qualitatively). The results were presented as three scientific papers and discussed at the symposium "Week of the Black Sea Pollution Assessment" in Istanbul, March 1996 [4, 23, 30].

During the two cruises organized between April 12 - May 5 and May 8-26 on the R/V "PROFESOR VODIANITSKYI" in 1997, Romanian specialists were in charge of dissolved O_2, nutrient and phytoplankton determinations. In the first cruise 42 oceanographic stations were carried out and over 200 water samples were analyzed for O_2, phosphate and silicate. The salinity gradients in front of the three Danube branches were determined. During the second cruise, 110 water samples from 25 oceanographic stations were analyzed for nutrient (NH_4 on board and nitrate, nitrite, phosphate and silicate at RMRI on refrigerated samples). Those analyses contributed to assess present eutrophication conditions in the NW part of the Black Sea. 150 phytoplankton samples were simultaneously collected for qualitative and quantitative analyses.

The data have been useful for developing a preliminary mathematical model of the Black Sea ecosystem.

5. IOC Black Sea Regional Programme in Marine Sciences and Services [21]. The IOC supported establishment of a regional mechanism for coordination of scientific improvement of observations and to extend the mutual data exchange, as an appropriate approach of the unique ecological problems of the Black Sea.

This initiative will take into account ongoing activities in the region avoiding overlapping and duplication. Two pilot projects have been launched:

- Black Sea observation and prediction research project: initiation of the Black Sea regional component of GOOS (Black Sea GOOS),
- Assessment of sediment flux mechanisms of formation, transformation, dispersion and ecological significance (Black Sea fluxes).

The Romanian participants from RMRI and NIMH will be involved in both pilot projects.

Two Romanian contributions were presented at the first scientific workshop organized within the framework of pilot project 2 in Istanbul, June 1997 [12, 29].

6. Environmental Management and Protection of the Black Sea Environmental Programme / GEF [16, 17]. This programme includes the Black Sea, Azov Sea, coastal zones of the riparian countries and drainage of all tributary rivers excepting the Danube (for which there is a special programme).

The major goal was creation of a framework for medium and long term regional co-operation, aiming at limiting the continual deterioration of the marine environment and resources and contributing to the rational utilization of resources.

To achieve this goal, a Programme Coordinating Unit (PCU) has been established with its headquarters in Istanbul (Turkey). National coordinators (RMRI for Romania) and six activity centers and focal points dealing with Black Sea priority problems have been established.

Within the framework of this programme, RMRI created the Fishery Activity Center in April 1994 including the technical aspects in connection with the objectives, functioning and communications with the fishery focal points in the Black Sea riparian countries. This Center has focused its activity on the following items: reports on the fishery status in each country, critical problems in regional fishery management, creation of databases including historical data regarding fish stock assessments, information on present resources and functional capacities of the fishing fleet and processing factories, background for a new Convention for Fisheries and Protection of Living Marine Resources in the Black Sea, strategy concerning the control on *Mnemiopsis leidyi* population, marine aquaculture and farms, creation of working groups for Resource Assessment, Modeling and Surveys (RAMS) and for fishery reconstruction.

RMRI specialists have provided logistical support for organization of two meetings at Constantza:

- the meeting of the RAMS working group (October, 1994), during which creation of a communication network between the Fisheries Activity Center and focal points, was discussed, the need for organizing a RAMSDATABASE was also discussed,

- the working meeting on fisheries (June-July 1995) for which the agenda included national reports on fishery activity in 1994, analysis of the GESAMP report (Geneva, March 1995) on the ctenophore *Mnemiopsis leidyi*, the analysis of the aquaculture mission report and results of the first RAMS meeting. Subsequently, appropriate experts from RMRI participated in specialized meetings which took place in Geneva, Istanbul and Varna in 1995.

Other achievements included participation in elaboration of the new text of the above mentioned Convention, creation of a regional database section concerning research cruises, elaboration of the list of Black Sea fish names in Romanian language for FAO, calculation of the growing parameters of the main commercial fish species at the Romanian littoral (for POPulation DYNamics (POPDYN) database), elaboration of the national report on research in the field of fishery resources in the Romanian sector and preparation of the following working meeting on fisheries (Constantza, 1996).

Remarkable results have also been obtained by each of the focal points which are co-ordinated at the national level by RMRI specialists:

Integrated Coastal Zone Management (ICZM)

- National Black Sea Environmental Priorities Study for Romania [31];
- participation in training course regarding implementation of the ICZM process and of some of its instruments (e.g. environmental impact assessment studies);
- elaboration of the ICZM National Report (2 volumes);
- elaboration of documentation concerning geographic limits of the coastal zone and the institutional framework necessary for ICZM process implementation;
- elaboration of a proposal for a pilot project regarding "The Plan for Investment and Management of the Coastal Zone on the Territory of Constantza County" and of related documentation;
- elaboration of some reports for the Activity Center:
  - report on the ICZM National Report;
  - report on coastal zone limits;
  - report on establishing ICZM institutional framework;
  - report on the pilot project and beginning of implementation.

Biodiversity

- elaboration of the National Report on Biodiversity of the Romanian Black Sea Sector [27];
- submission of three pilot project proposals, in accordance with the decisions of the working group at Varna, July 1995:
  - a programme for the rehabilitation of the biodiversity of Mamaia Bay which has been affected by anthropogenic impact;
  - a regional monitoring programme of the biodiversity in the north-western Black Sea;
  - a programme for the ecological reconstruction of Tekirghiol Lake;
- nomination of the RMRI representative for the subgroup "Conservation of the Black Sea Mammals" and participation in the working meeting (Istanbul, December 1995);
- presentation of the report on Romanian research regarding the Black Sea dolphins (Istanbul, December 1995);
- participation of Romanian specialists in the working meetings (Istanbul, February

1994; Batumi, November 1994; Varna, February 1995).

Routine Monitoring

- Participation in scientific courses and seminars regarding:
    - monitoring of beach and bathing water quality;
    - monitoring of drinking water quality;
    - methodology of assessment of land based pollution sources;
    - contaminants in the sediments;
    - collection and analysis methods for the oil contamination of the aquatic environment.
- participation in two joint meetings of the Activity Centers for routine and special monitoring aiming at finalizing the joint Black Sea monitoring programme.

Special Monitoring: Pollution Effects on Marine Organisms:

- participation in the seminar discussing convenient techniques for approaching the pollution effects on marine organisms;
- participation in the intensive course for achieving general knowledge on biomarkers and demonstrations of new methods;
- participation in the intensive course for achieving the necessary knowledge for choosing the sampling sites, number of samples and statistical analysis of results.

Emergency Response in Case of Accidental Marine Oil Pollution

- participation of RMRI specialists in the meetings of the Activity Center (Varna, 1994-1995) in order to elaborate the national contingency plan in accordance with the questionnaires and frameworks suggested by IMO for their unitary integration in the Regional Black Sea Programme;
- participation in the initiation course MEDPOL '95 organized by the Regional Mediterranean Center for the Emergency Response against Accidental Marine Pollutions (REMPEC) for the executives of this activity at the Mediterranean and Black Sea;
- elaboration of the draft National Contingency Plan (which is to be amended by ministries and organizations involved);
- starting of a cooperation project within the framework of the programme CEE/ECOS-OUVERTURE with CEDRE, Brest (France), aiming at the creation of a National Center for training specialists in marine pollution response at Constantza.

7. National Integrated Monitoring System / Coastal Monitoring [38]. The aim of PHARE Project ZZ.92.11/02.01/B001 "Black Sea Environmental Programme - Monitoring, Laboratory Analysis and Information Management" was to improve the environmental management in the Black Sea, strengthen environmental monitoring networks, improve the comparability of sampling technique and laboratory analyses and develop and implement an information management system.

RMRI has participated in the creation of the National Integrated Monitoring System with respect to defining the quality standards for the coastal and marine environment and with respect to specifying physical, chemical and biological indicators in these areas.

The contribution of RMRI consisted of:

- establishing the monitoring network taking into account pollution sources, hydrological patterns, different utilities of beaches and sea water;
- establishing parameters to be monitored, sampling frequency, analysis methods;
- setting up the collaboration framework with interested agencies of the Ministry of Environment in order to give efficient use of data and to avoid duplicating this effort.

References

1. Aubrey D G., Belberov Z., Bologa A.S., Eremeev V.N , Unluata U (1992) A coalition to diagnose the patient: CoMSBlack and the Black Sea, *MarTech*, **2**, 1, 5-8
2. Aubrey D.G., Moncheva S., Demirov E , Diaconu V., Dimitrov A. (1996) Environmental changes in western Black Sea related to anthropogenic and natural conditions, *Journal of Marine Systems*, **7**, 411-425.
3. Basturk O., Saydam C., Salihoglu I., Eremeeva L.V , Konovalov S.K., Stoyanov A., Dimitrov A., Cociasu A., Dorogan L , Altabet M. (1994) Vertical variations in the principle chemical properties of the Black Sea in autumn of 1991, *Marine Chemistry*, **45**, 149-165
4. Bequevort S., Bouvier T., Cauwet G., Popa L. (1996) Dynamics of the microbial food-web, personal communication.
5. Bodeanu N , Ruta G. (in press) Long-term evolution of the phytoplankton blooms in the Romanian Black Sea area, *NATO ASI Series*.
6. Bologa A.S. (1990) The Romanian Marine Research Institute at its 20th anniversary· Tradition, status, perspectives, *Cercetari* marine - *Recherches marines*, **23**, 7-13.
7. Bologa A.S. (1991-1992) International relations of the Romanian Marine Research Institute· necessity, achievements, prospects, *Cercetari marine - Recherches marines*, **24-25**, 5-9.
8. Bologa A S. (1994a) Marine research priorities in Romania, *Mediul inconjurator*, **5**, 1, 3-13.
9. Bologa A S. (1994b) Programe de cercetare-gestionare marina in România, *Marea noastra*, **4**, 12, 22-23
10. Bologa A.S. (1996) Rezultate semnificative obtinute de Institutul Român de Cercetari Marine in cadrul programelor regionale de cercetare si gestionare privind Marea Neagra in perioada 1991-1995. *Mediul inconjurator*, 7, 2, 54-68.
11. Bologa A.S , Frangopol P.T., Vedernikov V.I , Stelmakh L.V., Yunev O.A., Yilmaz A , Oguz T. (in press) Planktonic primary production in the Black Sea, *NATO ASI Series*.
12. Bondar C., Blendea V. (1997) Water and sediment transport of the Danube into the Black Sea, personal communication.
13. Cociasu A., Dorogan L., Humborg C., Popa L. (1996) Long-term ecological changes in the Romanian coastal waters. *Marine Pollution Buletin*, **32**, 32-38.
14. Cociasu A , Diaconu V., Popa L., Buga L., Nae I., Dorogan L., Malciu V. (1997) The nutrient stock of the Romanian shelf of the Black Sea during the last three decades in E Ozsoy and A Mikaelyan (Eds.), *Sensitivity to Change· Black Sea, Baltic Sea and North Sea*, NATO ASI Series, 2. Environment, 27, Kluwer Acad Publ , 49-63.
15. Diaconu V , Cociasu A. (in press) Physical and chemical processes in the water masses of the western shelf areas of the Black Sea, *NATO ASI Series*.

16 Global Environment Facility (GEF) (1994, 1995, 1996) Black Sea Environmental Programme Annual Report
17 GEF (1996) Strategic Action Plan for the Rehabilitation and Protection of the Black Sea, GEF, Istanbul, Turkey, 31 October 1996.
18 Humborg C , Ittekkat V., Cociasu A., Bodungen B. (1997) Effect of Danube River dam on Black Sea biogeochemistry and ecosystem structure *Nature*, 386, 385-388.
19. Intergovernmental Oceanographic Commission (IOC) (1991) Hydroblack '91 CTD Intercalibration Workshop, WHOI, Woods Hole, USA, 1-10 December, *UNESCO Workshop Report No. 91*
20 IOC (1993) CoMSBlack '92A Physical and Chemical Intercalibration Workshop, Erdemli, Turkey 15-29 January, *UNESCO Workshop Report No 98.*
21. IOC (1996) Circular Letter No 1495, Paris, 10 June, 5.
22. Lee H.A. (1994) Science for Stability, A NATO Programme, The first two phases 1979 to 1993, NATO Scientific Affairs Division, 1110 Belgium, 85-86.
23. Moncheva S., Ruta G , Bodeanu N. (1996) Dynamics of phytoplankton species distribution and biomass in the NW Black Sea in the summer 1995 conditions, personal communication.
24. NATO (1994) Newsletter from the Science Committee and the Committee on the Challenges of Modern Society, Issue No. 40 1st Quarter, 4.
25 Oguz T , Aubrey D.G , Latun V S., Demirov E , Koveshnikov L., Sur H.I., Diaconu V., Besiktepe S., Duman M., Limeburner R., Eremeev V. N. (1994) Mesoscale circulation and thermoshaline structure of the Black Sea observed during HydroBlack '91 *Deep-Sea Research*, **41**, 4, 603-628.
26. Panin N , Jipa D., Gomoiu M.-T., Secrieru D. (in press) Importance of sedimentary processes in Environmental changes, *NATO ASI Series.*
27 Petranu A. (Ed.) (1997) *Black Sea Biological Diversity Romania*, GEF Black Sea Environmental Programme, UN Publications, New York, 314 pp.
28 Petranu A , Apas M., Bologa A S , Bodeanu N., Dumitrache C , Moldoveanu M., Radu G., Tiganus V (in press) Status and evolution of the Romanian coastal ecosystem, *NATO ASI Series.*
29 Piescu V., Bologa A.S , Cociasu A , Cuingioglu E , Mihnea R , Patrascu V , Pecheanu I. (1997) Danube input of pollutants in sediments along the Romanian Black Sea coast, personal communication.
30. Popa L., Krastev A., Pencheva R., Ragueneau O., Verlinmeren J., Stoyanov A. (1996) The nutrient environment: general trends and the 1995 summer conditions, personal communication.
31 Postolache I (Ed.) (in press) GEF/BSEP: National Black Sea Environmental Priorities Study for Romania, Constanta, Romania, 1997
32 Remond A (1995) PECO 1994, Catalogue of participation in the Third Framework Programme in 1994, Project Summaries, 1-10, 1-13
33. Stelmakh L.V., Yunev O.A., Finenko Z.Z., Churilova T.Ja., Bologa A.S. (in press) Peculiarities of seasonal variability of primary production in the Black Sea, *NATO ASI Series.*
34 Unluata U., Aubrey D.G., Belberov Z , Bologa A.S., Eremeev V.N., Vinogradov M.E. (1993) International program investigates the Black Sea, *EOS*, American Geophysical Union, 74, 36, Sep. 7, 401-407-412.
35. Unluata U., Aubrey D.G., Belberov Z., Bologa A.S , Eremeev V.N., Vinogradov M.E. (1993) Progress report on the Co-operative Marine Science Program for the Black Sea 1992-1993,*UNESCO / IOC / Inf-924*, Paris, 27 Feb., 1-16.
36 Yilmaz A , Yunev O.A., Vedernikov V.I., Moncheva S , Bologa A.S , Cociasu A , Ediger D (in press) Unusual temporal variations in the spatial distribution of chlorophyll *a* in the Black Sea during 1990-1996, *NATO ASI Series.*
37. Yunev O.A., Vedernikov V.I., Yilmaz A., Bologa A.S. (in press) Biooptical data quality control, *NATO ASI Series*
38. World Bank (1992) Romania, Environment Strategy Paper, 31 July, 41-43

Working Groups Reports of the NATO ARW: "Environmental Degradation of the Black Sea: Challenges and Remedies"

Three working groups were established among the participants to review existing knowledge on the Black Sea's oceanography and further to identify needs for the future research. The working groups were Ecological Processes, Modeling and Physical Processes and Contamination. Each working group reviewed past efforts and addressed future research needs in the Black Sea. Recommendations for the future research needs were developed as a result of discussions in the working groups and afterwards finalized in the planetary sessions with common discussions of all participants. Working Groups stressed that international collaborative activities are about to terminate by the end of 1997 and/or 1998, it is of vital importance to pursue collaborations in the near future. Every efforts must therefore be made to allow the appropriate bridging of existing activities with the follow-on efforts so as not to loose the momentum and to ensure eventually the sustainable development of the Black Sea. The ARW specifically identified future research needs and a framework for a continuous ocean observing system and forecast capabilities for the Black Sea in parallel with developments in modern ocean sciences. Reports of the working groups are as follows;

1. Report of the Working Group 1: Ecological Processes

Observational ecological studies are needed to develop a predictive system for the Black Sea. Both local and basin scale studies are required. To achieve this goal Working Group 1 made the following recommendations for Process Studies.

1. Biological Methods need to be standardized between all Black Sea countries so that data from time series and regional studies are comparable. International Protocols (e.g. JGOFS) should be adopted. Standard methods for biomass and primary productivity should be the first goal.

2. Ecological data need to be collected to understand temporal and regional variability. Predictive ecological models need seasonal data for verification. When conducting these studies we strongly encourage different research groups to work together with the goal of obtaining more comprehensive data sets on the same samples. Surface time and regional surveys could be conducted from regularly schedules ferries and transport vessels. Oil drilling platforms might be useful for time series studies. In addition, each lab should initiate at least one time series station at a convenient coastal location where vertical profiles can be collected. Such time series process studies are necessary to document future environmental degradation of improvement.

Utilization of new technologies should be encouraged. Examples include The Continuous Plankton Recorder, Flow Cytometry for pico and nano plankton, High

S. Beşiktepe et al. (eds.),
Environmental Degradation of the Black Sea: Challenges and Remedies, 377–386.

Pressure Liquid Chromatography (HPLC) for pigment analyses of different phytoplankton populations and acoustical methods for zooplankton.

3. Environmental Risk Areas need to be identified and the local source of environmental stress identified. Monitoring of benthic biomass and species composition is the simplest approach for monitoring environmental quality because they are not as susceptible to seasonal and spatial variability. Pelagic studies should also be conducted. Species composition reflects the integrated pollution and environmental inputs. Proper baseline studies need to be conducted and a list of key indicator species identified. A recent GESAMP Report (1995) presented a list of suggested biological indicator species and indices.

4. We need to improve data available on river input and sediment quality for the major rivers. Good data is available for the Danube but similar studies are needed for the Dniester and Dnieper Rivers. The monitoring of the NW shelf region, started with EROS 2000 needs to be continued.

5. Shelf areas receive the main impact of these stresses and are the sites of changes in eutrophication and degradation of the benthic community. Little baseline data is available for the benthos, especially for the Turkish coast. Benthic flux data are also largely absent. Paleochemical studies on sediment cores should be used to document historical variability.

The benthic ecology and geochemistry of the sediments in contact with the oxic/anoxic interface in unknown. Benthic organisms around the methane seeps and vents on the shelf regions (especially the NW and SE shelves) should be investigated because unusual life forms have been identified at similar sites at other locations. Remote Operating Vehicles (ROVs) would be especially useful for these studies.

6. Specific Biological Process Field and experimental studies are required especially for key blooming species.

a) We know remarkably little about the life cycle of Aurelia and Mnemopsis in the Black Sea. Because of the lack of recent sampling we do not know if the populations are increasing, decreasing or stabilized. The seasonal variability, reproduction and feeding behavior are also unknown. For example, do Aurelia and Mnemopsis feed on each other?

The butterfish (Perilus triacanthus) has been proposed as a possible biological control. More data are needed about the food sources and preferences of this organism before this proposal can be evaluated. For example, what would these fish eat if gellatinous zooplankton were not available?

This WG3 did not object to pilot studies being conducted but recommended that a higher priority be given to building a better understand of the gellatinous species.

b) The key blooming species (phytoplankton and zooplankton) need to be better understood. The ecological modelers need Black Sea specific data for growth (m) and grazing (m) rates.

c) Specific studies should be conducted for

jelly organisms (e.g. Aurelia, Mnemopsis, Noctiluca and Rhizostoma)
(e.g., Noctiluca are thought to be a biological "dead end" but we don't know anything about their recycling)
Rhizosolenia
Ceratium
pico and nano plankton

7. Some studies suggest that inputs of atmospheric iron stimulate coccolith blooms. This hypothesis needs to be pursued.

8. The ecology of the oxic/anoxic interface is poorly known. Specific deficiencies are the anaerobic photosynthetic bacteria and nitrifying bacteria. Little is known about the bacteria involved in the sulfur and methane cycles.

9. Macroscopic variables reflecting the net effect of the ecological food web should be measured. These include new and regenerated primary production, the export flux of particulate organic carbon and the distribution, chemical nature and cycling of dissolved organic carbon. Changes in the ecological state of the euphotic zone could influence the export flux of organic carbon which could result in changes in the suboxic zone and sulfide distribution and production. The shelf regions would be especially susceptible to such changes.

10. We need to improve the regional (International) approach to Fisheries Studies. A regional approach is needed to understand reproduction, migration and wintering. A Fisheries Data Base should be constructed to aid in prediction of the evolution of fish stocks and controlling processes. The life cycles of key species (especially those with larvae grazed by jelly fish) need to be studied.

11. Fisheries Management needs evaluation of fish stocks for different regions. Genetic analysis is the optimum approach.

12. A science plan should be developed for monitoring by use of physiological and biochemical indicators and study of small commercial pelagic fish (e.g., sprat and anchovies). These small fish represent the boundary between the phytoplankton/zooplankton ecological models and the larger fish populations. The role of these fish in zooplankton dynamics and the physical forcing that causes long-term variability should be studied

13. Paleoceanographic studies (especially of the Quaternary) should be useful Integrative Tools for understanding variability in the recent past.

2. Report of the WG II: Modeling and Physical Processes

The existing efforts in the Black Sea addressed a series of important scientific and technical issues concerning with its environment, including development of infrastructure and improving the technical capabilities of the regional oceanographic institutions. Because these international collaborative activities are about to terminate by the end of 1997 and/or 1998, it is of vital importance to pursue such collaborations in the near-future. Every efforts must therefore be made to allow the appropriate bridging of existing activities with the follow-on efforts so as not to loose the momentum and to ensure eventually the sustainable development of the Black Sea.

Under the framework of the existing programs, significant achievements have been carried out on understanding the Black Sea environmental problems. For example, the Black Sea Environmental Program (BSEP) was initiated in 1993 to implement policy and a legal framework for the assessment, control and prevention of pollution and to maintain and to restore the biodiversity, to create and strengthen regional capabilities for managing the Black Sea ecosystem. The CoMSBlack program led to realization of a series of basin, sub-basin and regional scale multiship oceanographic surveys. The data collected during these surveys were instrumental in developing a new understanding of the physical-biogeochemical characteristics of the sea. The NATO-supported TU-Black Sea program opened a new horizon in enhancing scientific links between the major regional oceanographic institutions. It led to a series of scientifically important results, including development of a unique data base with its management system ,that will soon be accessible to the community .The EROS 2000 Project was launched in 1994 by the European Commission to strengthen the collaboration between Eastern and Western institutions. The project is focused on the eutrophication and contamination problems of the coupled River Danube-Northwestern shelf system through observations, process-oriented and modeling efforts. During the recent years, various groups developed and implemented, with a certain degree of success, the physical-biogeochemical-ecosystem and marine operational (storm surge, wind wave and oil spill) models having different levels of complexity. New technologies/methodologies are becoming available for monitoring and predicting the Black Sea more efficiently.

Needs for the Black Sea

On one hand, the modeling studies and the data collected within the framework of these activities have improved drastically our understanding of this complex ecosystem. On the other hand, they had some limitations related to insufficient data coverage and lack of understanding several crucial processes. Another important drawback of these studies was the insufficient coordination among the individual groups. The present critical economical situation in the riparian countries makes collaborative and cooperative research a must. Future research efforts at both national and international levels shall therefore take into account coordination of available activities in the field of research and operation services.

On the basis of our existing knowledge of the Black Sea, the future research initiatives should be oriented to explore, quantify and predict the marine environment through interdisciplinary sciences involving observations and modeling at a hierarchy of space and time scales.

Recommendations

It is recommended that the development and implementation of a forecasting and observations system should involve;

1) Prediction/Forecasting Modeling through

- *implementation of Ocean General Circulation Systems (OGCMs)* with different resolutions and spatial coverage. They form the physical component of the ecosystem for simulating and predicting the three dimensional structure of flow field, sea surface elevation, temperature and salinity distributions and their time evolution with mesoscale resolution.
- *development of an integrated network of physical-ecosystem-biogeochemical models (BGCMs)* for simulating and predicting the seasonal and longer term variability of the ecosystem and its related biogeochemical cycles forced by antropogenic and climate changes.
- *implementation of regional operational marine prediction system* including *interalia* wind waves, storm surge and oil spill prediction models.
- *development and implementation of data assimilation package* for the full set of predictive models, which increase the models' accuracy and to optimize relevant parameters.

2) Process modeling

The predictive modeling studies must be supported by parallel ongoing process-oriented modeling activities to understand physical-chemical and biological mechanisms governing the functioning the Black sea ecosystem and the related biogeochemical cycles. Knowledge gained by the process modeling studies is necessary to upgrade the forecasting capability.

2.a) Physical Process Modeling

The particular process-oriented physical modeling studies considered to be high priority are:

- dynamics of river-induced buoyant plumes and interaction with inner shelf waters,
- cross-shelf exchange processes between the outer shelf and the interior of the sea,
- instability characteristics of the Rim current jet,

- cold intermediate water mass formation at the northwestern shelf and cyclonic gyres of the basin interior, and their spreading over the basin,
- ventilation processes for the suboxic and anoxic layers,
- coastal upwelling,
- algorithm developments for satellite data processing.

2.b) Biogeochemical Process Modeling

The major process studies using the biogeochemical component of the system are:

- development of high trophic resolution ecosystem model including, in addition to phytoplankton, bacterioplankton and mesozooplankton, key biological components of the Black Sea ecosystem such as gelatinous omnivorous and carnivorous organisms and planktonic fishes (sprats, anchovy) to understand the relative roles of antropogenic activities and meteorological variations on ecosystem structure and functioning,
- development of numerical techniques to couple lower and higher resolution ecosystem model,
- models for estimating nutrient (N, P, S, O, C. Si) cycles ans sulphate reduction and denitrification,
- modeling redox processes across the oxic/anoxic interface by testing different hypotheses,
- modeling of anoxia in the shelf as a result of eutrophication, oxygen input and consumption, and physical processes,
- modeling the biogeochemical evolution of the sea from the begining of late quarternary period.

3) Black Sea Observing System

Development of an efficient observing system is a matter of highest priority as it is an inherent part of the prediction system and one of the most important requests for models development and verification. The observing system should be built with the following principles:

- correspondence to requirements of most crucial services and highest priority regional studies,
- ability to provide necessary accuracy and resolution of basic parameters,
- design basing on modern approaches involving observing system sensitivity studies,
- long-term systematic character,
- cost-efficiency,
- compliance with practices and standards available in the region and in the world,
- joint use of remote sensing and in-situ observations with account of future trends in observing techniques,

- avoidance of duplication of efforts and the system build up basing on already available systems and ongoing programs,
- account of the need to continue long-term series of data acquired in the result of past studies.

Efforts of research institutes and operational agencies should be joined in the development of the observing system. It is of primary importance to ensure the availablibility of data acquired in the course of research programmes to operational agencies in the most expeditious way.

The observational system which provide the data flow for the model initialization and updating should include;

- *ocean-atmosphere interface (meteorological) measurements* for the sea surface temperature, air temperature, surface wind velocity, wind waves, humidity and precipitation, surface radiation, cloud coverage.
- *In situ and remotely sensed measurements* for physical-chemical-biological properties of the sea.

The techniques to be used are continuous and/or discrete data sampling by means of *research vessels from the Black Sea institutions,*
ships of opportunity,
moored arrays,
satellite tracked drifters,
satellites,
shore-based and coastal stations.

In addition, there is a need for process-oriented field studies.

4) Data Management System

Efficiency of the data flow is an essential component of any operational forecasting system. the relevant data sets which are planned to be assimilated must be first received, assembled, quiality-controlled, processed and transmitted in near-real time. The sets and derived products must then be disseminated to the user community and archived. These requirements imply:

- to improve the existing communication capabilities. Two main potentials for more efficient data communication should be explored. Namely, the Internet connections between the neighboring countries and the use of the WMO GTS (Global Telecommunication System).
- to establish a network of centers specialized on the collection, processing and dissemination of particular set of data. The distribution of the work load between the oceanographic institutions allows to use the human and technical resources more effectively.

- every efforts must be devoted to include all available data into a distributed regional data base system.

TU-Black Sea data base system should be improved to manage additional data with additional variables. In particular, the historical data sets owned by individuals are under great risk of lost because of the lack of resources due to the economical difficulties of regional countries. After it is open to the regional community, it could potentially serve as a prototype regional data base.

3. Report of the WG III: Contamination

A generic Science Plan (SP) for science and technology (S&T) program development to foster capabilities for predicting and monitoring contaminantion of the Black Sea environment on a permanent basis is needed to assist in its protection and sustainable development. The SP should define the S&T research issues to be considered and strategies to be adopted for developing an ocean observation and prediction system for the entire Basin and in particular, its coastal and shelf seas.

Sensitivity analyses performed with the model will enable environmental managers to determine what natural and anthropogenic factors and what locations are most influential for changing conditions of the Black Sea.

The environmental managers can then better assess the relative impacts and, therefore, the relative merit of proposed projects throughout the Black Sea drainage basin. Managers will have for the first time, a scientifically based management tool for deciding which projects offer significant benefits for the Black Sea and which do not.

Needs for the Black Sea

Environmental Quality Objectives (EQOs)
1. Development of Environmental Criteria in order to set EQO's in relation to drinking water, food, industry, transport and tourism.

Specific Needs Training programme, funding & sustainability.

2. *Communication Network* Sustain a communication network in order to maximise the information pool to the
benefit of the riparian countries.

3. *Baseline Time Series Studies*

To develop and sustain cost efficient time series studies using a specific biomarkers of effect in order to target subsequent analytical chemistry and contaminant (including radionuclides) "hot spots", plumes and fates. This could benefit form data arising out of the proposed Black Sea Mussel Watch Programme.

<u>Specific Needs</u>
- Experimental programme for hypothesis testing.
- Determination of the health status of the biota to feed into mathematical models for forecasting and scenario testing.
- Remote sensing - result also to be fed into models as well as used in security.

4. *Environmental Security*

To prepare the Black Sea countries to respond to hazardous materials, oil (due to maritime accidents) or other chemicals which pose a threat to human health, property and the environment.

Specific Needs - Contingency planning to include sensitive zones, strategies and techniques of response and coordination.

5. Causes of Particular Concern

i Degradation products and xenoestrogens and their impact on the food chain and human health.

ii Ultra-violet B radiation impacts on the biota and surface chemistry and the consequences for the aquatic food web. It was considered advantageous to global monitoring of ozone depletion that a minimum of two laboratories within the Black Sea region be funded for UV-B monitorig equipment and connected into one of the international monitoring programmes such as ELDONET.

Recommendations

There is a requirement for a long term Science Plan for the Black Sea in order to implement the recommendations of the Strategic Action Plan 1996.

LIST OF PARTICIPANTS

Boris Alexandrov
Odessa Branch, Institute of Biology of the Southern Seas,
Ukrainian Academy of Science
37 Pushkinskaya Street
270011 Odessa
UKRAINE
Phone:
Fax:
e-mail: root@inbum.odessa.ua

Alexandru T. Balaban
Vicepresident
Academia Romana
Calea Victoriei 125
Bucuresti 711
Phone: 40 1 6594789
Fax: 40 1 2116608
e-mail: balaban@aix.acad.ro

William Bailey
Program Manager
GEO-CENTERS, Inc/US Navy
Suite 910
1755 Jefferson Davis Highway
Arlington, Virginia 22202 USA
Phone: 703/416-1023 ext.108
Fax: 703/416-1178
e-mail: baileywb@AOL.COM

Sukru T. Besiktepe
Middle East Technical University
Institute of Marine Sciences
P.O.Box 28, Erdemli, 33731
Icel-TURKEY
Phone: +90-324-5212406
Fax: +90-324-5212327
e-mail: sukru@deniz.ims.metu.edu.tr

Nicolae Bodeanu
Romanian Marine Research Institute
300, Bd. Mamaia 300 RO, 8700
Constantza 3
Phone: +40-41-650870 / ext. 33
Fax: +40-41-831274
e-mail: bodeanu@alpha.rmri.ro

Alexandru S. Bologa
Scientific Director
Romanian Marine Research Institute
300, Bd. Mamaia 300 RO, 8700
Constantza 3
Phone: +40-41-643288
Fax: +40-41-831274
e-mail: abologa@alpha.rmri.ro

Dexter Bryce
Senior scientist
GEO-CENTERS, Inc/US Navy
Suite 910
1755 Jefferson Davis Highway
Arlington, Virginia 22202 USA
Phone: 703/416-1023 ext.104
Fax: 703/416-1178
e-mail: dexbryce@aol.com

Francois Cabioch
CEDRE
Technopole Brest Iroise, Boite Postale 72
29280, Plouzane
FRANCE
Phone: 33 2 98 49 12 66
Fax: 33 2 98 49 64 46
e-mail: Fcabioch@ifremer.fr

Namik Cagatay
Institute of Marine Sciences and Management
Istanbul University
Muskule Sohak
Vefa 34470, Istanbul
TURKEY
Phone: 90-212-5282539
Fax: 90-212-5268433
e-mail:debien@superonline.com

Luis Veiga da Cunha
Science Administrator
NATO Scientific Affairs Division
B-1110 Brussels
BELGIUM
Phone: (32-2) 707 4229
Fax: (32-2) 707 4232
e-mail:

Adriana Cociasu
Romanian Marine Research Institute
300, Bd. Mamaia RO-8700 Constantza 3
ROMANIA
Phone: +40-41-650870 ext.47
Fax: +40-41-831274
e-mail: acociasu@alpha.rmri.ro

Vasile Diaconu
Romanian Marine Research Institute
300, Bd. Mamaia RO-8700 Constantza 3
ROMANIA
Phone: +40-41-650870 / ext. 46
Fax: +40-41-831274
e-mail: diaconu@alpha.rmri.ro

Dumitru Dorogan
Romanian Marine Research Institute
300, Bd. Mamaia RO-8700 Constantza 3
ROMANIA
Phone: +40-41-650870 ext.45
Fax: +40-41-831274
e-mail: dorogan@alpha.rmri.ro

Heleni Florou
National Centre for Scientific Research "Demokritos"
Institute of Nuclear Technology - Radiation Protection
Environmental Radioactivity Laboratory
Aghia Paraskevi 15310
P.O. Box 60228 - Athens
GREECE
Phone: +301-6517306
Fax: +301-6519180
e-mail: eflorou@cyclades.nrcps.ariadne-t.gr

Naci Gorur
Istanbul Technical University
Department of Geological Engineering
80626, Ayazaga, Istanbul
TURKEY
Phone: 90-212-2856211
Fax: 90-212-2856210
e-mail: debien@superonline.com

Paul F. Hankins
Environmental Program Manager
US Navy Supervisor of Salvage
2531 Jefferson Davis Highway
Arlington, VA 22242-5101
USA
Phone: 703/607-2758
Fax: 703/607-2757
e-mail:
hankins_paul_f@hq.navsea.navy.mil

Charles G. Hardin
Programe Manager, Bussines Development
Wasting House Savannah Rivin Co.
1359 Silver Bluff Rp. B-1
Aiken, SC 29 803
USA
Phone: 803 643 4845
Fax: 803 643 4850
e-mail: CGHARDIN@SCSCAPE.NET

Mikhail V. Ivanov
Institute of Microbiology
Academy of Sciences
117811, Prospect 60 Oktjabrja 7,
Moscow B-312
RUSSIA
Phone: 007(095) 1352139
Fax: 007(095) 1356530
e-mail: ivanov@imbran.msk.ru

Dan C. Jipa
Scientific Secretary
National Institute of Geology and Geoecology
Str. Caransebes 1
RO-78344 Bucharest
ROMANIA
Phone:
Fax:
e-mail:

David Kaftan
Institute of Microbiology
Laboratory of Photosynthesis
Academy of Sciences of the Czech Republic
37981 Trebon
CZECH REPUBLIC
Phone: +420-333-721101
Fax: +420-333-721246
e-mail: david@mbu.chvi.cz

Irakli Khomeriki
Head of the Oceanographic Centre
Tbilisi State University
Faculty of Geography and Geology
1 Charchavadze ave.
PO Box 380028 Tbilisi
GEOGIA
Phone: (995-32) 226420
Fax: (995-32) 221103

Sergey K. Konovalov
Marine Hydrophisical Institute
Ukrainian National Academy of Sciences
2, Kapitanskaya st.,
Sevastopol 335000
UKRAINE
Phone: +38(0692) 525276
Fax: +38(0692) 444253
e-mail: sergey@alpha.mhi.iuf.net

Genady Korotaev
Marine Hidrophysical Institute
Ukrainian National Academy of Sciences
2, Kapitanskaya St., Sevastopol 335000
UKRAINE
Phone: +38(0692) 520779
Fax: +38(0692) 444253
e-mail:
ipdop@fossil.ukrcom.sebastopol.ua

Ruben D. Kosyan
Director
Russian Academy of Sciences
Southern Branch of the
P.P.Shirshov Institute of Oceanology
Gelendzhik-7, 353470
RUSSIA
Phone: +7 86141 23261
Fax: +7 86141 23189
e-mail: KOSIAN@SDIOS.sea.ru /
KOSIAN@ONLINE.sea.ru

Christiane Lancelot
Universite Libre du Bruxelles
GMMA, CP 221, Bd. du Triomphe
B-1050 Bruxelles
BELGIUM
Phone: 32-26505988
Fax:
e-mail: lancelot@ulb.ac.be

David Lowe
Plymouth Marine Laboratory
Citadel Hill
Plymouth PL1 2PB
UNITED KINGDOM
Phone: 44-1752-633208
Fax: 44-1752-633102
e-mail: D.LOWE@pml.ac.uk

Vladimir Mamaev
GEF Dolmabahce sarayi
2.inci Harekat Kosru
80680 Besiktas Istanbul
TURKEY
Phone: (90-212) 227-9927
Fax: (90-212) 227-9933
e-mail: vmamaev@dominet.in.com.tr

Laurence Mee
GEF Dolmabahce sarayi
2.inci Harekat Kosru
80680 Besiktas Istanbul
TURKEY
Phone: (90-212) 227-9927
Fax: (90-212) 227-9933

Radu Mihnea
Romanian Marine Research Institute
300, Bd. Mamaia, RO-8700 Constantza 3
ROMANIA
Phone: +40-41-650870 / ext. 37
Fax: +40-41-831274
e-mail: ircm@alpha.rmri.ro

Alexandr S. Mikaelyan
Head, Phytoplankton group
P.P. Shirsov Institute of Oceanology
Nachimova prospect 23, 1171 Moscow
RUSSIA
Phone: +7-095-1247749
Fax: +7-095-1245483
e-mail: mikael@ecosys.sio.rssi.ru

Maria Mirza
Romanian Marine Research Institute
300, Bd. Mamaia, RO-8700 Constantza 3
ROMANIA
Phone: +40-41-650870
Fax: +40-41-831274
e-mail: mirzam@alpha.rmri.ro

Snejhana Moncheva
Marine Biology and Ecology Dept.
Bulgarian Academy of Sciences
Institute of Oceanology
P.O. Box 152, 9000 Varna
BULGARIA
Phone: +359 52 774549
Fax: +359 52 435015
e-mail: office@iobas.io-bas.bg

James W. Murray
School of Oceanography
Box 357940
University of Washington
Seattle WA 98195-7940
Phone: (206) 5434730
Fax: (206) 6853351
e-mail:
JMURRAY@U.WASHINGTON.EDU

Simion Nicolaev
Director
Romanian Marine Research Institute
300, Bd. Mamaia, RO-8700 Constantza 3
ROMANIA
Phone: +40-41-643288
Fax: +40-41-831274
e-mail: nicolaev@alpha.rmri.ro

Ulrich Niermann
Am Sackenkamp 37
23774 Heiligenhafen
GERMANY
Phone: 0049(0)4362-900237
Fax: 0049(0)4362-900238
e-mail: UNIERMANN@T-online.de

Temel Oguz
Middle East Technical University
Institute of Marine Sciences
P.O.Box 28, Erdemli, 33731
Icel-TURKEY
Phone: +90-324-5212406
Fax: +90-324-5212327
e-mail: oguz@deniz.ims.metu.edu.tr

Adriana Petranu
Romanian Marine Research Institute
300, Bd. Mamaia, RO-8700 Constantza 3
ROMANIA
Phone: +40-41-650870
Fax: +40-41-831274
e-mail: petranu@alpha.rmri.ro

Maria Popova
Deputy Director
National Institute of Meteorology
and Hydrology
Bulgarian Academy of Sciences
Sofia, 1784, "Tzarigradsko chaussee" 66
BULGARIA
Phone: 359-2-873805
Fax: 359-2-884494
e-mail: maria.popova@meteo.bg

Florica Porumb
Romanian Marine Research Institute
300, Bd. Mamaia, RO-8700 Constantza 3
ROMANIA

Ioan Porumb
Romanian Marine Research Institute
300, Bd. Mamaia, RO-8700 Constantza 3
ROMANIA

Ondrej Prasil
Institute of Microbiology
Laboratory of Photosynthesis
Academy of Sciences of the Czech
Republic
37981 Trebon
CZECH REPUBLIC
Phone: +420-333-721101
Fax: +420-333-721246
e-mail: prasil@.MBU.ENVI.CZ

Vladimir E. Ryabinin
Head
Marine Forecasting Research Laboratory
Hydrometcentre of Russia
9-13, Bol.Predtechensky Per.
Moscow, 123242
RUSSIA
Phone: +7-095-2552178
Fax: +7-095-2551582
e-mail: rusgmc@glasnet.ru (Subject
"Att.V.Ryabinin")

Ilkay Salihoglu
Assistant Director
Middle East Technical University
Institute of Marine Sciences
P.O.Box 28, Erdemli, 33731
Icel-TURKEY
Phone: +90-324-5212406
Fax: +90-324-5212327
e-mail: ilkay@deniz.ims.metu.edu.tr

Eugen Silviu Stanescu
Romanian Naval League (NGO)
Division Environmental Protection
Nicolae Titulescu Str. 13
RO-8700 Constantza
ROMANIA
Phone: 40-41-611836

Lyubomir Stoyanov
Emergency Response Centre
Ministry of Environment
1, Slavejkov Sj.
Varna 9000
BULGARIA
Phone: 359 52 221407
Fax: 359 52 602594
e-mail: riseco@mbox.digsys.bg

Anatoly Stolbov
Senior scientist
Departamental of Animal Physiology,
Institute of Biology of Southern Seas
335011, av.Nachimov, Sevastopol
UKRAINE
Phone: 380 692-592813
Fax:
e-mail: shulman@ibss.iuf.net

Fred Touchstone
General Manager, PCCI.Inc
1201 E.Abingdon Drive
Alexandria, Virginia 22314
USA
Phone: 703-684-2060
Fax: 703-684-5343
e-mail: ftouchstone@pccii.com

Umit Unluata
Director
Middle East Technical University
Institute of Marine Sciences
P.O.Box 28,
Erdemli - Icel 33731
TURKEY
Phone: +90-324-5212406
Fax: +90-324-5212327
e-mail: unluata@deniz.ims.metu.edu.tr

Zahit Uysal
Middle East Technical University
Institute of Marine Sciences
P.O.Box 28
Erdemli-Icel 33731
TURKEY
Phone: +90-324-5212406
Fax: +90-324-5212327
e-mail: uysal@deniz.ims.metu.edu.tr

Vladimir L. Vladimirov
Head
Data Base Laboratory
Marine Hydrophysical Institute
2 Kapitanskaya St.
Sevastopol, Crimea, 335000
UKRAINE
Phone: +380-692-525276
Fax: +380-692-444253
e-mail: vlvlad@alpha.mhi.iuf.net

Igor L. Volkov
P.P. Shirshov Institute of Oceanology
RAS, Nakhimovsky pr.36, Moscow
117218
RUSSIA
Phone: 7-095-1245949
Fax: 7-095-1245983
e-mail: geochem@geo.sio.rssi.ru

Evgeniy V. Yakushev
P.P. Shirshov Institute of Oceanology
RAS
Lab. of Biochemistry and
Hydrochemistry
Nakhimovsky pr.36, Moscow 117218
RUSSIA
Phone: 7-095-1247742
Fax: 7-095-1245983
e-mail: yakushev@fadr.msu.ru

Oleg A. Yunev
Department of Ecological Physiology
of Phytoplankton
Institute of Biology of Southern Seas
335011 Sevastopol, Nakhimov st.2
UKRAINE
Phone:
Fax:
e-mail: shulman@ibss.iuf.net

Victor E. Zaika
Director
Institute of Biology of Southern Seas
Nakhimov st., 335000 Sevastopol
UKRAINE
Phone:
Fax:
e-mail: vzaika@ibss.iuf.net